CRC SERIES IN AGRICULTURE

Editor-in-Chief

Angus A. Hanson, Ph.D.
Vice President
and
Director of Research
W-L Research, Inc.
Highland, Maryland

HANDBOOK OF SOILS AND CLIMATE IN AGRICULTURE

Editor
Victor J. Kilmer (Retired)
Chief
Soils and Fertilizer Research Branch
National Fertilizer Development Center
Tennessee Valley Authority
Muscle Shoals, Alabama

HANDBOOK OF PLANT SCIENCE IN AGRICULTURE

Editor
B. R. Christie, Ph.D.
Professor
Department of Crop Science
Ontario Agricultural College
University of Guelph
Guelph, Ontario, Canada

HANDBOOK OF PEST MANAGEMENT IN AGRICULTURE

Editor
David Pimentel, Ph.D.
Professor
Department of Entomology
New York College of Agricultural
and Life Sciences
Cornell University
Ithaca, New York

HANDBOOK OF ENGINEERING IN AGRICULTURE

Editor
R. H. Brown, Ph.D.
Chairman
Division of Agricultural Engineering
Agricultural Engineering Center
University of Georgia
Athens, Georgia

HANDBOOK OF TRANSPORTATION AND MARKETING IN AGRICULTURE

Editor
Essex E. Finney, Jr., Ph.D.
Assistant Center Director
Agricultural Research Center
U.S. Department of Agriculture
Beltsville, Maryland

HANDBOOK OF PROCESSING AND UTILIZATION IN AGRICULTURE

Editor
Ivan A. Wolff, Ph.D.
Director
Eastern Regional Research Center
Science and Education Administration
U.S. Department of Agriculture
Philadelphia, Pennsylvania

HANDBOOK OF ANIMAL SCIENCE IN AGRICULTURE

CRC Handbook of Transportation and Marketing in Agriculture

Volume II
Field Crops

Editor

Essex E. Finney, Jr., Ph.D.
Assistant Center Director
Agricultural Research Center
U.S. Department of Agriculture
Beltsville, Maryland

CRC Series in Agriculture

Editor-in-Chief

Angus A. Hanson, Ph.D.
Vice President
and
Director of Research
W-L Research, Inc.
Highland, Maryland

CRC Press, Inc.
Boca Raton, Florida

Library of Congress Cataloging in Publication data

Main entry under title:

CRC handbook of transportation and marketing in agriculture.

(CRC series in agriculture)
Includes bibliographical references and indexes.
CONTENTS: v. 1. Food commodities.—v. 2. Field crops.
1. Produce trade—Handbooks, manuals, etc.
2. Farm produce—Handbooks, manuals, etc. 3. Farm produce—Transportation—Handbooks, manuals, etc.
I. Finney, Essex E. II. Series.
HD9000 .C18 630'.68'8 80-27543
ISBN 0-8493-3851-4 (v. 1)
ISBN 0-8493-3852-2 (v. 2)

Direct all inquiries to CRC Press, Inc., 2000 N.W. 24th Street, Boca Raton, Florida 33431.

International Standard Book Number 0-8493-3851-4 (Volume I)
International Standard Book Number 0-8493-3852-2 (Volume II)

Library of Congress Card Number 80-27543
Printed in the United States

PREFACE

CRC SERIES IN AGRICULTURE

Agriculture, because of its pivotal role in the development of civilized societies, contributed much to the development of various scientific disciplines. Thus, agricultural pursuits led to the practical application of chemistry, and gave rise to such major disciplines as economics and statistics. The expansion of scientific frontiers, and the concomitant specilization within disciplines, has proceeded to the point where agricultural scientists classify themselves in an array of disciplines and subdisciplines, i.e., nematologist, geneticist, physicist, virologist, and so forth. Nevertheless, within the framework of these various disciplines and mission oriented agricultural research, information of primary interest and concern in the solution of agriculturally oriented problems is generated. Although some of the basic information finds its way into the plethora of reference books available within most disciplines, no attempt has been made to develop a comprehensive handbook series for the agricultural sciences.

It is recognized that there are serious difficulties in developing a meaningful handbook series in agriculture because of the range and complexity of agricultural enterprises. In fact, the single common denominator that applies to all agricultural scientists is their universal concern with at least some aspect of the production and utilization of farm products. The disciplines and resources that are called for in a specific investigation are either the same or similar to those utilized in any area of biological research, or in any one of several fields of scientific endeavor.

The sections in this handbook series reflect the input of different editors and advisory boards, and as a consequence, there is considerable variation in both the depth and coverage offered within a given area. However, an attempt has been made throughout to bring together pertinent information that will serve the needs of nonspecialists, provide a quick reference to material that might otherwise be difficult to locate, and furnish a starting point for further study.

The project was undertaken with the realization that the initial volumes in the series could have some obvious deficiencies that will necessitate subsequent revisions. In the meantime, it is felt that the primary objectives of the Section Editors and their Advisory Boards has been met in this first edition.

A. A. Hanson
Editor-in-Chief

PREFACE

Food and fiber are produced on millions of farms and ranches scattered around the globe. Many of these agricultural products are fragile, perishable, and vary naturally in quality. The transportation and marketing network which assembles these commodities, sorts them for quality, and puts them at the disposal of the world community of consumers is complex and expensive. A high degree of efficiency, therefore, is required to improve productivitiy, reduce losses in marketing channels, and deliver a safe, high quality product to the consumer at minimum cost.

This publication presents information and data on technological and operational aspects of the agricultural marketing system. Emphasis is placed upon techniques used to handle, process, sort, package, store, transport, and distribute food and fiber products.

It was not possible to include all of the information originally outlined for this section on transportation and marketing. In fact, only about one half of the topics originally envisioned have been included in this edition. No one publication can adequately cover all of the phases of marketing. Separate books could be written on each individual topic or phase. However, it was my intention to highlight a broad scope of significant information that is important and useful to those who are concerned with the handling, transportation, and marketing of agricultural commodities.

A reference book of this nature cannot be assembled without the cooperation and support of many authorities in the scientific community. I especially wish to acknowledge the members of the Advisory Board. Credit also is extended to the many contributors who displayed an unusual amount of persistence and patience in selecting and organizing the materials requested. Their cooperation and contributions are deeply appreciated.

Finally, readers are invited to offer suggestions for changes and additions in future editions and to report any errors that may be found. Comments and inquires should be directed to the Publisher.

Essex E. Finney, Jr.
Beltsville, Md.

EDITOR-IN-CHIEF

Angus A. Hanson, Ph.D., is Vice President and Director of Research, W-L Research, Inc., Highland, Maryland, and has had broad experience in agricultural research and development. He is a graduate of the University of British Columbia, Vancouver, McGill University, Quebec, and received the Ph.D. degree from the Pennsylvania State University, University Park, in 1951.

An employee of the U.S. Department of Agriculture from 1949 to 1979, Dr. Hanson worked as a Research Geneticist at Universtiy Park, Pa., 1949 to 1952, and at Beltsville, Md., serving successfully as Research Leader for Grass and Turf Investigations, 1953 to 1965, Chief of the Forage and Range Research Branch, 1965 to 1972, and Director of the Beltsville Agricultural Research Center, 1972 to 1979. He has been appointed to a number of national and regional task forces charged with assessing research needs and priorities, and has participated in reviewing agricultural needs and research programs in various foreign countries. As Director at Beltsville, he was directly responsible for programs that included most dimensions of agricultural research.

In his personal research, he has emphasized the improvement of forage crops, breeding and management of turfgrasses, and the breeding of alfalfa for multiple pest resistance, persistence, quality, and sustained yield. He is the author of over 100 technical and popular articles on forage crops and turfgrasses, and has served as Editor of *Crop Science* and the *Journal of Environmental Quality.*

THE EDITOR

Essex E. Finney, Jr., was born in May 1937, in Powhatan, Va. He received a B.S. degree (with honors) in Agricultural Engineering from Virginia Polytechnic Institute, Blacksburg, an M.S. from the Pennsylvania State University, University Park, and a Ph.D. in 1963 from Michigan State University, East Lansing. He was named a Princeton Fellow in Public Affairs from 1973 to 1974, and studied at the Woodrow Wilson School of Public and International Affairs, Princeton University, New Jersey.

From 1963 to 1965, Dr. Finney was a commissioned officer in the U.S. Army and served as a Branch Chief in the Special Projects Division of the Industrial Operations Directorate at Rocky Mountain Arsenal in Denver, Colo. Dr. Finney accepted a position as a Research Agricultural Engineer in the Agricultural Research Service of the U.S. Department of Agriculture in September 1965. He was assigned to the Instrumentation Research Laboratory of the Market Quality Research Division in Beltsville, Md. In 1972, he was appointed Chairman of the Agricultural Marketing Research Institute at Beltsville, and, in 1977, Assistant Director of the Beltsville Agricultural Research Center.

Dr. Finney is a member of the Institute of Food Technologists, Tau Beta Pi Engineering Association, and the American Society of Agricultural Engineers (ASAE). He has been particularly active in the Food Engineering Division of ASAE, having served as secretary, vice-chairman, and chairman of that Division between 1967 and 1970. He was elected to the Board of Directors of ASAE in 1970. In 1976, Dr. Finney was appointed to a study team for the World Food and Nutrition Study of the National Academy of Sciences.

Dr. Finney's early research dealt with methods for drying forages and cereal grains. His more recent research accomplishments relate to improved methods for measuring the quality of food and agricultural products. He has published data on the resistance of fruits and vegetables to mechanical damage. He has also developed new nondestructive methods for evaluating firmness and texture of food commodities. Dr. Finney has authored about 50 scientific publications and addressed professional groups throughout the U.S. and Canada. He received a Journal Paper Award from the American Society of Agricultural Engineers in 1969, and is currently the Editor of the *CRC Handbook of Transportation and Marketing in Agriculture.*

ADVISORY BOARD

CONTRIBUTORS

John A. Barton, M.S.
Rate Economist
Consumer Economics Department
Northeast Utilities
Hartford, Connecticut

Louis N. Bass, Ph.D.
Director
National Seed Storage Laboratory
Agricultural Research
Science and Education Administration
U.S. Department of Agriculture
Colorado State University
Fort Collins, Colorado

Donald B. Bechtel, Ph.D.
Research Chemist
Grain Marketing Research Laboratory
Agricultural Research
Science and Education Administration
U.S. Department of Agriculture
Mahhattan, Kansas

Albert H. Boyd, Ph.D.
Professor/Agronomist
Seed Technology Laboratory
Department of Agronomy
Mississippi Agricultural and Forestry Experimental Station
Mississippi State University
Mississippi State, Mississippi

Roger Brook, Ph.D.
Extension Agricultural Engineer
Agricultural Engineering Department
Michigan State University
East Lansing, Michigan

Dana P. Childs
Research Entomologist
Tobacco Storage Insects Research
Agricultural Research
Science and Education Administration
U.S. Department of Agriculture
Richmond, Virginia

Sergio Davila
Ph.D. Candidate
Agricultural Engineering
Agricultural and Biological Engineering Department
Mississippi State University
Mississippi State, Mississippi

Robert Davis, Ph.D.
Laboratory Director
Stored-Product Insects Research and Development Laboratory
Science and Education Administration
U.S. Department of Agriculture
Savannah, Georgia

George H. Foster
Professor
Agricultural Engineering Department
Purdue University
West Lafayette, Indiana

Warren E. Garner
Director (Retired)
Cotton Ginning Research Laboratory
Science and Education Administration
U.S. Department of Agriculture
Stoneville, Mississippi

Albert H. Graves
Industrial Engineer
Science and Education Administration
U.S. Department of Agriculture
Raleigh, North Carolina

Martin M. Kulik, Ph.D.
Research Plant Pathologist
Seed Research Laboratory
Plant Genetics and Germplasm Institute
Science and Education Administration
U.S. Department of Agriculture
Beltsville, Maryland

Y. Pomeranz, Ph.D.
Director
Grain Marketing Research Laboratory
Agricultural Research
Science and Education Administration
U.S. Department of Agriculture
Manhattan, Kansas

Beverly G. Reeves, Ph.D. (Deceased)
Agricultural Engineer
Cotton Mechanization and Ginning
Science and Education Administration
Extension
U.S. Department of Agriculture
Memphis, Tennessee

G. Burns Welch, Ph.D.
Professor
Agricultural Engineer
Agricultural and Biological Engineering
Department
Mississippi State University
Mississippi State, Mississippi

Marion E. Whitten, Ch.E.
Research Chemist (Retired)
Collaborator
Transportation and Packaging
Laboratory
Agricultural Marketing Service
U.S. Department of Agriculture
Beltsville, Maryland

Philip C. Williams, Ph.D.
Research Scientist
Grain Research Laboratory
Canadian Grain Commission
Canada Department of Agriculture
Winnipeg, Manitoba, Canada

Lowell W. Woodstock, Ph.D.
Research Plant Physiologist
Seed Research Laboratory
Plant Genetics and Germplasm Institute
Science and Education Administration
U.S. Department of Agriculture
Beltsville, Maryland

TABLE OF CONTENTS

TABLE OF CONTENTS

CRC HANDBOOK OF TRANSPORTATION AND MARKETING IN AGRICULTURE

Volume I

Introduction

MARKETING COSTS AND PRICE SPREADS

Essex E. Finney, Jr.

The system for marketing food and other agricultural products in the United States is complex and expensive.[1-4] In 1978, consumer retail expenditures for food from U.S. farms was estimated as $208 billion.[1] An estimated $141 billion, or slightly over two thirds of the total, represented "marketing costs", i.e., charges for transporting, storing, processing, packaging, and distributing food after it leaves the farm. Recent trends show that a significant proportion (25%) of the marketing cost is associated with the marketing functions performed by restaurants and other away-from-the-home eating places.[5] Between 1960 and 1974, for example, spending on food eaten outside the home doubled while food purchased for preparation in the home increased by 40%.[6]

Increases in the price of food are the result of both higher values of products sold from the farm and widening farm-retail marketing spreads. The difference between the price paid for a food product by consumers and the farm value of an equivalent quantity of farm product is the farm-retail price spread.[7] These price spreads vary for individual items (Figure 1) and are determined, to a large extent, by the amount of processing, packaging, and handling which occurs between the farmer and the consumer.

Breaking down price spreads into component costs is one step toward identifying specific functions for which increased efficiency might have a large impact on reducing marketing costs and retail prices, and thereby easing inflationary pressures in the overall economy.[8] Many approaches have been used to subdivide and classify marketing functions. Kohls and Downey[2] discuss some of the approaches; however, they emphasize that any subdivision or listing of functions must be recognized as being arbitrary. Nevertheless, commodity specialists within the U.S. Department of Agriculture have developed a systematic method to classify, monitor, analyze, and report marketing costs and price spreads.[1,7-9] Costs and margins associated with functions such as assembly, processing, wholesaling, and retailing were computed for 13 individual items for 1977 and 1978 (Table 1). For seven items, costs were allocated, to the extent possible, into the following major components: labor, packaging, transportation, business taxes, depreciation, rent, energy, interest, repairs, advertising, other costs, and profits (Tables 2 to 7). In general, the three principal individual cost factors are labor, packaging, and transportation (Figure 2).

These data have been developed by specialists in the Economics, Statistics, and Cooperatives Service (ESCS) of the U.S. Department of Agriculture. The information provided can serve as a general guide. However,the individual figures are subject to short-term fluctuations as well as long-range shifts. Figure 3, for example, illustrates the shifts which have occurred over the past 3 decades in the distribution of the tobacco user's dollar.[5] In this particular instance, the share which the farmer receives has decreased by 50% while the proportion for wholesaling and retailing has increased substantially. It is important, therefore, to consult commodity specialists within the U.S. Department of Agriculture for the most recent reports on marketing costs and spreads for individual items.

The percent of income spent for food in the U.S. (Figure 4) compares favorably with that of most other major industrial countries. There are striking differences, however, in the percent of family income spent on food by family income quintile. Gallo et al.[10] report, for example, that about one third of the income of low income families is spent on food. Thus, rising marketing spreads and retail costs create a special burden

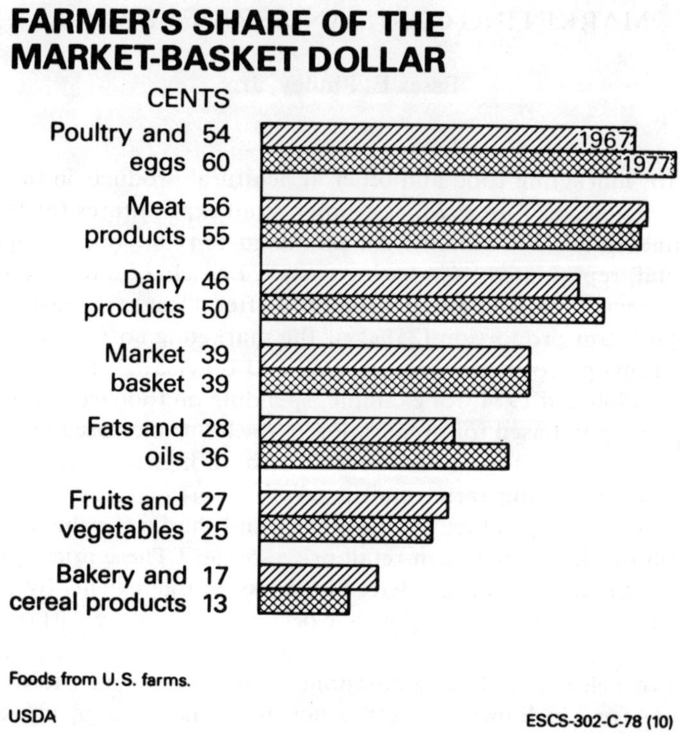

FIGURE 1. Proportion of retail costs accounted for by value for various commodity groupings.[5]

for those with low income.[8] In addition to marketing costs, the share of consumer income devoted to food expenditures depends on farm commodity prices, food consumption patterns, and changes in consumer income.

A significant volume of non-food items also moves through the agricultural marketing system. In 1973, for example, the total food and fiber marketing sector distributed products worth about $244 billion.[4] Farmers purchased $66 billion of production ingredients such as fertilizer, seed, chemicals, petroleum, machinery, and equipment to produce basic food and fiber products worth about $90 billion. These products were processed and distributed by a marketing sector which added $155 billion in costs. Of the $244 billion of goods marketed, $18 billion was exported. Considerable growth has occurred since 1973. In 1978, for example, agricultural exports were estimated as being over $26 billion.[8] The agricultural marketing system, therefore, represents an important segment of the nation's economy.

Table 1
RETAIL PRICES, FARM VALUES, AND MARKETING SPREADS BY FUNCTION FOR SELECTED FOODS, ANNUAL AVERAGES, 1977 AND 1978 (CENTS)

Food item and year	Farm value[a]	Assembly and procurement	Processing	Intercity transportation	Wholesaling	Retailing[b]	Retail price
Beef, choice (pound)							
1978	108.8	2.3	6.1	2.1	8.7	53.9	181.9
1977	83.3	2.2	6.3	2.0	6.8	47.8	148.4
Pork (pound)							
1978	74.5	2.1	28.9	2.2	6.3	29.6	143.6
1977	63.6	2.0	31.2	2.2	5.2	21.2	125.4
Broilers (pound)							
1978	37.2	1.0	8.7	1.4	3.8	14.4	66.5
1977	33.0	1.1	8.0	1.4	3.7	12.9	60.1
Eggs, grade A or AA large (dozen)							
1978	49.7	0.9	10.5	1.6	3.4	12.4	78.5
1977	53.8	0.9	10.3	1.5	3.5	12.3	82.3
Milk, sold in stores (½ gallon)							
1978	47.5[c]	3.0	14.1	[d]	13.4	9.2	87.3[e]
1977	45.1[c]	2.9	13.3	[d]	12.6	10.0	83.9
Butter (pound)							
1978	96.4	3.4	17.2	2.7	6.4	20.9	145.2[e]
1977	85.8	3.2	15.4	2.6	6.1	20.0	133.1
Bread, white (pound)							
1978	5.4[f]	1.5[h]	13.1	0.4[i]	16.7	[d]	37.1
1977	4.5[g]	1.4[h]	12.6	0.4[i]	16.6	[d]	35.5
Potatoes (fall) (10-lb bag)							
1978	41.4	[j]	20.0	22.8	16.3	72.6	173.1[k]
1977	44.3	[j]	16.9	24.1	13.2	66.6	165.1
Oranges, Calif. (dozen)							
1978	50.2	2.3	26.9	16.7	13.2	79.1	188.4[e]
1977	36.2	2.3	20.6	14.5	11.2	55.5	140.3

Table 1 (continued)
RETAIL PRICES, FARM VALUES, AND MARKETING SPREADS BY FUNCTION FOR SELECTED FOODS, ANNUAL AVERAGES, 1977 AND 1978 (CENTS)

Food item and year	Farm value[a]	Assembly and procurement	Processing	Intercity transportation	Wholesaling	Retailing[b]	Retail price
Lettuce, Calif. (head)							
1978	15.0	0.4	7.7	8.9	3.4	21.6	57.0[e]
1977	5.1	0.4	7.3	9.4	3.7	20.1	46.0
Orange juice, frozen (6-oz can)							
1978	19.1	0.6	8.8	1.8	4.3	8.1	42.7[l]
1977	10.7	0.6	12.1	1.7	3.0	5.9	34.0
Tomatoes, Calif., whole (303 can)							
1978	4.8	0.7	17.4	3.6	2.6	8.5	37.6[k]
1977	3.9	0.6	21.1	3.4	1.5	6.8	37.3
Tomato catsup, Calif., (14-oz bottle)							
1978	7.1	1.1	18.9	4.4	7.5	9.9	48.9[k]
1977	5.7	0.9	23.7	4.2	3.5	10.8	48.8

[a] The farm value is the gross return to farmers for the quantity of farm products equivalent to the unit sold at retail minus imputed value of byproducts. Because of losses from processing, waste, and spoilage, the farm value represents larger quantities than the retail unit.
[b] Instore costs only. Headquarters and warehousing expenses are included in wholesaling.
[c] Prices received by farmers are normally quoted for 3.5% butterfat (BF) at plant. This price has been adjusted for transportation from farm to plant and 20 ¢/100 lb to reflect the value of butterfat between 3.5% milk and the average of 3.2% milk sold in stores.
[d] Included in wholesaling.
[e] Annual average computed from BLS retail prices and price indexes.
[f] Includes wheat, 3.3 ¢ and other farm ingredients, 2.1¢.
[g] Includes wheat, 2.7¢ and other farm ingredients, 1.8¢.
[h] Assembly of wheat and milling.
[i] Flour only.
[j] Included in farm value.
[k] Based on January-June BLS prices.
[l] Estimated by Florida Department of Citrus.

From Dunham, D., *U.S. Dep. Agric. Agric. Econ. Rep.* No. 420, 1979.

Table 2
CHOICE BEEF AND PORK: COMPONENTS OF PRICE SPREADS PER POUND AT RETAIL, 1977 (CENTS)

Item	Beef	Pork
Farm value	83.3	63.6
Assembly of live animal	2.2	2.0
Slaughtering	6.3	31.2
Labor	1.7	13.5
Packaging	0.2	2.2
Transportation	0.7	1.7
Business taxes	0.1	0.2
Depreciation	0.3	0.6
Rent	0.2	0.1
Repairs	0.2	0.9
Advertising	0.1	0.6
Interest	0.2	0.4
Energy[a]	0.2	1.2
Other	2.0[b]	7.8[c]
Profit before taxes	.4	2.0
Wholesaling	8.8[d]	7.4[e]
Retailing	47.8	21.2
Retail price	148.4	125.4

[a] Includes water.

[b] Includes 0.6 ¢ for administrative expenses such as wages and salaries, insurance, payroll taxes, office supplies, data processing, and travel; 0.1 ¢ for inspection and grading; 0.4 ¢ for shipping and selling; 0.2 ¢ for procurement; and miscellaneous items.

[c] Includes 1.7 ¢ for administrative expenses such as wages and salaries, insurance, payroll taxes, office supplies, data processing and travel; 1.2 ¢ shipping and selling; 1.5 ¢ for procurement, and miscellaneous items.

[d] Includes transportation of 2 ¢ intercity and 1 ¢ intracity.

[e] Includes transportation of 2.2 ¢ intercity and 1 ¢ intracity.

From Dunham, D., *U.S. Dep. Agric. Agric. Econ. Rep.* No. 420, 1979.

Table 3
BROILERS: COMPONENTS OF PRICE SPREADS PER POUND AT RETAIL, 1977 (CENTS)

		Marketing functions				
Item	Farm value[a]	Assembly	Processing	Hauling, distributing[b]	Retailing	Retail price
Labor	—	0.69	3.60	2.51	7.5	—
Packaging	—	—	1.35	0.15	0.7	—
Transportation[c]	—	—	—	—	—	—
Business taxes	—	—	0.20	0.20	0.5	—
Depreciation	—	—	0.40	0.25	0.2	—
Rent	—	—	[d]	0.10	0.3	—
Repairs	—	—	0.30	0.15	0.1	—
Advertising	—	—	0.15	—	0.7	—
Interest	—	—	0.10	0.10	0.2	—
Energy	—	0.35	0.65	1.02	0.3	—
Other	—	—	0.65	0.38	1.5	—
Profit	—	—	0.60	0.24	0.9	—
Unallocated	—	0.06[e]	—	—	—	—
Total	33.0	1.10	8.00	5.10	12.9	60.1

Note: — = Not estimated.

[a] Farm value computed from U.S. Department of Agriculture live broiler equivalent prices. Prices converted to ready-to-cook equivalent by multiplying by 1.41 to allow for shrinkage and losses in processing and marketing.
[b] Includes long-distance transportation plus wholesaling and local delivery.
[c] Includes 1.1¢ for assembly, 1.4¢ for long-distance transportation, and 1.7¢ for local delivery.
[d] Included in depreciation.
[e] Includes overhead and profit.

From Dunham, D., *U.S. Dep. Agric. Agric. Econ. Rep.* No. 420, 1979.

Table 4
EGGS, GRADE A AND AA LARGE: COMPONENTS OF PRICE SPREADS PER DOZEN AT RETAIL, 1977 (CENTS)

		Marketing functions				
Item	Farm value[a]	Assembly	Processing	Hauling, distributing[b]	Retailing	Retail price
Labor	—	0.49	3.00	2.49	6.1	—
Packaging	—	—	4.10	0.12	0.3	—
Transportation[c]	—	—	—	—	—	—
Business taxes	—	—	0.40	0.20	0.4	—
Depreciation	—	—	0.50	0.20	0.1	—
Rent	—	—	[d]	0.10	0.1	—
Repairs	—	—	0.25	0.10	0.1	—
Advertising	—	—	0.30	—	0.9	—
Interest	—	—	0.20	0.10	0.3	—
Energy	—	0.33	0.65	1.13	0.1	—
Other	—	—	0.20	0.29	2.6	—

Table 4 (continued)
EGGS, GRADE A AND AA LARGE: COMPONENTS OF PRICE SPREADS PER DOZEN AT RETAIL, 1977 (CENTS)

		Marketing functions				
Item	Farm value[a]	Assembly	Processing	Hauling, distributing[b]	Retailing	Retail price
Profit	—	—	0.70	0.27	1.3	—
Unallocated	—	0.08[e]	—	—	—	—
Total	53.8	0.9	10.3	5.0	12.3	82.3

Note: — = Not estimated.

[a] Farm value calculated from prices received by producers. Average price multiplied by 1.03 to allow for 3% loss during marketing.
[b] Includes long-distance transportation plus wholesaling and local delivery.
[c] Includes 0.9¢ for assembly, 1.5 ¢ for long-distance transportation, and 1.6 ¢ for local delivery.
[d] Included in depreciation.
[e] Includes overhead and profit.

From Dunham, D., *U.S. Dep. Agric. Agric. Econ. Rep.* No. 420, 1979.

Table 5
GRADE A WHOLE MILK SOLD THROUGH RETAIL STORES: COMPONENTS OF PRICE SPREADS PER HALF-GALLON, 1977 (CENTS)

Item	Farm value[a]	Procurement and assembly[b]	Processing	Wholesaling	Retailing	Retail price
Labor	—	0.40	4.5	6.4	5.7	—
Packaging	—	—	4.6	—	0.2	—
Transportation	—	2.10	—	2.3	—	—
Business taxes	—	0.03	0.5	0.4	0.4	—
Depreciation	—	0.07	0.5	0.5	0.1	—
Rent	—	0.02	0.3	0.3	0.2	—
Repairs	—	0.04	0.4	0.3	0.1	—
Advertising	—	0.01	—	0.6[c]	1.0	—
Interest	—	0.05	0.1	0.1	0.1	—
Energy	—	0.04	0.9	0.3	0.3	—
Other	—	0.18	0.8	0.5	1.2	—
Profit before taxes	—	—	0.7	0.9	0.7	—
Total	45.1	2.94	13.3	12.6	10.0	83.9

Note: — = Not estimated.

[a] Prices received by farmers are normally quoted for 3.5% butterfat at plant of first receipt. This price has been adjusted for transportation from farm to first plant to give the farm price, which was adjusted 20 ¢/100 lb to reflect the value of butterfat between 3.5% butterfat and the average of 3.32% packaged whole milk. There are approximately 23 half-gal/100 lb.
[b] Includes laboratory and on-the-farm field service to assure quality, pickup at the farms, transportation, receiving and reloading as necessary, and final delivery to the processor.
[c] Dairy products, especially milk, are also advertised through cooperative ventures. This cost is not included in this estimate.

From Dunham, D., *U.S. Dep. Agric. Agric. Econ. Rep.* No. 420, 1979.

Table 6
BUTTER: COMPONENTS OF PRICE SPREADS PER POUND, 1977 (CENTS)[a]

Item	Procurement				Wholesaling			
	Farm value[b]	Contract hauling[c]	Other[d]	Manufacturing	Printing and packaging[e]	Hauling and warehousing	Retailing	Retail price
Labor	—	—	0.18	2.25	1.13	—	7.8	—
Packaging supplies	—	—	—	1.80	1.64	—	0.4	—
Transportation	—	2.91	—	—	—	2.58	—	—
Business taxes	—	—	0.01	0.10	0.06	—	0.5	—
Depreciation	—	—	0.02	0.39	0.06	—	0.2	—
Rent/lease	—	—	0.01	0.12	0.03	—	0.2	—
Repairs	—	—	[f]	0.19[g]	0.09	—	0.1	—
Advertising	—	—	—	0.03[h]	0.02	—	1.5	—
Interest	—	—	—	0.13	0.07	—	0.5	—
Energy	—	—	0.03	0.71	0.06	—	0.2	—
Other	—	—	0.07	0.81[i]	0.29	—	5.8	—
Profit before taxes	—	—	—	5.30	0.09	—	2.8	—
Unallocated	0	0	0	0	0	6.14	0	—
Total	85.77	2.91	.33	11.83	3.54	8.72	20.0	133.1

Note: — = Note estimated.

[a] 1-lb carton of quarter-lb wraps.
[b] Market value of 0.81103 lb of butterfat and 0.19 lb of skim milk at the plant less the producer hauling charge.
[c] Of the contract hauling charge, 2.58 ¢ was paid by producers and 0.33 ¢ paid by manufacturers.
[d] Costs to plants for hauling milk from farms in manufacturer-owned trucks, laboratory and field service to producers, and operating receiving and reload stations.
[e] Costs incurred by specialized butter wholesalers in printing and packaging 50% of the butter, assuming that the other half was printed by manufacturers.
[f] Less than 0.005%.
[g] Excludes plant labor used to repair buildings and equipment which is in labor.
[h] Dairy products are also advertised through cooperative ventures. This cost is not included.
[i] Salt and color (used in manufacturing), general supplies, office supplies, insurance, water and sewage, professional and other services, travel and related expenses, and other expenses. Cost of seasonal storage not included.

From Dunham, D., *U.S. Dep. Agric. Agric. Econ. Rep.* No. 420, 1979.

Table 7
WHITE PAN BREAD: COMPONENTS OF PRICE SPREADS PER 1-LB LOAF, 1977 (CENTS)

	Farm value				Baker-wholesaler			
Item	Wheat	Other ingredients	Assembly of wheat and flour milling[a]	Transportation of flour	Processing	Wholesaling	Retailing	Retail price
Labor	—	—	0.53	—	4.38	8.27	—	—
Packaging	—	—	0.03	—	2.41	0.13	—	—
Transportation	—	—	0.36	—	—	0.05	—	—
Business taxes	—	—	0.04	—	0.16	0.06	—	—
Depreciation	—	—	0.09	—	0.63	0.19	—	—
Rent	—	—	0.01	—	—	0.11	—	—
Repairs	—	—	0.02	—	0.12	0.10	—	—
Advertising	—	—	0.02	—	—	0.69	—	—
Interest	—	—	0.05	—	0.03	0.04	—	—
Energy	—	—	0.09	—	0.48	—	—	—
Other and unallocated	2.7	1.8	0.01[b]	0.43	3.53[b,c]	2.86[b]	—	—
Profit	—	—	.18	—	0.83	0.97	—	—
Total	2.7	1.8	1.43	0.43	12.57	13.47	3.1	35.5

Note: — = Note estimated.

[a] Price spread attributed to handling wheat at country receiving points, transporting it to flour mill, milling the wheat, and wholesaling the flour.

[b] Residual, includes items such as telephone service, janitorial supplies, contributions, and other miscellaneous items.

[c] Includes cost of nonfarm ingredients such as yeast, yeast food, dough conditioners and stabilizers, etc., (0.78 ¢) and merchandising, handling and processing of nonfarm ingredients other than wheat (0.32 ¢).

From Dunham, D., *U.S. Dep. Agric. Agric. Econ. Rep.* No. 420, 1979.

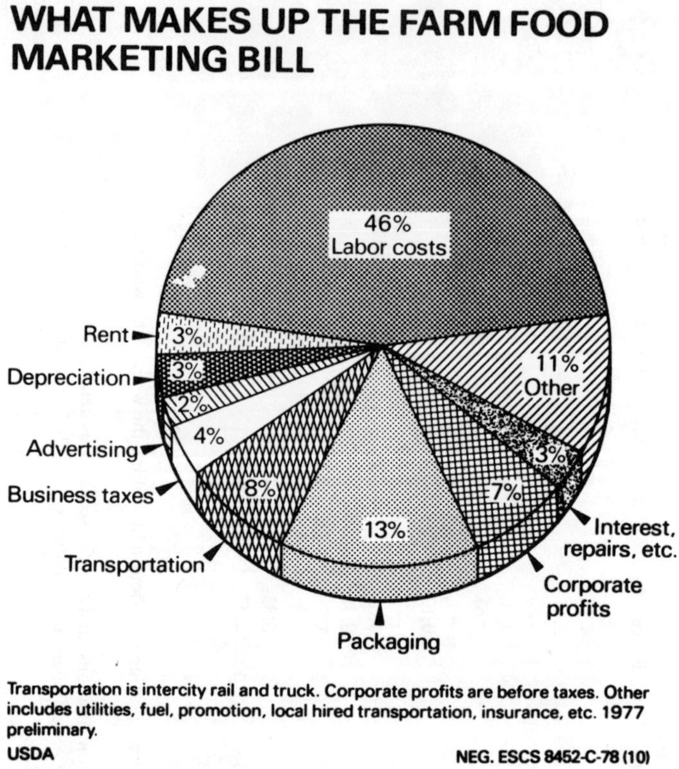

FIGURE 2. Distribution of the marketing bill by component costs.[5]

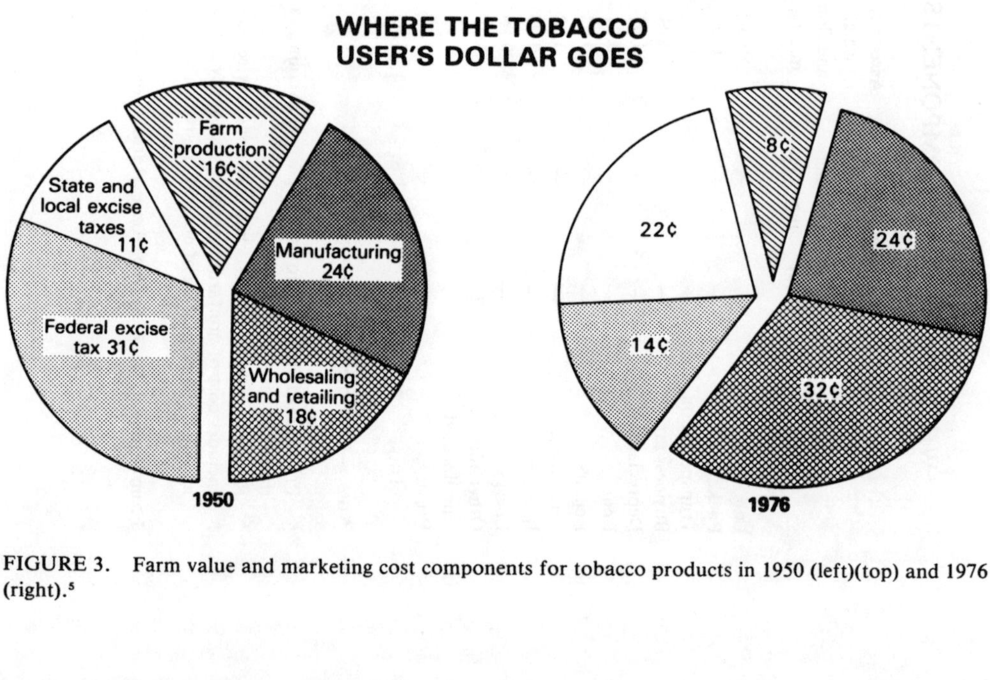

FIGURE 3. Farm value and marketing cost components for tobacco products in 1950 (left)(top) and 1976 (right).[5]

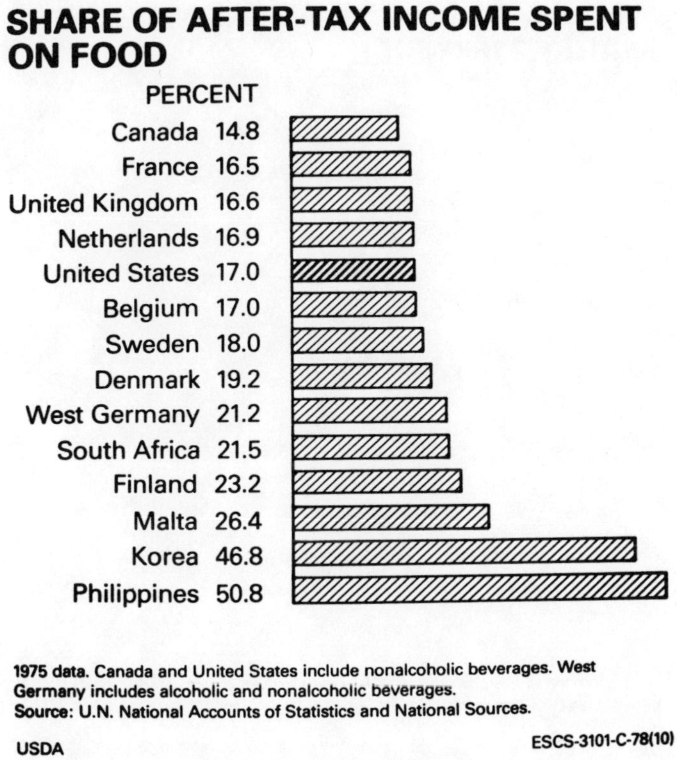

FIGURE 4. Proportion of income spent on food by selected industrial countries around the world.[5]

REFERENCES

1. **Dunham, D.,** Developments in marketing spreads for food products in 1978, *U.S. Dep. Agric. Agric. Econ. Rep.,* No. 420, 1979.
2. **Kohls, R. L. and Downey, W. D.,** *Marketing of Agricultural Products,* 4th ed., Macmillan, New York, 1972, 20.
3. **Linstrom, H. R.,** Farmer to consumer marketing, Rep. No. ESCS-01, U.S. Department of Agriculture, Washington, D.C., 1978.
4. **Manley, W. T.,** The food and fiber system — how it works, *U.S. Dep. of Agric. Agric. Info. Bull.,* 383, 1975.
5. **Anon.,** Handbook of agricultural charts, *U.S. Dep. Agric. Agric. Handb.* No. 551, 1978.
6. **LeBovit, C. and Gallo, A. E.,** Changing U.S. Food Spending Patterns, National Food Review, NFR-6, U.S. Department of Agriculture, Washington, D.C., 1979, p. 17.
7. **Anon.,** Cost components of farm-retail spreads for selected foods, *U.S. Dep. Agric. Agric. Econ. Rep.,* No. 343, 1976.
8. **Salathe, L. E. and Boehm, W. T.,** Food prices in perspective, *U.S. Dep. Agric. Agric. Info. Bull.,* 427, 1979.
9. **Anon.,** Distribution of the Food Dollar by Marketing Function and Expense Item, Marketing and Transportation Situation, MTS-195, U.S. Department of Agriculture, Washington, D.C., 1974.
10. **Gallo, A. E., Boehm, W. T., and LeBovit, C.,** Changes in Food Expenditures by Income Group, National Food Review, NFR-6, U.S. Department of Agriculture, Washington, D.C., 1979, p. 12.

ENERGY REQUIREMENTS FOR MARKETING FUNCTIONS

John A. Barton

The transportation and marketing system for agricultural products involves many steps and processes. Energy requirements for transportation and marketing functions vary by commodity and type of operation. In examining energy usage in transportation and marketing functions, it is useful to aggregate the multitude of activities into three major sectors: processing, wholesaling and retailing, and transportation. Processing is defined to include manufacturing processes which convert raw agricultural products into a form which consumers purchase; wholesaling and retailing involves the storage sale of manufactured food products and fresh produce, and transportation provides the linkage from production through retailing. Food preparation, both in households and out of home, not considered here to be a marketing function.

Total energy requirements for marketing and distribution functions are substantial, about 35% more than the energy used in agricultural production (Table 1).

Data for the wholesale-retail sector are incomplete because cold storage- and warehousing electricity-use data do not exist. Electricity use for these functions (primarily associated with wholesaling operations) is substantial, with the bulk associated with cold storage and freezer space. However, total consumption is small compared to retail supermarkets. The treatment of wholesale-retail energy use is therefore limited to retail supermarket establishments.

Energy requirements for processing, retailing, and transportation are discussed in greater detail in the following three sections. Data presented are for 1976 and 1977. However, because per unit energy requirements are leveling off or decreasing while volumes of agricultural products and manufactured foods are slowly increasing, these aggregate estimates should approximate current energy requirements.

ENERGY USE IN FOOD PROCESSING

Energy consumption in the food processing industry totaled 938 billion Btus in 1976. This amount was the equivalent of 7.4% of total energy use in all manufacturing. Natural gas accounted for about half of the total Btus consumed and electricity 14% (Table 2).

Of the nine three-digit Standard Industrial Classification (SIC) food processing industries,* the sugar and confectionery products industry (SIC 206) consumed the most energy — about 17%. The bakery products industry (SIC 205) used the smallest amount, around 5% after miscellaneous food and kindred products (SIC 209) (Table 3). With respect to fuel type, natural gas represented the largest share of total energy consumed in each of the nine three-digit industries. In four of the three-digit industries, natural gas represented over 50% of total Btu consumption. Actual consumption data for 1976 are shown in Table 4.

The bulk of energy use in food processing is associated with process heat applications. With the exception of the bakery products industry, which uses substantial amounts of natural gas in ovens, most fuel is used in boilers. In direct fuel applications, the use of natural gas is preferred over other fuels because of its clean-burning char-

* The Standard Industrial Classification (SIC) code system was developed by the Office of Management and Budget to classify industries by product type. Two-digit codes identify major product groups, such as food processing, three-digit codes break down food processing by food product group, and the four-digit codes provide a further breakdown.

Table 1
CURRENT AND DIRECT ENERGY REQUIREMENTS FOR AGRICULTURAL PRODUCTION, MARKETING, AND DISTRIBUTION

	Petroleum fuels (million gallons)	Liquid petroleum gas[a] (million gallons)	Natural gas (million mcf)	Coal (thousand short tons)	Electricity (million kWh)	Total Btus (billions)
Agricultural production[b]	6,530	1,435	165	34	31,527	1,279
Marketing and distribution	4,570	—	472	3,358	90,236	1,611
Processing	1,239	—	447	3,358	40,906	957
Retailing	28	—	25	—	49,330	198
Transportation	3,303	—	—	—	—	456

[a] LP gas is used in food processing and transportation, but the amounts are small and not quantifiable.

[b] Includes crop drying.

Table 2
FUELS AND ELECTRICITY CONSUMPTION IN THE FOOD PROCESSING INDUSTRY, 1976

	Physical units	Btus (billions)	Percent of total energy
Distillate oil (million gallons)	516	72,240	8
Residual oil (million gallons)	699	93,660	10
Coal (thousand short tons)	3,292	85,592	9
Natural gas (billion cubic feet)	438	459,900	49
Electricity (million kWh)	39,062	133,319	14
Other fuels and fuels not specified by kind	—[a]	92,789	10
Total	—	937,500	100

[a] — = Not applicable.

From Annual Survey of Manufacturers, 1976, Fuels and Electric Energy Consumed, U.S. Department of Commerce, Washington, D.C., 1978.

acteristics. In many cases, the use of fuel oil or coal would impart undesirable residues in the food product, particularly where the flame directly contacts the food (as in hog-singeing).

Energy consumed in actual packaging operations is not easily separated from total energy use. As a percentage of total energy, however, the energy directly associated with packaging operations is likely to be small. Most energy consumed would be elec-

Table 3
PERCENTAGE OF ENERGY CONSUMPTION IN THE FOOD PROCESSING INDUSTRY BY THREE-DIGIT SIC AND FUEL TYPE, 1976 (PERCENTAGE OF TOTAL BTUS)

SIC Code	Item	Purchased fuels and electricity	Distillate oil	Residual oil	Coal	Natural gas	Electricity	Other fuels and fuel not specified by type	Percentage of total energy consumed in food processing
SIC 20	Food and kindred products	100	8	10	9	49	14	10	100.0
SIC 201	Meat products	100	7	7	3	47	22	14	11
SIC 202	Dairy products	100	8	10	1	44	20	17	10
SIC 203	Canned and preserved fruits and vegetables	100	9	12	2	55	14	8	13
SIC 204	Grain mill products	100	6	8	18	49	14	5	16
SIC 205	Bakery products	100	5	6	—[a]	57	18	14	5
SIC 206	Sugar and confectionery products	100	6	16	26	44	4	4	17
SIC 207	Fats and oils	100	10	11	4	55	10	10	11
SIC 208	Beverages	100	11	9	7	48	15	10	11
SIC 209	Miscellaneous food preparations and kindred products	100	10	11	—	43	21	15	5

[a] — = Less than 0.5%.

From Annual Survey of Manufacturers, 1976, Fuels and Electric Energy Consumed, U.S. Department of Commerce, Washington, D.C., 1978.

Table 4
FUELS AND ELECTRICITY CONSUMPTION BY THREE-DIGIT SIC FOOD PROCESSING INDUSTRIES, 1976

SIC Code	Item	Distillate oil (million gal)	Residual oil (million gal)	Coal (1,000 short tons)	Natural gas (billion ft³)	Electricity (billion kWh)
SIC 20	Food and kindred products	516	699	3,292	438	39,062
SIC 201	Meat products	54	51	123	46	6,764
SIC 202	Dairy products	50	67	23	39	5,310
SIC 203	Canned and preserved fruits and vegetables	80	109	106	66	5,199
SIC 204	Grain mill products	64	85	1,016	69	6,085
SIC 205	Bakery products	17	21	1	28	2,709
SIC 206	Sugar and confectionery products	65	185	1,378	67	2,082
SIC 207	Fats and oils	73	82	164	55	3,194
SIC 208	Beverages	77	62	280	46	4,488
SIC 209	Miscellaneous food preparations and kindred products	37	39	1	21	3,230

From Annual Survey of Manufacturers, 1976, Fuels and Electric Energy Consumed, U.S. Department of Commerce, Washington, D.C., 1978.

tricity associated with mechanical operations, although some natural gas is used as a heat source in sealing food containers.

ENERGY USE IN TRANSPORTATION

Petroleum fuels are required for the transportation of agricultural inputs to farms, farm products to processors or directly to market, and processed foods to supermarkets. Annual fuel requirements for transporting agricultural inputs, farm commodities, and processed foods total about 2892 million gal of diesel fuel and 411 million gal of gasoline (Table 5). This compares with 2733 million gal of diesel fuel and 3535 million gal of gasoline used in farm production in 1978. Some liquid petroleum gas (derived mostly from natural gas) is used in trucking and some electricity is used in rail service, but the amounts of energy involved are small in comparison to gasoline and diesel fuel.

Agricultural Inputs

The seven major farm inputs considered are: fertilizers, farm machinery, pesticides, formula feeds, petroleum fuels, feeder livestock, and crop seeds. Other miscellaneous inputs such as contract combining and aerial applications of pesticides were not included because of data limitations and the small magnitude of transportation fuel re-

Table 5
TRANSPORTATION FUEL REQUIREMENTS IN THE FOOD SYSTEM, 1977[2]

	Ton-miles[a] (millions)	Diesel fuel (million gal)	Gasoline (million gal)
Agricultural inputs	82,359	505	114
Farm commodities[b]	141,098	1,119	297
Processed foods	115,033	1,268	—
Total	338,670	2,892	411

[a] A ton-mile is a measure of product shipment and is defined as the movement of a ton of product a distance of 1 mile. Hence, if 10 tons are shipped 10 miles, ton-miles would equal 100.

[b] Includes off-farm transportation only.

Table 6
ESTIMATED FUEL REQUIREMENTS TO TRANSPORT SELECTED FARM INPUTS[2]

Item	Ton-miles (millions)	Btus (billions)	Diesel fuel (million gal)	Gasoline (million gal)
Fertilizers	43,543	27,751	190	5
Farm machinery	1,905	4,372	27	0
Pesticides	506	490	4	0
Formula feed	18,506	26,058	164	30
Petroleum fuels	13,733	35,020	59	59
Feeder livestock	3,281	8,365	42	20
Crop seeds	1,065	2,716	19	0
Total	82,539	104,722	505	114

quirements. The distribution of fertilizers, petroleum products, and formula feeds account for about 80% of the fuel consumed in transporting farm inputs (Table 6).

Agricultural Commodities

Transportation of grains and soybeans accounts for about two thirds of total ton-miles for agricultural commodities and consumes around 60% of total gasoline requirements. The transportation of grains and soybeans and fruits and vegetables together accounts for more than 90% of total ton-miles and nearly 85% of total diesel fuel requirements (Table 7).

MANUFACTURED FOOD PRODUCTS

Manufactured food products are generally shipped from processing plants to packing houses or regional warehouses for distribution to retail stores. Deviations from this pattern may include aggregation of product at the wholesale level or direct shipment to the retail outlet, but the bulk of sales by the major retail chains are handled through large regional warehouses.

At the regional warehouse, product loads are broken down and generally reshipped to retail stores in mixed loads. Approximately 277 million tons of manufactured food products and about 20 million tons of fresh fruits and vegetables were shipped to warehouses in 1977. Fuel requirements for transporting fruits and vegetables to ware-

Table 7
ESTIMATED FUEL REQUIREMENTS TO TRANSPORT AGRICULTURAL COMMODITIES TO PROCESSORS[2]

Item	Ton-miles (millions)	Diesel fuel (million gal)	Gasoline (million gal)
Grains and soybeans	93,543	451	203
Fruits and vegetables	33,721	503	22
Livestock	3,706	54	14
Milk	3,172	38	23
Sugar	1,846	15	11
Poultry and eggs	1,117	26	16
Peanuts	79	1	1
Cotton	2,716	27	6
Tobacco	208	4	1
Total	141,098	1,119	297

houses are included in the estimates for agricultural commodities. Ton-miles and fuel requirements for movements of food products from processors and packing houses to regional warehouses, and from warehouses to retail outlets are shown below:[2]

	Tons (millions)	Average mileage (miles)	Ton-miles (millions)	Total Btus (billions)	Diesel fuel (million gallons)
To warehouse	277	765	101,227	139,774	998
To supermarket	297	50	14,850	37,868	270
Total	—[a]	—	116,077	177,642	1,268

[a] — = Not applicable.

Gasoline is also used in the wholesale/retail sector, but the amounts are small in comparison to diesel fuel use. Although gasoline is relied upon heavily for the local delivery of bakery, produce, and specialty items, the tonnages and low mileages involved would cause these requirements to be small.

Fuel requirements for transporting manufactured foods from processors to first destinations (primarily warehouses) are shown in Table 8.

FOOD RETAILING

Annual energy consumption in supermarkets totaled almost 200 trillion Btu in 1976. Most of the energy consumed, around 85% of the total, is electricity used for refrigeration, lighting, and mechanical equipment. Natural gas and fuel oils are used in significant quantities for space heating. Estimates of total fuels and electricity used in supermarkets in 1978 are shown in Table 9.

Of electricity use, refrigeration accounts for the largest share, with lighting accounting for much of the remainder. A relatively small amount of electricity use is associated with wrapping and packaging processes and other mechanical applications. Natural gas accounts for most of the non-electrical space heating and hot water requirements, with the remainder, fuel oil, used almost exclusively in the Northeast. The increased use of heat reclaim equipment, which efficiently utilizes the heat generated from the operation of refrigeration equipment, has resulted in a leveling off of natural gas use.

Table 8
TRANSPORTATION OF FOOD AND KINDRED PRODUCTS, 1977

TCC Code[a]	Ton-miles (millions)			Total Btus (millions)			Diesel fuel (million gal)			
	Rail	Truck	Barge	Rail	Truck	Barge	Rail	Truck	Barge	Total
Food and kindred products TCC 20[b]	53,234	41,175	6,818	37,453	100,332	1,989	269	716	13	998
Meat products TCC 201	4,033	11,401	31	3,045	29,734	5	22	212	—	234
Dairy products TCC 202	1,250	2,200	7	944	5,998	1	7	43	—	50
Canned and preserved fruits and vegetables TCC 203	12,709	9,486	247	8,858	24,190	43	64	173	—	237
Grain mill products TCC 204[c]	13,964	3,584	0	9,734	9,139	0	69	65	0	134
Bakery products TCC 205	268	760	13	187	1,060	21	1	8	—	9
Sugar — beet and cane TCC 206	4,586	989	3,227	3,197	2,521	1,371	23	18	10	51
Confectionery and related products TCC 207	376	1,714	32	262	4,370	5	2	31	—	33
Beverages and flavoring extracts TCC 208	5,709	5,850	153	3,979	12,945	26	29	92	—	121
Miscellaneous food preparations TCC 209	10,534	4,069	3,003	7,217	10,375	517	52	74	3	129

[a] Commodities are classified by the 1972 Transportation Commodity Code (TCC).
[b] May not equal column totals due to rounding.
[c] Excludes SIC 20421 — Prepared feed; animal, fish, and poultry in SIC 204. Transportation of processed feeds is included in the farm inputs section.

Table 9
ENERGY USE IN SUPERMARKETS

Item	Physical quantity	Btus (billions)	Percent of total energy (percent)
Electricity (million kWh)	49,330	168,364	85
Natural gas (million mcf)	24.52	25,746	13
Fuel oil (million gal)	28.3	3,960	2
Total	—[a]	198,070	100

[a] — = Not applicable.

From *Energy Usage in Supermarkets,* Super Market Institute, Spec. Res. Rep. No. 19, Chicago, 1976.

Virtually no new natural gas hook-ups have occurred since 1973. As more heat reclaim equipment is installed, the level of use of natural gas and fuel oil in retail supermarkets may begin to decline slightly. Total electrical usage and average heating energy requirements by region are shown in Table 10.

Table 10
ENERGY USAGE IN SUPERMARKETS BY REGION

Region	Heating Btus (Btus/100 ft² of selling area per day)	Total electricity usage (Kwh/100 ft² of selling area per day)
Northern	46,780	24.97
Central	36,748	25.91
Southern	19,871	23.00
Mountain/Northwest	36,201	29.49
Southwest	23,603	24.08
All stores	37,663	25.05

From *Energy Usage in Supermarkets,* Super Market Institute, Spec. Res. Rep. No. 19, Chicago, 1976.

In addition to the energy used in retail supermarkets, substantial amounts of electricity are used in warehousing operations, primarily for cold storage and freezers. Unfortunately, there are few data available which would indicate the magnitude of electricity requirements for refrigeration in wholesaling operations and supermarket chain warehouses. Aggregate estimates based on cubic footage of cold storage and freezer space are not reliable since electricity requirements vary as a result of weather conditions, size of unit, and frequency of inventory turnover.

REFERENCES

1. **Anon.,** Annual Survey of Manufacturers, 1976, Fuels and Electric Energy Consumed, U.S. Department of Commerce, Washington, D.C., 1978.
2. **Anon.,** Transportation Energy Requirements in the Food and Fiber System, Economics, Statistics, and Cooperatives Service, U.S. Department of Agriculture, unpublished data.
3. **Anon.,** Energy Usage in Super Markets, Super Market Institute, Spec. Res. Rep. No. 19, Chicago, 1976.

Cereals and Feed Grains Including Soybeans

STRUCTURE OF CEREAL GRAINS

Y. Pomeranz and D. B. Bechtel

The cereal grain is a one-seeded fruit called a caryopsis, in which the fruit coat is adherent to the seed. As the fruit ripens, the pericarp (fruit wall) becomes firmly attached to the wall of the seed proper. The pericarp, seed coats, nucellus, and aleurone cells form the bran. The embryo occupies only a small part of the seed. The bulk of the seed is taken up by the endosperm, which constitutes a food reservoir.[1-3]

The floral envelopes (modified leaves known as lemma and palea), or chaffy parts, within which the caryopsis develops, persist to maturity in the grass family. If the chaffy structures envelope the caryopsis so closely that they remain attached to it when the grain is threshed (as in rice and most varieties of oats and barley), the grain is considered to be covered. However, if the caryopsis readily separates from the floral envelopes on threshing, as with common wheats, rye, hull-less barleys, and the common varieties of corn, these grains are considered to be naked.

The structure of the wheat kernel is shown in Figure 1. The dorsal side of the wheat grain is rounded, while the ventral side has a deep groove or crease along the entire longitudinal axis. At the apex or small end (stigmatic end) of the grain is a cluster of short, fine hairs known as brush hairs. The pericarp, or dry fruit coat, consists of four layers: the epidermis, hypodermis, cross cells, and tube cells. The remaining tissues of the grain are the inner bran (seed coat and nucellar tissue), endosperm, and embryo (germ). The aleurone layer consists of large rectangular, heavy-walled, supposedly starch-free cells. Botanically, the aleurone is the outer layer of the endosperm, but as it tends to remain attached to the outer coats during wheat milling, it is shown in the diagram as the innermost bran layer.

The embryo (germ) consists of the plumule and radicle, which are connected by the mesocotyl. The scutellum serves as an organ for food storage. The outer layer of the scutellum, the epithelium, may function as either a secretory or absorption organ. In a well-filled wheat kernel, the germ comprises ∿2 to 3% of the kernel, the bran 13 to 17%, and the endosperm the remainder. The inner bran layers (the aleurone) are high in protein, whereas, the outer bran (pericarp, seed coats and nucellus) is high in cellulose, hemicelluloses, and minerals; biologically, the outer bran functions as a protective coating and remains practically intact when the seed germinates. The germ is high in proteins, lipids, sugars (chiefly sucrose), and minerals; the endosperm consists largely of starch grains surrounded by protein.

Grains of other cereals are similar in structure to wheat. The corn grain is the largest of all cereals (Figure 2). The kernel is flattened, wedge-shaped, and broader at the apex than at its attachment to the cob. The aleurone cells contain much protein and oil and also contain the pigments that make certain varieties appear blue, black, or purple. Two types of starchy endosperms — horny and floury — are found beneath the aleurone layer. The horny endosperm is harder and contains a higher level of protein. In dent corn varieties, the horny endosperm is found on the sides and back of the kernel and bulges in toward the center at the sides. The floury endosperm fills the crown (upper part) of the kernel, extends downward to surround the germ, and shrinks as dent corn matures, causing an indentation at the top of the kernel. In a typical dent corn, the pericarp comprises ∿6%, the germ 11%, and the endosperm 83% of the kernel. Flint corn varieties contain more horny than floury endosperm. A small amount of floury endosperm extends around the top of the kernel, with the bulk of the floury portion around the embryo.

The common varieties of oats have the fruit (caryopsis) enveloped by a hull com-

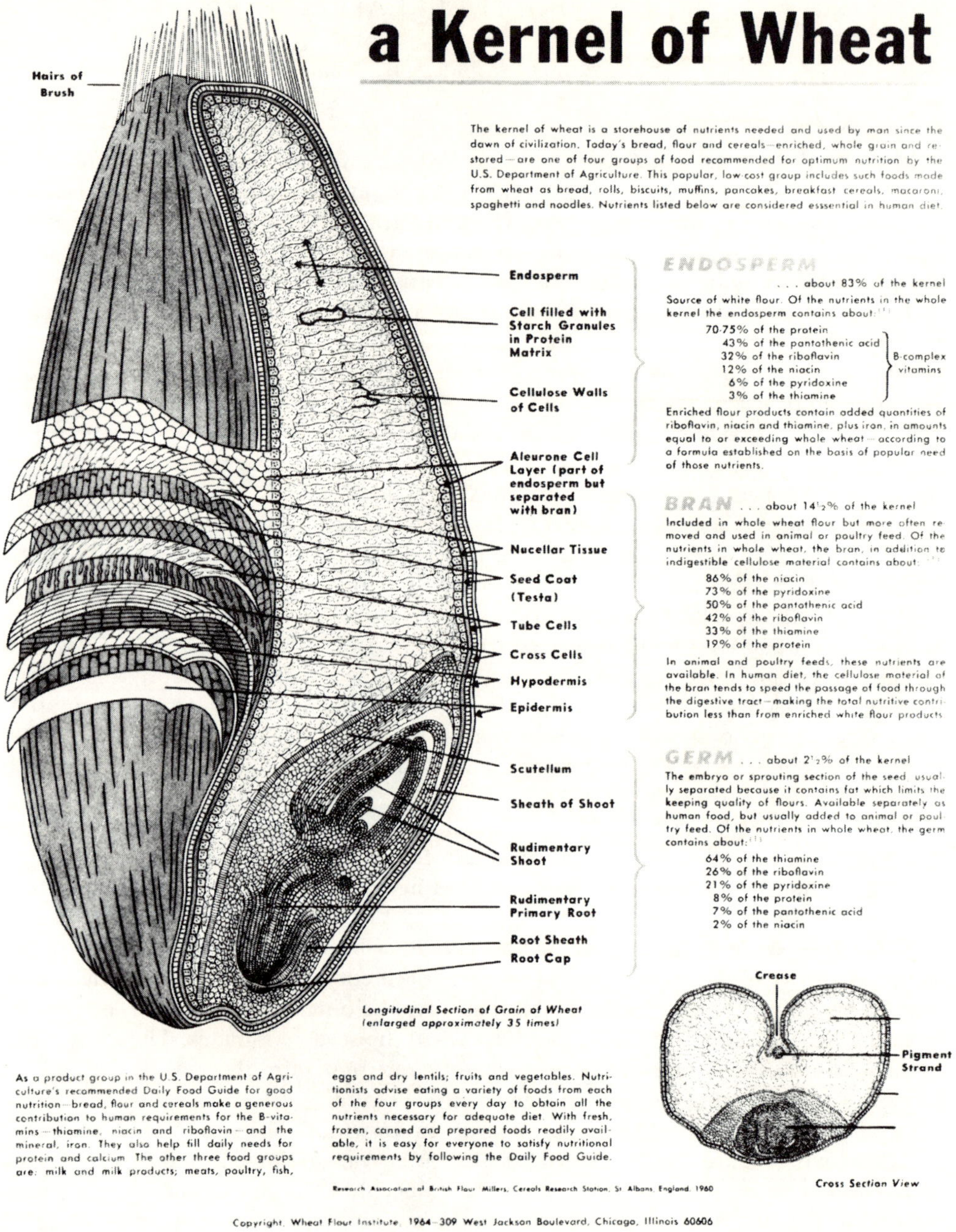

FIGURE 1. (A) Longitudinal; (B) cross section of a wheat kernel. (Magnification approximately ×35.) (Reproduced with permission of the Wheat Flour Institute, Chicago.)

posed of certain floral envelopes. Naked or hull-less oat varieties are not extensively grown. In light thin oats, hulls may comprise as much as 45% of the grain; in very heavy or plump oats, they may represent only 20%. The hull normally makes up ~30% of the grain. Oat kernels, obtained by removing hulls, are called groats.

Rice is a covered cereal; in the threshed grain (or rough rice), the kernel is enclosed in a tough siliceous hull, which renders it unsuitable for human consumption. When this hull is removed during milling, the kernel (or caryopsis), comprised of the pericarp

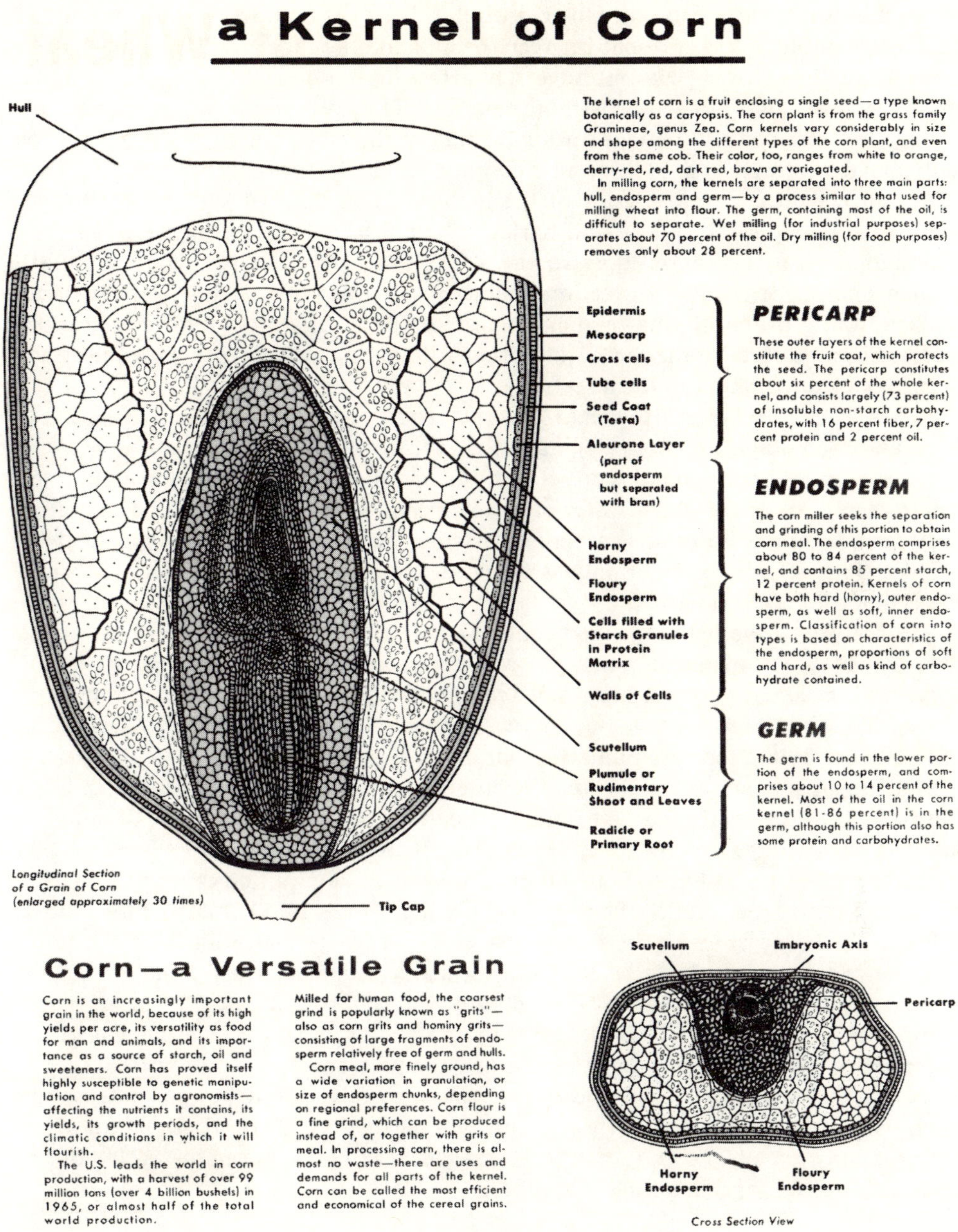

FIGURE 2. (A) Longitudinal; (B) cross section of a corn kernel. (Magnification approximately ×30.) (Reproduced with permission of the Wheat Flour Institute, Chicago.)

(outer bran) and the seed proper (inner bran, endosperm, and germ), is known as brown rice or sometimes as unpolished rice. Brown rice is in little demand as a food. Unless stored under favorable conditions, it tends to become rancid and is more subject to insect infestation than are the various forms of milled rice. When brown rice is subjected to further milling processes, the bran and germ are removed and the purified endosperms are marketed as white rice or polished rice.

The Hull and Bran Layers

The structure and adherence of the hull are important in protecting the germinating grain and in the malting process. One reason that barley is uniquely suited for malting is that a cementing layer is present between the hull and the caryopsis. The hull restricts excessive seedling growth without adversely affecting the desirable enzymic degradation of insoluble high molecular weight materials. The adhering hull also protects the seedling from mechanical damage during turning of the malt, and provides a filtration bed during the extraction of soluble malt components in the mashing process.

The hull, as such, or as a result of the high concentration of silica, can slow the attack of storage insects on rice and barley. The palea and lemma in barley and rice are held together by two hook-like structures. The bran (pericarp, seed coats, nucellus, and aleurone) affords little protection against insect infestations. Lack of resistance to infestation is probably due to the thinness of the bran, which allows easy insect penetration, and because the large quantities of lipid and protein present in the aleurone provide nourishment to the insects.

The outer pericarp layers of wheat (epidermis and hypodermis) have no intercellular spaces and are composed of closely adhering, thick-walled cells. The inner layers of the pericarp, on the other hand, consist of thinner-walled cells and often contain intercellular spaces, through which water can move rapidly and in which molds are commonly found. Similarly, molds can enter through the large intercellular spaces at the base of the kernel where the grain was detached from the plant at harvest and where there is no protective epidermis.

The structure of the pericarp, seed coats, and nucellus also explains how the kernel reacts to water. Following initial rapid water absorption, the rate decreases significantly. The seed coat offers more water resistance than the nucellus. The sponge-like pericarp absorbs water readily, but the water is prevented from readily entering the kernel interior by the seed coat and nucellar cuticles. Since the cuticles are deficient in the attachment region, water can readily enter the kernel, mainly via the germ. The ability of the germ to absorb and hold considerable amounts of water probably accounts, in part, for the susceptibility of the germ to attack by molds.

An intact grain stores better than a damaged or ground grain; deterioration changes (i.e., rancidity, off-flavors, etc.) occur slowly in the whole grain but quite rapidly after the grain has been ground. The hull, apparently, prevents the grain from becoming rancid by protecting the bran layers from mechanical damage during harvesting and subsequent handling. Once rough rice is dehulled, it rapidly becomes rancid, primarily resulting from free fatty acid production by the action of lipase. The fatty acids, in turn, become oxidized and cause rancidity. The lipids and lipase are normally compartmentalized in the aleurone and germ cells. Possibly, the dehulling process and subsequent handling of dehulled rice result in aleurone cell disruption and allow the rice to become rancid. The hull, therefore, is necessary to prevent cell disruption during harvesting, storage, and handling procedures.

The Germ

The site of the germ in the kernel and whether it protrudes or is well protected by adjacent layers will govern the extent to which the germ is retained intact during threshing and thus will determine the grain's usefulness for seeding or malting. The ease with which the germ is removed from the caryopsis during milling depends on several factors.

The germ is a separate structure and generally can be easily separated from the rest of the cereal grain. However, the scutellar epithelium (located next to the endosperm) has finger-like cells, which in wheat are attached to one another for about one third of their length.[4] The free ends protrude toward the adjacent starchy endosperm cells. The protruding epithelial cells may form an amorphous cementing layer between germ and endosperm. If part of the layer projects into the spaces between the finger-like cells of the scutellar structure, it may be difficult to separate the germ from the endosperm unless the cementing layer is softened. The softening may be accomplished by steeping in corn wet milling or by conditioning in wheat milling.

Germ separation is also enhanced by the fact that the germ takes up water faster and swells more readily than the endosperm. The strains resulting from differential swelling contribute to easy separation in milling.

Grinding and Milling

The starchy endosperm is the portion of the caryopsis that is used primarily as a food source. In the milling of wheat and rice, the bran layers and the germ are separated, thereby removing much of the lipid and protein.

Endosperm structure, composition, and associated texture govern both power requirements in milling and particle size of the resultant products. In wheat flour milling, the size and shape of the crease affect yield and composition of flour because the bran in the crease area is difficult to separate from the starchy endosperm. The thickness and irregularity of the aleurone around the starchy endosperm and in the crease make it impossible to increase flour yields in wheat milling by processes that successively "peel" outer layers.

Wheat varieties differ in vitreosity and hardness of their mature endosperms. The importance of grain hardness lies in its effect on the milling properties, and in particular on the amounts of semolina and farina and of damaged starch produced during milling.

Particles from the horny endosperm break along the wall of the cell and usually have sharp outlines. Floury endosperm breaks largely through the cell contents (rather than chiefly along the walls of the cells) and yields particles with irregular edges. The water-soluble proteins appear to be important in the genetic determination of wheat hardness.[5,6] According to Stenvert and Kingswood[7] however, differences in hardness are related to the continuity of the protein matrix and the strength with which the protein physically entraps starch granules.

Differences in the breakdown of wheats during grinding have important technological consequences. Thus, the clean separation of undamaged starch makes soft wheats the preferred material for air classification — process in which differences in specific gravity are used to separate finely pulverized wheat flours into protein-high and protein-low fractions. Air classification is the basis of numerous patents for producing "tailor-made" flours from a single flour.

Another consequence resulting from the variable amount of starch damaged during grinding is that mechanically damaged starch absorbs more water than does undamaged starch. That is of particular importance to bakers in countries in which the amount of water added to a dough is essentially limited only by the amount required to obtain a dough of adequate consistency for handling and for producing a well-developed bread of adequate shelf life. Damaged starch granules in flour from hard wheat are more susceptible to enzymatic degradation by amylases than are those from soft wheats; that is important in producing bread in sugar-free formulas in which fermentable sugars are primarily those produced by the action of malt on damaged starch, rather than by added sucrose.

Grain Drying

Changes in texture and structure during the drying of corn and rice are important in minimizing breakage during handling. Excessive cracks reduce the value of corn for producing foods such as breakfast cereals. Harsh heat treatment during grain drying may reduce starch yields, impair quality of the starch, and create difficulties in corn wet milling. Overheated corn also may cause difficulties in brewing beer and in distilling alcoholic beverages. The starch granules are imbedded in a proteinaceous matrix that hardens during overheating. The hardened matrix prevents the starch from being attacked by enzymes and converted to alcohol. In rice milling, harsh drying and accompanying structural cracks substantially reduce yields of head rice and increase the amounts of "brokens", and thus cause economic losses to the miller. The method of rice drying may also affect the texture and color of the milled rice and result in off-color, especially objectionable browning.

Nutritional Implications

The chemical composition of different cereal grains varies widely, since it is influenced by genetic, soil, cultural, and climatic factors. Variations are encountered in the relative amounts of proteins, lipids, carbohydrates, pigments, vitamins, and total ash; mineral elements present also vary widely. Cereals are characterized by relatively low protein and high carbohydrate contents; the carbohydrates consist essentially of starch (90% or more), pentosans, and sugars.

The various components are not uniformly distributed in the kernel. The hulls and bran are high in cellulose, pentosans, and ash; the germ is high in lipid and rich in proteins, sugars, and generally, ash. The endosperm contains the starch, has a lower protein content than the germ and the bran (in some cereals), and is low in fat and ash.

In recent years many studies have been concerned with the improvement of the nutritional value of cereal grains. Simple changes in grain morphology could be the basis of improvement. The embryo of cereal seeds is rich in protein (up to 38%), and the protein may contain about 7% lysine. Selection for larger embryos is particularly important if the whole seed (rather than starchy endosperm) is to be consumed. Variations in the number of aleurone cells of the endosperm exist in corn, rice, and barley. The aleurone layer is rich in protein having a good amino acid balance. Selection for a high aleurone cell number could be useful, provided the high number is associated with improved nutritional value. Increasing the relative surface area of the seed could lead to an increase in the aleurone layer. In early triticale (crosses between wheat and rye) and in *Avena sterilis* (a wild hexaploid oat), such an increase results from the otherwise undesirable development of long and thin, very small, or shrunken kernels. Both in wheat and in rice, much of the protein is concentrated in the aleurone and outermost subaleurone. Those tissues are diverted to feed during the milling and polishing of rice or during the milling of highly refined wheat flour. "Restructuring" cereal grains to obtain a more even distribution of protein throughout the whole endosperm would increase the protein content of milled products.

The hulls are rich in fibrous materials and low in protein. Many hull-less varieties of barley and oats have substantially more protein than the hulled varieties do. However, the low yield of the available hull-less varieties discourages their cultivation.

REFERENCES

1. **Pomeranz, Y. and MacMasters, M. M.,** Structure and composition of the wheat kernel, *Baker's Dig.,* 42(2), 24, 28, and 32, 1968.
2. **Pomeranz, Y. and MacMasters, M. M.,** Wheat and other cereals, in *Kirk Othmer's Encyclopedia of Chemical Technology,* John Wiley & Sons, New York, 1970, 22.
3. **Pomeranz, Y. and Bechtel, D. B.,** Structure of cereal grains, in *Cereals '78: Better Nutrition for the World's Millions,* Pomeranz, Y., Ed., American Association of Cereal Chemists, St. Paul, Minn., 1978, 85.
4. **MacMasters, M. M.,** Important aspects of kernel structure, *Trans. ASAE,* 5(2), 247, 1962.
5. **Barlow, K. K., Buttrose, M. S., Simmonds, D. H., and Vesk, M.,** The nature of the starch-protein interface in wheat endosperm, *Cereal Chem.,* 50, 443, 1973.
6. **Simmonds, D. H., Barlow, K. K., and Wrigley, C. W.,** The biochemical basis of grain hardness in wheat, *Cereal Chem.,* 50, 553, 1973.
7. **Stenvert, N. L. and Kingswood, K.,** The influence of the physical structure of the protein matrix on wheat hardness, *J. Sci. Food Agric.,* 28, 11, 1977.

COMPOSITION OF CEREAL GRAINS

Y. Pomeranz

Cereals comprise members of the grass family, Gramineae, that are grown for their edible grains. Most important are wheat, rice, maize (corn), grain sorghum, barley, oats, and rye (Table 1). Some of their physical properties are given in Table 2. Cereals are important foods because they supply the most calories per acre, can be stored safely for a long time, and can be processed into many foods acceptable throughout the world. Cereals are adapted to widely varying soil and climatic conditions, can be cultivated with only a small amount of labor, and have high food value. Most developing countries rely heavily on cereals as food sources. More than two thirds of the world's cultivated area is planted to grain crops. These crops provide more than half of the calories consumed for human energy; in developing areas of the world they provide about two thirds of the total food. They also contribute greatly to meat and livestock products.

The chemical composition of the dry matter of different cereal grains, in common with other foods of plant origin, varies widely. Proximate analysis of important cereal grains is summarized in Table 3. Variations are encountered in the relative amounts of proteins, lipids, carbohydrates, pigments, vitamins, and ash; mineral elements present also vary widely.

As a food group, cereals are characterized by relatively low-protein and high-carbohydrate contents; the carbohydrates consist essentially of starch (90% or more), dextrins, pentosans, and sugars.

The various components are not uniformly distributed in the different kernel structures. Table 4 gives approximate size and properties of the principal parts that comprise the mature kernel of different cereals; Table 5 compares the weight and composition of the main anatomical parts of the wheat kernel with the composition of flours that vary in milling extraction rate. The hulls and pericarp are high in cellulose, pentosans, and ash; the germ is high in lipid content and rich in proteins, sugars, and ash constituents. The endosperm contains the starch and is lower in protein content than the germ and, in some cereals, than bran; it is also low in crude fat and ash constituents. As a group, cereals are low in nutritionally important calcium, and its concentration and that of other ash constituents is greatly reduced by the milling processes used to prepare refined foods. In these processes, hulls, germ, and bran, which are the structures rich in minerals and vitamins, are more or less completely removed. Mineral contents of cereal grains and cereal products are given in Table 6.

All cereal grains contain vitamins of the B group but all are completely lacking in vitamin C (unless the grain is sprouted) and vitamin D. Yellow corn differs from white corn and the other cereal grains in containing carotenoid pigments (principally cryptoxanthin, with smaller quantities of carotenes), which are convertible in the body to vitamin A. Wheat also contains yellow pigments but they are almost entirely xanthophylls, which are not precursors of vitamin A. The oils of the embryos of cereal grains are rich sources of vitamin E. The relative distribution of different vitamins in various kernel structures is not uniform, although the endosperm invariably contains the least (Table 7).

Protein contents of wheat and barley are important indexes of their quality for manufacture of various food products. The bread-making potentialities of bread wheat are largely associated with the quantity and quality of its protein.

The cereal grains contain water-soluble proteins (albumins), salt-soluble proteins (globulins), alcohol-soluble proteins (prolamins), and acid- and alkali-soluble proteins (glutelins). The prolamins are characteristic of the grass family and, together with the glutelins, comprise the bulk of the proteins of cereal grains. The following are names given to prolamins in proteins of the cereal grains: gliadin in wheat, hordein in barley, zein in maize, avenin in oats, kafferin in grain sorghum, and secalin in rye. Distribution of cereal endosperm proteins according to their solubility is summarized in Table 8.

The various proteins are not distributed uniformly in the kernel. Thus, the proteins fractionated from the inner endosperm of wheat consist chiefly of a prolamin (gliadin) and glutelin (glutenin) apparently in approximately equal amounts. The embryo proteins consist of nucleoproteins, an albumin (leucosin), a globulin, and proteoses, while in wheat bran a prolamin predominates with smaller quantities of albumins and globulins. When water is added, the wheat endosperm proteins, gliadin and glutenin, form a tenacious colloidal complex known as gluten (see Figure 1). Gluten is responsible for the superiority of wheat over the other cereals for the manufacture of leavened products, since it makes possible the formation of a dough that retains the carbon dioxide produced by yeast or chemical leavening agents. The gluten proteins collectively contain about 17.55% nitrogen; hence, in estimating the crude protein content of wheat and wheat products from the determination of total nitrogen, the factor 5.7 is normally employed rather than the customary value of 6.25, which is based on the assumption that, on the average, proteins contain 16% nitrogen.

As a class, cereal proteins are not so high in biological value as those of certain legumes, nuts, or animal products (Tables 9 and 10). Zein, the prolamin of corn, lacks lysine and is low in tryptophan. The limiting amino acid in wheat endosperm proteins is lysine. While biological values of the proteins of entire cereal grains are greater than those of the refined mill products, which consist chiefly of the endosperm, the American diets normally comprise various cereals, as well as animal products. Under those conditions, different proteins tend to supplement each other, and the cereals are important and valuable sources of amino acids for the synthesis of body proteins.

The carbohydrate contents, including crude fiber of various grains, are shown in Table 11. The main form of carbohydrate is starch, which is the main source of calories provided by the grains. The major portion of the carbohydrates is in the starchy endosperm.

Fatty acids in cereals occur in three main types — neutral lipids, glycolipids, and phospholipids. Total lipids and individual fatty acids in cereal grains and some of their products are shown in Table 12. The lipids in cereals are relatively rich in the essential fatty acid, linoleic acid. Saturated fatty acids (mainly palmitic) represent less than 25% of the total acids for most grains.

In summary, cereal grains are a diversified and primary source of nutrients. Their high starch contents make them major contributors of calories; they also contribute to our needs for proteins, lipids, vitamins, and minerals. Vitamins and minerals lost during milling into refined food products (wheat flour or white rice) can be, and in many countries are, replaced by nutrient fortification. The composition of cereal grains and their milled products make them uniquely suited in the prduction of wholesome, nutritional, and consumer-acceptable foods.

Table 1
COMMON CEREAL SPECIES

Tribe	Genus and species	Common name
Aveneae	*Avena sativa*	Oats
Hordeae	*Secale cereale*	Rye
	Hordeum vulgare	Barley
	Triticosecale	Triticale
	Triticum aestivum	Wheat (bread wheats)
	T. durum	Durum wheat (bread wheat)
Oryzeae	*Oryza sativa*	Rice (indica, japonica, javanica)
	O. glaberrima	African rice
Zizanieae	*Zizania aquatica*	Wild rice
Paniceae	*Panicum miliaceum*	Common millet
	Echinocloa spp.	Japanese millet
Andropoganeae	*Sorghum vulgare*	Sorghum
	S. bicolor	
Maydeae	*Zea mays*	Corn

From Simmonds, D. H., in *Cereals '78: Better Nutrition for the World's Millions,* Pomeranz, Y., Ed., American Association of Cereal Chemists, St. Paul, Minn., 1978, 105.

Table 2
SOME PHYSICAL PROPERTIES OF CEREAL GRAINS

Name	Length (mm)	Width (mm)	Grain wt (mg)	Bulk density (kg/m³)	Unit density (kg/m³)
Rye	4.5 — 10	1.5 — 3.5	21	695	—
Sorghum	3 — 5	2.5 — 4.5	23	1360	—
Paddy rice	5 — 10	1.5 — 5	27	575—600	1370 — 1400
Oats	6 — 13	1 — 4.5	32	356 — 520	1360 — 1390
Wheat	5 — 8	2.5 — 4.5	37	790 — 825	1400 — 1435
Barley	8 — 14	1 — 4.5	37	580 — 660	1390 — 1400
Maize	8 — 17	5 — 15	285	745	1310
Bullrush millet	3 — 4	2 — 3	11	760	1322
Wild rice	8 — 20	0.5 — 2	22	388 — 775	—
T'of	1 — 1.5	0.5 — 1	0.3	880	—
Findi	1	1 — 1.5	0.4	790	—
Finger millet	1.5	1.5	—	—	—

From Muller, H. G., in *Industrial Uses of Cereals,* Pomeranz, Y., Ed., American Association of Cereal Chemists, St. Paul, Minn., 1973, 24.

Table 3
PROXIMATE ANALYSIS OF IMPORTANT CEREAL GRAINS (% DRY WEIGHT)

Cereal	Nitrogen	Protein[a]	Fat	Fiber	Ash	N.F.E.[b]	Ref.
Barley							
Grain	1.2—2.2	11	2.1	6.0	3.1	—	3
Kernel	1.2—2.5	9	2.1	2.1	2.3	78.8	11
Maize	1.4—1.9	10	4.7	2.4	1.5	72.2	8, 11
Millet	1.7—2.0	11	3.3	8.1	3.4	72.9	8, 11
Oats							
Grain	1.5—2.5	14	5.5	11.8	3.7	—	1
Kernel	1.7—3.9	16	7.7	1.6	2.0	68.2	11
Rice							
Brown	1.4—1.7	8	2.4	1.8	1.5	77.4	6, 7, 11
Milled			0.8	0.4	0.8	—	
Rye	1.2—2.4	10	1.8	2.6	2.1	73.4	2, 8, 11
Sorghum	1.5—2.3	10	3.6	2.2	1.6	73.0	8, 11
Triticale	2.0—2.8	14	1.5	3.1	2.0	71.0	4, 11
Wheat (bread)	1.4—2.6	12	1.9	2.5	1.4	71.7	5, 11, 12
Wheat (durum)	2.1—2.4	13			1.5	70.0	9
Wild rice	2.3—2.5	14	0.7	1.5	1.2	74.4	8

[a] Typical or average figure.
[b] N.F.E. = Nitrogen-free extract. This is an approximate measure of the starch content.

From Simmonds, D. H., in *Cereals '78: Better Nutrition for the World's Millions,* Pomeranz, Y., Ed., American Association of Cereal Chemists, St. Paul, Minn., 1978, 107.

Table 4
APPROXIMATE GRAIN SIZE AND PROPORTIONS OF THE PRINCIPAL PARTS COMPRISING THE MATURE KERNEL OF DIFFERENT CEREALS

Cereal	Grain weight (mg)	Embryo (%)	Scutellum (%)	Pericarp (%)	Aleurone (%)	Endosperm (%)
Barley	36—45 (41)	1.85	1.53	18.3	79.0	79.0
Bread wheat	30—45 (40)	1.2	1.54	7.9	6.7—7.0	81—84
Durum wheat	34—46 (41)	1.6	1.6	12.0	86.4	86.4
Maize	150—600 (350)	1.15	7.25	5.5	5.5	82
Oats	15—23 (18)	1.6	2.13	28.7—41.4	28.7—41.4	55.8—68.3
Rice	23—27 (26)	2—3	1.5	1.5	4—6	89—94
Rye	15—40 (30)	1.8	1.73	12.0	85.1	85.1
Sorghum	8—50 (30)	7.8—12.1	7.8—12.1	7.3—9.3	7.3—9.3	80—85
Triticale	38—53 (48)	3.7	3.7	14.4	81.9	81.9

From Simmonds, D. H., in *Cereals '78: Better Nutrition for the World's Millions,* Pomeranz, Y., Ed., American Association of Cereal Chemists, St. Paul, Minn., 1978, 109.

Table 5
WEIGHT, ASH, PROTEIN, LIPID, AND CRUDE FIBER CONTENTS OF MAIN ANATOMICAL PARTS OF THE WHEAT KERNEL AND OF FLOURS OF DIFFERENT MILLING EXTRACTION RATE

Parameter (%)	Wheat kernel fractions				Milling extraction		
	Pericarp	Aleurone layer	Starchy endosperm	Germ	75	85	100
Weight	9	8	80	3	75	85	100
Ash	3	16	0.5	5	0.5	1	1.5
Protein	5	18	10	26	11	12	12
Lipid	1	9	1	10	1	1.5	2
Crude fiber	21	7	>0.5	3	>0.5	0.5	2

From Pomeranz, Y. and MacMasters, M. M., *Baker's Dig.*, 42(4), 24, 1978. With permission.

Table 6
MINERAL CONTENTS OF CEREAL GRAINS AND CEREAL PRODUCTS[4-9]

	Ca (mg/100 g)	Fe (mg/100 g)	Mg (mg/100 g)	P (mg/100 g)	K (mg/100 g)	Na (mg/100 g)	Cu (mg/100 g)	Mn (mg/100 g)	Zn (mg/100 g)
Barley	80	10	120	420	560	3	0.76	1.63	1.53
Buckwheat	110	4	390	330	450	—	0.95	3.37	0.87
Maize									
Grain	30	2	120	270	280	1	0.21	0.51	1.69
Bran	30	—	260	190	730	—	—	1.61	—
Germ	90	90	280	560	130	—	1.10	0.90	—
Millet (proso)	50	10	160	280	430	—	2.16	2.91	1.39
Oats	100	10	170	350	370	2	0.59	3.82	3.4
Rice									
Grain	40	3	60	230	150	9	0.33	1.76	1.8
White	30	1	20	120	130	5	0.29	1.09	1.3
Rye	60	10	120	340	460	1	0.78	6.69	3.05
Sorghum	40	4	170	310	340	—	0.96	1.45	1.37
Wheat									
Grain	50	10	160	360	520	3	0.72	4.88	3.4
Bran	140	70	550	1170	1240	9	1.23	11.57	9.8
Triticale	20	4	—	—	385	—	0.52	4.26	0.02

Table 7
VITAMIN CONTENTS OF CEREAL GRAINS AND CEREAL PRODUCTS[4,7,9-16]

	Thiamin (mg/100 g)	Riboflavin (mg/100 g)	Niacin (mg/100 g)	Vitamin B_6 (mg/100 g)	Folic acid (mcg/100 g)	Pantothenic acid (mg/100 g)	Biotin (mcg/100 g)	Vitamin E (IU/100 g)
Barley (pearled)	0.23	0.13	4.52	0.26	67	0	6	1
Buckwheat	0.60		4.4			1		
Maize	0.37	0.12	2.2	0.47	26	1	21	2
Oats	0.67	0.11	0.8	0.21	104	1	13	3
Millet	0.73	0.38	2.3					1
Rice								
Brown	0.34	0.05	4.7	0.62	20	2	12	2
Polished	0.07	0.03	1.6	0.04	16	1	5	1
Rye	0.44	0.18	1.5	0.33	34	1	—	2
Sorghum	0.38	0.15	3.9					
Wheat	0.57	0.12	7.4	0.35	78	1	6	1
Wheat germ	2.01	0.68	4.2	0.92	328	2		
Wheat bran	0.72	0.35	21.0	1.38	258	3	14	—
Wheat flour (patent)	0.13	0.04	2.1	0.05	25	1	1	—

Table 8
DISTRIBUTION OF CEREAL ENDOSPERM PROTEINS ACCORDING TO THEIR SOLUBILITY[a]

Cereal	Protein range (% dwb)	Albumins (water-soluble)	Globulins (salt-soluble)	Prolamin (alcohol-soluble)	Residue and glutelin (alkali-soluble)
Barley	10—16	3—4	10—20	35—45	35—45
Maize	7—13	2—10	10—20	50—55	30—45
Oats	8—20	5—10	50—60	10—15	5
Rice	8—10	2—5	2—8	1—5	85—90
Rye	9—14	20—30	5—10	20—30	30—40
Sorghum	9—13	Trace	Trace	60—70	30—40
Triticale	12—18	20—30	5—10	20—30	30—40
Wheat (HRS)	10—15	5—10	5—10	40—50	30—40
Wheat (durum)	12—16	10—15	5—10	40—50	30—40

[a] Expressed as percentage of total protein.

From Simmonds, D. H., in *Cereals '78: Better Nutrition for the World's Millions,* Pomeranz, Y., Ed., American Association of Cereal Chemists, St. Paul, Minn., 1978, 121.

Table 9
PROTEIN QUALITY OF CEREAL GRAINS (PER)[4,17-21]

	Actual	Estimate[a]		Actual	Estimate[a]
Barley	—	1.6	Sorghum	1.8	0.9
Buckwheat	—	1.8	Wheat	1.5	1.3
Foxtail Millet	—	1.0	Wheat germ	2.5	2.5
Job's Tears	—	1.0	Wheat gluten	—	0.7
Maize					
Normal	1.2	1.2	Wheat flour 80—90% extraction	—	1.1
Opaque-2	2.3	1.9			
Pearl millet	1.8	1.6	Wheat flour 70—80% extraction	—	1.0
Proso millet	—	1.4			
Rogi (finger) millet	0.8	—	Wheat flour 60—70% extraction	—	0.8
Rice					
Brown	1.9	1.8			
Polished	1.7	1.7			
Rye	1.6	1.6	Bulgur wheat	—	1.2
Oats	1.9	1.7	Triticale	1.6	1.4

[a] Estimated from amino acid content assuming availability of amino acids the same as the amino acids in casein.

Table 10
AMINO ACID COMPOSITION OF CEREALS (% BY WEIGHT)

Amino acid	Rice (brown)	Wheat (HRS)	Maize (field)	Sorghum	Pearl millet	Barley	Oats	Rye	Triticale	WHO requirement (1973)[a]
Tryptophan	1.08	1.24	0.61	1.12	2.18	1.25	1.29	1.13	1.08	1.0
Threonine	3.92	2.88	3.98	3.58	4.00	3.38	3.31	3.70	3.11	4.0
Isoleucine	4.69	4.34	4.62	5.44	5.57	4.26	5.16	4.26	3.71	4.0
Leucine	8.61	6.71	12.96	16.06	15.32	6.95	7.50	6.72	6.87	7.0
Lysine	3.95	2.82	2.88	2.72	3.36	3.38	3.67	4.08	2.77	5.5
Methionine	1.80	1.29	1.86	1.73	2.37	1.44	1.47	1.58	1.44	3.5
Cystine	1.36	2.19	1.30	1.66	1.33	2.01	2.18	1.99	1.55	
Phenylalanine	5.03	4.94	4.54	4.97	4.44	5.16	5.34	4.72	5.26	6.0
Tyrosine	4.57	3.74	6.11	2.75	—	3.64	3.69	3.22	2.14	
Valine	6.99	4.63	5.10	5.71	5.98	5.02	5.95	5.21	4.39	—
Arginine	5.76	4.79	3.52	3.79	4.60	5.15	6.58	4.88	4.99	—
Histidine	1.68	2.04	2.06	1.92	2.11	1.87	1.84	2.28	2.48	—
Alanine	3.56	3.50	9.95	—	—	4.60	6.11	5.13	3.53	—
Aspartic acid	4.72	5.46	12.42	—	—	5.56	4.13	7.16	5.00	—
Glutamic acid	13.69	31.25	17.65	21.92	—	22.35	20.14	21.26	31.80	—
Glycine	6.84	6.11	3.39	—	—	4.55	4.55	4.79	4.05	—
Proline	4.84	10.44	8.35	—	—	9.02	5.70	5.20	12.06	—
Serine	5.08	4.61	5.65	5.05	—	4.65	4.00	4.13	4.70	—
Total protein (%)	7.5	14.0	10.0	11.0	11.4	12.8	14.2		17.3	—

[a] Nutritionally essential amino acids.

From Simmonds, D. H., in *Cereals '78: Better Nutrition for the World's Millions,* Pomeranz, Y., Ed., American Association of Cereal Chemists, St. Paul, Minn., 1978, 120.

Table 11
CARBOHYDRATE CONTENTS OF CEREAL GRAINS AND THEIR PRODUCTS[4,9]

	Carbohydrate	
	Total (g/ 100 g)	Fiber (g/ 100 g)
Barley, pearled	78.8	0.5
Buckwheat, whole grain	72.9	9.9
Buckwheat flour		
Dark	72.0	1.6
Light	79.5	0.5
Bulgur		
Club wheat	79.5	1.7
Hard red winter wheat	75.7	1.7
White wheat	78.1	1.3
Maize		
Field corn	72.2	2.0
Sweet corn, raw	22.1	0.7
Maize flour (corn flour)	76.8	0.7
Malt, dry	77.4	5.7
Malt extract, dried	89.2	trace
Millet, proso	72.9	3.2
Oatmeal, dry	68.2	1.2
Popcorn		
Unpopped	72.1	2.1
Popped, plain	76.7	2.2
Rice		
Brown	77.4	0.9
White	80.4	0.3
Rice bran	50.8	11.5
Rice polish	57.7	2.4
Rye	73.4	2.0
Rye flour		
Light	77.9	0.4
Medium	74.8	1.0
Dark	68.1	2.4
Sorghum grain	73.0	1.7
Wheat		
Hard red spring	69.1	2.3
Hard red winter	71.7	2.3
Soft red winter	72.1	2.3
White	75.4	1.9
Durum	70.1	1.8
Wheat flour		
80% extraction	74.1	0.5
Straight, hard wheat	74.5	0.4
Straight, soft wheat	76.9	0.4
Patent, all purpose	76.1	0.3
Wheat bran	61.9	9.1
Wheat germ	46.7	2.5

Table 12
FATTY ACID COMPOSITION OF CEREALS AND RELATED PRODUCTS[4,22]

Food	Water	Total lipid	Fatty acid (gm/100 g food, edible portion)[a,b]									
			Saturated					Unsaturated				
			Sum	14:0	16:0	18:0	20:0	Sum	16:1	18:1	18:2	18:3
Barley *(Hordeum vulgare)*												
Whole grain	14	2.8	0.48	0.01	0.45	0.02	0	1.52	0.01	0.24	1.14	0.13
Buckwheat, domestic *(Fagopyrum* spp.)												
Whole grain	11	2.4	0.46	0.01	0.36	0.05	0.04	1.66	0	0.80	0.74	0.12
Corn *(Zea mays)*												
Whole grain, raw	13.8	4.1	0.47	0	0.40	0.06	0.01	3.07	0.01	0.91	2.12	0.03
Flour	12	2.6	0.30	0	0.25	0.04	0.01	2.00	—	0.64	1.34	0.02
Germ	0	30.8	3.93	0	3.30	0.54	0.09	25.57	0.04	7.58	17.7	0.25
Grits, degermed, enriched or unenriched dry form	12	0.8	0.09	0	0.08	0.01	—	0.60	—	0.18	0.41	0.01
Corn oil (commercial)	0	100	12.73	0	10.7	1.74	0.29	82.96	0.14	24.6	57.4	0.82
Cornmeal, white or yellow												
Bolted (nearly whole grain)	12	3.4	0.39	0	0.33	0.05	0.01	2.54	—	0.75	1.76	0.03
Degermed, enriched, or unenriched dry form	12	1.2	0.14	0	0.12	0.02	—	0.90	—	0.27	0.62	0.01
Farina, enriched, regular												
Dry form	10.3	1.5	0.24	—	0.22	0.01	0.01	0.83	—	0.11	0.69	0.03
Millet, pearl millet *(Pennisteum glaucum)*												
Whole grain	11.8	4.1	0.86	0	0.68	0.16	0.02	2.67	0.02	0.83	1.69	0.13
Oatmeal or rolled oats *(Avena sativa)*												
Dry form	8.3	7.4	1.37	0.02	1.21	0.10	0.04	5.65	0.02	2.60	2.87	0.16
Rice *(Oryza sativa)*												
Brown dry form	12	2.3	0.62	0.03	0.54	0.04	0.01	1.36	0.01	0.54	0.78	0.03
White, fully milled or polished, enriched dry form	12	0.8	0.21	0.01	0.19	—	—	0.47	—	0.19	0.27	0.01

Bran oil	0	100	19.50	0.69	16.7	1.58	0.53	74.56	0.26	39.2	33.5	1.60
Rye *(Secale cereale)*												
Whole grain	12.1	2.2	0.27	—	0.25	0.02	0	1.30	0.01	0.22	0.95	0.12
Sorghum *(Sorghum vulgare)*												
Whole grain	11	3.3	0.48	0.01	0.44	0.03	0	2.74	0.04	1.15	1.46	0.09
Triticale *(Triticum × Secale)*												
Whole grain flour	14	3.4	0.49	0.02	0.45	0.02	0	1.91	0	0.28	1.48	0.15
Wheat *(Triticum aestivum)*												
Whole grain												
Hard red spring	14	2.7	0.37	—	0.36	0.01	—	1.56	0.01	0.25	1.20	0.10
Hard red winter	14	2.5	0.35	—	0.33	0.02	—	1.47	0.01	0.28	1.08	0.10
Soft red winter	14	2.4	0.35	—	0.33	0.02	0	1.40	0.01	0.25	1.07	0.07
White	14	2.0	0.30	—	0.28	0.02	0	1.14	0.01	0.18	0.88	0.07
Flours												
Hard red spring	14	1.5	0.23	—	0.21	0.01	0.01	0.84	—	0.12	0.69	0.03
Hard red winter	14	1.5	0.20	—	0.19	0.01	—	0.74	—	0.10	0.64	0.03
Soft red winter	14	1.4	0.22	—	0.20	0.01	0.01	0.76	—	0.08	0.65	0.03
All purpose	14	1.4	0.23	—	0.22	0.01	—	0.72	—	0.11	0.58	0.03
Bran	14	4.6	0.74	0.01	0.69	0.04	—	3.09	0.02	0.71	2.20	0.16
Germ	14	10.9	1.88	—	0.01	1.81	0.06	8.18	0.04	1.54	5.86	0.74
Wheat, durum *(Triticum durum)*												
Whole grain	14	3.3	0.54	—	0.51	0.03	0	1.88	0.01	0.40	1.36	0.11
Semolina	14	1.8	0.33	—	0.31	0.02	0	0.90	0.01	0.17	0.68	0.04

[a] 14:0 = myristic; 16:0 = palmitic; 18:0 = stearic; 20:0 = arachidic; 16:1 = palmitoleic; 18:1 = oleic; 18:2 = linoleic; 18:3 = linolenic.

[b] — = <0.005 g.

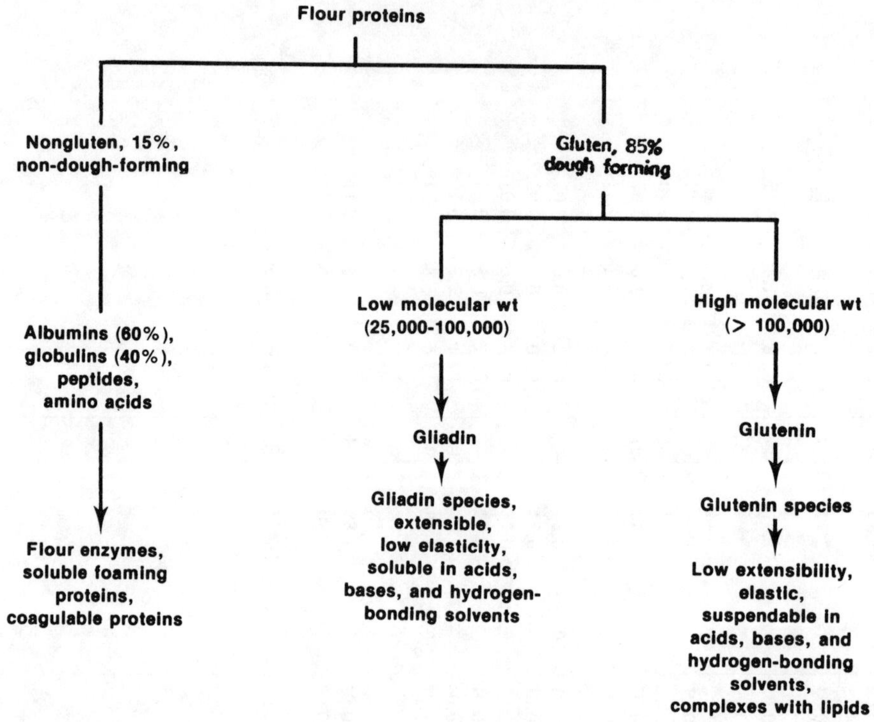

FIGURE 1. Wheat flour proteins. (Courtesy of J. Holme, Ogilvie Flour Mills, Montreal, Canada.)

REFERENCES

1. **Simmonds, D. H.,** Structure, composition, and biochemistry of cereal grains, in *Cereals '78: Better Nutrition for the World's Millions,* Pomeranz, Y., Ed., American Association of Cereal Chemists, St. Paul, Minn., 1978, 105; 107; 109; 120; and 121.
2. **Muller, H. G.,** Some physical properties of cereals and their products as related to potential industrial utilization, in *Industrial Uses of Cereals,* Pomeranz, Y., Ed., American Association of Cereal Chemists, St. Paul, Minn., 1973, 24.
3. **Pomeranz, Y. and MacMasters, M. M.,** Structure and composition of the wheat kernel, *Baker's Dig.,* 42(4), 24, 1978.
4. **Lockhart, H. B. and Nesheim, R. O.,** Nutritional quality of cereal grains, in *Cereals '78: Better Nutrition for the World's Millions,* Pomeranz, Y., Ed., American Association of Cereal Chemists, St. Paul, Minn., 1978, 207; 210; 212; 215; and 217.
5. **Altman, P. L. and Dittmer, D. S.,** *Metabolism,* Federation of American Societies for Experimental Biology, Bethesda, Md., 1968.
6. **Murphy, E. W., Willis, B. W., and Watt, B. K.,** Provisional tables on the zinc content of foods, *J. Am. Dietet. Assoc.,* 66, 345, 1975.
7. National Academy of Sciences, *Atlas of Nutritional Data on United States and Canadian Feeds,* National Academy of Sciences, Washington, D.C., 1971.
8. **Lorenz, K. and Reuter, F. W.,** Mineral composition of developing wheat, rye, and triticale, *Cereal Chem.,* 53, 683, 1976.
9. **Watt, B. K. and Merrill, A. L.,** Composition of Foods, Raw, Processed, Prepared, Agricultural Handbook No. 8, Agricultural Research Service, U.S. Department of Agriculture, Washington, D.C., 1963.
10. Vitamin E Content of Foods and Feeds for Human and Animal Consumption, Bull. 435, Agriculture Experiment Station, University of Wyoming, Laramie, 1965.
11. **Ross, L. and Lockhart, H. L.,** *Nutritive Values,* The Quaker Oats Co., Chicago, 1977.
12. **Harding, M. G. and Crooks, H.,** Lesser known vitamins in food, *J. Am. Dietet. Assoc.,* 38, 240, 1961.
13. *Heinz Nutritional Data,* 6th ed., H. J. Heinz, Pittsburgh, Pa., 1963.
14. **Miller, D. F.,** Composition of Cereal Grains and Forage, Publ. No. 585, National Academy of Science, National Research Council, Washington, D.C., 1958.
15. **Toepfer, E. W., Polansky, M. M., Eheart, J. F., Slover, H. T., Morris, E. R., Hepburn, F. N., and Quackenbush, F. W.,** Nutrient composition of selected wheats and wheat products. XI. Summary, *Cereal Chem.,* 49, 173, 1972.
16. **Butterfield, S. and Calloway, D. H.,** Folacin in wheat and selected foods, *J. Am. Dietet. Assoc.,* 60, 310, 1972.
17. Food and Agriculture Organization, Amino Acid Content of Foods and Biological Data on Proteins, FAO Nutritional Studies No. 24, Food and Agriculture Organization, Rome, Italy, 1970.
18. Proceedings of the High Lysine Corn Conference, Mertz, E. T. and Nelson, O. E., Eds., Corn Refiners Association, Inc., Washington, D.C., 1966.
19. **Hensley, G. W. and Lockhart, H. B.,** unpublished data, 1975.
20. Food and Agriculture Organization, Maize and Maize Diets, FAO Nutritional Studies No. 9, Food and Agriculture Organization, Rome, Italy, 1953.
21. **Shukla, T. P.,** Cereal proteins; chemistry and food applications, *CRC Crit. Rev. Food Sci. Nutr.,* 6, 1, 1975.
22. **Weihrauch, J. L., Kinsella, J. E., and Watt, B. K.,** Comprehensive evaluation of fatty acids in foods. VI. Cereal products, *J. Am. Dietet. Assoc.,* 68, 335, 1976.

ANALYTICAL TESTS FOR GRAIN QUALITY

P. C. Williams

THE ANALYTICAL EVALUATION OF CEREALS

There are four principal reasons for the chemical analysis of a commodity. The first is that of screening early generations for basic quality evaluation in breeding programs. The second, and probably the most widespread, is the analysis of the commodity to establish its market value, or its functional or nutritional potential. The third reason for analysis is that of quality control and monitoring, and the final reason is that of research investigation. Due to the large numbers to be processed, screening procedures usually call for rapidity and simplicity rather than absolute precision. Quality evaluation and control calls for a higher degree of accuracy and reproducibility. Tests to establish market value, and research investigations usually involve the most sophisticated techniques, and equipment.

THE PROXIMATE ANALYSIS OF CEREAL COMMODITIES

The term "proximate analysis" encompasses the determination of moisture, protein, total ash, oil (fat), "crude" fiber, and "nitrogen-free extract". The latter term includes sugars, other soluble carbohydrates, and starch. Having determined the moisture, protein, oil, ash, and crude fiber contents, a sizeable discrepancy exists between the original weight of material, and the constituents which can be accounted for by the above determinations. This is presumed lost during the acid and alkaline digestions involved in the crude fiber determination. The total discrepancy, less the protein content (which has been determined), has come to be referred to as the "nitrogen-free extract". Most of the literature concerning both the chemical composition and analytical procedures applicable to sorghums and millets have their origin in the U.S. Table 1 summarizes the results of a very large number of observations made on a number of different types of cereals.

Analytical Procedures for Cereals

Sampling

Accurate sampling is the first essential for accurate and precise results. In the case of analytical tests carried out for the screening of early generation breeding material, there is usually only a very small amount of material available, so that sampling is not the responsibility of the chemist. For most analytical operations the amount of material to be analyzed is too large to be accommodated or tested, and some form of subdivision is necessary. A sample is a subportion of a population. A good sample is representative of the population to the extent that were a series of subsequent samples to be withdrawn from the population at random, the mean result of testing all subsequent samples should agree with the first sample to the same degree that the results of testing the individual subsamples agree with each other.

The subject of sampling has been well-documented in a recent monograph.[20] In most cases, the samples arriving at analytical laboratories will have already been withdrawn. A Boerner subdivider can be used to divide samples into two equal halves, and the subdivision can be repeated until the subsample size is sufficiently reduced for the laboratory.

Table 1
PROXIMATE COMPOSITION OF SOME COMMON CEREALS[1-19]

	Protein		Oil		Fiber		Carbohydrate		Ash	
	$\bar{x}$	SD	$\bar{x}$	SD	$\bar{x}$	SD	$\bar{x}$	SD	$\bar{x}$	SD
Wheat										
Hard[a]	14.1	1.83	2.1	0.81	2.6	1.80	77.5	4.1	2.0	0.61
Soft	10.6	1.34	2.0	0.19	2.6	0.30	82.5	1.90	1.9	0.21
Durum	16.3	1.78	2.3	0.32	2.4	0.24	76.7	1.81	1.9	0.13
Barley	12.3	1.67	2.1	0.57	6.0	1.26	74.0	4.67	3.0	0.66
Oats	12.6	1.47	5.3	1.01	12.9	2.59	64.7	3.37	4.1	0.80
Oat groats	16.9	0.82	7.5	0.91	2.3	0.70	0.85	3.43	2.3	0.17
Rye	13.2	1.38	1.8	0.24	2.3	0.36	80.7	1.54	1.9	0.20
Corn	10.5	0.49	4.7	0.50	2.1	0.19	81.1	0.98	1.5	0.17
Sorghum	12.4	1.61	3.5	0.49	2.0	0.84	76.8	5.17	1.8	0.37
Millet	12.5	2.22	4.5	1.21	8.6	5.47	65.2	7.80	3.2	1.26
Buckwheat	12.2	1.09	2.8	0.31	11.6	1.45	71.0	1.74	2.4	0.10
Rice	8.5	0.82	1.8	0.13	9.7	0.59	74.6	1.32	5.3	0.62
Rice (milled)	8.7	0.59	1.2	0.76	1.1	0.86	88.0	2.17	1.1	0.43

Notes: All results on dry matter basis.
SD = standard deviation.

[a] Wheat protein = n × 5.7; all other protein = N × 6.25. For this table recent data in which more modern N: P factors were quoted for certain commodities, data were recalculated using the 6.25 factor.

Sample Preparation

Sample preparation usually involves two and sometimes three processes. The first is cleaning, which means the removal of foreign material. Foreign material includes grains of types other than that which the sample supposedly represents, also weed seeds, damaged kernels, pieces of plant material such as straw and chaff, stones, soil, and metal objects. These can be removed manually, in the case of small samples, or mechanically, using aspiration and screens. Cleaning is also discussed in Reference 20. The second step involves preliminary air-drying. This is only necessary in the case of high moisture grain and must precede the third process, that of grinding to a powdery form. Excessive moisture affects the grinding characteristics of a grain. Allow the samples to stand for 12 to 24 hr in a well-aerated location, preferably at a temperature of above 30°C (86°F). The moisture level will fall to a level which will not impede pulverization.

The final stage in sample preparation concerns pulverization, or grinding. Many types and models of grinder are available. The most important criteria of a laboratory grinder are uniformity of mean particle size and particle size distribution, insensitivity to variables such as grain size and moisture level, simplicity of use, durability, and the capacity for high throughput without excessive heat build-up. Grinders fall into five main types, namely burr mills, cutting-action mills, impeller mills, hammer mills, and pin mills. The various types and models of grinder available have recently been summarized.[21] Hammer mills and impeller mills are the most suitable for the pulverization of small grains such as sorghums and millets. Because of their small grain size, it is difficult to produce a uniform product by grinding millets on a burr mill, for example.

Following the pulverization stage, the powdered sample is usually stored for a certain period awaiting analysis. Storage containers vary widely in size, shape, and composition. Paper and plastic envelopes are convenient to store, but are difficult receptacles for mixing and withdrawing the ground sample. They are also more susceptible to infestation. Paper may be damaged by excessive moisture, and the stored sample may acquire or lose moisture, depending on the relative humidity. The best storage container for all-round ease of handling, stacking, and durability, is the flat, round metal can. As an alternative to metal cans plastic vials with a "snap-cap" are suggested. More recently round plastic vials have become widely adopted as an inexpensive storage receptacle. It is easier to mix a ground sample thoroughly in a vial than in a can, if spillage is to be avoided. Thorough mixing of pulverized or ground material is an essential prerequisite to precise analysis, since grinding usually introduces some degree of stratification of the material.

Moisture Determination

Standard Methods

Single-Stage Air Oven Method, SSAO[22]

Equipment — Oven capable of maintaining 130°C ± 1°C, preferably with air-circulating fan; analytical balance with precision of ±0.001 g; aluminum moisture dishes, 50 mm in diameter × 20 mm height; desiccator.

Procedure — Weigh accurately 1 to 2 g of well-mixed, ground material into a previously-weighed moisture dish. Place uncovered in oven for 60 min after temperaure has regained 130°C. Cover, cool in desiccator for 20 to 30 min, and reweigh. Calculate % SSAO moisture or % dry matter using the formula:

$$\%H_2O = \frac{(W_2 - W_3) \times 100}{(W_2 - W_1)}$$

$$\text{or \% dry matter} = 100 - \%H_2O$$

$$\text{or \% dry matter} = \frac{(W_3 - W_1) \times 100}{(W_2 - W_1)}$$

where W_1 = weight of empty sample can, W_2 = weight of sample can plus sample, and W_3 = weight of sample can plus sample after drying. Ideally, the procedure should be repeated until the dried sample attains constant weight. In practice, the initial 60-min drying period is sufficient, since the losses recorded after several periods of drying may result from charring of the sample rather than from additional moisture loss. The precision, or reproducibility of the test, is about ±0.10 − 0.15% moisture on a single-well-mixed pulverized sample. Grinding error increases this figure to about 0.2% moisture. Typical SSAO moisture figures for cereals range from 8 to 15%, depending upon the relative humidity of the atmosphere, length and method of storage, and method of grinding.

The Two-Stage Air Oven Method, 2SAO[23]

This is the standard method for the determination of moisture in intact grain kernels. Moisture is usually lost during the pulverization of grain, due to the heat developed in the grinder, and the air displacement, which may reach the equivalent of a wind velocity of up to 200 km/hr. The 2SAO procedure gives an accurate measurement of moisture in whole grains, regardless of the initial moisture level. The first stage involves air-drying a known weight of grain in tared dishes to a level of moisture below which no significant loss in moisture occurs during grinding. The grain is then ground, and the moisture content of the pulverized grain is determined by the SSAO method. The calculation of the overall moisture content involves both stages of drying.

Equipment — Is as for the SSAO test, plus flat, open dishes, capable of holding 25 g of whole grains in a single layer.

Procedure — Stage A: weigh 25 g of grain into an open tared dish. Place in a warm, well aerated atmosphere for at least 24 hr or until there is no appreciable further loss in moisture. Reweigh the dishes and record the loss in weight as "stage A" moisture. Stage B: pulverize the partially dried grain from stage A, using a burr mill or a cutting-action mill. These exert a relatively gentle action on the grain, and minimize undetected moisture loss. Carry out the SSAO test on a 2 g sample of the well-mixed ground grain. Calculate the overall moisture content of the original sample using the formula:

$$\%H_2O = H_2OA + \frac{(100 - H_2OA) \times H_2OB}{100}$$

The 2SAO test has a standard error of about 0.18 to 0.20% moisture. It is applicable to any type of grain and it is the only method for the determination of moisture in grain which is unaffected by either the moisture level itself, the temperature, or the bulk density (bushel or hectoliter weight) of the grain. As a result, provided that the correct precautions are taken to eliminate moisture loss during the grinding for stage B, the 2SAO test is the most accurate of all means of testing whole grain for moisture.

Electronic Methods

Electronic Moisture-Testing Instruments

During the past three decades, a large number of electronic instruments have become available for the testing of whole grains for moisture. The chief advantages of the instruments are the rapidity of testing (usually less than 1 min), together with their large sample weight, (100 to 250 g) which reduces sampling error. Also, they are non-destructive, and following the test, the grain can be used as seed if necessary. The disadvantages include sensitivity to temperature both of the grain and the atmosphere, and sensitivity to grain bulk density. In addition, the efficiency of capacitance meters is markedly reduced at moisture levels of less than 9% and above 25%. Finally because of the large sample size, most electronic meters are not suitable for use in early generation breeding work.

To calibrate an electronic moisture meter, assemble at least 50, and preferably 80 to 100 samples of the required grain. The sample size should exceed the sample size called for by the instrument by 100 g. Determine the moisture content of several of the samples by the 2SAO method. Samples of different moisture level can then be prepared by tempering. Increments of about 1% moisture should be included up to 3%, above and below the moisture range anticipated in future crops. Use distilled or demineralized water for tempering and store the tempered samples in sealed metal containers or heavy duty (4 to 6 mil) plastic bags for at least 7 days with daily agitation, to ensure even and thorough incorporation of moisture. In hot climates, the samples should be tempered for the first 24 hr at room temperature. Thereafter, store at 5 to 10°C (40 to 50°F) for the remainder of the period, to inhibit mold development. Low moisture samples are prepared by exposing the grain to the atmosphere for a few hours, or by standing overnight in a warm atmosphere, such as on top of an air-oven, if the ambient relative humidity tends to be high.

Determine the 2SAO moisture in all samples in duplicate. Next, analyze the samples by means of the instrument, again in duplicate. Calculate the means of all readings, both for 2SAO moisture, and instrumental reading, together with the respective standard deviations. Discard any duplicates which differ by twice the standard deviation, or more.

Linear regression is used to estimate the degree to which the dependent variable varies with respect to the independent variable. The formula for the calculation is

$$y\ (\text{dependent}) = a + bx\ (\text{independent})$$

$$\text{where } b = \frac{\Sigma xy - (\Sigma x \times \Sigma y)/N}{\Sigma x^2\ (\Sigma x)^2/N}$$

$$\text{and } a = \bar{y} - b\bar{x}$$

The term "b" is referred to as the regression coefficient. The magnitude of the intercept made on the origin by the line of slope bx is given by a.

Regress the 2SAO results against the meter readings, to obtain the regression equation $y = a + b_x$, where y = the true moisture content, and x = the meter reading. Test the efficiency of the new meter calibration by analyzing about 20 "unknown" samples, which have had moisture determined by the 2SAO method, but have not been included in the computation of the regression. If a consistent bias occurs this can be offset by adjusting the intercept (a) of the regression equation followed by re-analysis of the unknowns, to verify the accuracy of the analysis. Construct a chart to cover the entire range of meter readings encountered in the exercise, plus an additional 10% of the range at both ends.

Near Infrared Reflectance Spectroscopy (NIRS)

This technology was first introduced during 1971.[24] It is suitable for the simple, rapid testing of pulverized grains for moisture. The calibration procedure for an NIRS instrument is similar to that described above for an electronic moisture meter. It is described in detail in the manufacturers' operating manuals, but most modern NIRS instruments are delivered precalibrated for the more common grains and commodities.

Equipment — Neotec GQA®, Model 31EL; Technicon InfraAlyzer®, or Dickey-john® GAC, Model III; grinding mill, such as Udy Cyclotec, or Christie-Norris 5 in. hammer mill. Access to computer is preferable, although manufacturers will compute constants on request.

Procedure — To calibrate, assemble 50 to 80 samples and insure moisture range of 5 to 10% starting at about 8% moisture, as for moisture meter calibration. Pulverize about 20 to 25 g of sample, using the grinder. Mix and determine moisture by SSAO method. Fill the sample cell of the instrument, and level the surface. Cap the cell, and place in instrument, which has been switched to "calibrate" mode. Record the reflectance values. Repeat the operation for all samples, discarding the pulverized material after reading. Regress the reflectance values against the SSAO moisture results and compute the calibration constants, using multiple linear regression. Set calibration constants into instrument, according to operators' manual. Analyze 12 to 20 "unknown" samples. Calculate mean NIRS and SSAO results for unknowns, and adjust intercept, if necessary. NIRS instruments require recalibration for each new crop and season, and whenever the analysis of check samples reveals increasing standard error. To operate, pulverize a sample of about 20 to 25 g, mix thoroughly, load cell, insert into instrument which has been set to "analyze" mode, and record the result, which is displayed directly on the digital meter. The standard error of NIRS moisture determination in pulverized grain is usually 0.20 to 0.25, for freshly ground samples. Note: The NIRS method is only suitable for the determination of moisture in pulverized commodities.

Protein Determination

Standard Methods

Standard methods for the estimation of protein in grains involve the determination of total nitrogen, followed by the multiplication of the total nitrogen content by a

Table 2
NITROGEN TO PROTEIN (N:P) CONVERSION FACTORS FOR SOME COMMON CEREALS AND COMMODITIES

Commodity	N:P factor
Wheat (HRS)	5.61
Wheat flour	5.59
Wheat germ	5.45
Wheat bran	5.26
Rye flour	5.64
Triticale	5.76
Barley	5.67
Oats, hulled	5.50
Buckwheat	5.53
Millet	5.68
Sorghum	5.60
Soybean	5.69
Corn	5.36[a]

[a] Tkachuk, R., personal communication.

From Tkachuk, R., *Cereal Chem.*, 46, 419, 1969. With permission.

factor which converts the nitrogen figure to a protein figure. This is the nitrogen to protein (N:P) conversion factor. A summary of N:P factors for a number of crops and commodities has been documented by the World Health Organization.[25] Tkachuk, whose calculations were based upon amino-acid analysis, also determined the N:P factor recommended for use with several types of grain.

The Macro-Kjeldahl Test for Total Nitrogen[27]

The Macro-Kjeldahl test for the determination of total nitrogen in grains[28] is the standard method for the determination of nitrogen. It involves the reduction of all amino-acid and amide nitrogen to ammonium salts. The ammonia is liberated by neutralization and steam-distilled. The distillation is quantitative, and the ammonia is determined by direct titration. The result is converted to protein by application of the appropriate N:P factor. A single test takes an overall time of about 90 to 110 min.

Equipment — Macrodigestion and distillation equipment, similar to LabConco or Precision Scientific, equipped with gas heaters or 650 W electric heaters; 800-mℓ Pyrex® Kjeldahl flasks, 500-mℓ Erlenmeyer flasks; aspirators for bulk liquid storage; dispensers for 20, 25, and 50 mℓ; 20-mℓ pipettes; 25-mℓ burettes.

Reagents — These may be

1. Concentrated sulfuric acid, H_2SO_4, 96% or 66 Baume
2. Catalyst mix containing potassium sulfate, titanium dioxide, cupric sulfate ($CuSO_4$, $5H_2O$) and pumice (100 mesh) in the ratio 10.0:0.3:0.3:0.1. Alternatively use mercuric oxide in place of titanium oxide. Mercury is a more efficient catalyst, but is a highly toxic, cumulative poison

3. Sodium hydroxide, NaOH (40%)
4. Boric acid solution, 5% containing 4 mℓ of 1% bromocresol green indicator per liter
5. Methyl red solution, 2 mℓ of alcoholic 1% methyl red solution per liter
6. Standardized H_2SO_4 solution for titration. Standard 0.1 *N*H_2SO_4 is used to determine the total nitrogen, using the formula 1 mℓ 0.1 *N*H_2SO_4 = 1.4 mg N. Alternatively the H_2SO_4 can be standardized so that 1 mℓ = 1% protein (using a 1.0 g sample weight). The strength of acid depends on the N factor used, and is calculated using the formula

$$\text{Acid normality} = \frac{1}{(1.4 \times \text{N:P factor})}$$

$$\text{e.g., if N:P factor} = 5.8$$

$$\text{Acid normality} = \frac{1}{1.4 \times 5.8} = 0.1218$$

Procedure — Standardization of H_2SO_4. Dissolve 9.5 to 10.0 g of pure NaOH pellets in 100 mℓ distilled or demineralized water and dilute to 2 ℓ in a volumetric flask. Dry 3 to 4 g of pure potassium hydrogen phthalate for 2 hr at 130°C. Cool in a desiccator. Accurately weigh at least four 0.5 g portions into clean 500-mℓ Erlenmeyer flasks. Dissolve in about 100 mℓ distilled (demineralized) water. Add two drops of 1% phenolphthalein indicator. Rinse a clean burette with dilute NaOH solution and titrate the phthalate samples. Calculate the normality of the NaOH using the formula:

$$\text{Normality} = \frac{1000 \times \text{wt. of phthalate (g)}}{\text{vol. of titration (m}\ell\text{) } 204.23 \text{ (equiv. of phthalate)}}$$

Replicate results must agree to within 0.0001 *N*. Pipette 20-mℓ portions of dilute H_2SO_4 into Erlenmeyer flasks, add 100 mℓ of the methyl red indicator solution, and titrate with the standardized NaOH. Calculate the normality of the H_2SO_4 using the formula:

$$\text{Normality} = \frac{\text{vol. of titration (m}\ell\text{)} \times \text{normality if NaOH}}{20}$$

Duplicate titrations should agree within 0.0001 *N*. Adjust the strength of the standard H_2SO_4 to exactly 0.1142 *N* by the addition of water or concentrated H_2SO_4 as necessary. Note: For 5ℓ, 0.1 mℓ of demineralized H_2SO_4 reduces normality by 0.0007504 N. For 5 ℓ, 5mℓ of concentrated H_2O reduces normality by 0.00011 *N*. Analysis: accurately weigh 1.0 g of sample into the Kjeldahl flask; add 10 g catalyst and 20 mℓ concentrated H_2SO_4. Digest until clear, followed by a further 20 min. Cool, dilute with 300 mℓ water. Add 50 mℓ of 40% NaOH, connect to distillation apparatus, and distill into 25 mℓ of boric acid solution plus 100 mℓ methyl red solution. This combination provides the mixed indicator. Collect 250 to 280 mℓ of distillate and titrate against standardized H_2SO_4 until the green color just turns purple. Record the results: 1 mℓ of titration = 1% protein (N × 6.25). Multiply the result × 2 if 0.5 g sample is used. Reagent blanks should be determined daily, and the result subtracted from the sample titration. For the reagent blank, analytical grade dextrose is substituted for the sample.

Total Protein: Micro-Kjeldahl[29]

Equipment — Microdigester: this should be fitted with naked electric heating elements — Silex burners are suitable; distillation equipment such as Parnos/Wagner, or Markham stills; 10-mℓ pipettes and burettes; 10- and 20-mℓ dispensers; 50- or 100-mℓ Pyrex® Kjeldahl flasks; 250-mℓ Erlenmeyer flasks; plastic aspirators for liquid storage, analytical balance capable of weighing up to at least 1 g ± 0.0001 g.

Reagents — As for macro-Kjeldahl test, but accurately dilute standard $H_2SO_4 \times 5$, using pipette and volumetric flask. Note: The ammonia liberated after digestion can be determined colorimetrically provided that the equipment includes a spectrophotometer, or a Technicon AutoAnalyzer. References are cited in the literature summary.[30-32] The Tecator block digester is satisfactory for the simultaneous digestion of large numbers of samples.

Total Protein: Dye-Binding Procedure (DBC)[33-35]

Equipment — Udy DBC equipment — alternatively, spectrophotometer similar to Bausch and Lomb® Spectronic 20; shaker; dispenser for 40 mℓ; filtering apparatus using fiberglass filter pads; analytical balance with precision of ±0.001 g; plastic containers for dye solutions.

Procedure — Accurately weigh 0.3 to 1.0 g of sample into reaction vessel. Add 40 mℓ dye solution, shake 3 min (high speed) or 40 min (low speed). Filter into colorimeter cell (after allowing reaction vessel [sd] to stand upright for about 5 min). Record protein using table provided with equipment.

Notes: Adjust weight of sample according to anticipated protein content of material. Calibrate test against Kjeldahl test for individual crops if necessary. Select 30 to 40 samples covering the expected range of protein. Analyze by Kjeldahl and DBC methods. Compute regression equation $Y = a + bx$, where Y = % protein DBC, and x = % protein Kjeldahl. Construct a table for future DBC analysis. A recommended text for reference to statistical computation is found in Reference 34. If other than the Udy equipment, the sample cell is likely to be of too great a width for the employment of standard Udy dye solution. Dilute the dye by 10 to 100, depending on the width of the cell used, and construct a calibration chart.

The procedure is applicable to a wide range of commodities, but due to variations in kernel texture, and the color of the testa/pericarp, separate calibration is recommended for different cereals. Statistical correlation coefficients between DBC meter readings and Kjeldahl protein data should exceed 0.95 for the calibration, for most accurate results.

Total Protein: Biuret Method

The biuret reaction has been successfully applied to the determination of protein in various cereals (e.g., References 6 to 40), and a semiautomatic instrument is commercially available.[40] The chief source of error in the biuret procedure lies in the clarification of the extract prior to colorimetry. Nevertheless the test is rapid and economical and when applied to wheat is equal to the Kjeldahl test in precision. Off-colors may be introduced as a result of fiber components and certain tannins, and calibrations must be carried out against the Kjeldahl test for individual grain types.

Equipment — Balance capable of accuracy to within 0.001 g; centrifuge capable of at least 6000 rpm spectrophotometer, such as Bausch and Lomb® Spectronic 20; associated glassware; plastic centrifuge tubes.

Reagents. — For solution A — 173 g sodium citrate + 100 g anhydrous sodium carbonate dissolved in 600 mℓ demineralized water (heating to 60°C advisable). For solution B — Dissolve 17.3 g $CuSO_4$, $5H_2O$ in 150 mℓ demineralized water. For solution C — Add B to A and dilute to 1000 mℓ. Biuret working reagent: dissolve 140.25 g potassium hydroxide pellets in demineralized water and dilute to 250 mℓ. Add 20 mℓ of KOH solution to 280 mℓ of Solution C and dilute to 2 ℓ. This solution is stable for at least 21 days.

Procedure — Weigh 250 mg pulverized grain into 50 mℓ centrifuge tubes. Add 0.25 g CCl_4 + 25 mℓ of biuret working solution. Stopper and shake 10 min. Stand for 30 min, agitate briefly and centrifuge for 10 min at 6000 rpm. Decant supernatant and compare against reagent blanks at 550 nm. Calibrate the test against the Kjeldahl test using at least 50 samples. Calculate the regression equation and construct a chart for the direct reading of protein from absorbance values, for future samples. Correlations of biuret absorbance values against Kjeldahl protein should be at least 0.98.

Total Protein: Near-Infrared Reflectance Spectroscopy (NIRS)

Near-infrared reflectance spectroscopy was first introduced as a simple rapid method for the determination of protein, moisture and oil in cereals and oilseeds in 1971.[24] NIRS has subsequently been successfully applied to the routine analysis of cereals and oilseeds,[24,43] legumes,[44] sorghums,[45,46] and millets,[46] plus a large number of other commodities. Protein, oil, and moisture are the constituents most commonly measured using the technique, but more recently its application to the determination of fiber and related parameters in forages[47] and essential amino-acids in wheat[48] have been reported. Using a single portable instrument a single worker can process 250 samples in the average working day, with accuracy and precision equivalent to wet chemical methods. NIRS reqires no chemicals, minimal bench space, and does not call for the employment of highly-skilled laboratory technicians. It is a technique which is ideally suited to the screening of the large numbers of samples generated in plant-breeding programs, and also to the rapid on-the-spot analysis of marketable commodities at elevator sites.

Equipment — Model 31EL Grain Quality Analyzer (Neotec Corp.), InfraAlyzer® (Technicon Corp.), Grain Analysis Computer Model III (Dickey-john Corp.), or other suitable NIRS instrument; grinding mill such as Udy Cyclone sample mill, or equivalent.

Procedure — Calibration: proceed as described for moisture meters, using 50 samples, and operating procedure as described in respective manufacturers' operating manuals. Protein, cereal oil and fiber calibrations should be carried out on an "as is" basis with respect to moisture content. Analyze the samples by standard methods, e.g., Kjeldahl protein, and record reflectance values for all samples. Regress the reflectance values against the required standard analytical values to generate calibration constants. Set constants into the instrument as described in the operators' manuals. Analysis: prepare samples in exactly the same way as for those used in calibration, set instruments to "%", or "Analysis" modes, and proceed according to manufacturers' operating manuals.

Total Amino Acids by Ion Exchange Chromatography (IEC)[49-51]

This analysis is usually applied only to the more advanced progeny of breeding programs and new food and feed commodities. Efficient hydrolysis is essential. Some amino acids may be destroyed or modified during hydrolysis, and others may be incompletely released. The recommended procedure for most amino acids involves hydrolysis with 6 N hydrochloric acid for 24 hr at 110°C, with precautions to maximize recovery and minimize degradation. Individual amino acids such as tryptophan require special precautions during hydrolysis.

Equipment — Beckman® Spinco Model 120 Amino Acid Analyzer, Durrum® Model Amino-Acid Analyzer, or other suitable equipment; mercury diffusion pump.

Procedure — Hydrolysis: transfer 20 to 40 mg accurately weighed sample to 18-mm Pyrex® test tube. Add 40 mℓ 6*N*HCl, freeze to about −80°C (liquid nitrogen) and evacuate to 50 μm. Allow contents to melt, seal tube. Hydrolize at 110°C for 24 and 72 hr (two tubes). Remove HCl under vacuum (0.1 mm) in a desiccator containing

NaOH. This takes about 12 hr. Add 25 mℓ of 0.2 N Na citrate buffer, pH 2.2, containing Brig-35 detergent, to the residue. Filter from insoluble material, using Whatman® No. 52 paper (or equivalent) using vacuum. Use portions of filtrate for amino acid determination.[26] Analysis: proceed according to respective manufacturers' operating manuals.

Individual Amino Acids

Lysine is the most important limiting amino acid in most cereals. Methionine, tryptophan and threonine are other limiting amino acids.

Determination of Lysine (Ion-Exchange Chromatography)

Hydrolyze as for total amino acids. For routine analysis the chromatography column may be shortened and the colorimetry modified to record only lysine. This allows the determination of lysine in 3 to 4 hydrolysates per hour.[52]

Lysine by Dye-Binding Capacity[35]

Equipment — As for total protein by DBC.

Procedure — Determine total protein by DBC method. Verify protein content of highest testing samples by Kjeldahl. Reanalyze highest 25% of population, weighing samples on a constant protein basis (usually 60 mg protein per sample). The highest values represent the lines highest in basic amino acids, which are highly correlated to lysine itself. The procedure should be standardized by including samples of known (IEC) lysine content.

Fiber

The fiber in a diet is important because of its influence on the digestibility of protein, and other carbohydrates, on certain alimentary diseases. The term "crude fiber" refers to the insoluble residue remaining after digestion with 2 *N* sulfuric acid, followed by digestion with 2 *N* sodium hydroxide solution. Other fiber parameters include acid detergent and neutral detergent fiber (ADF and NDF) and lignin. Crude fiber data are usually lower than those for ADF and NDF, due to the more rigorous conditions employed. Procedures are outlined for crude fiber and ADF, which are the values most frequently sought.

Crude Fiber[53]

Equipment — These are

1. Extraction apparatus with condenser to fit 600-mℓ tallform lipless beaker and hot plate adjustable to temperature that will bring 200 mℓ water at 25° to rolling boil in 15 ± 2 min
2. Tall-form lipless beakers, 600-mℓ
3. Alundum crucibles No. 5204, porosity RA766 (min wt 15 g, max 20 g)
4. Air oven maintained at 130° ± 2°
5. Electric muffle with rheostat or automatic control and pyrometer maintained at 550 to 600°
6. Desiccator with efficient desiccant. (Ignited CaO is satisfactory)
7. Sintered glass filters or filtering crucibles; suction flask with suitable holder for alundum crucible (filter ring Sargent No. 2 with 60° fluted funnel)
8. Apparatus designed to preheat acid, alkali, and wash water, equipped with efficient condenser

Reagents — These are

1. Sulfuric acid solution containing 1.25 g H_2SO_4 per 100 mℓ
2. Sodium hydroxide solution, containing 1.25 g actual NaOH per 100 mℓ free, from Na_2CO_3. Strength of acid and alkali solutions must be accurately checked.
3. Ethyl alcohol, 95%
4. Ethyl ether, ACS grade
5. 10% H_2SO_4 solution

Procedure — Steps to follow are

1. Extract 2 g dry material with ethyl ether, or use residue from crude fat determination and transfer residue to digestion flask. If material is difficult to wet, add 3 drops diluted Dow-Corning Antifoam A emulsion (1 + 3).
2. Add 200 mℓ boiling H_2SO_4 solution; immediately connect digestion flask to condenser and heat. (Contents of flask must come to boiling within 1 min and boiling must continue briskly exactly 30 min.) Rotate flask frequently until sample is thoroughly wetted. Take care to keep material from remaining on sides of flask out of contact with solution. (Blast of air conducted into flask serves to reduce frothing of liquid.)
3. After 30 min remove flask, immediately filter through sintered glass and wash with boiling water until washings are no longer acid.
4. Bring quantity of NaOH solution to boiling and keep at this temperature under reflux condenser until used. Wash charge back into flask with 200 mℓ boiling NaOH solution, using wash bottle marked to deliver 200 mℓ. Connect flask with reflux condenser and boil exactly 30 min, timing boiling with alkali so that contents of different flasks reach boiling point about 3 min apart, which permits enough time for filtration.
5. After 30 min remove flask and immediately filter through sintered glass. Thoroughly wash residues with boiling water and transfer to alundum crucibles.
6. For materials difficult to filter, after 30 min boiling remove flask, add 10 to 50 mℓ hot 10% K_2SO_4 solution and immediately filter through sintered glass, using vacuum. Wash with hot 10% K_2SO_4 solution.
7. Return residue to digestion flask, thoroughly washing all residue with hot K_2SO_4 solution. Filter into alundum crucible.
8. After thorough washing with boiling water, wash with about 15 mℓ alcohol. Dry crucible and contents at 130° to constant weight.
9. Cool in efficient desiccator and weigh.
10. Ignite contents of crucible in electric muffle or over Meker burner at dull red heat until carbonaceous matter is consumed (approximately 20 min).
11. Cool in desiccator and weigh.
12. Report loss in weight as crude fiber.

$$\text{Calculation:}\quad \text{Crude fiber, \%} = \frac{\text{loss in wt.} \times 100}{\text{wt. of sample}}$$

Acid-Detergent Fiber[57]

Equipment — Includes the following:

1. Refluxing apparatus — any conventional apparatus suitable for crude fiber determinations. Berzelius beakers (600 mℓ) and condensers made from 500 mℓ 4-b flasks are also satisfactory.

2. Fritted glass crucibles. Use coarse porosity Pyrex® crucible 40 to 50 mℓ capacity. Wash and ash new crucibles at 400° before using. Remove and place in 100° forced draft oven ⩾ 1 hr. Cool in desiccator over P_2O_5 or Mg $(C1O_4)_2$ and weigh in same order samples are to be weighed. Hold length of time from oven to balance pan as constant as possible and always weigh crucibles in same order.

Reagents — These are

1. Sulfuric acid (72% by weight) — Standardize reagent grade H_2SO_4 to sp gr 1.634 at 20° or 24.00 *N:* Add 1200 g H_2SO_4 to 440 mℓ H_2O in 1 L MCA volumetric flask with cooling. Standardize to 1634 g/L at 20° by removing solution and adding H_2O or H_2SO_4 as required.
2. Acid-detergent solution — Add 20 g cetyl trimethylammonium bromide (tech. grade) to 1 L 1.00 *N* H_2SO_4, previously standardized. Agitate to aid solution.

Procedure — Weigh 1 g air-dried sample ground to pass 1 mm screen, or approximate equivalent amount wet material, into refluxing container. Add 100 mℓ acid-detergent solution at room temperature. Heat to boiling in 5 to 10 min; reduce heat to avoid foaming as boiling begins. Reflux 60 min from onset of boiling, adjusting boiling to slow, even level. Remove container, swirl, and fill weighed (W_1) fritted glass crucible and complete filtration, using min suction. Increase vacuum only as needed. Shut off vacuum. Break up filtered mat with rod and fill crucible 2/3 full with hot (90 to 100°) H_2O. Stir and let soak 15 to 30 sec. Dry with vacuum and repeat H_2O washing, rinsing sides of crucible. Wash twice in same manner with acetone. Repeat acetone washings until no more color is removed, breaking up all lumps so that solvent comes into contact with all particles of fiber. Remove residual acetone with vacuum. Dry 3 hr or overnight in 100° forced draft oven and weigh (W_2). Calculate % acid-detergent fiber $= 100 (W_2 - W_1)/S$, where S = grams of sample × grams of oven-dried matter per gram of air dried or wet matter, determined on separate sample.

Oil

By Extraction

Equipment — Soxhlet or Goldfisch extraction equipment, with associated glassware; balance capable of weighing up to 250 g ± 0.001 g. Air oven as for SSAO moisture testing.

Reagents — Anhydrous ether.

Procedure — Dry about 5 to 10 g of finely ground grain for 1 hr at 100°C ± 1°C. Accurately weigh 4 to 5 g of sample into soxhlet thimble, (or filter paper in the case of the Goldfisch procedure.) Place the sample inside the extraction equipment. Add anhydrous ether until the syphon empties the extractor, in the case of the soxhlet equipment, then add a further 30 to 40 mℓ. Use 40 mℓ for Goldfisch extraction. Collect the condensed solvent in tared soxhlet flask or Goldfisch beaker. Extract overnight, remove sample and evaporate off excess solvent. Complete solvent removal by drying at 100°C for 1 hr in a vacuum oven. Cool, weigh, and calculate % oil.

By N.I.R.S.

Equipment — As for protein.

Procedure — As for protein. For calibration the reflectance values are regressed against standard oil content values determined by extraction.

The Determination of Kernel Texture

Kernel texture or hardness exerts a very significant effect on the milling, processing

Table 3
TESTS FOR THE DETERMINATION OF FUNCTIONAL PROPERTIES OF VARIOUS COMMODITIES

Test	Determines	Commodity	Ref.
Farinograph	Rheological properties, water absorption	Wheat	59—61, 63
Alveograph	Rheological properties	Wheat	59, 63
Extensograph	Rheological properties	Wheat	59, 63
Mixograph	Rheological properties	Wheat	59, 62, 63
Amylograph	Starch pasting properties, alpha-amylase activity	Wheat, rye	60, 63
Baking	Baking quality[a]	Wheat, rye, triticale	63
Gassing power	Gassing power (6 hr)	Wheat	63
Cooking quality	Cooking quality	Legumes, rice, sorghums	65—68
Relative NPR	Biological value of proteins	Cereals, legumes[b]	68

[a] Many different types of baking test exist, some specific to local market needs.
[b] Includes derived products.

and cooking behavior of all cereals, pulses, and many other commodities.[55] For texture measurement the grinding/sieving, or particle size index test, the Strong-Scott barley pearler, or direct visual inspection of endosperms can be used. The particle size index (PSI) test is recommended on the basis of its simplicity, precision and speed.

Grind 20 g of kernels in a burr-mill such as a Falling Number KT-30 mill. Collect the powdered grain in a beaker or other suitable receptacle. Brush all fragments remaining in the grinding chamber into the beaker and mix thoroughly. Accurately weigh 10 g of well-mixed powdered sample, add about 70 g of whole wheat kernels to prevent clogging of sieve apertures and sieve for 10 min through a 200 mesh screen (74-μm apertures). Weigh the throughs and record the PSI as a percentage. In the case of wheat the PSI ranges from 6 to 7 for hard durum wheats to 45 to 50 for the softest club wheats. Barleys range from 15 to 35, and ryes and diploid wheats from 40 to 50. Other cereals such as rice and corn and most legumes display wide ranges in texture, which exerts a significant influence on cooking quality.

The Determination of Functional Properties

The determination of functional properties of commodities calls for complicated tests such as the baking of breads, and the preparation of pasta products, dough rheology and cooking quality. A detailed description of these procedures lies beyond the scope of this chapter. Table 3 identifies some of the more commonly employed tests, together with sources of further information.

REFERENCES

1. **Chamberlain, J. S.**, The Feeding Value of Cereals, Chem. Bull. No. 120, U.S. Department of Agriculture, Washington, D.C., 1909.
2. **Ball, C. R. and Rothgeb, B. E.**, Uses of Sorghum Grain, Farmers' Bull. No. 686, U.S. Department of Agriculture, Washington, D.C., 1915.
3. **Morrison, F. B.**, *Feeds and Feeding*, Morrison Publishing, Ithaca, N.Y., 1940.
4. **Baird, R. O. and Francis, C. K.**, Chemical composition of Kaffir corn, *J. Ind. Eng. Chem.*, 2, 531, 1910.
5. **Miller, D. F.**, Composition of Cereal Grains and Forages, No. 585, National Academy of Sciences, National Research Council, Washington, D.C., 1958.
6. Annual Reports, Grain Research Laboratory, Canadian Grain Commission, Winnipeg, 1960—1976.
7. **Wall, J. S. and Ross, W. M.**, *Sorghum Production and Utilization*, AVI Publishing, Westport, Conn., 1970.
8. **Kent-Jones, D. W. and Amos, A. J.**, *Modern Cereal Chemistry*, 5th ed. Northern Publishing, Liverpool, 1957.
9. National Academy of Sciences, *Atlas of Nutritional Data on U.S. and Canadian Feeds*, National Academy of Sciences, Washington, D.C., 1971.
10. **Baird, R. O.**, The chemistry of the kafir corn kernel, *Okla. Agric. Exp. Stn. Bull.*, No. 89, 1910.
11. **Kadkol, S. B. and Swaminathan, M.**, Chemical composition of different varieties of Ragi, E. coracana, *Bull. Central Food Tech. Res. Inst. Mysore*, 4, 12, 1954.
12. **Woll, F. W.**, Miscellaneous Chemical Work. II. Analyses of Feeding Stuffs, Wis. Agric. Exp. Sta. 16th Annu. Report, 1899, 271.
13. **Bersch, W.**, Millet and millet by-products, *Landswirtsch. Vers. Stn.*, 46, (2-3), 103, 1895.
14. **Rickett, R. C. and Whitehead, W. L.**, *Sorghum Research in Indiana, Sorghum Newsletter*, 3, 20, 1960.
15. **Biely, J.**, The Nutritive Value of Wheat in Poultry Rations, Canadian Wheat Board, Winnipeg, 1973.
16. **Coffman, F. A.**, *Oats and Oat Improvement*, American Society of Agronomy, Madison, Wis., 1961.
17. **Inglett, G. E.**, *Corn: Culture, Processing, Products*, AVI Publishing, Westport, Conn., 1970.
18. **Matz, S. A.**, *The Chemistry and Technology of Cereals as Food and Feed*, AVI Publishing, Westport, Conn., 1959.
19. **Matz, S. A.**, *Cereal Science*, AVI Publishing, Westport, Conn., 1969.
20. **Christensen, C. M., Ed.**, *Storage of Cereal Grains and Their Products*, 2d ed., American Association of Cereal Chemists, St. Paul Minn., 1974.
21. **Williams, P. C.**, A survey of grinders used in grain sample preparation, *J. Flour Feed Milling*, in press.
22. *Cereal Laboratory Methods*, No. 44-15A, American Association of Cereal Chemists, St. Paul, Minn., 1969.
23. *Cereal Laboratory Methods*, No. 44-18, American Association of Cereal Chemists, St. Paul, Minn., 1969.
24. **Rosenthal, R. D.**, *Introducing — the Neotec® Grain Quality Analyzer*, Annu. Meet. Kansas Assoc. Wheat-growers, Hutchinson, December 1971.
25. Food and Agriculture Organization, FAO Nutrition Reports Series No. 52, Food and Agriculture Organization, Rome, 1973.
26. **Tkachuk, R.**, Nitrogen to protein conversion factors for cereals and oilseed meals, *Cereal Chem.*, 46, 419, 1969.
27. *Cereal Laboratory Methods*, No. 46-12, American Association of Cereal Chemists, St. Paul, Minn., 1969.
28. **Kjeldahl, J. T.**, Neue Methodé zur Bestimmung des Stickstoffs in organischen Korpern, *Anal. Chem.*, 22, 366, 1883.
29. *Cereal Laboratory Methods*, No. 46-13, American Association of Cereal Chemists, St. Paul, Minn., 1969.
30. **Varley, J. A.**, Automatic methods for the determination of nitrogen, phosphorus and potassium in plant material, *Analyst (London)*, 91, 119, 1966.
31. **Crooke, W. M. and Simpson, W. E.**, Determination of ammonium in Kjeldahl digests of crops by an automated procedure, *J. Sci. Food Agric.*, 22, 9, 1971.
32. **Williams, P. C.**, The colorimetric determination of total nitrogen in feeding stuffs, *Analyst (London)*, 89, 276, 1964.
33. **Udy, D. C.**, Estimation of protein in wheat and flour by ion-binding, *Cereal Chem.*, 33, 190, 1956.
34. **Mossberg, R.**, Evaluation of protein quality and quantity by dye-binding capacity; a tool in plant-breeding, in *New Approaches to Breeding for Improved Plant Protein* (series 12), International Atomic Energy Agency, Vienna, 1969, 151.

35. **Munck, L.**, Improvement of nutritional value in cereals, *Hereditas*, 72(1), 1, 1972.
36. **Pinckney, A. J.**, The biuret test as applied to the estimation of wheat protein, *Cereal Chem.*, 38, 501, 1961.
37. **Jennings, A. C.**, Determination of the nitrogen content of cereal grain by colorimetric methods, *Cereal Chem.*, 38, 467, 1961.
38. **Williams, P. C.**, The determination of proteins in whole wheat meal and flour by the Biuret procedure, *J. Sci. Food Agric.*, 12, 58, 1961.
39. **Noll, J. S., Simmonds, D. H., and Bushuk, W.**, A modified biuret reagent for determination of protein, *Cereal Chem.*, 51, 610, 1974.
40. **Johnson, R. M. and Craney, C. W.**, Rapid biuret method for protein content in grains, *Cereal Chem.*, 48, 276, 1971.
41. **Simmonds, D. H. and Ronalds, J. A.**, Rapid protein determination in cereal grain using the biuret reaction, *Baker's Digest*, 49(4), 36, 1975.
42. **Williams, P. C.**, Application of Near-infrared reflectance spectroscopy to analysis of cereal grains and oilseeds, *Cereal Chem.*, 52, 561, 1975.
43. **Watson, C. A., Carville, D., Dikeman, E., Daigger, D., and Booth, D. G.**, Evaluation of two infrared instruments for determining protein content of hard red winter wheat, *Cereal Chem.*, 53, 214, 1976.
44. **Williams, P. C., Stevenson, S. G., Hawtin, G. C., and Starkey, P. M.**, The application of near-infrared reflectance spectroscopy to protein-testing in pulse-breeding programmes, *J. Sci. Food Agric.*, 29, 285, 1978.
45. **Williams, P. C. and Stevenson, S. G.**, Application of near-infrared reflectance spectroscopy to the analysis of cereals other than common wheat for protein, moisture and oil, *Cereal Foods World*, 23 (Abstr. 51-6), 462, 1978.
46. **Pomeranz, Y., Davis, G. D., Stoops, J. L., and Hubbard, J. D.**, Estimation of Protein Content and Quality in Grain Sorghum, Abstr. 112, Am. Assoc. Cereal Chemists 62nd Annu. Conf., San Francisco, 1977.
47. **Norris, K. H., Barnes, R. F., Moore, J. E., and Shenk, J. S.**, Predicting forage quality by infrared reflectance spectroscopy, *J. Anim. Sci.*, 43(4), 889, 1976.
48. **Norris, K. H., Williams, P. C., and Mattern, P. J.**, Rapid Estimation of Amino-Acids in Wheat by Near-Infrared Reflectance Spectroscopy, Abstr. 41, 62nd Annu. Conf., Am. Assoc. Cereal Chemists, San Francisco, 1977.
49. **Tkachuk, R. and Irvine, G. N.**, Amino acid compositions of cereals and oilseed meals, *Cereal Chem.*, 46, 206, 1969.
50. **Blackburn, S.**, *Amino Acid Determination*, Marcel Dekker, New York, 1978.
51. **Spackman, D. H., Stein, W. H., and Moore, S.**, Automatic recording apparatus for use in the chromatogaphy of amino acids, *Anal. Chem.*, 30, 1190, 1958.
52. **Mattern, P. J., Schmidt, J. W., and Johnson, V. A.**, Screening for high lysine content in wheat, *Cereal Sci. Today*, 15, 409, 1970.
53. *Cereal Laboratory Methods*, No. 32-15, American Association of Cereal Chemists, St. Paul, Minn., 1969.
54. **van Soest, P. J.**, Use of detergents in analysis of fibrous feeds. II. A rapid method for the determination of fibre and lignin, *JAOAC*, 46, 829, 1963.
55. **Williams, P. C.**, Relation of starch damage and related characteristics to kernel hardness in Australian wheat varieties, *Cereal Chem.*, 44, 383, 1967.
56. **Bloksma, A. and Hlynka, I.**, Basic considerations of dough properties, in *Wheat Chemistry and Technology*, Hlynka, I., Ed., American Association of Cereal Chemists, St. Paul, Minn., 1964.
57. **Pratt, D. B.**, Criteria of flour quality, in *Wheat Chemistry and Technology*, Hlynka, I., Ed., American Association of Cereal Chemists, St. Paul, Minn., 1964.
58. **Locken, L., Loska, S., and Shuey, W.**, *The Farinograph Handbook*, American Association of Cereal Chemists, St. Paul, Minn., 1960.
59. **Larmour, R. K., Working, E. B., and Offelt, C. W.**, Quality tests on hard red winter wheats, *Cereal Chem.*, 16, 733, 1939.
60. *Cereal Laboratory Methods*, 7th ed., American Association of Cereal Chemists, St. Paul, Minn., 1962.
61. **Houston, D. F.**, Ed., *Rice Chemistry and Technology*, American Association of Cereal Chemists, St. Paul, Minn., 1972.
62. **Dendy, D. A. V.**, Ed., *Sorghum and Millets for Human Food*, Int. Assoc. Cereal Chem. Symp. Vienna, 1976.
63. **Ackroyd, W. R. and Doughty, J.**, Legumes in Human Nutrition, *F.A.O.-U.N. Nutr. Stud.*, No. 19, 1966.

64. **Hulse, J. H., Rachie, K. O., and Billingsley, L. W., Eds.,** Nutritional Standards and Methods of Evaluation for Food Legume Breeders, Industrial Development Research Council, Ottawa, 1977.
65. Protein-Calorie Advisory Group of the United Nations System, *Guideline No. 16 on protein methods for cereal breeders as related to human nutritional requirements. PAG Bull.*, 5(2), 22, 1975.

DRYING, CLEANING, AND CONDITIONING

Roger C. Brook and George H. Foster

INTRODUCTION

Proper conditioning and storage of grains permits a fairly constant supply of grain products throughout the entire year. Drying removes excess moisture from freshly harvested grain, thereby achieving an acceptable moisture content for safe storage.

Definitions

Moisture content expresses the amount of water in grain either as wet basis (w.b.) moisture or dry basis (d.b.) moisture. Wet basis moisture content is defined as the weight of water per unit weight of wet grain. Wet basis moisture content is used in grading and marketing of grain. Dry basis moisture, often used in engineering and scientific research, is defined as the weight of water per unit weight of grain dry matter. Dry basis to wet basis moisture content conversions are given in Figure 1 or can be calculated using the following equation:

$$\text{Moisture content, \% w.b.} = \frac{100\ (\%\ \text{d.b.})}{100 + \%\ \text{d.b.}} \qquad (1)$$

Equilibrium moisture content is important in grain drying since it limits the moisture content which can be attained when grain is exposed to air of a specified temperature and relative humidity. It is the point when no moisture transfer occurs between the grain and the drying air. The equilibrium moisture content of grain increases with increasing relative humidity and with decreasing temperature, as illustrated in Figure 2. Equilibrium moisture contents for several grains are given in Table 1, or they can be calculated using the equations presented in Tables 2 and 3.

Standard bushel is the basis for determining market price paid for grain. A bushel, although a volume measure, has a specified weight and moisture content for grain. For example, a standard bushel of shelled corn or grain sorghum is defined as 56 lb at 15.5% w.b. moisture content. Similarly, a standard bushel of wheat or soybeans is 60 lb at 14% w.b. moisture; oats is 32 lb at 14.5% w.b. moisture.

Shrink is the weight reduction which occurs when excess moisture is evaporated from grain. The percentage shrink occurring due to moisture removal is given in Figure 3, or it can be calculated using the following equation:

$$\%\ \text{Shrink} = \frac{M_o - M_f}{100 - M_f} \times 100 \qquad (2)$$

where M_o is the initial and M_f is the final moisture content, wet basis.

Moisture Content Measurement

Moisture content of grain is routinely determined by using electronic moisture meters which measure the dielectric properties of grain that vary with moisture content. Indirect measurement methods are calibrated against a direct method, such as a standard air oven procedure, which measures the moisture removed from the grain. The oven test procedure is basically a four-step process (ASAE Standard S352; AACC Method 44-15A):

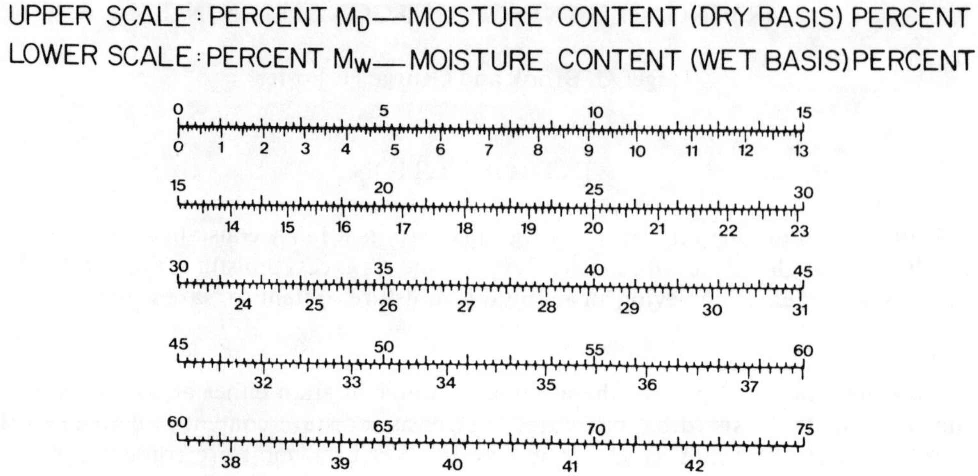

FIGURE 1. Equivalent moisture percentages — dry basis and wet basis. (From ASAE, Agricultural Engineers Yearbook, American Society of Agricultural Engineers, St. Joseph, Mo., 1978. With permission.)

1. Place a minimum of 15 g of a representative sample in each of two or more tared moisture dishes (for high moisture grain over 25% w.b. use 100-g portions). Weigh and record the weight of the dishes and contents.
2. Place the uncovered dishes and their covers in the oven. The oven temperatures and heating period depend on the grain type as specified in Table 4.
3. At the end of the heating period, cover the dishes and place them in a desiccator. Weigh and record the weight of the dishes and contents after they reach room temperature.
4. Calculate the wet basis moisture percentage by dividing the weight lost by the original sample weight and multiply by 100. Replicates should agree within 0.2% moisture.

DRYING AIR PROPERTIES

Drying of grain is affected by physical properties of the drying air. Relationships between the various properties of moist air are collectively called psychrometrics.

Psychrometrics

Psychrometrics are important to the understanding of grain drying and storage. The important psychrometric properties include:

1. Temperature — dry bulb, wet bulb, dew point (°F; °C)
2. Humidity — absolute, relative
3. Vapor pressure (lb/in.^{2}; Pa)
4. Specific volume (ft^3/lb; m^3/kg)
5. Enthalpy (Btu/lb; J/g)

A good review of these properties was presented by Brooker et al.[14] The mathematical relationships which describe the psychrometric properties are summarized in Table 5. Computer subroutines in English units and SI units are also available for calculating psychrometric properties. Graphical representation of these properties in the form of a psychrometric chart is often more convenient. Typical psychrometric charts, in English units, are presented in Figures 4 and 5. Other charts are also available for higher atmospheric pressure and other temperature ranges.[5]

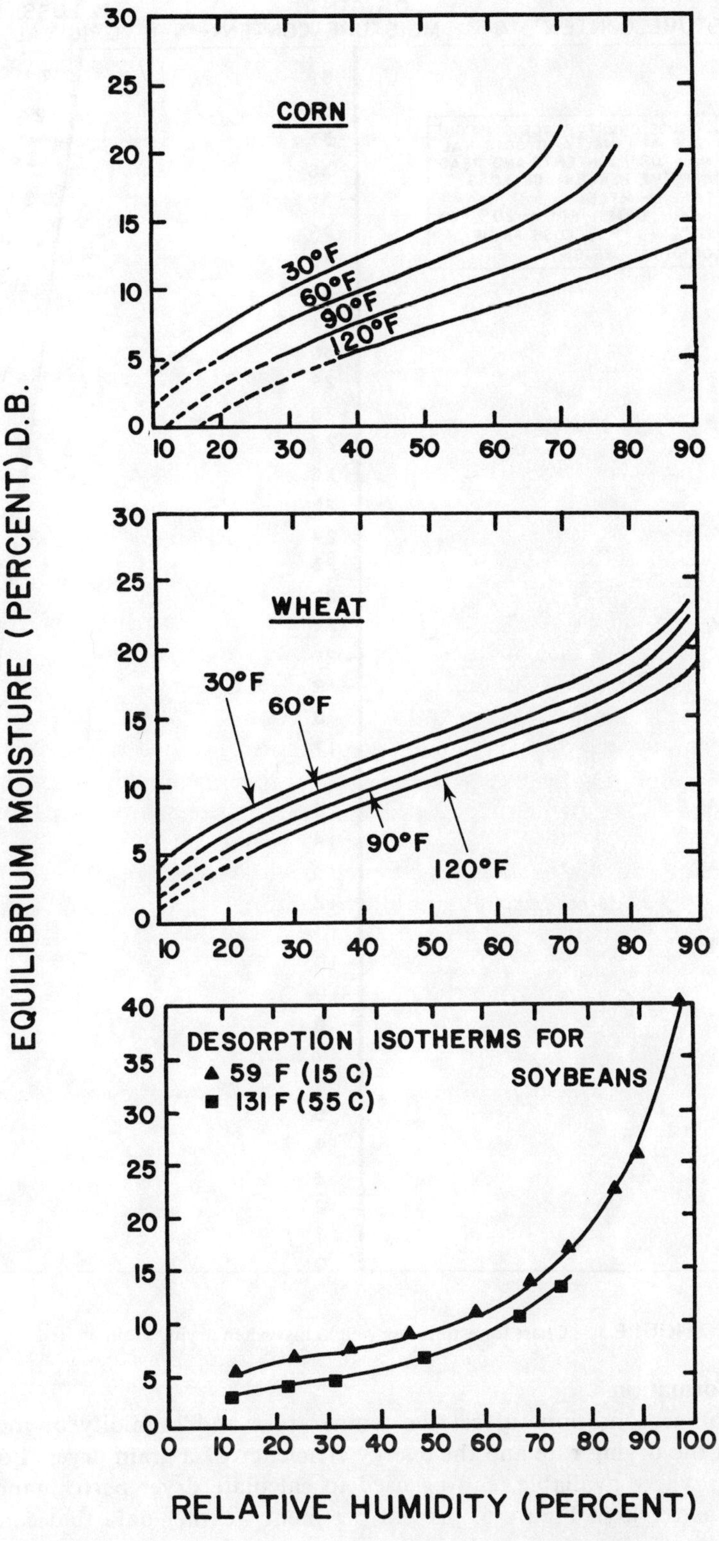

FIGURE 2. Equilibrium moisture content curves for corn, wheat, and soybeans. (From ASAE, Agricultural Engineers Handbook, American Society of Agricultural Engineers, St. Joseph, Mo., 1978. With permission.)

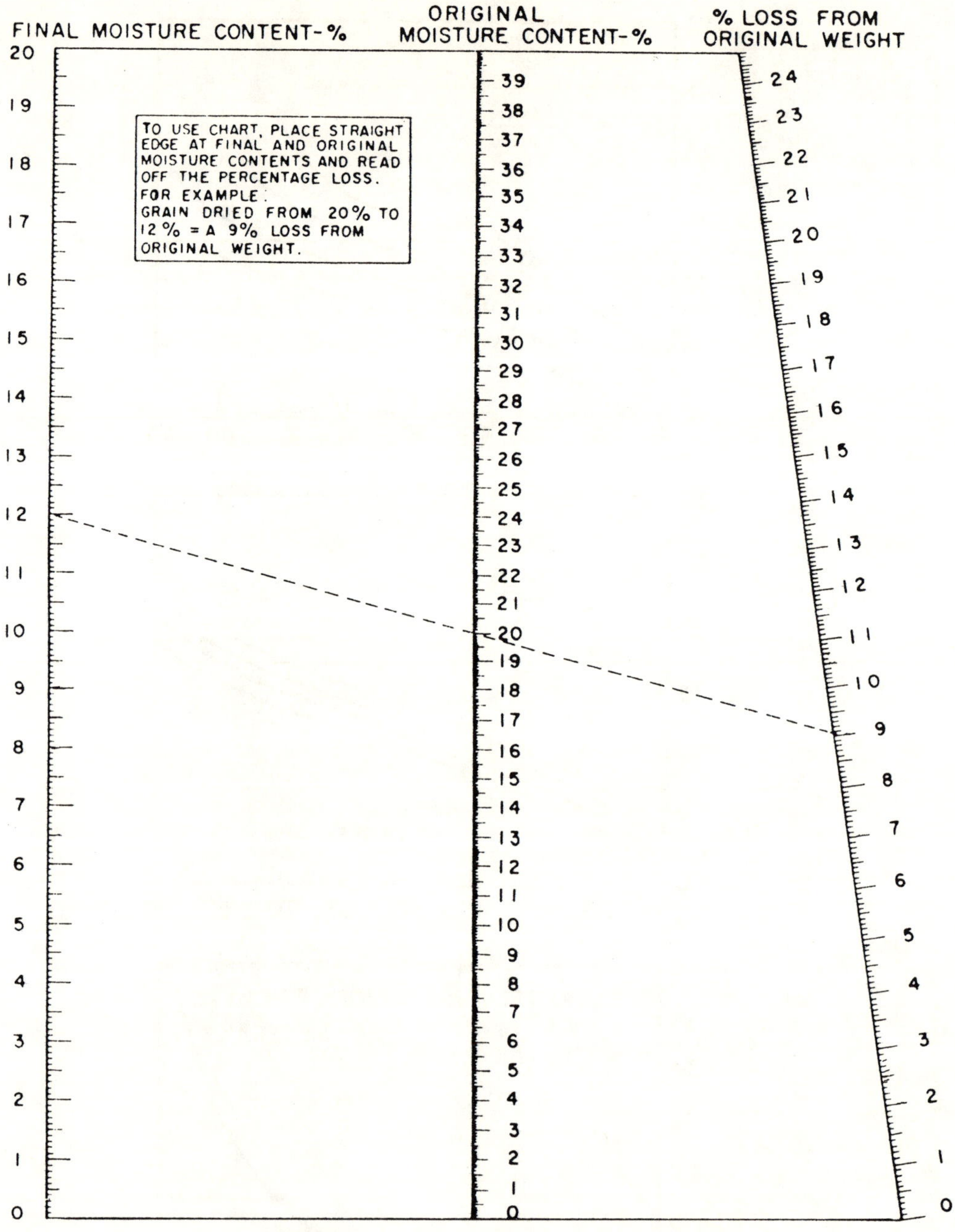

FIGURE 3. Chart for estimating weight loss when drying grain or seed.

Weather Information

Local weather conditions affect the temperature and humidity of the drying air which affect the drying rate and the energy efficiency of a grain dryer. Local weather information, where available, can be used to calculate dryer performance. Weather maps, such as those in Figure 6, present pertinent weather data that can be used in estimating dryer performance in the U.S.

Short range forecasting of weather conditions is important for making management and control decisions when operating a drying and storage facility. Regression equa-

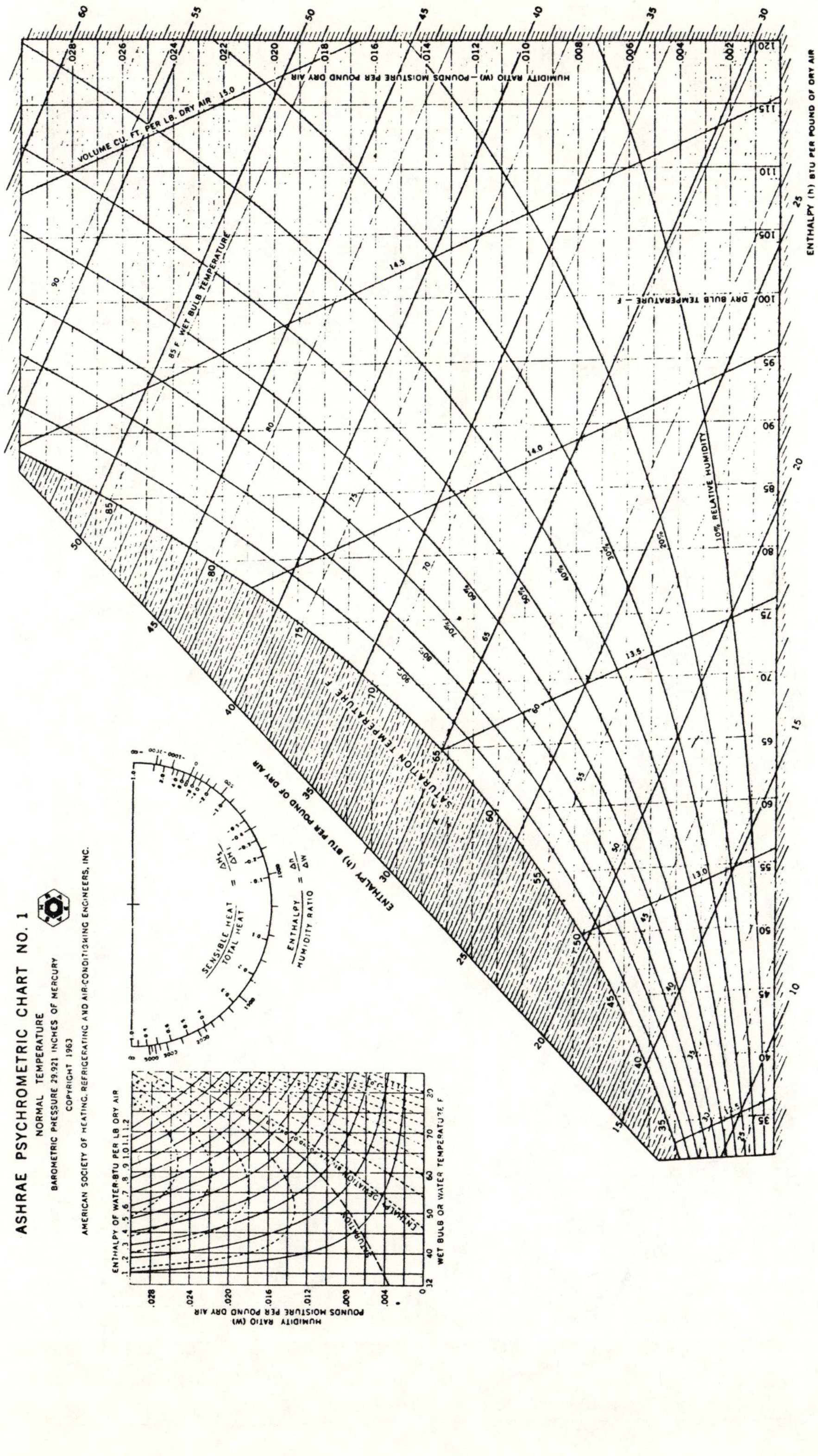

FIGURE 4. ASHRAE psychrometric chart for normal temperature range.

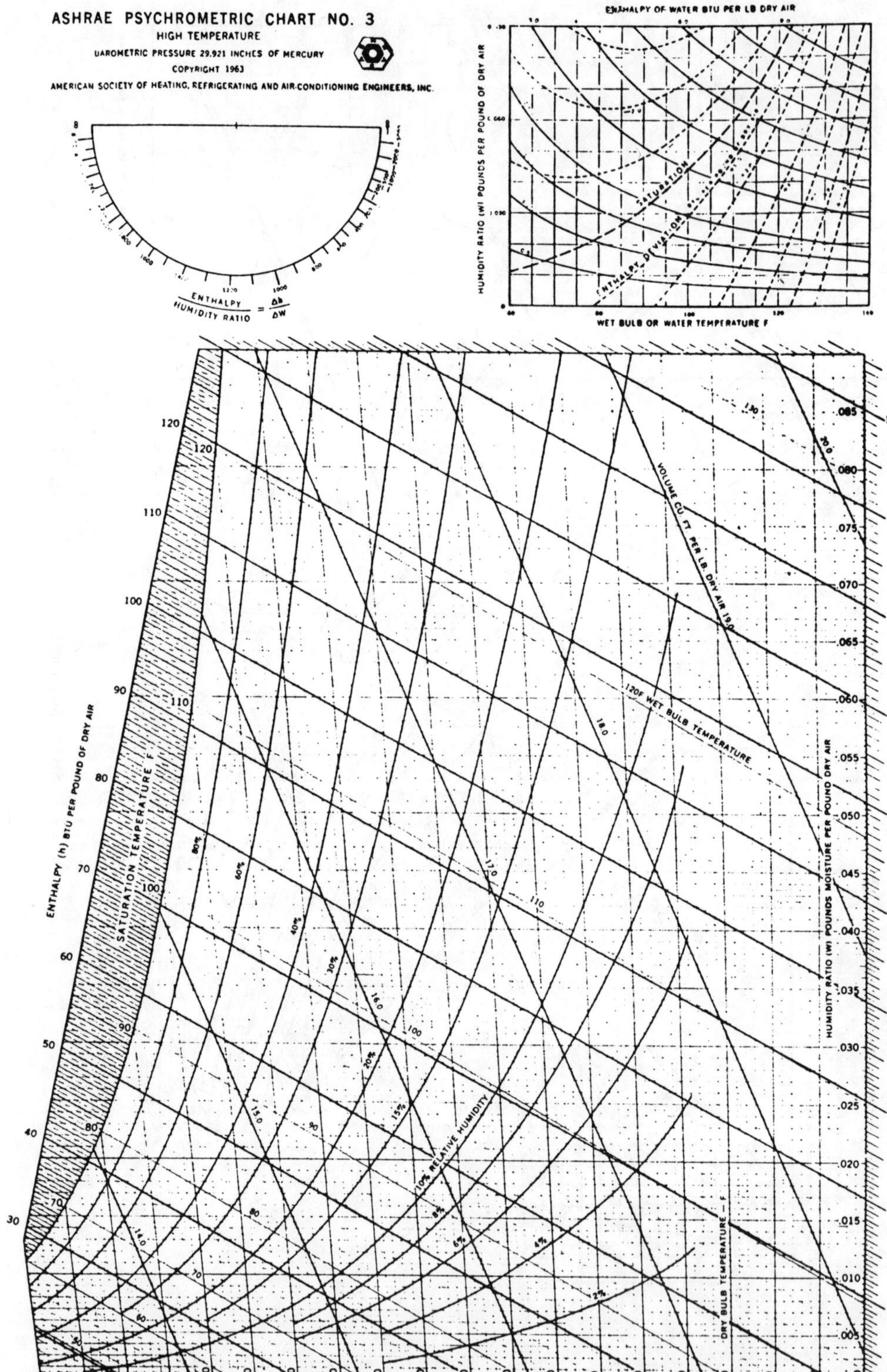

FIGURE 5. ASHRAE psychrometric chart for high temperature range.

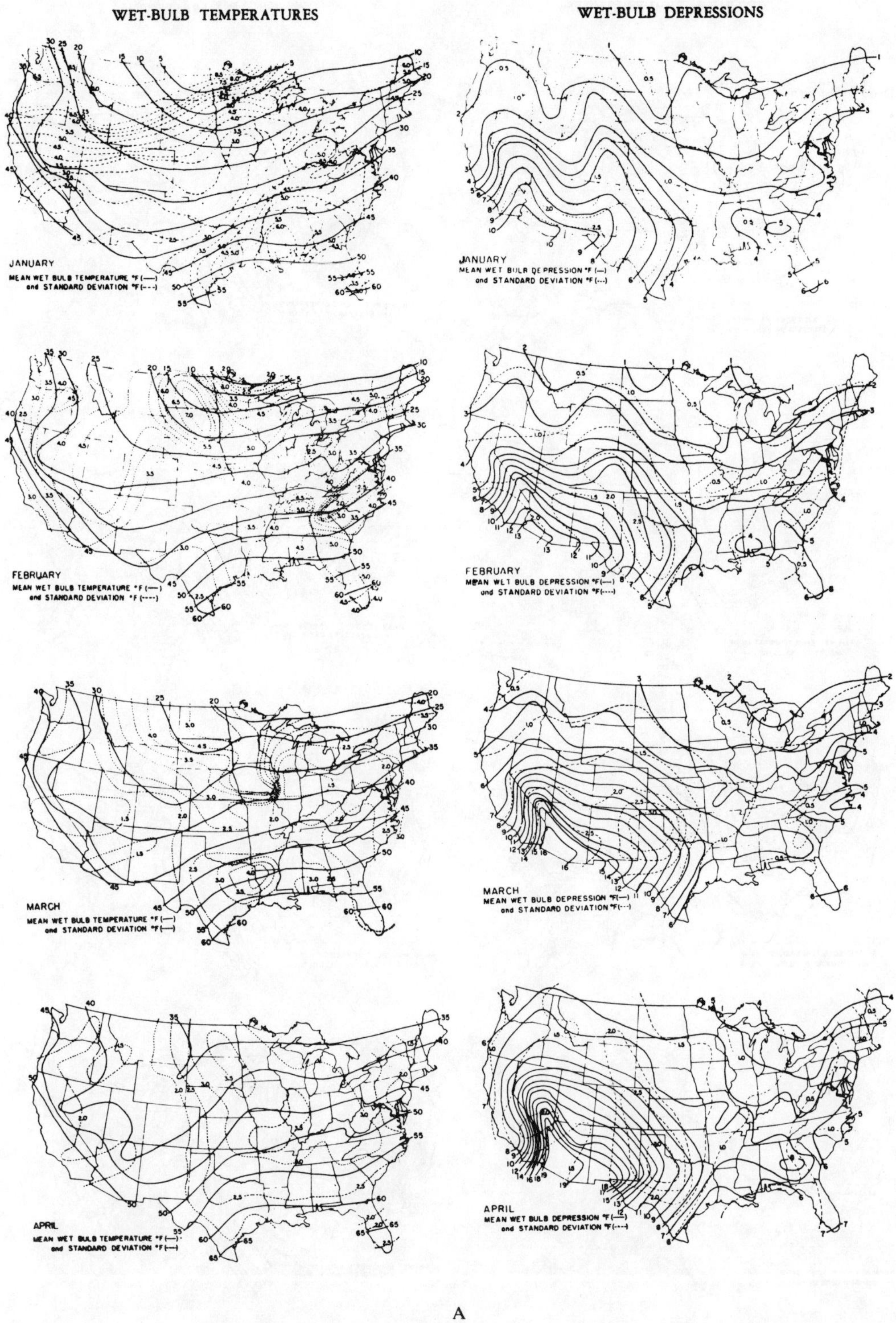

A

FIGURE 6. Wet-bulb temperatures and wet-bulb depressions — U.S. monthly averages. (A) January to April; (B) May to August; (C) September to December.

WET-BULB TEMPERATURES

WET-BULB DEPRESSIONS

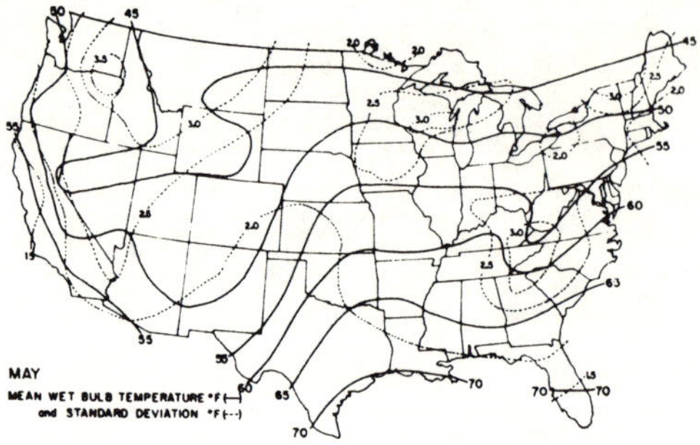

MAY
MEAN WET BULB TEMPERATURE °F (—)
and STANDARD DEVIATION °F (---)

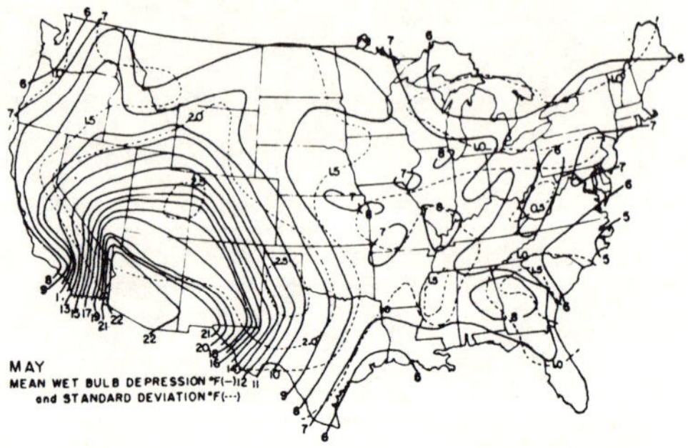

MAY
MEAN WET BULB DEPRESSION °F (—)
and STANDARD DEVIATION °F (---)

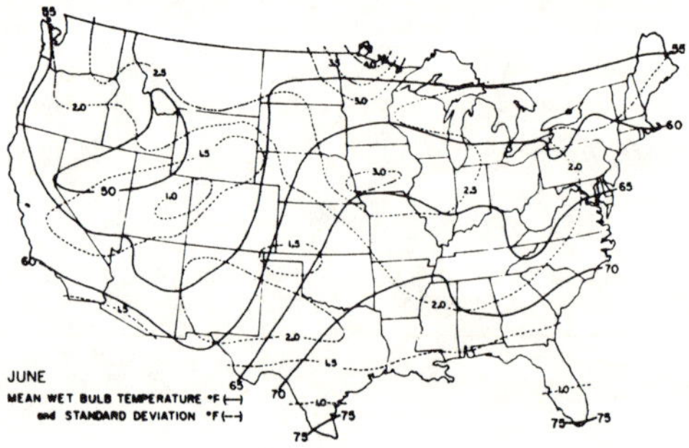

JUNE
MEAN WET BULB TEMPERATURE °F (—)
and STANDARD DEVIATION °F (---)

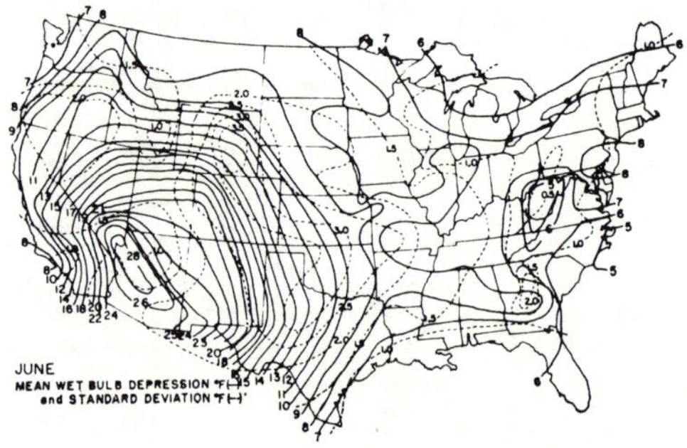

JUNE
MEAN WET BULB DEPRESSION °F (—)
and STANDARD DEVIATION °F (---)

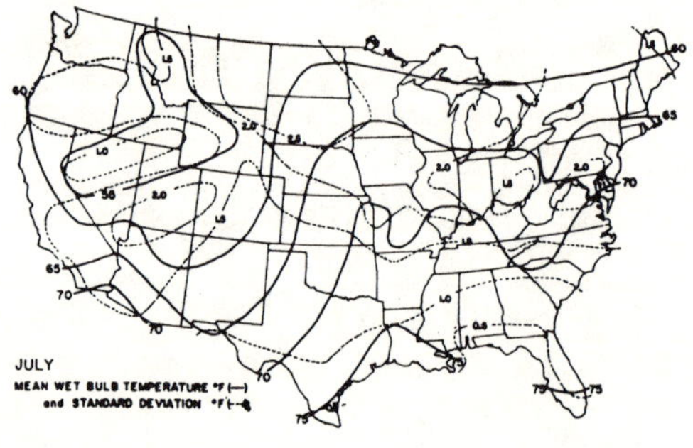

JULY
MEAN WET BULB TEMPERATURE °F (—)
and STANDARD DEVIATION °F (---)

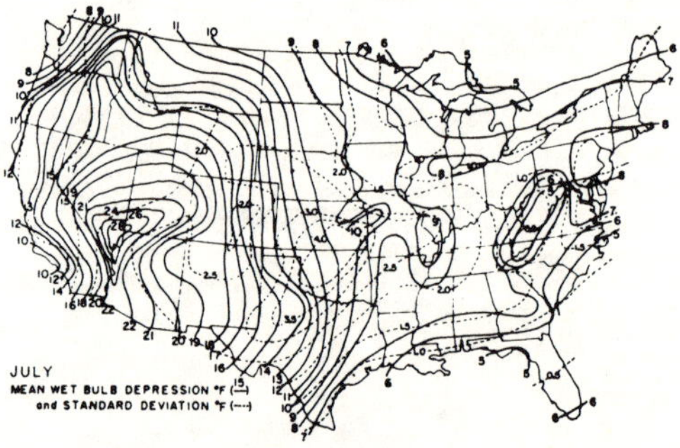

JULY
MEAN WET BULB DEPRESSION °F (—)
and STANDARD DEVIATION °F (---)

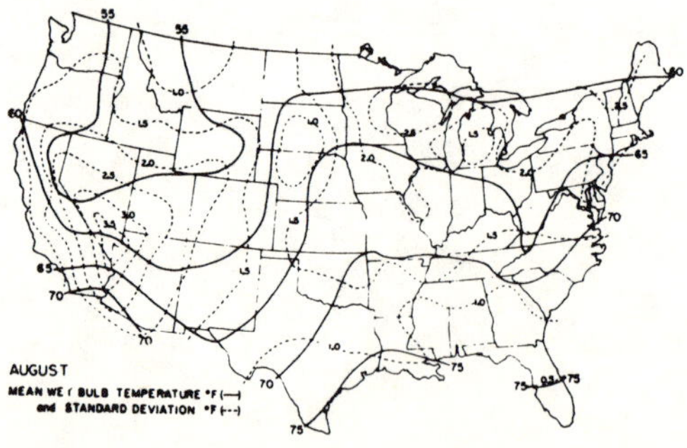

AUGUST
MEAN WET BULB TEMPERATURE °F (—)
and STANDARD DEVIATION °F (---)

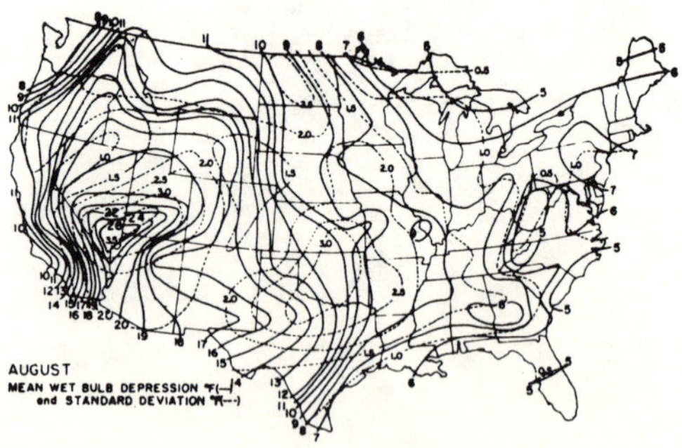

AUGUST
MEAN WET BULB DEPRESSION °F (—)
and STANDARD DEVIATION °F (---)

B

WET-BULB TEMPERATURES

WET-BULB DEPRESSIONS

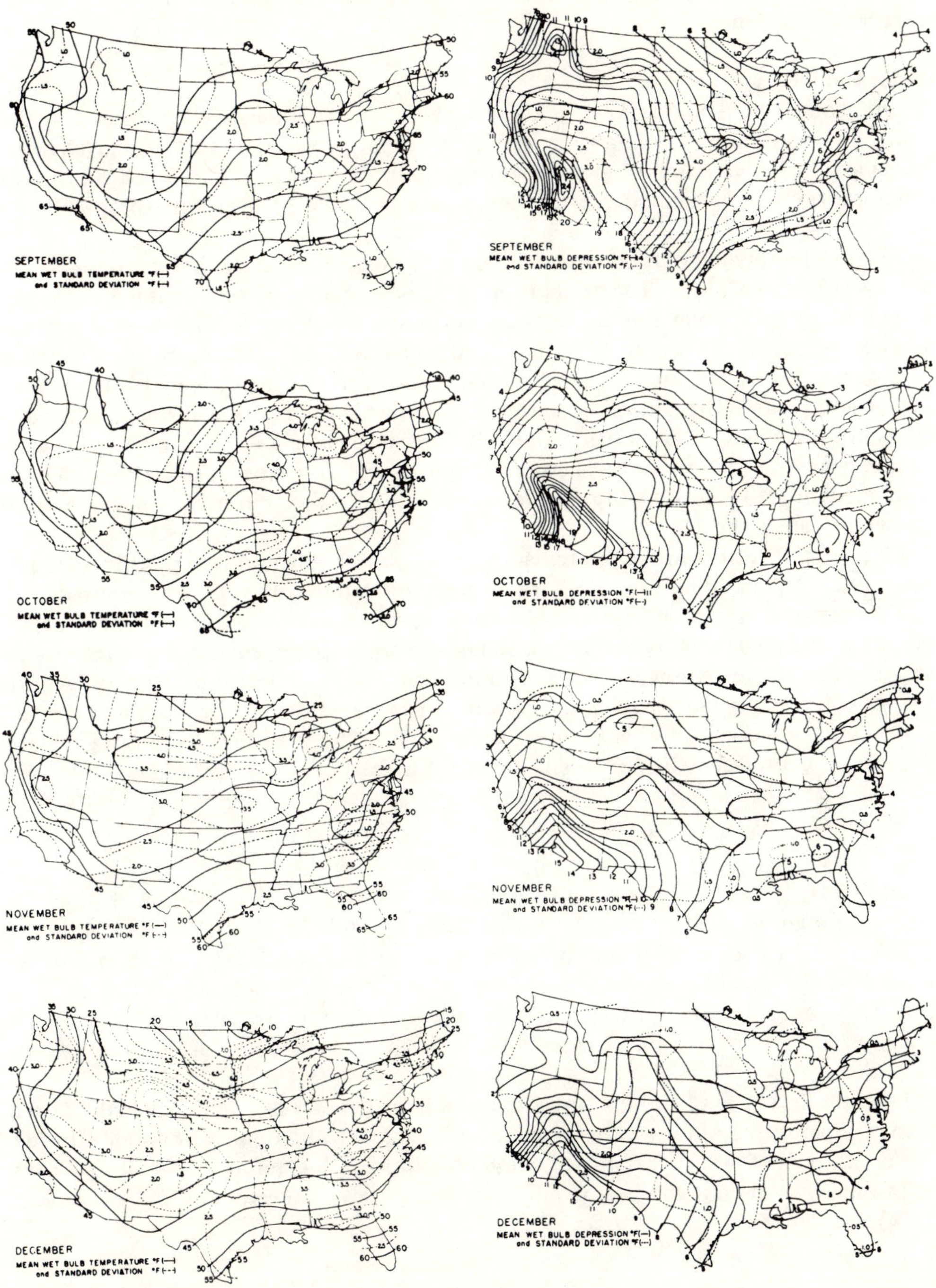

C

tions for dry bulb and wet bulb temperatures have been determined from 20 years of weather data for Lafayette, Ind. (see Table 6). These types of equations can be useful over a short term planning horizon, and can be updated periodically using current weather conditions.

Solar Radiation

One source of heat for drying grain is solar energy. Energy is captured by solar collectors and added to the drying air in the form of sensible heat. There has been considerable activity and progress in solar grain drying since 1973.[19] For the most part, solar energy has been used for low temperature drying of grain in storage bins on farms.

The availability of solar energy varies by geographical area, by time of the year, and with weather conditions. The amount of cloudiness and ground reflection is important in determining the solar flux on flat surfaces, especially those tilted toward the sun. A method of predicting the total radiation, direct beam and diffuse, on tilted surfaces was developed by Lui and Jordan,[38,39] and is the basis for the data in Tables 7 to 10. Baker and Klink[7] reported total measured radiation on horizontal surfaces at 15 locations in the North Central states in the U.S. They also calculated the radiation expected at the measuring stations at selected probability levels, based on 11 to 20 years of data. The median (50% probability) values reported by Baker compare well with those shown in Tables 7 to 10.

Flat plate collectors are used almost exclusively for applications such as low temperature grain drying. Since air is used for drying, air heating collectors are preferred.

The performance of flat plate collectors depend on the radiation incident on the collector, the angle factor, the transmittance-absorptance product of the cover or glazing, the rate of heat transfer to the working fluid in the collector, energy losses from the collector, energy losses from the collector from radiation, convection, and conduction, and the storage of energy in the collector.[16a] The total useful energy gain of a collector is expressed as a function of the inlet fluid temperature, as follows:

$$Q_u = A_c F_R [\overline{\tau\alpha} I_T - U(T_i - T_a)] \qquad (3)$$

where Q_u = total useful energy, Btu/hr-ft², W/m²; A_c = area of collector, ft², m²; I_T = total radiation incident on collector surface, Btu/hr-ft², W/m²; $\overline{\tau\alpha}$ = average collector transmittance-absorptance product, %; F_R = heat removal efficiency of transfer fluid (for water collectors used in residential heating, $F_R \approx 0.9$; air collectors, $F_R \approx 0.8$); U = overall heat transfer coefficient, Btu/hr−ft²−°F, W/m²−°C; T_i = temperature of entering heat transfer fluid, °F, °C; and T_a = temperature of ambient air, °F, °C.

In grain-drying applications, the fluid entering the collector (air) is usually at ambient temperature, and since the temperature rise is low, high mass flow rates through the collector may be used. Both of these conditions are conducive to high collection efficiencies, or permit the use of relatively inexpensive collectors at moderate (30 to 50%) efficiencies.[19]

DRYING GRAINS

This section presents information on grain properties important to understanding and design of grain drying systems. Standard work in the field of grain drying include Hall,[24] Luikov and Mikhailov,[40] Krischer,[36] Houston,[30] Brooker et al.,[14] and Bakker-Arkema et al.[a]

Physical Properties

Kernel size and weight are important in determining several of the physical properties discussed below. Representative dimensions for grain kernels are presented in Table 11. When available, the standard deviation represented by the data is also included.

The equivalent diameter in Table 11 is the diameter of a sphere which contains the same volume as the grain kernel. The percent sphericity indicates how well the grain kernel approximates a spherical shape. Equivalent ellipsoid dimensions are also included in Table 11. Moisture contents are included, since physical size and weight varies with the amount of moisture in the grain kernels. Closely related to these dimensions are the kernel weight and volume density as presented in Table 12. Additional data for shelled corn are presented in Table 13.

Bulk density (lb/ft^3, kg/m^3) is the weight per unit volume of bulk grain. An equivalent term often used in commercial market channels is *test weight* (lb/bu). (In this case, a bushel is defined as 1.25 ft^3.) Table 12 presents bulk density data at several different moisture contents. Moisture content has a sizable effect on bulk density or test weight, and is related to shrink during drying. Test weight adjustments for shelled corn as affected by harvest moisture content and percent of kernels that are mechanically damaged are given in Table 14.

Specific surface area (ft^2/ft^3, m^2/m^3) is the amount of kernel surface area per unit volume of grain kernels. The greater the specific surface area, the greater the area for heat and moisture exchange during drying. Representative values for specific surface area are presented in Table 15.

Porosity is the percent air space in bulk lots of grain and is an important parameter associated with the resistance encountered when moving air through grain. A lower porosity leads to greater friction as the air moves through the grain mass. Representative porosity or void ratio values are presented in Table 16. Porosity is affected by moisture content and particle size, and is also related to bulk density.

Resistance to Airflow

The energy necessary to force air through a mass of grain depends on the mass flow rate of air, the surface and shape characteristics of the product, the porosity, the size distribution of particles and the depth of the grain mass in the direction of airflow. The resistance to airflow is commonly referred to as pressure drop or static pressure and is commonly expressed in height of water column displaced (inches of water) or more correctly in pressure units (lb/in.^{2}, Pa). Static pressure results from energy lost through friction and turbulence.

Pressure drop data for several grains and other seeds are presented in Figure 7. The data are reported for clean, relatively dry grain which was loose filled (not packed). Experience has shown that using a packing factor or multiplier of 1.3 to 1.5 with the data in Figure 7 represents typical bin fill situations fairly well. Some of the data in Figure 7 can also be represented by the equation and constants given in Table 17.

Increasing the percentage of fines or broken kernels (fm) in grain will increase the pressure drop. The pressure drop per unit depth (P) for shelled corn can be corrected to account for fines:

$$P_{corrected} = P_{clean}\ [32.1 + (467.0 - 0.236Q)\ fm] \quad (4)$$

The correction, as presented by Haque et al.,[25] is applicable for airflows (Q) in the range 0.076 to 0.2m^3/m^2-sec (15 to 40 cfm/ft^2) through shelled corn with fines up to 20%. Fines were defined as material passing through a 4.76mm (12/64 in.) round hole sieve.

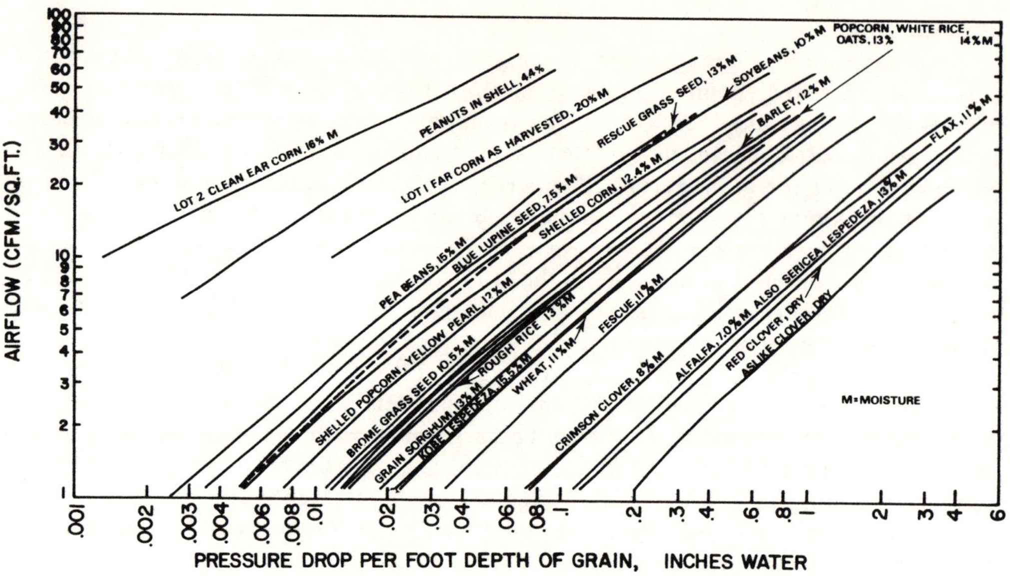

FIGURE 7. Resistance of grains and seeds to air flow. (Courtesy of Shedd, C. K., *Agric. Eng.*, 34, 616, 1953. With permission.)

Thermal Properties

Heat of vaporization (Btu/lb; J/g) the amount of heat necessary to vaporize moisture from the grain is moisture and temperature dependent, as illustrated in Table 18. The increase in heat with decreasing moisture content is significant when considering energy requirements for drying grain. Equilibrium moisture content curves provide the necessary data to determine this product-temperature-moisture dependent property. A plotting technique[49a] plots equilibrium vapor pressures vs. saturation vapor pressures for air on log-log axis, illustrated for shelled corn in Figure 8. The slope of the moisture content isoteres are equal to the ratio of the heat of vaporization of free water compared to that of the grain. Heat of vaporization can also be calculated using a method presented by Gallaher.[22] The resulting equation and constants are given in Table 19 for grain vapor pressures as calculated from Table 3.

Thermal conductivity (Btu/hr-ft-°F; J/sec-m-°C) values for grain are moisture and temperature dependent. Representative values for several grains are given in Table 20.

Specific heat capacity (Btu/lb-°F; J/g-°C) values for grains are primarily moisture dependent. Representative values for several grains are given in Table 20. Specific heat capacity values, as affected by moisture content, can be calculated as given in Table 21.

Convective heat transfer coefficient (Btu/hr-ft^2-°F; J/sec-m^2)is a property affecting heat transfer from flowing air to the grain particles. A relationship between spheres in a packed bed and a flowing gas[41] can be utilized to calculate the convective heat transfer coefficient for grain kernels represented by a sphere of equivalent radius. The corresponding equation and constants are presented in Table 22.

Moisture transfer rate can be expressed in several ways. Empirical equations based on exposure of thin layers of grain to large volumes of air are often used to describe moisture loss as a function of drying time. Representative equations for shelled corn for different temperature ranges are presented in Table 23. Similar equations for other products are presented in Table 24. Alternatively, moisture loss can be related through a diffusion process, with the apparent diffusion coefficient (ft^2/hr; m^2/sec) available for several grains as presented in Table 25. Additional comparisons of the relative drying rates of various agricultural products is presented in Table 26.

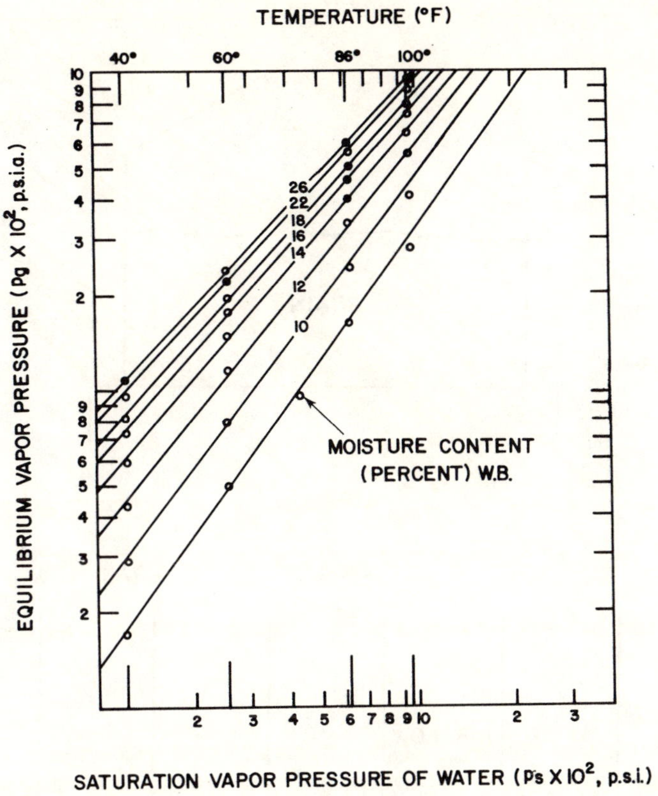

FIGURE 8. Othmer plot to determine equilibrium moisture content for shelled corn.

Grain Dryer Types

There are four basic types of grain dryers, classified according to the relative directions of airflow and grain flow. The four types are

1. Fixed bed — the grain is stationary and the air is forced through the bed from one side (typically the bottom, as illustrated in Figure 9).
2. Crossflow — the grain flows downward between two perforated sheets, while the air is forced through the grain perpendicular to the grain flow (illustrated in Figure 9).
3. Concurrent flow — the grain and air both flow in a downward direction, the air velocity being three to five times greater than the grain velocity (illustrated in Figure 9).
4. Counterflow — the grain flows downward while the air is forced upward through the grain (illustrated in Figure 9).

The drying process in all types can be described using heat and mass balances on a small element of grain (Spencer,[61] O'Callaghan et al.,[49] Thompson,[66] Ingram et al.,[31] and Meiring and Hofkes,[42]). A good review of the describing differential equations was presented by Brooker et al.,[14] with the resulting equations as presented in Table 27. A heat balance can also be useful for estimating the performance of a grain drying system:

$$\frac{cfm}{v} (C_a) (T - T_e) t = h_{fg} DM (M_o - M_f) \quad (5)$$

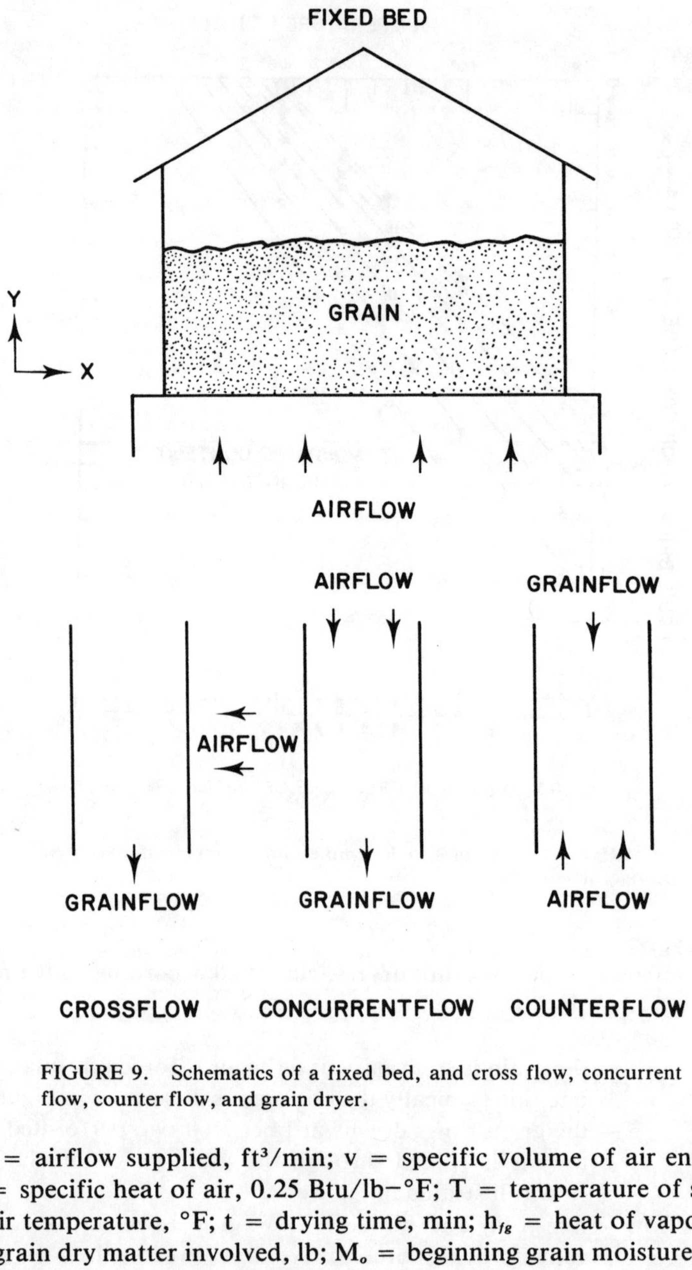

FIGURE 9. Schematics of a fixed bed, and cross flow, concurrent flow, counter flow, and grain dryer.

where cfm = airflow supplied, ft^3/min; v = specific volume of air entering the fan, ft^3/lb; C_a = specific heat of air, 0.25 Btu/lb−°F; T = temperature of supply air, °F; T_e = exit air temperature, °F; t = drying time, min; h_{fg} = heat of vaporization, Btu/lb; DM = grain dry matter involved, lb; M_o = beginning grain moisture, decimal d.b.; and M_f = final grain moisture, decimal d.b. The temperature of the supply air may be obtained from weather information presented previously. The exit temperature of the air will be influenced by the equilibrium moisture properties of the grain being dried. A good estimate of fixed-bed drying performance can be obtained by assuming air exiting at 85% relative humidity.

Grain Drying Equipment

When selecting grain drying equipment, primary considerations are drying capacity, reliability, operating efficiency, and initial cost. The effect of drying on grain quality is also important. General characteristics of commonly available drying systems are given below.

Deep Bin Drying (Fixed Bed Type)

Grain is placed in the bin to depths of 12 to 20 ft. Successful operation of a bin drying system depends largely on adequate airflow. Drying must be completed before the wet grain on top of the bin begins to mold. Recommended airflow rates for drying corn of a range of moisture contents follows.

Initial moisture content (% w.b.)	Airflow rate (cfm/bu)
24	3
22	2
20	1
17	1/2

Supplemental heat can be added (5 to 10°F) and, in high humidity areas, increases the probability of the grain reaching a storable moisture content and reduces drying time. Supplemental heat may cause overdrying. Added heat also increases mold activity. Stirring devices are used to give more even drying throughout the grain depth and break up pockets of fines in the bin. Deep bin drying systems are low cost but have low drying capacity. Seasonal volumes of 5000 to 8000 bu/bin are normal. Low energy cost and excellent grain quality are advantages.

Batch-In-Bin Drying (Fixed Bed Type)

Bin type dryers can be set up to dry grain in batches 30 to 50 in. deep. Higher airflow, 10 to 25 cfm/bu, used along with drying air temperatures of 100 to 140°F, increases drying capacity compared to deep bin drying. Batch-in-bin dryers have moderate initial investment and moderate energy costs. Good to excellent grain quality can be expected. Seasonal volumes of 8000 to 15,000 bu are normal.

Counterflow Bin Drying

A common variation of in-bin drying uses controlled bottom unloading on a semi-continuous basis. Counterflow drying is efficient in heat utilization, but may cause some grain quality degradation if high drying temperatures are used. Drying capacity is relatively high and initial costs are greater than in-bin batch drying.

Column Batch or Continuous (Crossflow Type)

Column thickness in these type dryers varies from 10 to 22 in. Recently, column thickness has been increased in a number of dryers in order to increase fuel use efficiency. High airflow rates (70 to 90 cfm/bu) and high drying air temperatures (up to 220°F) permit high drying capacities. Manual batch dryers have a moderate initial investment. Automated batch and continuous crossflow dryers are considered high investment dryers. The energy cost for these types of dryers are also higher than any of the types previously discussed. Grain quality problems with nonuniform drying and brittle kernels commonly result from the high drying rate. Grain turning or mixing devices are used in some dryers to increase uniformity of drying, especially in those using thicker grain columns. Crossflow dryers can handle seasonal volumes from 15,000 to 60,000 bu and up.

Many crossflow dryer designs exhaust the cooling air. Through changes in dryer design, the energy in the cooling air can be recycled to the drying fan, thus resulting in an energy saving. Some recent dryers also recycle up to half of the drying air. This permits 15% saving in energy consumption, with only a minor reduction in drying capacity.

Concurrent Flow Dryers

In a concurrent flow dryer, the air and the product both flow downward though the dryer. The hottest air encounters the wettest grain. The air is cooled rapidly due to the high rate of evaporation from the wet grain, allowing the use of air temperatures of 300 to 500 °F. Typically energy consumption for a concurrent flow dryer is 20 to 25% less than for a comparable crossflow dryer.

Concurrent flow dryers are considered high investment dryers. The concurrent flow dryer subjects all grain kernels to the same drying treatment thus avoiding nonuniformity in moisture content inherent in crossflow dryers. Grain quality, with respect to kernel brittleness, is usually better than in crossflow dryers. Seasonal volumes of 40,000 bu and up are normal.

Dryeration

With dryeration, corn is dried to 16 to 18% moisture content in a high-temperature dryer and transferred hot to a separate bin where, following a 4- to 10-hr tempering period, the grain is cooled slowly using an airflow rate of 1/2 to 1 cfm/bu. Dryeration results in a 20 to 25% decrease in energy use, 30 to 35% higher dryer capacity and a better quality product. About two percentage points of additional moisture is removed as the grain is cooled slowly following the tempering period.

Combination Drying

Combination or partial high temperature drying is similar to dryeration except the corn is removed from the dryer at a moisture content of about 20%. The remaining moisture is removed using a bin dryer. Combination drying results in a 25 to 30% reduction in energy requirements while doubling the apparent capacity of the drying system. The slow drying and cooling below 20% moisture insures a high quality dried product. Disadvantages of combination drying systems include the extra handling involved and the extra management required by the low temperature drying portion of the process.

Drying Effects On Grain Quality

Product quality and its determination are important for the marketing and processing of cereal grain. Desirable properties of high-quality cereal grains are

1. Appropriately low and uniform moisture content
2. Low percentage of broken and damaged kernels
3. Low susceptibility to breakage
4. High test weight
5. High starch yield
6. High oil recovery
7. High protein quality
8. High viability
9. Low mold count
10. High nutritive value

An exhaustive determination of grain quality is not a simple task as is clear from the list given above. The grain dealer is especially interested in 1, 2, 3, and 4; the wet or dry miller in 5, 6, and 7; the farmer seeding corn in 8; the animal feeder in 9 and 10. While not all quality factors are important to every user, they are all affected by the cereal grain drying process.

Stress Cracks and Broken Kernels

Drying cereal grains with hot air causes stress cracks. Stress cracks are in the starchy

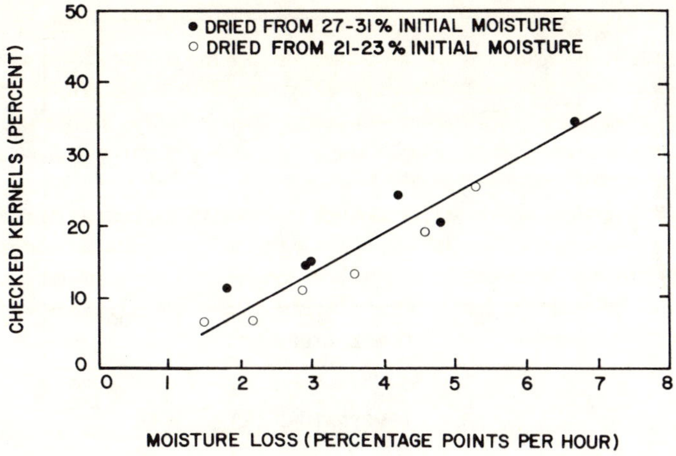

FIGURE 10. Effect of drying speed on stress cracks in corn. (From Thompson, R. A. and Foster, G. H., *U.S. Dep. Agric. Mark. Res. Rep.*, 631, 1963.)

endosperm inside the kernel and do not rupture the seed coat. The extent of stress-crack development in yellow corn has been related to the overall drying rate,[64] expressed as moisture loss in percentage points per hr as represented in Figure 10. The relation between stress cracks and drying rate has been confirmed for white corn[73] and for eastern European yellow corn varieties.[33]

Grain is often cooled as the last step in the drying process. Rapid cooling of high-temperature corn causes much of the stress-crack development. Delayed slow cooling results in a significant reduction in the percentage of stress-cracked kernels. A conventional crossflow dryer with rapid cooling may produce as many as 90% stress-cracked kernels, delayed cooling (dryeration) reduces this to 40% while natural air drying produces only about 8% stress-cracked kernels.[18]

Other factors also influence the formation of stress cracks during the drying process. Beginning the artificial drying process at a lower kernel moisture content tends to reduce the percentage of stress cracks observed.[54,56] The moisture content of the grain during drying also affects the degree of stress cracking. The maximum development of stress cracks occurred while corn was drying through the moisture range from 19 to 14%.[64] Relative humidity in the range of 10 to 30% has little effect on stress-crack development but relative humidity above 30% shows decreasing occurrence of stress cracks with increasing relative humidity.[72]

Milling Properties

Kernel temperature during drying has a significant effect on starch quality and starch yield. Figure 11 illustrates the effect of corn-kernel temperature on starch yield. Lasseran[37] reported, for European corn varieties, that a single-stage dryer has a marked and depressive effect on starch quality for operating temperatures above 90 °C (194 °F). They further indicated that two-stage drying with intermediate storage using aeration may be acceptable. Jansen and Friedrich[32] reported a fourfold increase in maltose content (from 9.4 to 35.5 mg) during hydrothermal treatment of corn (100 to 142 °C).

Oil recovered by wet milling of corn is entirely from the germ. A regression analysis of oil recovery vs. drying temperature has shown that recoverable oil decreases with increasing temperature at the rate of 0.6%/°F ($r = 0.95$) in the temperature range 49 to 149°C (120 to 300°F).[70] Freeman[21] stated that as much as 90% of the oil in damaged germs may be absorbed by the gluten prior to or during processing. Glogorcan et al.[23]

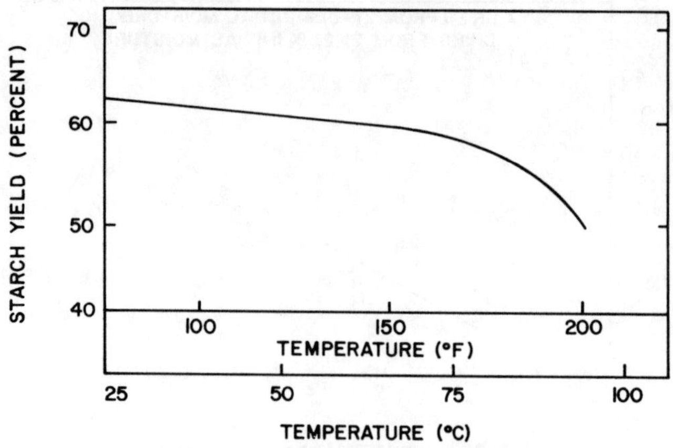

FIGURE 11. Effect of corn kernel temperature on starch yield expressed as a percentage of dry weight. (From Brooker, D. B., Bakker-Arkema, F. W., and Hall, C. W., *Drying Cereal Grains*, AVI Publishing, Westport, Conn., 1974. With permission.)

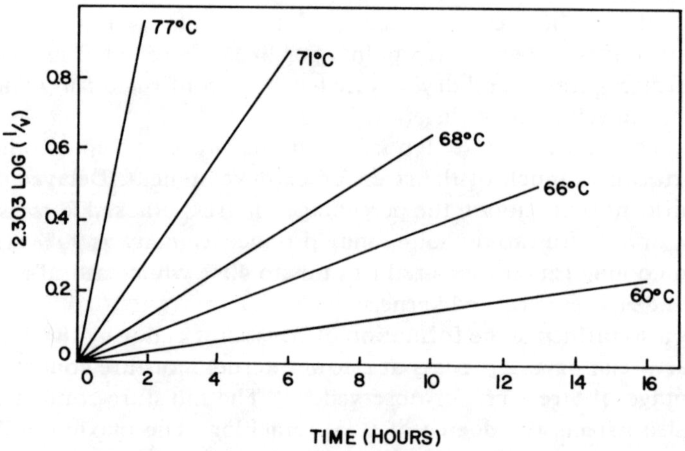

FIGURE 12. The denaturation of gluten in flour at 13.9% moisture at various temperatures. (From Becker, H. A. and Sallans, H. R., *Cereal Chem.*, 33, 254, 1956.)

found no diffusion of oil in corn dried at 40 to 160°C (104 to 320°F) and stored up to 25 months. Overhults et al.[50] reported that the "overall quality" of the oil obtained from soybeans did not significantly deteriorate in beans dried at temperatures up to 100°C (212°F). Some interior breakdown was evident at the higher temperatures, indicating that at temperatures greater than 100°C the soybean may begin to show a deterioration in oil quality.

Artificial drying affects protein content and quality in grain. High drying temperatures may destroy certain amino acids.[62] The amount of protein which can be extracted using NaCl is reduced markedly as corn is heated to 143°C (290°F).[71] The indication is that higher drying temperatures affect the solubility of certain proteins and sugars normally present in corn.[70]

Becker and Sallans[11a] reported that the denaturation of gluten in wheat flour is a first-order reaction. This is illustrated in Figure 12 where a function of the bread loaf volume, V, is plotted vs. time at various temperatures. The denaturation of the gluten in wheat flour directly affects the loaf volume. The quality of Michigan soft wheat dried in a single-stage concurrent-flow dryer has been investigated. A summary of results is contained in Table 28.

Grain Viability

Seed grain requires a high germination percentage. Kreyger[35] determined the critical kernel temperature for a number of small grains as a function of equilibrium relative humidity, as given in Table 29. It is obvious that there is a significant difference between the various small-grain species. Wheat germination is the least heat sensitive, followed by oats, corn, and rye. Germination of all grains is more heat sensitive at high than at low moisture contents.

Nutritive Value

A large part (about 90%) of the shelled corn harvested and consumed in the U.S. is used as animal feed. Williamson[74] discussed several physical characteristics that are important when considering corn for animal feed:

1. Consistency
2. Energy content
3. Palatability
4. Mold
5. Protein and amino acid profile

Assessing the effect of overheating on the nutritive value of corn is at best a difficult task. Different animals have different requirements. Energy contained in the starch of a corn kernel provides a consistent energy level in most ruminant rations. However, corn energy is not adequate for all purposes. Mink and broilers require a more concentrated energy source in the form of fat for optimum performance.[74]

Mühlbauer and Christ[46] investigated the influence of kernel temperature and time at that temperature on the nutritive value of corn used for animal feed. The quality criterion used was the content of the essential (and limiting) amino acids methionine, cystine, lysine, and their combination. Rapid drying at 180°C resulted in better quality protein than longer drying at 140 or 160°C, as illustrated in Table 30.

Quality changes in grain due to drying can be determined by a combination of analytical methods and by animal feeding trials. A rapid method to evaluate the effect of drying on the nutritive value of grain is needed. Mühlbauer et al.[47] conducted an extensive investigation using various drying conditions and concluded that changes in available lysine content could be correlated linearly ($r = 0.74 - 0.82$) with color changes in the kernels. The time-temperature effect of drying on color changes in shelled corn is illustrated in Figure 13.

STORAGE OF GRAINS

Grain will store satisfactorily if it is dry, cool, and clean.[20] Cooling can sometimes replace drying, as when moist grain is held through the winter months in cold climates. Grain that is clean will resist mold growth and insect infestation problems at one to two percentage points higher moisture content than grain containing broken kernels and trash. Most cereal grains, if clean and undamaged, and when properly cooled and managed, will store for 1 year at 14% moisture content. Soybeans and other oilseeds need to be maintained at about two percentage points lower moisture content than cereal grains to assure successful storage.

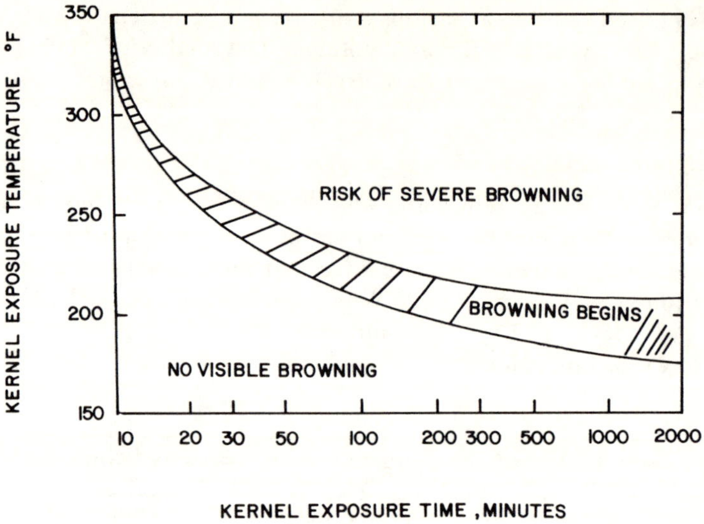

FIGURE 13. Heat damage (browning) risk for corn kernels exposed to various air temperatures.

Aeration, or mechanical ventilation of grain in storage, is used to cool grain and maintain its temperature at desired levels.[29] Aeration is also needed to maintain uniform grain temperatures to prevent air convection that results from temperature differences that develop in large bulks of stored grain. Convected air carries moisture from the warm parts of the grain bulk and deposits it in cooler grain. The problem is most severe in regions such as the U.S. cornbelt, where there are large differences between summer and winter temperatures.

Aeration is also used to hold wet grain until it can be mechanically dried, to hold moist grain at near freezing temperatures during cold weather, and to cool hot grain from a heated air drying system either as it is put into storage, or after tempering as used in the dryeration process. Airflow rates recommended for the various applications of aeration are given in Table 31. For bins with grain depths of over 50 ft, it is sometimes necessary to reduce airflow rates below those recommended in order to stay within practical fan power limits. This applies mostly to aerating smaller sized grains such as wheat.

Fan horsepower required to aerate corn at various depths are given in Table 32. The same data for wheat are in Table 33. These data are based on resistance pressures of various grains as reported by Shedd[60] and shown in Figure 7. The loose fill data are increased 30% to adjust for packed filling of bulk bins and a static pressure of 0.5 in. water gage is added for fan entrance and duct losses. Thus, fan horsepower is calculated as follows:

$$\text{Fan HP} = \frac{\text{cfm} \times [(\text{Shedd's loose fill data} \times 1.3) + 0.5]}{3000} \qquad (6)$$

The data in Tables 32 and 33 are given for each 100 ft^2 of bin floor or cross-sectional area. The cross-sectional areas of common-sized bins are given in Table 34 along with a multiplier to convert the data in Tables 32 and 33 to the requirements for the bin size under consideration.

CLEANING OF GRAINS

Grain may be cleaned before drying, storage, or marketing to remove small pieces of broken kernels and foreign material. This material, if left with the whole, sound kernels, accelerates spoilage in storage, and, if more than 2 to 3% of the total, lowers the market grade. Cleaning and sorting is based on either the size, shape or specific gravity of the material.

Screens

Screens are commonly used to separate grain materials by particle size and shape. The American Society for Testing and Materials has established Standard E-11, Specifications for Wire-Cloth Sieves for Testing Purposes. The specifications for these screens or sieves are presented in Table 35.

The Agricultural Marketing Service, USDA, has established sieve sizes to be used for determining dockage and foreign material in grains. The specifications are presented in Table 36. Sieving may be done by hand or with vibrating or shaking devices.

Specific Gravity Separators

Specific gravity separators utilize gravity and an upward airflow to achieve a highly controlled degree of separation. Though expensive and low in capacity, gravity separators can make separations not possible with other devices. They are particularily useful in the seed industry for separating light-weight, infertile seed from good seed. Representative specific gravity values are presented in Table 37.

Table 1
GRAIN EQUILIBRIUM MOISTURE CONTENT (% WET BASIS)[14]

		Relative humidity (%)									
Material	Temp °F	10	20	30	40	50	60	70	80	90	100
Barley	77	4.4	7.0	8.5	9.7	10.8	12.1	13.5	15.8	19.5	26.8
Buckwheat	77	5.0	7.6	9.1	10.2	11.4	12.7	14.2	16.1	19.1	24.5
Flaxseed	77	3.3	4.9	5.6	6.1	6.8	7.9	9.3	11.4	15.2	21.4
Oats	77	4.1	6.6	8.1	9.1	10.3	11.8	13.0	14.9	18.5	24.1
Rye	77	5.2	7.6	8.7	9.9	10.9	12.2	13.5	15.7	20.6	
Shelled corn	20				10.4	11.8	13.3	15.0	16.6		
YD	32				11.0	12.5	14.0	15.8	18.0	21.8	
	40	6.3	8.6	9.8	11.0	12.4	13.8	15.7	17.6	21.5	
	50	6.6	8.0	9.3	10.8	12.2	13.8	15.2	17.5	21.8	
	60	7.5	7.8	9.0	10.3	11.3	12.4	13.9	16.3	19.8	
	77	5.1	7.0	8.3	9.8	11.2	12.9	14.0	15.6	19.6	23.8
	86	4.4	7.4	8.2	9.0	10.2	11.4	12.9	14.8	17.4	
	90	4.9	6.6	7.7	9.3	10.8	12.4	14.0	16.2	19.3	
	100	4.0	6.0	7.3	8.7	9.0	11.0	12.5	14.2	16.7	
	120				8.6	10.0	11.2	13.1	14.9		
	155				7.4	8.4	10.0	11.5	12.2		
	160	3.9	6.2	7.6	9.1	10.4	11.9	13.9	15.2	17.9	
Sorghum	30		8.2	10.1	11.2	12.3	13.5	14.5	15.8		
	60		7.5	9.5	10.7	11.8	12.9	14.0	15.5		
	77	4.4	7.3	8.6	9.8	11.0	12.0	13.8	15.8	18.8	21.9
	90		7.0	8.7	10.2	11.8	12.2	13.1	14.8		
	120		6.6	8.0	9.4	10.7	11.6	12.7	14.3		
Soybeans	77		5.5	6.5	7.1	8.0	9.3	11.5	14.8	18.8	
Wheat											
Soft red winter	20				11.3	12.8	14.1	15.3	17.0		
	32				11.0	12.2	13.5	14.7	16.2		
	50				10.2	11.7	13.1	14.4	16.0		
	70				9.7	11.0	12.4	14.0			
	77	4.3	7.2	8.6	9.7	10.9	11.9	13.6	15.7	19.7	25.6
Hard red winter	77	4.4	7.2	8.5	9.7	10.9	12.5	13.9	15.8	19.7	25.0
Hard red spring	77	4.4	7.2	8.5	9.8	11.1	12.5	13.9	15.9	19.7	25.0
White	77	5.2	7.5	8.6	9.4	10.5	11.8	13.7	16.0	19.7	26.3
Durum	77	5.1	7.4	8.5	9.4	10.5	11.5	13.1	15.4	19.3	26.7
Wheat	30		6.9	9.2	10.4	11.8	13.2	14.5	16.3		
	50		7.9	9.2	10.6	12.7	14.2	15.0	17.3		
	60		6.1	7.8	9.6	10.7	12.7	13.8	15.3		
	68	5.6	7.1	8.3	9.6	10.9	12.2	13.5	15.0		
	77	5.8	7.6	9.1	10.7	11.6	13.0	14.5	16.8	20.6	
	90		5.3	7.0	8.6	10.3	11.5	12.9	14.3		
	104	4.6	6.2	7.4	8.6	10.0	11.3	12.3	14.2		
	120			6.2	7.4	9.6	10.4	11.9	13.6		
	122	4.0	5.8	6.7	8.1	10.0	10.8	12.6	15.1	19.4	
	176	2.5	3.7	4.8	5.7	6.7	8.0	9.8	11.5		

Table 2
DESORPTION EQUILIBRIUM MOISTURE CONTENT EQUATION AND CONSTANTS FOR THE HENDERSON-THOMPSON EQUATION[55]

$$Me = \frac{\left[\frac{\ln(1 - rh)}{-K(T + C)}\right]^{1/N}}{100}$$

Me = equilibrium moisture; decimal d.b.
rh = relative humidity; decimal
T = temperature, °C

Product	K	N	C	SE moisture
Barley	0.000022919	2.0123	195.267	0.0080
Beans, edible	0.000035953	1.7639	193.091	0.0178
Corn, yellow dent	0.000086541	1.8634	49.810	0.0127
Rice, rough	0.000019187	2.4451	51.161	0.0097
Sorghum	0.000008532	2.4757	113.725	0.0087
Soybean	0.000503633	1.3628	43.016	0.0165
Wheat				
Durum	0.000025738	2.2110	70.318	0.0068
Hard	0.000023008	2.2857	55.815	0.0071
Soft	0.000012300	2.5558	64.346	0.0122

Table 3
DISORPTION EQUILIBRIUM MOISTURE CONTENT EQUATION AND CONSTANTS FOR THE CHUNG-PFOST EQUATION[55]

$$M_e = B - C \ln[-1.98 (T + A) \ln(rh)]$$

M_e = equilibrium moisture; decimal d.b.
rh = relative humidity; decimal
T = temperature, °C

Product	A	B	C	SE moisture
Barley	91.323	0.368149	0.042787	0.0055
Beans, edible	120.098	0.480920	0.066826	0.0174
Corn, yellow dent	30.205	0.379212	0.058970	0.0121
Rice, rough	35.703	0.325535	0.046015	0.0133
Sorghum	102.849	0.391444	0.050970	0.0086
Soybean	24.576	0.375314	0.066816	0.0175
Wheat				
Durum	112.350	0.415593	0.055318	0.0057
Hard	50.998	0.395155	0.056788	0.0061
Soft	35.662	0.308163	0.042360	0.0147

Table 4
OVEN TEMPERATURE AND HEATING PERIOD FOR MOISTURE CONTENT DETERMINATIONS

Seed	Oven temp ± 1°C	Heating time		Ref.
		hr	min	
Barley	130	20	0	26
Corn	103	72	0	68
Fescue	130	3	0	26
Flax	103	4	0	68
Oats	130	22	0	26
Rape	130	4	0	26
Rye	130	16	0	26
Ryegrass	130	3	0	26
Safflower	130	3	0	26
Sorghum	130	18	0	26
Wheat	130	19	0	26

Table 5
EQUATIONS AND CONSTANTS RELATING PSYCHROMETRIC PROPERTIES OF AIR IN ENGLISH AND SI UNITS

	Constants	
Equation[a]	**English**[4]	**SI**[9,57]
Saturation line		
$\ln(P_s) = A - \frac{B}{T} - C\ln(T)$	$A = 23.3924$	$A = 31.9602$
	$B = 11286.6489$	$B = 6270.3605$
	$C = 0.46057$	$C = 0.46057$
	$459.69 < T < 491.69°R$	$255.38 < T < 273.16°K$
$\ln(P_s/R) = (A + BT + CT^2 + DT^3 + ET^4)/(FT - GT^2)$	$R = 3206.18$	$R = 22105847.39$
	$A = -27405.526$	$A = -27405.526$
	$B = 54.1896$	$B = 97.5413$
	$C = -0.045137$	$C = -0.146244$
	$D = 0.215321 \times 10^{-4}$	$D = 0.12558 \times 10^{-3}$
	$E = -0.462027 \times 10^{-8}$	$E = -0.48502 \times 10^{-7}$
	$F = 2.41613$	$F = 4.34903$
	$G = 0.00121547$	$G = 0.39381 \times 10^{-2}$
	$491.69 < T < 959.69°R$	$273.16 < T < 533.16°K$
Latent heat at saturation		
$h_{fg} = A - B(T - C)$	$A = 1220.844$	$A = 2839776.184$
	$B = 0.05077$	$B = 212.56384$
	$C = 459.69$	$C = 255.38$
	$459.69 < T < 491.69°R$	$255.38 < T < 273.16°K$
$h_{fg} = A - B(T - C)$	$A = 1075.8965$	$A = 2502535.259$
	$B = 0.56983$	$B = 2385.76424$
	$C = 491.69$	$C = 338.72$
	$491.69 < T < 609.69°R$	$273.16 < T < 338.72°K$
$h_{fg} = (A - Bt^2)^{1/2}$	$A = 1354673.214$	$A = 7329155978000$
	$B = 0.9125275587$	$B = 15995964.08$
	$609.69 < T < 959.69°R$	$338.72 < T < 533.16°K$
Wet bulb line		
$P_{swb} - P_v = B(T_{wb} - T)$	$A = 0.2405$	$A = 1006.9254$
where	$459.69 < T < 959.69°R$	$255.38 < T < 533.16°K$
$B = \dfrac{A(P_{swb} - P_{atm})(1.0 + 0.15577P_v/P_{atm})}{0.62194\,h'_{fg}}$		
Humidity ratio		
$H = 0.6219P_v/(P_{atm} - P_v)$	$459.69 < T < 959.69°R$	$255.38 < T < 533.16°K$
	$P_V < P_{atm}$	$P_V < P_{atm}$
Specific volume		
$V_{sa} = AT/(P_{atm} - P_v)$	$A = 0.370486$	$A = 287.0$
	$459.69 < T < 959.69°R$	$255.38 < T < 533.16°K$

Table 5 (continued)
EQUATIONS AND CONSTANTS RELATING PSYCHROMETRIC PROPERTIES OF AIR IN ENGLISH AND SI UNITS

	Constants	
Equation[a]	English[4]	SI[9,57]
Enthalpy		
$h = A(T-B) - H[C + D(E - T_{dp})] + h''_{fg} H + FH(T - T_{dp})$	A = 0.2405	A = 1006.9254
	B = 459.69	B = 255.38
	C = 143.35	C = 333432.1
	D = 0.485	D = 2030.598
	E = 491.69	E = 273.16
	F = 0.448	F = 1875.6864
	$459.69<T<491.69°R$	$225.38<T_{dp}<273.16°K$
$h = A(T-B) + CH(T_{dp} - D) + h''_{fg} H + EH(T - T_{dp})$	A = 0.2405	A = 1006.9254
	B = 459.69	B = 273.16
	C = 1.0	C = 0.0
	D = 491.69	D = 4186.8
	E = 0.448	E = 273.16
		F = 1875.6864
	$491.69<T<671.69°R$	$273.16<T_{dp}<373.16°K$

[a] List of symbols:
h = enthalpy (Btu/lb; J/kg), h_{fg} = latent heat at saturation (Btu/lb; J/kg), h'_{fg} = latent heat at T_{wb} (Btu/lb; J/kg), h''_{fg} = latent heat at T_{dp} (Btu/lb; J/kg), H = humidity ratio (lb/lb; kg/kg), ln = natural logarithm (base e), P_{atm} = atmospheric pressure (psi; Pa), P_s = saturation vapor pressure at T (psi; Pa), P_{swb} = saturation vapor pressure at T_{wb} (psi; Pa), P_v = vapor pressure (psi; Pa), T = dry bulb temperature (°R; °K), T_{dp} = dew point temperature (°R, °K), T_{wb} = wet bulb temperature (°R; °K), and V_{sa} = air specific volume (ft³/lb; m³/kg).

Table 6
SHORT TERM DRY BULB AND DEWPOINT TEMPERATURE FORECAST EQUATIONS[16]

Dry Bulb Prediction

T1 = 0.244 + 1.198 (X1) − 0.262(X3) + 0.0578(X18) R square = 0.980, SD = 2.115
T2 = 0.714 + 1.357(X1) − 0.470(X3) + 0.167(X18) − 0.0713(X24) R square = 0.956, SD = 3.143
T3 = 1.292 + 1.416(X1) − 0.587(X3) + 0.295(X18) − 0.155(X24) R square = 0.926, SD = 4.073
T4 = 1.970 + 1.399(X1) + 0.436(X18) − 0.625(X3) − 0.258(X24) R square = 0.897, SD = 4.801
T5 = 2.732 + 1.322(X1) + 0.574(X18) − 0.366(X24) − 0.595(X3) R square = 0.873, STD der = 5.338
T6 = 3.571 + 1.193(X1) + 0.694(X18) − 0.467(X24) − 0.504(X3) R square = 0.850, SD = 5.786

Dew Point Prediction

T1 = 0.522 + 1.148(X1) − 0.161(X3) R square = 0.982, SD = 1.716
T2 = 0.726 + 1.286(X1) − 0.354(X3) + 0.136(X18) − 0.0859(X24) R square = 0.965, SD = 2.373
T3 = 1.247 + 1.360(X1) − 0.467(X3) + 0.238(X18) − 0.164(X24) R square = 0.943, SD = 3.023
T4 = 1.846 + 1.380(X1) − 0.525(X3) + 0.349(X18) − 0.252(X24) R square = 0.919, SD = 3.589
T5 = 2.494 + 1.356(X1) + 0.457(X18) − 0.341(X24) − 0.537(X3) R square = 0.896, SD = 4.072
T6 = 3.179 + 1.292(X1) + 0.550(X18) − 0.422(X24) − 0.502(X3) R square = 0.872, SD = 4.508

Note: These equations predict the temperature (°F) T (−) hours in advance when the respective dry bulb and dewpoint temperatures X(−) hours in the past are given.

Table 7
DAILY TOTAL[a] SOLAR RADIATION AT COLUMBUS, OHIO — 40.0°N. LATITUDE[53a] (MONTHLY AVERAGES ON SURFACES TILTED AT SELECTED ANGLES TOWARD THE SUN AT SOLAR NOON)

	Btu/ft²/day at specified tilt angles				
Month	0° (Horizontal)	18.4° (4/12 slope)	40.0° (Latitude)	55.0° (Lat. + 15)	90.0° (Vertical)
January	486	651	788	838	791
February	747	936	1073	1106	972
March	1113	1285	1368	1343	1040
April	1481	1569	1528	1411	909
May	1839	1838	1683	1485	837
June	2111	2042	1804	1549	803
July	2041	1986	1766	1524	801
August	1573	1593	1482	1324	779
September	1189	1284	1277	1197	811
October	920	1088	1180	1170	922
November	479	584	656	667	572
December	430	581	703	744	686

[a] Direct, diffuse and reflected.

Table 8
DAILY TOTAL[a] SOLAR RADIATION AT MADISON, WIS. — 43.1°N. LATITUDE[53a] (MONTHLY AVERAGES ON SURFACES TILTED AT SELECTED ANGLES TOWARD THE SUN AT SOLAR NOON)

	Btu/ft²/day at specified tilt angles				
Month	0° (Horizontal)	18.4° (4/12 slope)	43.1° (Latitude)	58.1° (Lat. + 15)	90.0° (Vertical)
January	565	856	1146	1246	1235
February	812	1087	1339	1410	1331
March	1232	1480	1641	1632	1349
April	1455	1562	1533	1418	979
May	1745	1759	1600	1411	855
June	2032	1983	1733	1489	840
July	2047	2009	1767	1522	861
August	1740	1786	1655	1475	909
September	1444	1608	1631	1533	1082
October	993	1224	1379	1375	1125
November	556	743	903	938	847
December	496	758	1020	1110	1099

[a] Direct, diffuse and reflected.

Table 9
DAILY TOTAL[a] SOLAR RADIATION AT COLUMBIA, MO. — 39.0°N. LATITUDE[53a] (MONTHLY AVERAGES ON SURFACES TILTED AT SELECTED ANGLES TOWARD THE SUN AT SOLAR NOON)

	Btu/ft²/day at specified tilt angles				
Month	0° (Horizontal)	18.4° (4/12 slope)	39.0° (Latitude)	54.0° (Lat. + 15)	90.0° (Vertical)
January	651	912	1124	1210	1153
February	941	1209	1402	1459	1288
March	1316	1536	1644	1624	1252
April	1631	1729	1689	1563	988
May	2000	1993	1829	1612	873
June	2129	2053	1820	1563	788
July	2149	2084	1857	1599	807
August	1953	1981	1846	1645	914
September	1690	1857	1869	1762	1158
October	1203	1455	1597	1598	1256
November	840	1126	1337	1405	1249
December	590	839	1039	1119	1056

[a] Direct, diffuse and reflected.

Table 10
DAILY TOTAL[a] SOLAR RADIATION AT DODGE CITY, KAN. — 37.8°N. LATITUDE[53a] (MONTHLY AVERAGES ON SURFACES TILTED AT SELECTED ANGLES TOWARD THE SUN AT SOLAR NOON)

	Btu/ft²/day at specified tilt angles				
Month	0° (Horizontal)	18.4° (4/12 slope)	37.8° (Latitude)	52.8° (Lat. + 15)	90.0° (Vertical)
January	953	1404	1753	1910	1815
February	1186	1552	1803	1885	1645
March	1566	1842	1976	1961	1492
April	1976	2103	2062	1910	1161
May	2127	2113	1943	1710	887
June	2460	2359	2092	1784	826
July	2401	2317	2067	1772	834
August	2211	2238	2090	1859	978
September	1842	2023	2037	1922	1231
October	1421	1736	1910	1922	1500
November	1065	1461	1743	1849	1647
December	874	1316	1659	1815	1729

[a] Direct, diffuse and reflected.

Table 11
THE PHYSICAL DIMENSIONS OF GRAIN KERNELS

Product	Equiv sphere diameter[a] (in.)	% Sphericity	Ellipsoid dimensions[a] Major (in.)	Interm. (in.)	Minor (in.)	Moisture % w.b.	Ref.
Barley	0.155 (0.008)	44.5	0.411 (0.035)	0.137 (0.010)	0.108 (0.058)	7.8	44
	0.141 (0.005)					8.4	65
	0.180 (0.011)					12.0	10
Flax	0.088 (0.002)	46.4	0.194 (0.008)	0.092 (0.005)		6.4	44
Milo	0.137	84.7	0.170 (0.010)	0.160 (0.007)	0.110 (0.012)	9.2	44
Navy bean	0.266 (0.016)					12.0	10
Oats	0.142 (0.004)	34.2	0.508 (0.058)	0.114 (0.012)	0.090 (0.011)	8.7	44
	0.136 (0.004)					9.5	65
	0.232 (0.022)					12.0	10
Rice	0.132 (0.004)	46.7	0.333 (0.017)	0.123 (0.004)	0.089 (0.004)	8.9	44
Rice							
Medium	0.179	46.2				15.0	75
Long	0.180	37.5				15.0	75
Rye	0.122					7.8	65
Shelled corn	0.315	99.4	0.640 (0.036)	0.798 (0.042)	0.504 (0.028)	6.7	44
	0.305 (0.006)					10.0	65
	0.386 (0.053)					12.0	10
Sorghum	0.138 (0.003)					10.2	65
Soybeans	0.249 (0.005)					<7.0	65
	0.253 (0.14)					14.6	65
Wheat							
Hard	0.152 (0.008)	61.5	0.258 (0.017)	0.126 (0.010)	0.122 (0.010)	7.7	44
Soft	0.141 (0.004)					10.6	65
	0.158 (0.010)					12.0	10
	0.175 (0.013)					12.0	10

[a] Values in parentheses are standard deviations.

Table 12
WEIGHT, VOLUME, AND DENSITY OF GRAIN KERNELS

Product	Weight[a] (× 10⁵ lb)	Volume[a] (× 10⁴ in.³)	Unit density (lb/ft³)	Bulk density[a] (lb/ft³)	Moisture (% wb)	Ref.
Barley	9.93	20.32	88.44	37.8	7.8	10
	(1.47)	(2.10)	—	(2.0)		
	7.56	—	—	38.6	12.0	44
Flax	1.46	3.47	72.71	43.6	6.4	10
	(0.16)	(0.45)	—	(0.5)		
Milo	6.34	13.43	81.57	48.1	9.2	44
	(0.21)	(0.44)	—	(0.2)		
Navy bean	46.60	—	—	38.6	12.0	10
	—	—	—	—		
Oats	7.50	15.18	85.38	29.0	8.7	44
	(0.73)	(1.22)	—	(3.8)		
	9.05	—	—	25.6	12.0	10
	—	—	—	—		
Rice	5.94	11.99	85.61	36.1	8.9	44
	(0.45)	(1.07)	—	(0.7)		
Rice, medium	9.80	20.48	82.67	37.4	12.0	75
	10.20	21.12	83.47	38.6	14.0	75
	10.70	21.87	84.53	39.6	16.0	75
	11.70	23.61	85.64	40.5	18.0	75
Rice, long	11.20	22.76	85.05	36.6	12.0	75
	11.30	22.81	85.59	36.7	14.0	75
	11.70	23.51	86.00	37.8	16.0	75
	12.00	24.02	86.34	38.4	18.0	75
Shelled corn	74.20	—	—	41.2	12.0	10
Wheat						
hard	9.48	18.47	88.69	50.0	7.7	44
	(1.44)	(2.78)	—	(0.7)		
soft	8.17	—	—	48.0	12.0	10
	(0.78)	—	—	—		

[a] Values in parentheses are standard deviations.

Table 13
RELATIONSHIP BETWEEN THREE AXIAL DIMENSIONS AND VOLUME, IN 50-KERNEL SAMPLES OF SHELLED CORN[44]

Sample and statistics	Major diameter (mm)	Intermediate diameter (mm)	Minor diameter (mm)	Weight (g)
Single variety middle of ear				
Mean	10.78	7.58	4.57	0.27
Variance	0.475	0.214	0.151	0.001
Correlation with volume-r	0.66	0.59	0.27	—
Regression equation $\ln V = -7.14 + 1.18 \ln a + 1.04 \ln b + 0.48 \ln c$				
Single variety whole ear				
Mean	11.15	7.30	4.86	0.27
Variance	1.071	0.670	0.586	0.003
Correlation with volume-r	0.75	0.78	0.30	—
Regression equation $\ln V = -6.19 + 0.62 \ln a + 1.13 \ln b + 0.6 \ln c$				
Mixture of varieties				
Mean	12.66	8.5	5.24	0.35
Variance	1.72	0.77	0.5	0.004
Correlations with volume-r	0.35	0.41	0.32	—
Regression equation $\ln V = -6.18 + 0.9 \ln a + 0.71 \ln b + 0.68 \ln c$				

Table 14
ADJUSTMENT TO BE ADDED TO WET HARVEST TEST WEIGHT OF SHELLED CORN TO OBTAIN EXPECTED TEST WEIGHT LEVEL AFTER DRYING TO 15.5% MOISTURE[24a]

Damage (%)	% Moisture at Harvest (w.b.)							
	30	28	26	24	22	20	18	16
	lb/bu							
45	0.3	0						
40	0.7	0.2						
35	1.3	0.7						
30	1.8	1.3	0.8					
25	2.4	1.9	1.4	0.9	0.3			
20	3.1	2.6	2.0	1.5	1.0	0.5		
15	3.8	3.2	2.7	2.2	1.7	1.2	0.6	0.2
10	4.5	4.0	3.5	2.9	2.4	1.9	1.4	0.8
5	5.3	4.7	4.2	3.7	3.2	2.7	2.1	1.6
0	6.1	5.6	5.0	4.5	4.0	3.5	2.9	2.4

Table 15
REPRESENTATIVE SPECIFIC SURFACE AREA FOR SEVERAL GRAINS AT 12% W.B. MOISTURE

Product	Specific surface area (ft^2/ft^3) Mean	SD	Ref.
Barley	452	58	10
Navy beans	477	20	10
Oats	334	63	10
Rice, medium	345	—	75
Shelled corn	239	66	10
Wheat, soft	360	50	10

Table 16
VOID RATIO OF GRAIN KERNELS IN BULK

Product	Void ratio %	Moisture content %w.b.	Ref.
Barley	56.0	8.4	65
Navy bean	45.6	14.9	52
	46.9	18.3	52
	47.8	25.0	52
Oats	40.8	9.5	65
Rice	59.0	12.0	75
	57.9	14.0	75
	56.5	16.0	75
	55.0	18.0	75
Shelled corn	45.9	9.4	65
	39.4	13.7	65
	44.7	16.0	52
	51.0	19.1	52
	53.4	23.7	52
Sorghum	58.3	10.2	65
Soybeans	57.2	<7.0	65
	54.3	14.6	65
Wheat, soft	60.8	10.2	65
	58.0	13.2	65

Table 17
CALCULATION OF STATIC PRESSURE FOR AIRFLOW THROUGH GRAIN[4]

$$P = \frac{aQ^2}{\ln(1 + bQ)}$$

P = static pressure (Pa/m)
Q = airflow (0.01–0.20 m^3/m^2 sec)

Product	Value of a	Value of b
Barley	2.18×10^4	8.01
Oats	2.53×10^4	14.60
Shelled Corn	2.06×10^4	30.70
Soybeans	1.14×10^4	12.80
Wheat	2.91×10^4	9.84

Table 18
HEAT OF VAPORIZATION (BTU/LB) OF SELECTED GRAINS[28]

Product and MC (% d.b.)	Temp (°F) 32	50	70	100	150
Wheat					
5	1289	1277	1265	1243	1208
10	1230	1218	1207	1186	1153
15	1174	1163	1152	1133	1101
20	1101	1091	1081	1062	1032
Corn					
5	1474	1460	1446	1422	1382
10	1384	1371	1359	1335	1298
15	1304	1292	1280	1258	1223
20	1202	1191	1180	1160	1127
Sorghum					
5	1299	1287	1275	1253	1218
10	1226	1214	1203	1182	1149
15	1165	1154	1144	1124	1093
20	1107	1097	1087	1068	1038
Water	1075	1065	1055	1037	1008

Table 19
EQUATIONS AND CONSTANTS FOR CALCULATING LATENT HEAT OF VAPORIZATION FOR VARIOUS GRAINS

Latent heat[a] = (1094.0 − 0.57T) [1 + A exp (−BM)]

$$\text{Latent heat} = (1094.0 - 0.57T)\,[1 + A\exp(-BM)]$$

Grain	A	B
Barley	0.8266	19.901
Beans, edible	0.4922	14.965
Corn, yellow	1.2925	16.961
Rice, rough	2.0692	21.739
Sorghum	1.0060	19.650
Soybean	0.7000	14.969
Wheat		
Durum	0.7405	18.010
Hard	1.2287	17.612
Soft	2.5348	23.628

[a] T = grain temperature, °F;
M = grain moisture, decimal d.b.

Table 20
REPRESENTATIVE THERMAL PROPERTIES FOR VARIOUS GRAINS

	Moisture content	Temperature (°F)	Specific heat (Btu/lb−°F)	Thermal conductivity (Btu/hr−ft−°F)	Ref.
Corn, yellow	0.9	54.0—83.8	0.366	0.0812	34
	5.1	—	0.404	0.0847	34
	9.8	—	0.438	0.0878	34
	14.7	—	0.484	0.0919	34
	20.1	—	0.531	0.0945	34
	24.7	—	0.567	0.0982	34
	30.2	—	0.588	0.0996	34
Oats	0.0	—	0.305	—	27
	9.1	—	—	0.0370	8
	27.7	—	—	0.0537	8
Rapeseed	5.5	66.9	0.354	—	45
	6.1	39.9—80.1	—	0.1125	12
	8.0	39.9—80.1	—	0.1215	12
	10.1	66.9	0.447	—	45
	11.5	39.9—80.1	—	0.1345	12
	12.8	39.9—80.1	—	0.1450	12
	15.5	66.9	0.433	—	45
Rice, rough	12.0	—	0.382	0.0590	75
	14.0	—	0.405	0.0607	75
	16.0	—	0.429	0.0624	75
	18.0	—	0.452	0.0642	75
	20.0	—	0.476	0.0650	75
Soybeans	0.0	53.5—82.5	0.377	—	3
	8.7	—	0.443	—	3
	15.0	—	0.480	—	3
	17.1	—	0.493	—	3
	21.2	—	0.518	—	3
Wheat, hard	0.0	—	0.310	—	6
	1.4	—	0.319	0.0706	6
	5.3	—	0.357	0.0755	6
	7.4	—	0.407	0.0874	6
	13.6	—	0.460	0.0967	6
Wheat, soft	0.7	51.2—89.9	0.347	0.0679	34
	5.5	—	0.375	0.0706	34
	10.3	—	0.428	0.0747	34
	14.4	—	0.500	0.0786	34
	20.3	—	0.522	0.0798	34

Table 21
EQUATION AND CONSTANTS FOR CALCULATING THE SPECIFIC HEAT OF GRAINS AS AFFECTED BY MOISTURE CONTENT

Equation: Specific heat (Btu/lb-°F) = A + B (Moisture, % w.b.)

Grain	Mean temp °F	A	B	Ref.
Corn, yellow	68.9	0.350	0.00851	34
	95.3	0.325	0.00949	34
Oats	—	0.305	0.0078	27
Rapeseed	35.1	0.308	0.00678	45
	66.9	0.324	0.00764	45
Rice	—	0.220	0.01301	75
Rough				
	—	0.265	0.01070	27
Shelled	—	0.287	0.00910	27
Finished	—	0.282	0.00900	27
Sorghum	75.0	0.332	0.00770	59a
Soybeans	68.0	0.391	0.00461	3
Wheat	—	0.283	0.00724	6
Hard white				
Soft white	—	0.068	0.00065	34

Table 22
EQUATIONS AND CONSTANTS FOR CALCULATING THE HEAT TRANSFER COEFFICIENT FOR VARIOUS GRAINS

Equation:[a] $h = A\,(G_aT/B)^c$

Grain	A	B	C	Ref.
Barley	0.5738	0.25	0.601	13
Corn, yellow	0.0934	0.69T	0.850	58
Navy bean	0.2781	0.90T	0.660	48
Wheat	0.5738	0.25	0.601	13

[a] Definition of symbols: G_a = airflow, lb/hr-ft²; h = convective heat transfer, BTU/hr-ft²-°F; T = air temperature, °R.

Table 23
THINLAYER DRYING EQUATIONS FOR SHELLED CORN[a]

Equation	Constants	Ref.
$MR = \frac{6}{\Pi^2} \exp(-Kt)$	$K = 1944.0 \exp - \frac{5023}{T + 459.69}$	51
$t = A\, ln MR + B(ln MR)^2$	$A = 1.86178 + 0.00488T$	66
	$B = 427.364 \exp(-0.03301T)$	
$\frac{t}{60} = P_i(M - M_e)^{q_i} - P_i(M_0 - M_e)^{q_i} + t_i$	$i = 1: M_0 \geq M \geq M_1$	67
	$M_1 = 0.4(M_0 - M_e) + M_e$	
	$t_1 = 0$	
	$P_1 = \exp(-2.45 - 6.42M_0^{1.25}$	
	$-3.15rh + 9.62M_0\sqrt{rh}$	
	$+0.030T - 0.12\, V_a$	
	$q_1 = 3.98 + 2.87\, M_0$	
	$-[0.19/(rh + 0.015)]$	
	$+0.016T$	
	$i = 2: M_1 \geq M \geq M_2$	
	$M_2 = 0.12(M_0 - M_e) + M_e$	
	$t_2 = [P_1(M_1 - M_e)^{q_1} - P_1(M_0 - M_e)^{q_1}]/60$	
	$P_2 = \exp[2.82 + 7.49(rh + 0.01)^{0.67} - 0.0179T]$	
	$q_2 = -\exp(0.810 - 3.11\, rh)$	
	$i = 3: M_2 \geq M \geq M_e$	
	$t_3 = [P_2(M_2 - M_e)^{q_2} - P_2(M_1 - M_e)^{q_2}]/60 + t_2$	
	$P_3 = 0.12(M_0 - M_e)^{(q_2 - q_3)}(P_2 q_2/q_3)$	
	$q_3 = -1.0$	
$MR = \exp(-Rt^c)$	$R = \exp(-xt^y)$	17
	$x = [6.0142 + 1.453 \times 10^{-4}(rh)^2]^{0.5}$	
	$-T[3.353 \times 10^{-4} + 3.0 \times 10^{-8}(rh)^2]^{0.5}$	
	$y = 0.1245 - 2.197 \times 10^{-3}(rh)$	
	$+2.3 \times 10^{-5}(rh)T - 5.8 \times 10^{-5}T$	
	$C = 0.664$	
	or $R = 4.74 \times 10^{-2} + 7.62 \times 10^{-4} rh$	16
	$+5.73 \times 10^{-2} T - 2.03 \times 10^{-5}(rh)T$	
	$C = 0.76 + 5.23 \times 10^{-4} rh - 4.57 \times 10^{-3} T$	
	or $R = -3.47 \times 10^{-2} + 2.87 \times 10^{-3} T$	43
	$C = 0.54 + 3.24 \times 10^{-3} rh$	

[a]Definition of Symbols:

- rh = relative humidity, decimal
- t = drying time, hr
- MR = moisture ratio = $(M_t - M_e)/(M_0 - M_e)$
- M_e = equilibrium moisture, decimal d.b.
- M_0 = moisture at time t = 0, decimal d.b.
- M_t = moisture at time t, decimal d.b.
- T = drying temperature, °F
- V_a = air velocity, ft/hr

Table 24
THINLAYER DRYING EQUATIONS FOR SELECTED GRAINS

Equation	Product	Constants	Range	Ref.
$MR = \exp(-kt)$	Barley	$k = 139.3\exp\left[-\frac{7976}{T + 460}\right]$		49
	Wheat	$k = 2000\exp\left[-\frac{9179}{T + 460}\right]$		49
$MR = \exp(-xt^y)$	Rough rice	$x = -0.19 - 0.45\ rh + 6.75 \times 10^{-3}\ T$	$90 \leq T \leq 125$ F	
		$y = -3.47 \times 10^{-2} + 1.94\ rh - 1.77\ rh^2 + 5.26 \times 10^{-3}\ T$	$0.20 \leq rh \leq 0.85$	
		Initial moisture content	$M_0 = 24\%$ w.b.	
		Air velocity	$V_a = 118$ ft/min	
$t = A\ ln\ MR + B\ (ln\ MR)^2$	Sorghum	$A = -25.87 + 0.34T - 1.08 \times 10^{-3}\ T^2$	$80 \leq T \leq 160$F	53
		$A = 0.54 - 1.70 \times 10^{-3}\ T$	$160 \leq T \leq 240$F	
		$B = 30.35\ \exp\ (-1.80 \times 10^{-2}\ T)$	$80 \leq T \leq 240$F	
		Air velocity	$20 \leq V_a \leq 60$ ft/min	
		Initial moisture content	$M_0 = 18$–19%w.b.	
$MR = 1.0 - 24.82(Dt)^{0.5} + 142.29Dt$	Wheat	$D = 76.3\exp\left[-\frac{11080}{T + 460}\right]$	$(Dt)^{0.5} \leq 0.034$	11
$MR = 0.509\exp(-629.68Dt)$	Wheat		$(Dt)^{0.5} \geq 0.034$	11

Definition of symbols: rh = relative humidity, decimal; t = drying time, hr; MR = moisture ratio = $(M_t - M_e)\ (M_o - M_e)$; M_e = equilibrium moisture, decimal d.b.; M_o = moisture at time t = 0, decimal d.b.; M_t = moisture at time t, decimal d.b.; T = drying temperature, °F; V_z = drying air velocity, ft/min.

Table 25
EQUATION AND CONSTANTS FOR CALCULATING APPARENT DIFFUSION COEFFICIENTS FOR ESTIMATING THE DRYING RATE IN SEVERAL GRAINS[a]

Product	Diffusion coefficient	Range	Ref.
Corn	$D = 1.63 \times 10^{-3} \exp\left[(2.5 \times 10^{-2} T + 6.02) M - \frac{4523}{T + 460}\right]$	$120° \leq T \leq 160°F$ $0.1 \leq rh \leq 0.7$	15
	$D = 6.3 \times 10^{-9} \exp\left[-\frac{6946}{T + 460}\right]$	$0.005 \leq M \leq 0.35$ 100-170°F	51
Rough rice			
Hull	$D = 2.77 \times 10^{-2} \exp\left[-\frac{5184}{T + 460}\right]$	—	63
Bran	$D = 8.58 \exp\left[-\frac{9198}{T + 460}\right]$		63
Endosperm	$D = 5210 \exp\left[-\frac{13284}{T + 460}\right]$		63
Soybean	$D = 3.88\left[0.49 \exp\left(-\frac{0.59}{M_o - M_e}\right) + 0.018(M - M_e) \exp\left(-\frac{5648}{T + 460}\right)\right]$	—	59
Wheat	$D = 76.8 \exp\left[-\frac{11080}{T + 460}\right]$	—	61

[a] Definition of symbols: r = kernel radial coordinate, ft; r_o = equivalent diameter, ft; r^h = relative humidity, decimal; D = diffusion coefficient, ft^2/hr; M = moisture content, dec. d.b.; M_e = equilibrium moisture, dec. d.b.; M_o = moisture content at time $t = 0$, decimal d.b.; T = kernel temperature, °F.

Table 26
RELATIVE DRYING RATES OF DIFFERENT PRODUCTS AS COMPARED TO WHEAT AT 18% W.B. and 35°C (95°F)[35]

	% Moisture content w.b.				
	22	20	18	16	14
Peas	95	55	35	23	15
Corn	135	90	50	30	15
Wheat	210	150	100	65	35
Rye	225	175	115	70	35
Oats	450	250	150	100	60
Sugar beet seed	1650	1200	800	500	300
Rapeseed	—	1500	1150	800	500

Table 27
DIFFERENTIAL EQUATIONS DESCRIBING THE DRYING PROCESS FOR FOUR COMMON DRYER TYPES[a]

Fixed bed

$$\frac{\partial T}{\partial x} = \frac{-ha}{G_a c_a + G_a c_v W}(T - \theta)$$

$$\frac{\partial \theta}{\partial t} = \frac{ha}{p_p c_p + p_p c_w M}(T - \theta) + \frac{h_{fg} + c_v (T - \theta)}{p_p c_p + p_p c_w M} G_a \frac{\partial W}{\partial x}$$

$$\frac{\partial W}{\partial x} = -\frac{P_p \partial M}{G_a \partial t}$$

$$\frac{\partial M}{\partial t} = \text{an appropriate thin-layer equation}$$

Cross flow

$$\frac{\partial T}{\partial x} = \frac{-ha}{G_a c_a + G_a c_v W}(T - \theta)$$

$$\frac{\partial \theta}{\partial y} = \frac{ha}{G_p c_p + G_p c_w M}(T - \theta) - \frac{h_{fg} + c_v (T - \theta)}{G_p c_p + G_p c_w M} G_a \frac{\partial W}{\partial x}$$

$$\frac{\partial W}{\partial x} = -\frac{G_p}{G_a}\frac{\partial M}{\partial y}$$

$$\frac{\partial M}{\partial t} = \text{an appropriate thin-layer equation}$$

Table 27 (continued)
DIFFERENTIAL EQUATIONS DESCRIBING THE DRYING PROCESS FOR FOUR COMMON DRYER TYPES[a]

Concurrent flow

$$\frac{dT}{dx} = \frac{-ha}{G_a c_a + G_a c_v W} (T - \theta)$$

$$\frac{d\theta}{dx} = \frac{ha}{G_p c_p + G_p c_w M} (T - \theta) - \frac{h_{fg} + c_v (T - \theta)}{G_p c_p + G_p c_w M} G_a \frac{dW}{dx}$$

$$\frac{dW}{dx} = - \frac{G_p}{G_a} \frac{dM}{dx}$$

$$\frac{dM}{dt} = \text{an appropriate thin-layer equation}$$

Counter flow

$$\frac{dT}{dx} = \frac{ha}{G_a c_a + G_a c_v W} (T - \theta)$$

$$\frac{d\theta}{dx} = \frac{ha}{G_p c_p + G_p c_w M} (T - \theta) + \frac{h_{fg} + c_v (T - \theta)}{G_p c_p + G_p c_w M} G_a \frac{dW}{dx}$$

$$\frac{dW}{dx} = \frac{G_p}{G_a} \frac{dM}{dx}$$

$$\frac{dM}{dt} = \text{an appropriate thin-layer equation}$$

[a] Definition of symbols: a = specific surface area ft^2/ft^3; c = specific heat, Btu/lb — °F; h = heat transfer coefficient, Btu/hr-ft^2-°F; h_{fg} = heat of vaporization, Btu/lb; p = dry weight density, lb/ft^3; t = time, hr; x = grain depth coordinate, ft; y = column length coordinate, ft; G = dry weight flow rate, lb/hr-ft^2; M = grain moisture content, decimal d.b.; T = air temperature, °F; W = air humidity ratio, lb/lb; θ = grain temperature, °F. Subscripts: a = air; p = product; v = water vapor; w = liquid water.

Table 28
RESULTS OF FOUR QUALITY FACTORS OF SOFT WHEAT DRIED AT DIFFERENT INLET AIR TEMPERATURES IN A ONE-STAGE CONCURRENT-FLOW DRYER[a,2]

Drying temp °C	Germination %	Ash %	Protein %	AWR[a] %
121	71	0.37	8.2	51.5
149	71	0.38	8.6	52.9
204	51	0.39	9.0	59.0
232	28	0.40	8.9	66.3
Control	93	0.37	8.2	51.6

[a] AWR = Alkaline water retention.

Table 29
CRITICAL KERNEL TEMPERATURE (°C) OF SOME SMALL GRAINS AS A FUNCTION OF EQUILIBRIUM RELATIVE HUMIDITY (CRITERION: LESS THAN 5% VIABILITY DECREASE)[35]

	% Relative Humidity			
	60	70	80	90
Oats	59	55	50	—
Wheat	63	62	58	52
Corn	52	51	48	46
Rye	53	50	45	41

Table 30
LYSINE CONTENT OF CORN AS AFFECTED BY DRYING TEMPERATURE TO DRY FROM 26 TO 14% W.B.[46]

Drying temp (°C)	Drying time (min)	Lysine content (mg/16 mg N)
80	107	2.4
120	35	2.4
140	21	2.0
160	14	1.8
180	7	2.1

Table 31
RECOMMENDED AERATION AIRFLOW RATES

Purpose of aeration	Airflow rate (cfm/bu)
Maintaining dry grain	1/20 to 1/10
Holding moist grain	1/4 to 1/2
Cooling hot grain (from dryer)	1/2 to 1

Table 32
FAN HORSEPOWER FOR AERATING SHELLED CORN[a] — REQUIREMENTS PER 100 ft² OF BIN CROSS-SECTION OR FLOOR AREA

Corn depth (ft)	Aeration air flow rate (cfm/bu)					
	1/20	1/10	1/4	1/2	3/4	1
	(hp/100 ft²)					
10	0.007	0.015	0.04	0.10	0.19	0.31
15	0.011	0.025	0.06	0.24	0.54	1.0
20	0.016	0.037	0.10	0.48	1.2	2.3
25	0.021	0.05	0.21	1.0	2.3	4.3
30	0.029	0.07	0.36	1.7	4.3	8.0
40	0.048	0.14	0.83	4.3	—[b]	—[b]
50	0.076	0.26	1.6	1.00	—[b]	—[b]
60	0.12	0.40	3.0	—[b]	—[b]	—[b]
80	0.24	0.99	7.8	—[b]	—[b]	—[b]
100	0.45	1.88	—[b]	—[b]	—[b]	—[b]

Note: Fan horsepower values shown for shelled corn may be used for soybeans.

[a] Horsepower based on fan static efficiency of 47%.
[b] Above recommended maximum of 10 hp/100 ft².

Table 33
FAN HORSEPOWER FOR AERATING WHEAT[a] — REQUIREMENTS PER 100 ft² OF BIN CROSS-SECTION OR FLOOR AREA

Wheat depth (ft)	Aeration air flow rate (cfm/bu)					
	1/20	1/10	1/4	1/2	3/4	1
	(hp/100 ft²)					
10	0.009	0.018	0.068	0.21	0.42	0.72
15	0.015	0.040	0.12	0.61	1.32	2.31
20	0.024	0.071	0.34	1.31	3.01	5.88
25	0.039	0.12	0.65	2.71	6.42	—[b]
30	0.059	0.19	1.11	4.41	—[b]	—[b]
40	0.12	0.42	2.49	—[b]	—[b]	—[b]
50	0.21	0.76	5.15	—[b]	—[b]	—[b]
60	0.34	1.33	8.62	—[b]	—[b]	—[b]
80	0.79	3.21	—[b]	—[b]	—[b]	—[b]
100	1.45	6.03	—[b]	—[b]	—[b]	—[b]

Note: Fan horsepower values shown for wheat may be used for sorghum, rough rice, barley and oats.

[a] Horsepower based on a fan static efficiency of 47%.
[b] Above recommended maximum of 10 hp/100 ft.

Table 34
CROSS-SECTIONAL OR FLOOR AREAS OF ROUND BINS

Bin diameter (ft)	Floor area (ft²)	Multiplier[a]
18	250	2.5
21	346	3.5
24	452	4.5
27	572	5.7
30	706	7.1
33	855	8.6
36	1017	10.2
40	1256	12.5
42	1385	13.9
48	1808	18.1

[a] Multiplier times horsepower values from Tables 32 or 33 is the fan horsepower required for the bin size selected.

Table 35
STANDARD SPECIFICATIONS FOR WOVEN-WIRE CLOTH SIEVES

U.S. standard sieve no.	Nominal sieve opening	
	mm	in.
4	4.76	0.187
6	3.36	0.132
8	2.38	0.0937
12	1.68	0.0661
16	1.19	0.0469
20	0.841	0.0331
30	0.595	0.0234
40	0.420	0.0165
50	0.297	0.0117
70	0.210	0.0083
100	0.149	0.0059
140	0.105	0.0041
200	0.074	0.0029
270	0.053	0.0021

Table 36
RECOMMENDED SIEVES FOR DETERMINING DOCKAGE AND FOREIGN MATERIAL IN GRAINS[69]

	Description of sieve			
		Size of perforation		
Common name of sieve	Shape of perforation[a]	Common designation (in.)	Decimal equivalent of diameter or width (in.)	Approximate number of perforations/ft²
Grain sorghums dockage	Round hole	2 ½/64 diam.	0.0391	27,970
Corn sieve	Round hole	12/64 diam.	0.1875	2,640
Fine seed	Round hole	1/12 diam.	0.0833	9,820
Flaxseed dockage	Round hole	4 ½/64 diam.	0.0703	13,795
Soybean				
Foreign material	Round hole	8/64 diam.	0.1250	4,736
Splits[b]	Slotted	8/64 diam. × 3/4	0.1250	638
		9/64 diam. × 3/4	0.1406	590
		10/64 diam. × 3/4	0.15625	557
Small chess	Slotted	0.064 × 3/8	0.0640	2,633
Large chess	Slotted	0.070 × 1/2	0.0700	1,655
Barley sizing				
Class I, barley	Rectangular	5/64 × 3/4	0.0781	865
Class III, western	Rectangular	5 ½/64 × 3/4	0.0858	856
Small buckwheat	Equilateral triangle	Inscribed circle 5/64 diameter	0.0781	2,845
Flaxseed sieve	Slotted	3/64 × 3/8	0.0469	2,690
Flaxseed dockage	Wire mesh	4 × 16 meshes to in.²		

[a] Perforations in each row are staggered with respect to perforations in adjacent rows.
[b] The soybean splits sieves may be used in making a partial separation of the "splits" in soybeans.

Table 37
REPRESENTATIVE VALUES OF SPECIFIC GRAVITY FOR SELECTED GRAINS

Product	Specific gravity	Ref.
Barley	1.39	44
White hulless	1.33	76
Coast	1.13	76
Buckwheat	1.10	76
Corn	1.29	44
	1.19	76
Flax	1.14	44
Millet	1.11	76
Milo	1.32	44
Oats	1.37	44
	0.99	76
Rice	1.37	44
	1.12	76
Rye	1.23	76
Sorghum, grain	1.24	76
Soybeans	1.18	76
Wheat	1.42	44
	1.30	76

REFERENCES

1. **Agrawal, Y. C. and Singh, R. P.**, Thin-Layer Drying Studies on Short-Grain Rough Rice, ASAE Tech. Paper 77-3531, American Society of Agricultural Engineers, St. Joseph, Mich., 1977.
2. **Ahmadnia, A.**, Quality of Soft Wheat Dried in a Concurrent-Counter Flow Dryer, M. S. thesis, Michigan State University, East Lansing, 1977.
3. **Alam, A. and Shove, G. C.**, Hygroscopicity and thermal properties of soybeans, *Trans. ASAE,* 16, 707, 1973.
4. American Society of Agricultural Engineers, *Agricultural Engineers Yearbook,* American Society of Agricultural Engineers, St. Joseph, Mich., 1978.
5. American Society of Heating, Refrigeration, and Air-conditioning Engineers, *Handbook of Fundamentals,* American Society of Heating, Refrigeration, Air-conditioning Engineers, New York, 1977.
6. **Babbit, E. A.**, The thermal properties of grain in bulk, *Can. J. Res.,* F23, 388, 1945.
7. **Baker, D. G. and Klink, J. C.**, Solar Radiation Reception, Probabilities and Areal Distribution in the North Central Region, Tech. Bull. 300, University of Minnesota, St. Paul, 1975.
8. **Bakke, A. L.**, Thermal conductivity of stored oats with different moisture content, *Plant Phys. J.,* 10, 521, 1935.
9. **Bakker-Arkema, F. W., Brook, R. C., and Lerew, L. E.**, Cereal grain drying, in *Advances in Cereal Science and Technology,* Vol. 2, Pomeranz, Y., Ed., American Association of Cereal Chemists, St. Paul, Minn., 1978.
10. **Bakker-Arkema, F. W., Rosenau, J. R., and Clifford, W. H.**, Measurement of grain surface area and its effects on the heat and mass transfer rates in fixed and moving beds of biological products, *Trans. ASAE,* 14, 864, 1971.
11. **Becker, H. A.**, A study of diffusion in solids of arbitrary shape, with application to the drying of the wheat kernel, *J. Appl. Polym. Sci.,* 1, 212, 1959.
11a. **Becker, H. A. and Sallans, H. R.**, A study of the relationship between time, temperature, moisture content, and loaf volume by the bromate formula in the heat treatment of wheat and flour, *Cereal Chem.,* 33, 254, 1956.
12. **Bilanski, W. K. and Fisher, D. R.**, Thermal conductivity of rapeseed, *Trans. ASAE,* 19, 788, 1976.
13. **Boyce, D. A.**, Grain moisture and temperature changes with position and time during drying, *J. Agric. Eng. Res.,* 10, 333, 1965.
14. **Brooker, D. B., Bakker-Arkema, F. W., and Hall, C. W.**, *Drying Cereal Grains,* AVI Publishing Westport, Conn., 1974.
15. **Chu, S. T. and Hustrulid, A.**, Numerical solution of diffusion equations, *Trans. ASAE,* 11, 705, 1968.
16. **Colliver, D. G., Brook, R. C., and Peart, R. M.**, Optimal Management, Procedures for Solar Grain Drying, ASAE Tech. Paper 78-3512, American Society of Agricultural Engineers, St. Joseph, Mich., 1978.
16a. **Duffie, J. A. and Beckman, W. A.**, *Solar Energy Thermal Processes,* John Wiley & Sons, New York, 1974.
17. **Flood, C. A., Sabbah, M. A., Meeker, D., and Peart, R. M.**, Simulation of natural air corn drying system, *Trans. ASAE,* 15, 156, 1972.
18. **Foster, G. H.**, Heated air drying and corn quality, in *Preservation of Wet Harvested Grains,* Multon, J. L. and Guilbot, A., Eds., Institute National de la Recherche Agronomique, Paris, 1973.
19. **Foster, G. H. and Peart, R. M.**, Solar Grain Drying — Progress and Potential, Agr. Inf. Bull. 401, Agricultural Research Service, U.S. Department of Agriculture, Washington, D.C., 1976.
20. **Foster, G. H. and McKenzie, B. A.**, Managing Grain for Year-Round Storage, Bull. AE-90, Coop. Ext. Ser., Purdue University, Lafayette, Ind., 1979.
21. **Freeman, J. E.**, Quality factors affecting value of corn for wet milling, *Trans. ASAE,* 16, 671, 1973.
22. **Gallaher, G. L.**, A method for determining the latent heat of agricultural crops, *Agric. Eng.,* 32, 34, 1951.
23. **Glogorcan, M.**, Distribution of oil in corn in relation to harvesting, drying and storage, *Muehle Mischfuttertechn.,* 112, 131, 1974.
24. **Hall, C. W.**, *Drying Farm Crops,* AVI Publishing, Westport, Conn., 1957.
24a. **Hall, G. E. and Hill, L. D.**, Test Weight as a Grading Factor for Shelled Corn, No. AERR 124, University of Illinois, Urbana, 1973.
25. **Haque, E., Foster, G. H., Chung, D. S., and Lai, F. S.**, Static Pressure Drop Across a Bed of Corn Mixed with Fines, ASAE Tech. Paper 78-3512, American Society of Agricultural Engineers, St. Joseph, Mo., 1978.
26. **Hart, J. B., Feinstein, L., and Golumbic, C.**, Oven Method for Precise Measurement of Moisture in Seeds, Market Res. Rep. 304, U.S. Department of Agriculture, Washington, D.C., 1959.
27. **Haswell, G. A.**, A note on the specific heat of rice, oats and their products, *Cereal Chem.,* 31, 341, 1954.

28. **Haynes, B. C.**, Vapor Pressure Determination of Seed Hygroscopicity, Tech. Bull. 1229, U.S. Department of Agriculture, Washington, D.C., 1961.
29. **Holman, L. E.**, Aeration of Grain in Commercial Storages, Market Res. Rep. 178, U.S. Department of Agriculture, Washington, D.C., 1966.
30. **Houston, D. F., Ed.**, *Rice-Chemistry and Technology*, American Association of Cereal Chemists, St. Paul, Minn., 1972.
31. **Ingram, G. W.**, Deep bed dryer simulation with inter-particle moisture diffusion, *J. Agric. Eng. Res.*, 21, 263, 1976.
32. **Jansen, H. D. and Friedrich, W.**, Modification of corn starch by heat treatment, *Muehle Mischfuttertechn.*, 112, 65, 1975.
33. **Katic, Z.**, Maiskornbeschadigung bei kunstlicher Trocknung, *Muehle Mischfuttertechn.*, 110, 534, 1973.
34. **Kazarian, E. A. and Hall, C. W.**, Thermal properties of grains, *Trans. ASAE*, 8, 33, 1965.
35. **Kreyger, J.**, Drying and Storing Grains, Seeds and Pulses in Temperate Climates, Inst. Stor. Proc. Agric. Prod., Wageningen, Netherlands, 1972.
36. **Krischer, O.**, *Die Wissenschaftlichen Grundlagen der Trocknungstechnik*, Springer-Verlag, Berlin, 1969.
37. **Lasseran, J. C.**, Incidence of drying and storage conditions of corn on its quality for the starch industry, *Staerke*, 25, 257, 1975.
38. **Lui, B. Y. H. and Jordan, R. C.**, The interrelationship and characteristic distribution of direct, diffuse and total solar radiation, *Sol. Energy*, 4, 1, 1960.
39. **Lui, B. Y. H. and Jordan, R. C.**, Availability of solar energy for flatplate solar heat collectors, in Applications of Solar Energy for Heating and Cooling of Buildings, GRP 170, American Society of Heating and Ventilating Engineers, New York, 1967.
40. **Luikov, A. V. and Mikhailov, Y. A.**, *Theory of Energy and Mass Transfer*, Pergamon, New York, 1965.
41. **McAdams, W. H.**, *Heat Transmission*, McGraw-Hill, New York, 1954.
42. **Meiring A. G. and Hofkes, H. J.**, The determination of performance characteristics and energy use of grain dryers through computer simulation, *Grundl. Landtech.*, 26, 1, 1977.
43. **Misra, M. K. and Brooker, D. B.**, Thin-Layer Drying of Shelled Corn, ASAE Tech. Paper 78-3002, American Society of Agricultural Engineers, St. Joseph, Mo., 1978.
44. **Mohsenin, N. N.**, *Physical Properties of Plant and Animal Materials*, Gordan & Breach, New York, 1970.
45. **Moysey, E. B., Shaw, J. T., and Lampman, W. P.**, The effect of temperature and moisture on the thermal properties of rapeseed, *Trans. ASAE*, 20, 768, 1977.
46. **Mühlbauer, W. and Christ, W.**, Die zulassige Einwirkungszeit verschiender Korn temperaturen bei der Trocknung von Mais fur die tierische Ernahrung, *Grundl. Landtech.*, 24, 161, 1974.
47. **Mühlbauer, W., Huss, W., and Kuppinger, H.**, Fabmetrische Schnellmethode zum Erkennen einer Proteinschadiguing bei der Trocknung von Maiskornern mit hohen Luft temperaturen, *Grundl. Landtech.*, 26, 128, 1976.
48. **Ngoddy, P. O., Bakker-Arkema, F. W., and Bickert, W. G.**, Heat transfer in a deep bed of pea beans, *Q. Bull. Mich. Agric.*, 49, 1966.
49. **O'Callaghan, J. R., Menzies, D. J., and Bailey, P. H.**, Digital simulation of agricultural dryer performance, *J. Agric. Eng. Res.*, 16, 223, 1971.

49a. **Othmer, D. F.**, Correlating vapour pressure and latent heat data, *J. Ind. Eng. Chem.*, 32, 841, 1940.

50. **Overhults, D. G., White, G. M., Hamilton, H. E., Ross, I. J., and Fox, J. D.**, Effect of heated air drying on soybean oil quality, *Trans. ASAE*, 18, 942, 1975.
51. **Pabis, S. and Henderson, S. M.**, Grain drying theory. II. A critical analysis of the drying curve for shelled maize, *J. Agric. Eng. Res.*, 6, 272, 1961.
52. **Patterson, R. J., Bakker-Arkema, F. W., and Bickert, W. G.**, Static pressure-airflow relationships in packed beds of granular biological materials such as grain — II, *Trans. ASAE*, 14, 172, 1971.
53. **Paulsen, M. R. and Thompson, T. L.**, A Drying Analysis of Grain Sorghum, ASAE Tech. Paper 72-819, American Society of Agricultural Engineers, St. Joseph, Mich., 1972.

53a. **Pederson, J. H., Ed.**, Low Temperature and Solar Grain Drying Handbook, Midwest Plan Service, Iowa State University, Ames, 1980.

54. **Peplinski, A. J., Brekke, O. L., Griffin, E. L., Hall, G., and Hill, L. D.**, Corn quality as influenced by harvest and drying conditions, *Cereal Foods World*, 20, 145, 1975.
55. **Pfost, H. B., Mauer, S. G., Chung, D. S., and Milliken, G. A.**, Summarizing and Reporting Equilibrium Moisture Data for Grains, ASAE Tech. Paper 76-3520, American Society of Agricultural Engineers, St. Joseph, Mich., 1976.

55a. **Rodriguez-Arias, J. H.**, Desorption Isotherms and Drying Rates of Shelled Corn in the Temperature Range of 40 to 140°F, Ph.D. thesis, Michigan State University, East Lansing, 1956.

56. **Ross, I. J. and White, G. M.** Discoloration and stress cracking of white corn as affected by overdrying, *Trans. ASAE,* 15, 327, 1972.
57. **Rugumayo, E. W.,** personal communication, 1979.
58. **Sabbah, M. A.,** Convective Heat Transfer and Tempering Effects on Cooling Rate of Shelled Corn, Ph.D. thesis, Purdue University, West Lafayette, Ind., 1971.
59. **Sabbah, M. A., Meyer, G. E., Keener, H. M., and Roller, W. L.,** Reversed Air Drying for Fixed Bed of Soybean Seed, ASAE Tech. Paper 76-3023, American Society of Agricultural Engineers, St. Joseph, 1976.

59a. **Sharma, D. K. and Thompson, T. L.,** Specific heat and thermal conductivity of sorghum, *Trans. ASAE,* 16, 114, 1973.

60. **Shedd, C. K.,** Resistance of grains and seeds to air flow, *Agric. Eng.,* 34, 616, 1953.
61. **Spencer, H. B.,** A mathematical simulation of grain drying, *J. Agric. Eng. Res.,* 14, 226, 1969.
62. **Stefanov, L. P.,** Optimum drying temperature for maize, *Oli Grassi Deriv.,* 7, 12, 1971.
63. **Steffe, J. F.,** Moisture Diffusion and Tempering in the Drying of Rough Rice, Ph.D. thesis, University of California, Davis, 1979.
64. **Thompson, R. A. and Foster, G. H.,** Stress cracks and breakage in artificially dried corn, *U.S. Dep. Agric. Mark. Res. Rep.,* 631, 1963.
65. **Thompson, R. A. and Isaacs, G. W.,** Porosity determinations of grains and seeds with an air-comparison pycnometer, *Trans. ASAE,* 10, 693, 1967.
66. **Thompson, T. L.,** Predicted Performances and Optimal Designs of Convection grain dryers, Ph.D. thesis, Purdue University, Lafayette, Ind., 1967.
67. **Troeger, J. M. and Hukill, W. V.,** Mathematical description of the drying rate of fully exposed corn, *Trans. ASAE,* 14, 1153, 1971.
68. Grain Division, Oven method for determining moisture content of grain and related agricultural commodities, in Equipment Manual, U.S. Department of Agriculture, Washington, D.C., 1971.
69. U.S. Department of Agriculture, Grain Inspection Manual, Instruction 918(GR)-6, Agricultural Marketing Service, U.S. Department of Agriculture, Washington, D.C., 1974.
70. **Vojnovich, C., Anderson, R. A., and Griffin, E. L.,** Wet milling properties of corn after field shelling and artificial drying, *Cereal Foods World,* 20, 333, 1975.
71. **Wall, J. S., James, C., and Donaldson, G. L.,** Corn proteins: chemical and physical changes during drying of grain, *Cereal Chem.,* 52, 779, 1975.
72. **Westerman, P. W., White, G. M., and Ross, I. J.,** Relative humidity effect on the high temperature drying of shelled corn, *Trans. ASAE,* 16, 1136, 1973.
73. **White, G. M. and Ross, I. J.,** Discoloration and stress cracking in white corn as affected by drying temperature and cooling rate, *Trans. ASAE,* 15, 504 1972.
74. **Williamson, J. L.,** Nutritional requirements of livestock as related to corn quality, in *Corn Quality in World Markets,* Hill, L. D., Ed., Interstate, Danville Ill., 1975.
75. **Wratten, F. T., Pode, W. D., Chesness, J. L., Bal, S., and Ramarao, V.,** Physical and thermal properties of rough rice, *Trans. ASAE,* 12, 801, 1969.
76. **Zink, F. J.,** Specific gravity and air space of grains and seeds, *Agric. Eng.,* 16, 439, 1935.

STORED-GRAIN INSECTS AND THEIR CONTROL

Robert Davis

INTRODUCTION

Insect control in stored cereals and feed grains, including soybeans, is based on the need to maintain both quality and quantity as demanded by processors and consumers and to meet rules and regulations promulgated by governmental action agencies. Control of insects attacking postharvest stored cereals and feed grains often requires a program of both passive and active approaches. The former involves prevention through the use of a good sanitation program and other good storage or manufacturing practices. The latter includes the use of some sampling techniques for the purpose of detection followed by a physical manipulation or chemical pesticide application if required. The use of passive programs such as good sanitation is not only a well recognized practice in the food storage and processing industries but a requirement. The Food, Drug, and Cosmetic Act, Section 402 (a)(4), states a food is adulterated "if it has been prepared, packed, or held under insanitary conditions whereby it may become contaminated with filth."

The control of insects in stored cereals and feed grains offers some unique challenges for consideration by the responsible pest controller or sanitarian. Indeed, this is true for most agricultural commodities in the marketing channels. Many of these considerations relate to the quality of the end product while many others relate to health and safety. Economics will always be a primary consideration because of the increased inherent value of a harvested commodity through the accumulation of the input in manpower and energy. Examples of some of these considerations would be as follows:

1. What is the pest species and where is it to be controlled?
2. What will be the value lost if control is not attempted or effected?
3. Is there time at this point in the marketing channel to effect a control procedure?
4. What control options are available, legal, and safe at this point in the marketing channel?
5. What effect will the control action have on the commodity immediately and on the end product?

COMMON INSECT AND MITE PESTS

The number of pest species that are commonly encountered in cereals, feed grains, and soybeans is not extensive. For the reader who is interested in having one or more readily available references, there are several good text books as well as state, federal, and international publications that will serve adequately.[1-5] However, specifics on the biology, ecology, and behavior of each pest species will only be found by reading the scientific literature as it relates to each species. Table 1 lists the common insects and mites attacking cereals, feed grains, and soybeans. Wherever a positive identification is desired or required, the services of a taxonomist should be secured. The extension entomologists located at the nearest Land Grant College can provide assistance in getting insect identification.

Insect pests of stored grains are often divided into groupings based on their feeding habits. Primary invaders, the first to attack an unprocessed whole kernel, usually are never pests of milled or processed grain. The major exception to this rule is the lesser

Table 1
COMMON AND SCIENTIFIC NAMES OF INSECTS AND MITES COMMONLY FOUND ATTACKING CEREALS, FEED GRAINS, AND SOYBEANS

Common name	Scientific name
Weevils	
Granary weevil[a]	*Sitophilus granarius* (L.)
Rice weevil[a]	*Sitophilus oryzae* (L.)
Maize weevil[a]	*Sitophilus zeamais* (Motschulsky
Broadnosed grain weevil	*Caulophilus oryzae* (Gyllenhal)
Coffee bean weevil	*Araecerus fasiculatus* (De Geer)
Bean weevil[b]	*Acanthoscelides obtectus* (Say)
Cowpea weevil[b]	*Callosobruchus maculatus* (F.)
Borers	
Lesser grain borer[a]	*Rhyzopertha dominica* (F.)
Larger grain borer[a]	*Prostephanus truncatus* (Horn)
Moths	
Angoumois grain moth[a]	*Sitotroga cerealella* (Olivier)
Pink scavenger caterpillar[a]	*Pyroderces rileyi* (Walsingham)
European grain moth[a]	*Nemapogen granella* (L.)
Rice moth[a]	*Corcyra cephalonica* (Stainton)
Indian meal moth[c]	*Plodia interpunctella* (Hubner)
Mediterranean flour moth[c]	*Anagasta kuehniella* (Zeller)
Meal moth[c]	*Pyralis farinalis* L.
Almond moth[c]	*Cadra cautella* (Walker)
Tobacco moth[c]	*Ephestia elutella* (Hubner)
Raisin moth[c]	*Cadra figuliella* (Gregson)
Beetles	
Cadelle	*Tenebroides mauritanicus* (L.)
Sawtoothed grain beetle[c]	*Oryzaephilus surinamensis* (L.)
Merchant grain beetle[c]	*Oryzaephilus mercator* (Fauvel)
Squarenecked grain beetle[c]	*Cathartus quadricollis* (Guerin-Meneville)
Foreign grain beetle[c]	*Ahasverus advena* (Waltl)
Mexican grain beetle[c]	*Pharoxonotha kirschi* (Reitter)
Siamese grain beetle[c]	*Lophocateres pusillus* (Klug)
Flat grain beetle[c]	*Cryptolestes pusillus* (Schonherr)
Rusty grain beetle[c]	*Cryptolestes ferrugineus* (Stephens)
Confused flour beetle[c]	*Tribolium confusum* Jacquelin duVal
Red flour beetle[c]	*Tribolium castaneum* (Herbst)
American black flour beetle[c]	*Tribolium audax* Halstead
Longheaded flour beetle[c]	*Latheticus oryzae* Waterhouse
Slenderhorned flour beetle[c]	*Gnatocerus maxillosus* (F.)
Smalleyed flour beetle[c]	*Palorus ratzeburgi* (Wissmann)
Depressed flour beetle[c]	*Palorus subdepressus* (Wollaston)
Broadhorned flour beetle[c]	*Gnatocerus cornutus* (F.)
Larger black flour beetle	*Cynaeus angustus* (LeConte)
Yellow mealworm[a]	*Tenebrio molitor* L.
Dark mealworm[a]	*Tenebrio obscurus* F.
Lesser mealworm[c]	*Alphitobius diaperinus* (Panzer)
Redhorned grain beetle[c]	*Platydema ruficorne* (Strum)
Black carpet beetle[c]	*Attagenus megatoma* (F.)
Warehouse beetles[a]	*Trogoderma* spp.
Hairy spider beetle[a]	*Ptinus villiger* (Reitter)
Whitemarked spider beetle[a]	*Ptinus fur* (L.)
Brown spider beetle[a]	*Ptinus clavipes* Panzer
Twobanded fungus beetle[c]	*Alphitophagus bifasciatus* (Say)
Hairy fungus beetle[c]	*Typhaea stercorea* (L.)
Corn sap beetle[a]	*Carpophilus dimidiatus* (F.)
Cigarette beetle[a]	*Lasioderma serricorne* (F.)
Drugstore beetle[a]	*Stegobium paniceus* (L.)

Table 1 (continued)
COMMON AND SCIENTIFIC NAMES OF INSECTS AND MITES COMMONLY FOUND ATTACKING CEREALS, FEED GRAINS, AND SOYBEANS

Common name	Scientific name
Mites	
Flour mite	*Acarus siro* L.
	Tyrophagus spp.
	Glycyphagus spp.
Insect parasites	*Anisopteromalus calandrae* (Howard)
	Choetospila elegans Westwood
	Bracon hebetor Say
	Venturia canescens (Gravenhost)
	Scenopinus fenistralis (L.)
	Scenopinus glabrifions Meigen
Insect predators	*Xylocoris flavipes* (Reuter)

[a] Primary feeder.
[b] Pest of stored soybeans primarily.
[c] Secondary feeder.

grain borer which readily attacks both wheat and flour. Secondary invaders of stored grain are usually the primary invaders of milled and processed cereals and often attack grain which is no longer sound after attack by primary invaders. Tertiary invaders, such as the fungus beetles, are often found in grain and grain products which have a high moisture content and is supporting various fungus on which these beetles feed. Some authorities will also group these insect pests by their food preference; i.e., grain weevils, bean weevils, grain borers, grain moths, meal moths, and flour beetles.

INSECT SURVEYS, SAMPLING, AND DETECTION METHODS

Insect surveys, including methods of sampling and associated detection methods, can be found in most food industry sanitation programs as a part of their good storage or manufacturing practices. Some industry associations have established sanitation committees that recommend to their membership good storage or manufacturing practices which include insect survey methods. State and federal action agencies also have their published survey, sampling, and detection methods which they use in grading commodities and regulating quality. A discussion of the merits and disadvantages of the many available methods is beyond the scope of this presentation. However, it can be stated that the prerequisites of any insect survey of cereals and feed grains would involve the following: (1) sufficient number of samples to provide confidence that the whole is represented, (2) sample size is large enough to provide meaningful results, (3) each sample is selected at random but within the habitat of the insect species being surveyed, and (4) sample method does not exclude the species being sought because of behavioral characteristics.

The examination of the samples is an area that is often critical because of the time required. Visual examination is the simplest and perhaps the most common approach. Often when sample numbers are small, a visual examination of grain can be quite satisfactory for some situations such as on-the-farm storage. However, in cereal grains to be milled or processed this is not always satisfactory, particularly if the infestation is light or the examination is to assure a minimum of insect fragments following processing. In such instances, more sophisticated methods of examination may be necessary.[6,7]

When examination of large numbers of grain samples reveals a significant portion of the insect population may be hidden inside the individual kernels, it may be desirable to resort to other than visual methods of examination. Some success has been made of X-ray radiography in examining samples for hidden internal insect infestation in grain.[8,14] Modern X-ray equipment is now designed to provide Polaroid pictures which eliminate the time-consuming developing process. The use of a machine to crush the grain onto ninhydrin-treated filter paper strips was developed by Dennis and Decker[15] and further modified by Ashman et al.[16] into a hand-carried unit (Ashman-Simon infestation detector) for field application which has found some degree of usage around the world. Detection techniques involving aural monitoring of in-kernel insect activity and the subjection of samples to nuclear magnetic resonance have been reported by Street.[17,18] Both would appear to offer promising approaches to insect detection in limited situations but neither has been developed for practical use. The use of food- and pheromone-baited traps has also seen limited use in research but still lacks perfection in most situations.[19,20]

An insect detection device using an infrared CO_2 analyzer has been developed by Bruce and Street[21,22] that can be used in a nondestructive manner on both grain and processed packaged products. Using infrared spectrography, it measures the amount of CO_2 produced by insect respiration. Commercial prototypes are being produced by Horiba Instrument Company and Technicon Instrument Company.

PREVENTIVE ACTIONS

Following the securing of clean grain, sanitation is the first and oftentimes the only procedure needed to maintain the grain clean and free of insect infestation and/or contamination. A good sanitation program can develop good customer relationships as well as better employer-employee relationships. Sanitation programs are never static but are always being modified to allow for newer techniques and changing rules and regulations involving storage, milling, and other good manufacturing practices. Each industry's pest control operator or responsible sanitarian will have to evolve his own program that best meets his own storage situations. The recommended USDA procedures[3,4] for clean, safe storage of grain following harvest include the essential considerations. The following steps are recommended: (1) store only dry grain in weather-tight, rodentproof bins, (2) clean out all bins before loading with grain, (3) spray walls and floors of wooden bins and around the door frames of metal bins, (4) clean up and dispose of litter, waste grain, and feed accumulations, (5) apply protective dust or spray directly to grain as it is binned or fumigate promptly after binning, (6) inspect the bins at least monthly and fumigate if an infestation is discovered.

PHYSICAL CONTROLS

These controls vary widely in their application and approach to insect control from the circulation of cool ambient air to the use of sophisticated techniques involved in gamma radiation. The two most commonly employed techniques are drying and lowering the grain temperature. Both are effective techniques and are soundly based on the biologies of the pest insects. Low moisture grain does not provide sufficient water for insect development, and cooler temperature of the grain will slow the insect's developmental rate.

The periodic turning of grain will often effectively reduce an insect population through the tumbling and abrasive actions of the grain, particularly those insect stages feeding outside the grain kernels. Turning grain has in some incidences resulted in a 100% reduction in an insect population.[23,24] This approach to insect control is currently being researched. One interesting research approach is the utilization of fluidized bed drying — moving the grain over a bed of rising hot air — which effectively dries the grain and controls the insects in one step.[25]

The use of ionizing radiations is an effective means of controlling stored-grain insects.[26-30] Dosages of 20 to 50 krad from the gamma and electron accelerators are approved in the U.S. for disinfestation of wheat and wheat products.[31] Indications from extensive research on insects attacking other postharvest commodities conducted by the U.S. Department of Agriculture-Savannah are that this level of ionizing radiation is effective.[32-51] The utilization of this means of disinfesting grain and grain products has never been accepted commercially because of the large initial capital outlay and the stationary aspects of the required facility. The use of electron accelerators to disinfest grain and grain products has proven to be as effective as gamma radiation and may be more acceptable commercially because of their lower initial costs and greater mobility.[52]

Hermetic storage and modified atmospheric gas mixtures have been used successfully, the latter receiving increased interest in recent years. Hermetic storage, burying grain stores in sealed chambers so that the grain respiration reduces the O_2 concentration where it will not sustain insect life, has been used since the earliest times. It is still practiced in some parts of the world by subsistence farmers. It has also been used quite successfully in parts of Africa and Argentina to hold the carry-over grains from year to year when exporting their surplus was not possible.[53]

Recently, research in Australia and in the U.S. has shown that reduction by purging with or otherwise increasing the percentage of N_2 or CO_2 concentrations in grain storage can effectively control grain insects with no adverse effect on grain quality and germination. For the details on the application and utilization of these control methods, the reader is referred to the excellent discussions by Banks and Annis,[54] Jay,[55] and Storey.[56]

Heat has also been used successfully in limited situations, such as mills, retail stores, or storerooms in dwellings. Raising the ambient temperatures to a range between 120 to 150°F for several hours has been reported as a means of controlling exposed insects.[57,58] Infrared and microwave radiations have shown potential as a means to control grain insects both exposed and internal in a fraction of a minute on moving or falling grain streams.[23,59-63]

Other mechanisms of physical control involve food and pheromone traps, and light traps; electrocuting grills; air curtains; screens; and repellents (insect-resistant packaging). All of these mechanisms need to be further evaluated; however, some may have limited application presently in sanitation or control programs.

CHEMICAL CONTROL

The use of chemical pesticides still remains the mainstay in the arsenal of those attempting to effect insect control in stored grain. However, the use of these pesticides requires that they be applied in strict accord with their EPA-approved manufacturer's or formulator's label and labeling materials. Also, some pesticides may be designated as "restrictive use pesticides" and thereby carry the additional requirement that they be applied by or under the supervision of an individual certified to purchase and use these pesticides. The certification for purchase and use of these restrictive pesticides is vested in the States by EPA. Those interested in this certification program should direct their inquiries to their state's extension service located at the Land Grant Colleges. All other questions regarding the use of any pesticide that is not answered on the pesticide label or labeling should be directed to the state's extension entomologists.

The use of pesticides to control stored-grain insects can be divided into either a preventive or a corrective approach. Prevention involves the use of pesticides to limit the spread of existing infestation and limit the possibility of acquiring an infestation from outside sources. Correction involves the use of a pesticide to kill or eliminate an existing infestation but usually does not provide protective value against possible reinfestation.

Preventive Approach

This approach will usually involve one or more of the following: (1) spraying a residual-type pesticide onto the walls and floor of the empty storage facility before it is filled, (2) spraying or dusting a residual-type pesticide on the grain as it is being placed in storage, (3) dispensing a pesticidal vapor or aerosol into the head space over the grain during storage, or (4) spraying or dusting a residual-type pesticide onto the grain surface during storage. Several pesticides are approved and have been used successfully in protecting stored grain (Table 2). The reader is referred to U.S. Department of Agriculture Handbook No. 554[64] for recommended pesticides and guidelines on their use.

The use of a residual-type pesticide sprayed onto the walls and floor of empty storage facilities before they are filled is highly recommended. This practice should be considered as a part of any good sanitation program and be applied after the facility has been physically cleaned. Application methods for utilizing residual treatments in empty storage facilities are discussed in U.S. Department of Agriculture Leaflet No. 345,[65] Chapman and Schmitt,[66] and Campbell.[4] Recommended formulations and dosages will vary depending upon the pesticide used and the construction of the storage facility.

The treatment of the grain as it goes into storage with a pesticidal protectant is highly recommended if the grain is to be stored for over 1 month. Even if storage is to be short term, such a treatment may be warranted to assure relatively insect-free grain during subsequent storage, transportation, and processing. Most pesticides used as grain protectants have a limited time of effectiveness. In uninterrupted storage they can be effective for 9 to 12 months and the insect pests are not resistant to the pesticide used. Application methods and equipment for utilizing grain protectants are discussed by Campbell[4] and in U.S. Department of Agriculture Marketing Bulletin No. 20.[67] U.S. Department of Agriculture Handbook No. 129[68] gives information on the use of this technique on rice.

The dispensing of pesticidal vapors and aerosols into the headspace over stored grain is directed at controlling flying insects and against any that are exposed on the grain surface or on surfaces of the storage facility. This technique should be implemented only if needed and is not usually considered a routine control procedure. However, if for some reason the grain surface became infested with insects, particularly moth species, this technique can be a valuable addition to the overall preventive approach to insect control. The application of vapors and aerosols can be either continuous through the use of the impregnated vapor strips or intermittent through the use of mechanical or thermal foggers, timed aeosols, or ultralow volume (ULV) applicators.

The dusting or spraying of a residual-type pesticide on the surface of stored grain can be used as a substitute to treating the grain as it goes into storage or as an adjunct to such a treatment. If there exists a choice, the latter would be recommended. Campbell[4] discusses the use of surface pesticide treatments in stored grain. Unless the storage facility is clean, sound, and tight and only insect-free grain is placed in storage, surface treatments alone will usually be inadequate. Most uses of surface treatments are made during the storage period to assure that the grain surface is adequately protected to provide a good barrier against subsequent infestation. Grain surfaces through their exposure to the extremes of the storage environment will often require additional pesticide treatment as the original residual treatment degrades.

Table 2
SUGGESTED PESTICIDES FOR PREVENTIVE CONTROL IN STORED GRAIN

Insecticide or treatment	Tolerance (ppm)	Dosage (active ingredients per 1000 ft^3 unless otherwise stated)	How, where, and when to apply
		Facility Treatment	
Malathion (premium grade)	8 on grain	0.45 lb/ 1000 ft^2	Mix with water; at least 2—4 weeks before grain is binned, spray inside walls and floor of bin at rate of 2 gal/ 1000 ft^2
Methoxychlor	2 on grain	0.4 lb/ 1000 ft^2	
Pyrethrins + piperonyl butoxide	3 + 20	0.013 − 0.13 lb/ 1000 ft^2	Mix with water and apply to walls and floor of empty storage at the rate of 2 gal/ 1000 ft^2
		Grain Treatment	
Malathion (premium grade)	8	0.63 lb/ 1000 bu	Mix with water 3—5 gal/1,000 bu; spray on grain stream as it goes into storage
		0.6 lb/ 1000 bu	Mix dust into grain before storing
Pyrethrins + piperonyl butoxide	3 + 20	0.06 + 0.6 lb/ 1000 bu	Mix with water 3—5 gal/1000 bu; apply as protective spray to grain before it is stored
		0.06 + 0.83 lb/ 1000 bu	Mix dust into grain before storing
		Headspace Treatment	
Dichlorvos	0.5	1 — strip	Suspend strip over commodity before moths emerge in spring; replace when effectiveness diminishes
Pyrethrins + piperonyl butoxide	3 + 20	0.006 + 0.06 lb/ 1000 ft^3 of air space	Apply with thermal aerosol generator
		Residual Surface Treatment	
Malathion (premium grade)	8	0.32 lb in 1—2 gal water/ 1000 ft^2	Surface spray; will not control insects established beneath the surface

Corrective Approach

When assurance is needed that grain is free of live insect infestations, the use of fumigants offers the only practical approach. Fumigants, when properly handled and applied, are a safe, an efficient, and an effective means of eliminating an infestation of live insects. However, in the hands of an inexperienced person, they can be very hazardous and quite often ineffectively used. Efficacy will depend upon achieving fumigant gas distribution throughout the grain mass and then retaining the required concentration of the gas for the proper time interval. The serious reader is referred to the excellent manual by Monro[69] for a complete discussion on most aspects of fumigation.

Next to retaining the fumigant gas with the grain for the required time interval, which is largely a matter of sealing the structure with gasproof tarpaulins, plastic films, tapes, etc. the selection of an appropriate fumigant is of prime consideration. This selection will depend on a number of factors, some of which will only be gained through experience with particular situations. However, some factors that can and should be considered are (1) can the fumigant legally be used on the grain, (2) type of grain and its quality, (3) temperature of grain, (4) kind or type of facility, (5) spot or bulk treatment, (6) time available to effect fumigation, (7) allowable residue tolerances, (8) how will fumigant distribution be achieved, and (9) how will aeration be accomplished.

The legality of using a particular fumigant is entirely within the providence of federal, state, and local regulatory agencies. In most instances, the U.S. Environmental Protection Agency's approved label and labeling material will provide information on all Federal requirements and restrictions. However, other governmental agencies may have additional restrictions. Also, the use of an approved fumigant may be also vested in individuals or firms involved in the ownership of the grain or storage facility. It is always a good practice to inform all parties involved, even with a remote interest, that the grain has been or is to be fumigated.

Most other factors to be considered in selection of an appropriate fumigant relate directly or indirectly to the physical characteristics and behavior of the fumigants themselves. Table 3 lists some of the physical characteristics of the chemicals commonly used as grain fumigants and in fumigant mixtures, their residue tolerances, and relative toxicity to grain insects. A brief discussion of some common fumigants follows to illustrate the importance of considering a fumigant's physical characteristics. A thorough knowledge of fumigants cannot be overemphasized, and the individual anticipating their use should learn as much about them as possible before using them.

Carbon disulfide (CS_2) is moderately toxic to stored-grain insects. In the past it has been used extensively as a fumigant, but because of its high flammability and resulting fires and explosions it is no longer commonly used alone as a fumigant. Its use in present fumigation technology is as a constituent of fumigant mixtures, usually with carbon tetrachloride (CCl_4), to eliminate the hazard of fire and explosion. A common fumigant mixture containing CS_2 is the one often referred to as "80—20". This is a mixture by volume but by weight it contains about 16.5% CS_2, 82.5% CCl_4, and 1% inert ingredients. Fumigant mixtures containing CS_2 are usually sprayed onto the grain surface or some absorbent material to increase the rate of vaporization. This method seems to work quite well in large flat storages of grain. In upright storages where the depth of grain is greater, the fumigant mixture is often pumped into the storage and placed just below the surface of the grain.

Carbon tetrachloride (CCl_4) is rarely if ever used as a fumigant by itself because of its low toxicity to insects. However, it is a major constituent in many fumigant mixtures with CS_2 and ethylene dichloride (EDC) because it is nonflammable and serves to reduce the hazard of fire and explosion. CCl_4 is slow to vaporize and when it does the resulting gas is quite heavy. There are indications that it acts as a vehicle to carry its companion liquids and gases to greater depths.[70]

Chloropicrin (CCl_3NO_2) is one of the most toxic fumigants to stored-grain insects. Unfortunately, it is a strong lachrymator, difficult to vaporize and aerate, and corrosive to metals which has resulted in a voluntary restriction of its use.

Ethylene dibromide ($CH_2 \cdot Br \cdot CH_2 \cdot Br$) is highly toxic to stored-grain insects, but it has only minor uses as a grain fumigant. It is difficult to vaporize, is readily sorbed,

Table 3
COMMON CHEMICALS USED AS FUMIGANTS OR IN FUMIGANT MIXTURES ON GRAIN

Fumigant	Boiling point	Sp gravity (air = 1)	Flammability	Relative toxicity to grain insects	Residue tolerance (ppm)
Carbon disulphide (CS_2)	46.3°C	2.64	Yes	Moderate	Exempt
Carbon tetrachloride (CCl_4)	76.8°C	5.32	Nonflammable	Low	Exempt
Chloropicrin (CCl_3NO_2)	112°C	5.68	Nonflammable	Very low	Exempt
Ethylene dibromide ($CH_2Br \cdot CH_2 \cdot Br$)	131.6°C	6.487	Nonflammable	High	50 inorganic bromide 125 inorganic bromide in rice 240 inorganic bromide in popcorn
Ethylene dichloride ($CH_2Cl \cdot CH_2 \cdot Cl$)	83.5°C	3.42	Yes	Low-moderate	Exempt
Hydrocyanic acid	26°C	0.9	Yes	High	
Methyl bromide (CH_3Br)	3.6°C	3.27	Nonflammable	Very high	50 inorganic bromide 125 inorganic bromide in rice 240 inorganic bromide in popcorn
Phosphine	−87.4°C	1.21	Yes	Very high	0.1

and therefore does not penetrate or aerate well. In mixtures with CCl_4 and methyl bromide (MB) and in mixtures with both EDC and CS_2, it has found application in both bulk and spot fumigants of grain.

Ethylene dichloride ($CH_2 \cdot Cl \cdot CH_2 \cdot Cl$) is not very toxic to insects attacking stored grain. However, because of its flammability, it is always used in a mixture usually in a 3:1 ratio with CCl_4. Such mixtures have worked well under warm conditions at exposures of 2 weeks or more.

Hydrocyanic acid (HCN) is a very toxic fumigant to insects attacking stored grain. Unfortunately, it is readily sorbed on most commodities and therefore does not penetrate well. HCN is no longer used extensively as a grain fumigant.

Methyl bromide ($CH_3 \cdot Br$) is one of the more commonly used fumigants. It penetrates well and has a low boiling point and therefore can be used at lower temperatures than many other fumigants. Its major drawback is its weight and therefore it tends to be sorbed or stratify unless recirculation is used. This fact has limited its usefulness in most grain fumigations unless used in a mixture with another fumigant such as EDB.

Phosphine (PH_3) is a fumigant that is being used more extensively in recent years. It is available in a solid form as aluminum phosphide which lends ease in handling and applying. The aluminum phosphide reacts with the atmospheric moisture in the storage and produces phosphine gas.

$$2\,AlP + 6\,H_2O \rightarrow 2\,PH_3 + 2\,Al(OH)_3$$

Formulations of aluminum phosphide are commonly available in 1-g pellets, 3-g tablets, and 34-g bags (sachets) which release 0.3 g, 1 g, and 11 g of phosphine gas, respectively. Phosphine is very toxic to insects, but it has two drawbacks in that it is flammable and fumigation times of at least 3 days are required. The fumigation period can be extended considerably with low grain temperatures. The flammability of phosphine under recommended fumigation dosages and approved application procedures is never a problem as gas concentrations are held well below the level of flammability.

SAFE USE OF FUMIGANTS

Fumigant use requires that the applicator always be cognizant that he is working with toxic chemicals that can be lethal to man as well as to insects. A fumigation should always involve more than one trained person so they can assist each other in the event of an accident. However, accidents can be prevented by a thorough knowledge of the fumigant being used and proper application techniques. The most important consideration in the use of fumigants is to minimize the contact with the gas. This involves proper sealing of the facility to be treated and, whenever possible, applying the fumigant from a remote location. When there is a possibility of gas exposure, the fumigator should be familiar with all the protective and warning devices available to him and make use of them. The threshold limit values (TLVs) published by the American Conference of Government Industrial Hygienists[71] is a comprehensive list of airborne concentrations of toxic gases and their safe exposure times. Measurements of these TLVs may be difficult in many actual situations, particularly in relation to exposure. However, glass detector tubes are available from several manufacturers that can be used to detect concentration levels. If entry into the immediate area is required and safe reentry after aeration, periodic use of these detector tubes should be considered a requirement. The detection of any fumigant gas will demand the consideration to use personal protective devices such as gas masks and self-contained unit breathing apparatus (SCUBA). Canister-type masks have both fumigant concentration and time limitations that must be strictly adhered to. It also must always be remembered that all fumigants are not only toxic by the respiratory route, but some may be absorbed through the skin and gas masks will offer little protection. Most field detection equipment used for the measurement of gas concentration for the purpose of effecting a control within the fumigated area is not adequate for safety purposes. For a complete discussion of the safe application of fumigants to stored grain the reader is referred to several U.S. Department of Agriculture publications on the subject[65-68] and to Monro.[69]

REFERENCES

1. **Metcalf, C. L. and Flint, W. P.,** *Destructive and Useful Insects,* 4th ed., McGraw-Hill, New York, 1962, 1087.
2. **Hughes, A. M.,** The Mites of Stored Food, Tech. Bull. 9, Ministry of Agriculture, Fisheries, and Food, Her Majesty's Stationery Office, London, 1961, 287.
3. Agricultural Research Service, Stored Grain Insects, *U.S. Dep. Agric. Agric. Handb.,* 500, 57, 1978.
4. **Campbell, W. V.,** Stored Grain Insects and Their Control in the Middle Atlantic States, Southern Coop. Ser. Bull. No. 75, Agricultural Experiment Station, Raleigh, N.C., 1969, 30.
5. **Cotton, R. T.,** *Pests of Stored Grain and Grain Products,* Burgess Publishing, Minneapolis, 1963, 318.
6. Approved Methods Committee, *Approved Methods of the American Association of Cereal Chemists,* 8th ed., American Association of Cereal Chemists, St. Paul, Minn., 1975.
7. **Gorham, J. R., Ed.,** Training Manual for Analytical Entomology in the Food Industry, FDA Tech. Bull. No. 2, Association of Official Analytical Chemists, Washington, D.C., 1977, 174.

8. **Dobie, P.**, An investigation into the use of an x-ray technique in the study of pre-emergent stages of *Sitophilus oryzae* (L.) developing in Manitoba wheat, *J. Stored Prod. Res.*, 9, 7, 1973.
9. **Festus, I.**, Detection and estimation of internal pest infestation in seeds by the application of x-ray techniques, *OEPP/EPPO Bull.*, 3, 65, 1972.
10. **Kirkpatrick, R. L. and Wilbur, D. A.**, The development and habits of the granary weevil, *Sitophilus granarius*, within the kernel of wheat, *J. Econ. Entomol.*, 58, 979, 1965.
11. **Mills, R. B. and Wilbur, D. A.**, Radiographic studies of Angoumois grain moth development in wheat, corn, and sorghum kernels, *J. Econ. Entomol.*, 60, 671, 1967.
12. **Milner, M., Lee, M. R., and Katz, R.**, Application of x-ray technique to the detection of internal insect infestation of grain, *J. Econ. Entomol.*, 43, 933, 1950.
13. **Pedersen, J. R. and Brown, R. A.**, X-ray microscope to study behavior of internal-infesting grain insects, *J. Econ. Entomol.*, 53, 678, 1960.
14. **Stermer, R. A.**, Automated x-ray inspection of grain for insect infestation, in Quality Detection in Foods, J. J. Caffney Company, Electric Power and Processing Division of American Society of Agricultural Engineers, St. Joseph, 1976, 110.
15. **Dennis, N. M. and Decker, R. W.**, A method and machine for detecting living internal insect infestation in wheat, *J. Econ. Entomol.*, 55, 199, 1962.
16. **Ashman, F., Elias, D. G., Ellison, J. F., and Spratley, R.**, An instrument for detecting insects within food grains, *Milling*, 151, 32, 1969.
17. **Street, M. W., Jr.**, A method of aural monitoring of in-kernel insect activity, *J. Ga. Entomol. Soc.*, 6, 72, 1971.
18. **Street, M. W., Jr.**, Nuclear magnetic resonance for detecting hidden insect infestation in stored grain, *J. Ga. Entomol. Soc.*, 6, 249, 1971.
19. **Burkholder, W. E.**, Application of Pheromones for Manipulating Insect Pests of Stored Products, Proc. Symp. Insect Pheromones and Their Application, Ministry of Agriculture and Forestry, Tokyo, 1976, 111.
20. **Burkholder, W. E.**, Application of pheromones and behavior-modifying techniques in detection and control of stored-product insects, in Proc. 2nd Int. Working Conf. Stored-Product Entomology, 1979, 56.
21. **Bruce, W. A. and Street, M. W., Jr.**, Infrared CO_2 detection of hidden insects, *J. Ga. Entomol. Soc.*, 9, 260, 1974.
22. **Street, M. W., Jr. and Bruce, W. A.**, CO^2 analyzer detects insects hidden in foods, *Food Eng.*, 48, 94, 1976.
23. **Kirkpatrick, R. L. and Cagle, A.**, Controlling insects in bulk wheat with infrared radiation, *J. Kans. Entomol. Soc.*, 51, 386, 1978.
24. **Cogburn, R. R., Tilton, E. W., and Brower, J. H.**, Bulk-grain gamma irradiation for control of insects infesting wheat, *J. Econ. Entomol.*, 65, 818, 1972.
25. **Dermott, T. and Evans, D. E.**, An evaluation of fluidized-bed heating as a means of disinfesting wheat, *J. Stored Prod. Res.*, 14, 1, 1978.
26. **Golumbic, C. and Davis, D. F.**, Radiation disinfestation of grain and seeds, in *Food Irradiation*, International Atomic Energy Agency, Vienna, 1966, 473.
27. **Schroeder, H. W. and Tilton, E. W.**, Infrared Radiation for the Control of Immature Insects in Kernels of Rough Rice, AMS-445, U.S. Department of Agriculture, Washington, D.C., 1961, 10.
28. **Tilton, E. W. and Brower, J. H.**, Status of U.S. Department of Agriculture research on disinfestation of grain and grain products, in *Radiation Preservation of Food*, International Atomic Energy Agency, Vienna, 1973, 295.
29. **Tilton, E. W., Brower, J. H., and Cogburn, R. R.**, Critical evaluation of an operational bulk-grain and packaged product irradiator, *Int. J. Radiat. Eng.*, 1, 49, 1971.
30. **Tilton, E. W., Brower, J. H., and Cogburn, R. R.**, Insect control in wheat flour with gamma irradiation, *Int. J. Appl. Radiat. Isot.*, 25, 301, 1974.
31. Code of Federal Regulations, Title 21, Section 121.3003, 1968, *Fed. Regist.*, March 2, 1967.
32. **Tilton, E. W., Burkholder, W. E., and Cogburn, R. R.**, Effects of gamma radiation on *Rhyzopertha dominica, Sitophilus oryzae, Tribolium confusum,* and *Lasioderma serricorne, J. Econ. Entomol.*, 59, 1363, 1966.
33. **Burkholder, W. E., Tilton, E. W., and Cogburn, R. R.**, Effects of gamma radiation on the grain mite, *Acarus siro, J. Econ. Entomol.*, 59, 976, 1966.
34. **Tilton, E. W., Burkholder, W. E., and Cogburn, R. R.**, Effect of gamma radiation on *Trogoderma glabrum* and *Attagenus piceus, J. Econ. Entomol.*, 59, 944, 1966.
35. **Brower, J. H. and Tilton, E. W.**, Comparative gamma radiation sensitivity of the sawtoothed grain beetle and the merchant grain beetle, *Environ. Entomol.*, 1, 735, 1972.
36. **Brown, G. A., Brower, J. H., and Tilton, E. W.**, Gamma radiation effects on *Sitophilus zeamais* and *S. granarius, J. Econ. Entomol.*, 65, 203, 1972.

37. **Brower, J. H. and Tilton, E. W.,** Gamma radiation effects on *Trogoderma inclusum* and *T. variabile*, *J. Econ. Entomol.*, 65, 250, 1972.
38. **Brower, J. H. and Scott, H. C.,** Gamma radiation sensitivity of the spider beetle, *Gibbium psylloides* (Coleoptera: Ptinidae), *Can. Entomol.*, 104, 1551, 1972.
39. **Brower, J. H. and Mahany, P. G.,** Gamma radiation sensitivity of the cadelle, *Tenebroides mauritanicus* (Coleoptera: Ostomidae), and the flat grain beetle, *Cryptolestes pusillus* (Coleoptera: Cucujidae), *J. Ga. Entomol. Soc.*, 8, 174, 1973.
40. **Brower, J. H., Miller, G. L., and Edenfield, J. E.,** Gamma radiation sensitivity of the corn sap beetle, *Carpophilus dimidiatus* (Coleoptera: Nitidulidae), *J. Ga. Entomol. Soc.*, 8, 55, 1973.
41. **Brower, J. H.,** Depressed flour beetle: sensitivity to gamma irradiation, *J. Econ. Entomol.*, 66, 1318, 1973.
42. **Cogburn, R. R., Tilton, E. W., and Brower, J. H.,** Almond moth: gamma radiation effects on the life stages, *J. Econ. Entomol.*, 66, 745, 1973.
43. **Brown, G. A. and Davis, R.,** Sensitivity of red flour beetle eggs to gamma radiation as influenced by treatment age and dose rate, *J. Ga. Entomol. Soc.*, 8, 153, 1973.
44. **Brower, J. H.,** Sensitivity of *Tenebrio molitor* and *T. obscurus* to gamma irradiation, *J. Econ. Entomol.*, 66, 1175, 1973.
45. **Brower, J. H.,** Inability of populations of *Callosobruchus maculatus* to develop tolerance to exposures of acute gamma irradiation, *Ann. Entomol. Soc. Am.*, 67, 287, 1974.
46. **Brower, J. H.,** Radiosensitivity of the slenderhorned flour beetle, *Gnathocerus maxillosus* (Coleoptera: Tenebrionidae), *Fla. Entomol.*, 57, 91, 1974.
47. **Brower, J. H.,** Radioresistance of the red flour beetle, *Tribolium castaneum* (Coleoptera: Tenebrionidae), exposed to sublethal doses of gamma irradiation for 25 generations, *Can. Entomol.*, 106, 241, 1974.
48. **Brower, J. H.,** Radiosensitivity of the squarenecked grain beetle, *Cathartus quadricollis* (Guerin-Meneville) (Coleoptera: Cucujidae), *J. Kans. Entomol. Soc.*, 47, 254, 1974.
49. **Brower, J. H.,** Radiosensitivity of the longheaded flour beetle, *J. Econ. Entomol.*, 68, 220, 1975.
50. **Brower, J. H.,** Radiosensitivity of *Tribolium destructor* Uyttenboogaart (Coleoptera: Tenebrionidae), *J. Stored Prod. Res.*, 11, 223, 1975.
51. **Brower, J. H.,** Radiosensitivity of stored-product insects: *Sitophagus hololeptoides* (Coleoptera: Tenebrionidae), *J. Ga. Entomol. Soc.*, 10, 341, 1975.
52. **Watters, F. L.,** Potential of accelerated electrons for insect control in stored grain, in Proc. 2nd Int. Working Conf. Stored-Product Entomology, 1979, 278—86.
53. **Bailey, S. W.,** Air-tight storage of grain: its effects on insect pests. I. *Calandra granaria* L. (Coleoptera, Curculionidae), *Aust. J. Agric. Res.*, 6, 33, 1955.
54. **Banks, H. J. and Annis, P. C.,** Suggested Procedures for Controlled Atmosphere Storage of Dry Grain, Division of Entomology Tech. Paper No. 13, Commonwealth Scientific and Industrial Research Organization, Canberra, 1977, 23.
55. **Jay, E. G.,** Suggested Conditions and Procedures for Using Carbon Dioxide to Control Insects in Grain Storage Facilities, ARS Bull. No. 51-46, U.S. Department of Agriculture, Washington, D.C., 1971.
56. **Storey, C. L.,** Exothermic inert-atmospheric generators for control of insects in stored wheat, *J. Econ. Entomol.*, 66, 511, 1973.
57. **Dean, G. A., Cotton, R. T., and Wagner, G. B.,** Flour mill insects and their control, *U.S. Dep. Agric. Circ.*, 390, 40, 1937.
58. **Cotton, R. T. and Wilbur, D. A.,** Insects, in *Storage of Cereal Grains and Their Products*, Christensen, C. M., Ed., American Association of Cereal Chemists, St. Paul, Minn., 1974, 211.
59. **Tilton, E. W. and Schroeder, H. W.,** Some effects of infrared irradiation on the mortality of immature insects in kernels of rough rice, *J. Econ. Entomol.*, 56, 727, 1963.
60. **Kirkpatrick, R. L. and Tilton, E. W.,** Elevated temperatures to control insect infestation in wheat, *J. Ga. Entomol. Soc.*, 8, 264, 1973.
61. **Kirkpatrick, R. L.,** Infrared radiation for control of lesser grain borers and rice weevils in bulk wheat (Coleoptera: Bostrichidae & Curculinoidae), *J. Kans. Entomol. Soc.*, 48, 100, 1975.
62. **Kirkpatrick, R. L.,** The use of infrared and microwave radiation for control of stored-product insects, *Proc. 1st Int. Working Conf. Stored-Product Entomology*, 1975, 431.
63. **Kirkpatrick, R. L.,** Controlling insects in bulk wheat with infrared, *J. Kans. Entomol. Soc.*, 51, 386, 1978.
64. **Neal, J. W., Jr., Ed.,** Guidelines for control of insect and mite pests of foods, fibers, feeds, ornamentals, livestock, forests and forest products, *U.S. Dep. Agric. Agric. Handb.*, 554, 822, 1979.
65. U.S. Department of Agriculture, Insects in Farm-Stored Wheat; How to Control Them, Leaflet 345, U.S. Department of Agriculture, Washington, D.C., 1966, 8.

66. **Chapman, H. C. and Schmitt, J. B.,** Stored-grain insects and their control in the Northeastern states, *N.J. Agric. Exp. Stn. Bull.,* 791, 16, 1959.
67. Agricultural Marketing Service, Methods and equipment for bulk treatment of grain against insects, *U.S. Agric. Mark. Serv. AMS Ser.,* 20, 7, 1962.
68. Agricultural Research Service, Controlling insect pests of stored rice, *U.S. Dep. Agric. Agric. Handb.,* 129, 1971, 19.
69. **Monro, H. A. U.,** Manual of Fumigation for Insect Control, 2nd ed. rev., FAO Agricultural Studies No. 79, Food and Agriculture Organization, Rome, Italy, 1969, 381.
70. **Berck, B.,** Distribution and persistence of fumigant mixtures applied to grain, in *Proc. 10th Int. Cong. Entomol.,* 4, 99, 1958.
71. American Conference of Government Industrial Hygienists, Threshold limit values, *AMA Arch. Ind. Hyg. Occup. Med.,* 9, 545, 1964.

Seeds

CLEANING AND CONDITIONING SEEDS FOR MARKETING

G. Burns Welch, A.H. Boyd, and Sergio Davila

When seed is brought in from the field, it generally contains a variety of undesirable materials such as stems, weed seeds, dust, and sometimes other crop seeds. Before the major crop seed can meet acceptable standards as required by law and demanded by progressive farmers, the undesirable materials must be separated from the crop seed. To accomplish this, the seed is conditioned through a machine or a series of machines which are capable of removing the contaminating materials. Thus, seed conditioning is an important link in the business of making available an ample supply of high quality, varietally pure seeds in a viable condition, free of noxious matter.

There are many different types of machines and special procedures used in conditioning seed. To describe all of these would require a volume in itself. However there are certain machines that are used more commonly than others for conditioning seeds of grains, soybean, (*Glycine Max* L. Merr) grasses, and clovers. These are the ones which will be covered in this chapter. Once an operator understands the material presented here, he should have little trouble in becoming familiar with other machines and processes.

PRINCIPLES OF SEED SEPARATIONS

The removal of undesirable material from crop seed requires conditioning the seed through one or more specialized machines. These machines are designed to remove the undesirable materials by exploiting differences in some physical property between the crop seed and undesirable material. Thus, it is important that the personnel responsible for conditioning seed have a thorough understanding of the differences in physical properties by which seed can be separated. Once a processor understands these principles, he is then able to examine a seed mixture and determine the separation problems and the necessary machines and conditioning sequence to clean and upgrade a seed lot to acceptable standards.

To affect a separation, the seed must be basically different in one or more physical characteristics from the other seed or contaminating material. Thus, it is by some difference in a physical characteristic that a machine is able to sort out the undesirable material from the crop seed. Since there are often several differences between seed of different crops or other contaminants, the conditioner should select the one difference that is great enough to allow mechanical separation by a machine at a reasonable capacity and a high degree of precision.

The following physical characteristics are the ones most commonly exploited in cleaning and separating seed.

Size

Size is a term often used in comparing one of the physical properties of seed. It is true that some seeds are larger or smaller than others. Where there is a great difference in the overall size of two seeds, such as soybeans and morning glory (*Ipomoea* spp.), it is acceptable to think in terms of separating the seeds on the basis of overall size; but when comparing two seeds that approach being the same size, we can be more precise by thinking of seeds as having the three dimensions — width, thickness, and length. This is easy to see when examining a lima bean (*Phaseolus lunatus* L.) or a kernel of corn (*Zea mays* L.) taken from the middle section of the ear. Seed differing in any one or two of these dimensions can be separated as follows.

1. Width — Seed having the same length and thickness but differing in width can be separated with screens that have round perforations. These may be a flat-type of screen such as used in the air-screen cleaners or indented cylindrical screens as used in width graders.
2. Thickness — Seeds having the same length and width but of different thicknesses can be separated with screens that have oblong perforations. The width of the opening is associated with the thickness of the seed. Again the screens may be the flat-type screen as used in air-screen cleaners, or the spiral-grooved cylindrical screen as used in thickness graders.
3. Length — Seed having nearly the same width and thickness but differing in length can be separated with an indented disc or indented cylinder separator. These machines are often referred to as length separators.

Weight

Seed of different weight or specific gravity can be separated with machines in which the seed passes through an airstream and by a gravity table which utilizes a flotation process on a vibrating deck.

Shape

Round seed can be separated from seed having a flat side by allowing the seed to roll down the flights of a vertical spiral. The machine for making this type of separation is known as a spiral separator.

Surface Texture

Seed having a rough surface can be separated from seed of similar size but having a smooth surface with revolving, fabric-covered rolls as in a roll mill.

Color

Seed of the same size but differing in color can be separated by a color sorter, which utilizes a photoelectric cell that has been calibrated to sort out off-colored seed.

Electrical Conductivity

A few varieties of seed can be separated to a very limited degree by their difference in electrical conductivity. However, this principle of separation is limited to only a few applications and to research investigations with small or light-weight seed. At the present time, this method of separating seed has not been developed sufficiently for widespread application.

Modification of Physical Properties to Enable Separations

The surfaces of some weed seed become sticky when wetted with oil or water. Seed with cracked seedcoats or rough surfaces will adhere more to liquid than seed with a smooth and/or unbroken seedcoat. If a small amount of oil or water is first added to a mixture where some seeds have this characteristic, and then sawdust or iron powder is added to the seed mixture and thoroughly mixed, the seed with sticky or rough seedcoats will take on more of the added material than the seed with smooth seedcoat. If sawdust is added, it will form a pellet around the undesirable particles and become large enough that it can be separated from the crop seed on the basis of size and/or weight. If iron powder is added, the powder adheres to the sticky seed; then if the seed mixture is passed over a magnet the sticky or rough seed which took on some of the iron power is attracted to the magnet and directed into a different receptacle than the smooth seed. This is a very specialized process and is used mainly for removing certain weed seed from small legume seed.

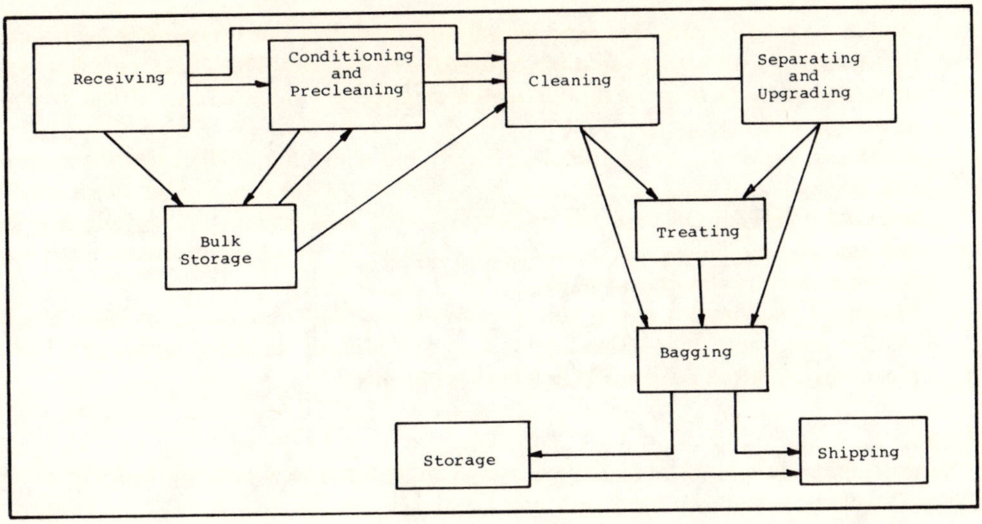

FIGURE 1. Flow diagram showing various steps in seed processing. The route that a particular batch of seed would follow during processing depends upon condition of seed upon arrival, kind of seed, cleaning and upgrading problems, seed treatment requirements, and shipping schedule. (Courtesy of the Seed Technology Laboratory, Mississippi State University, Mississippi State, 1979.)

STEPS IN SEED CONDITIONING

A seed conditioning sequence can be divided into steps as shown in Figure 1. It may not be necessary for some particular crop to be processed through all of these steps. The steps that seed must pass through from the time it is received until it is finished or ready for sale depend upon the moisture content and the undesirable material in that particular seed lot when it is received. For example, soybean would probably require only cleaning, upgrading, and bagging, whereas rice might require precleaning, drying, cleaning, upgrading, treating, and bagging. A well-designed seed conditioning plant is so arranged that any of these steps can be bypassed without interruption of the flow of seed through the conditioning plant. A brief description of the different conditioning steps is given below.

Receiving and Bulk Storage

A convenient means of unloading seed from the trucks when it is brought into the seed processing plant should be provided. If the seed is transported in bulk, a dump pit large enough to accommodate a truck can be driven over the pit to unload. An elevator connected with the pit is needed to move the seed into the bulk storage areas.

Drying

When seed is harvested at a moisture content too high for safe storage, it must be dried immediately to prevent loss of germination due to mold growth and heating. After the seed is dried to a moisture content safe for storage, it can be cleaned at a later date.

Precleaning

If seed contains a large amount of trash, it is advantageous to remove part of this undesirable material before drying, placing in storage, or cleaning. This precleaning operation in a seed conditioning plant is usually done with an aspirator, a screen machine having one or two vibrating screens, or rotating cylindrical screen (Figure 2).

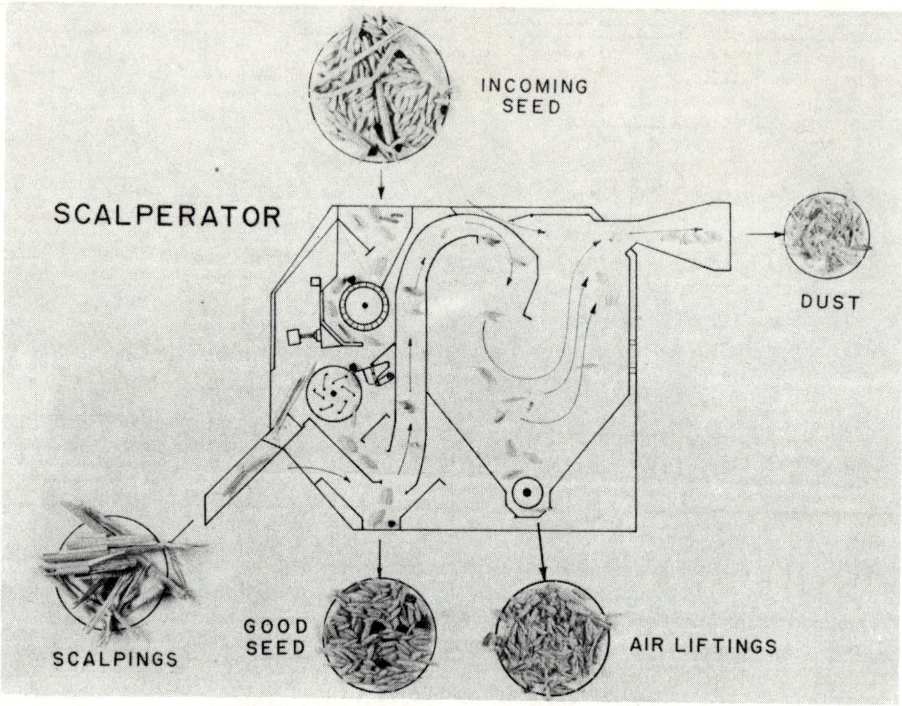

FIGURE 2. Scalperator used for precleaning seed. Large material removed by rotating cylindrical screen, small light material removed by airstream. (Courtesy of the Seed Technology Laboratory, Mississippi State University, Mississippi State, 1979.)

This precleaning operation in a seed conditioning plant is usually done with an aspirator, a screen machine having one or two vibrating screens, or rotating cylindrical screen (Figure 2). The screen machines may or may not have a fan to induce an airstream to remove the dust and light chaff. The machines used to perform this precleaning are commonly called precleaners or scalpers. Scalpers are machines that remove only material larger than the seed being handled. Other types of precleaners may utilize both scalping and screening (removal of a fine fraction) for a rough cleaning process either with or without assistance of an air separation.
Some of the advantages gained from precleaning are

1. Removes the large trash that will not move easily through cup elevators and feed hoppers on cleaning machines.
2. Better separations can be made with the air-screen cleaner when the large and small trash have been first removed. The seed flows more uniformly from the feed hopper and down the screens, as the screens are not flooded with trash. Screen perforations closer to be size of the seeds can be used to give a more accurate separation.
3. Increases the capacity of the air-screen cleaner.
4. Removes high-moisture green material which would increase time and cost of artificial drying.
5. Reduces volume to be stored; thus, trash will not take up valuable storage space.

Some crop seed requires preconditioning before starting through the cleaning and upgrading steps. For example, corn requires shelling before it can be cleaned. Grasses, some grains, and small-seeded legumes often require some type of special conditioning, such as a debearder, to break apart seed pods before they can be finished (Figure 3).

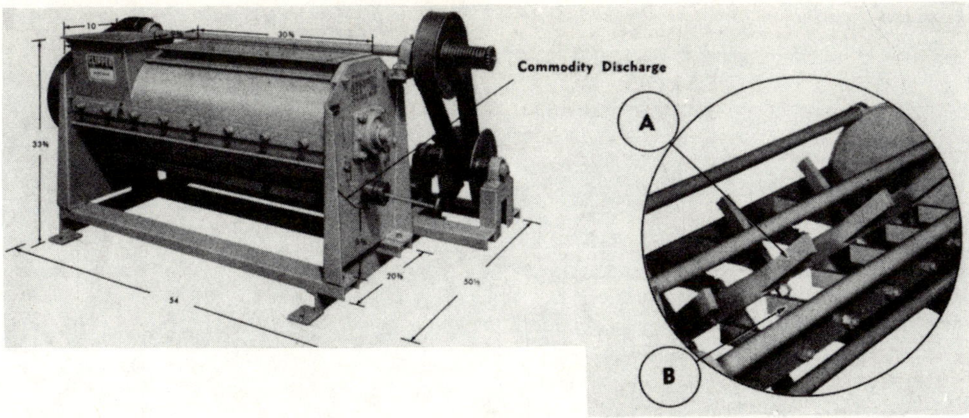

FIGURE 3. Debearder used for removing awns and breaking apart seed clusters. (A) Rotating beater arm with special hard coated edges; (B) Stationary beater arm. (Courtesy of the Ferrell-Ross Company, Saginaw, Mich., 1978.)

Also, hullers or scarifiers may be needed to remove some seed covering and/or scratch the seed coat to allow moisture imbibition where hard seed is a consideration.

Basic Cleaning — Air-Screen Cleaner

Almost all seed lots are cleaned first or after precleaning with an air-screen machine to remove the light trash, dust, and most of the undesirable material, both larger and smaller than the crop seed. Many seed lots can be cleaned and made into an acceptable product with only this process. Other seed lots may require further upgrading, and the upgrading machines can function properly only when the seed has first been cleaned. Thus the air-screen cleaner is usually considered a necessary and basic machine in most all seed conditioning plants.

The air-screen cleaner uses three cleaning principles: aspirations or airstreams to remove light material, top screens (scalpers) to remove the large material, and bottom screens (graders) to remove the small material. These machines vary in size from one-screen models with one airstream to the large commercial machines with up to seven screens and three airstreams. Two-, three-, four-, and five-screen model cleaners are most commonly used in the industry.

To understand the principles of operation and successfully operate any seed conditioning machine, the operator should know the principal parts of the machine and their functions, the flow path of the seed through the machine, and the adjustments which affect the precision of the separation and capacity of the machine. A diagram of a four-screen air-screen cleaner with two scalper and two grader screens is shown in Figure 4.

Function and Adjustment of the Air-Screen Cleaner Parts

Feed Hopper

The feed hopper (Figure 4) is a box-like device at the top of the machine which provides a uniform metered flow of seed to the screens. In the bottom of the feed hopper, there is an adjustable gate to regulate the flow of the seed from the hopper. Also the bottom is a revolving roller just above the adjustable gate to assist in forcing the seed to discharge from the hopper. On some machines the speed of the roller can be varied, which will regulate the amount of seed discharging from the hopper. Therefore the rate of feed can be controlled by the position of the feed gate and also by the speed of the feed roller. Generally a more accurate rate of feed can be obtained by using a combination of the two adjustments.

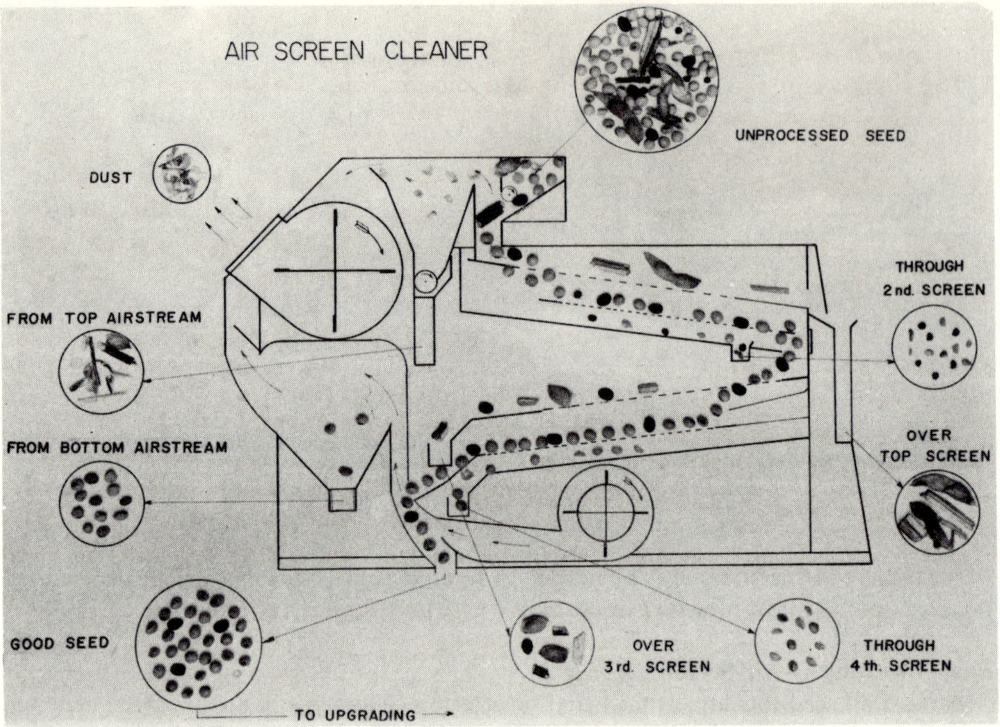

FIGURE 4. Flow diagram for four-sceen air-screen cleaner. (Courtesy of the Seed Technology Laboratory, Mississippi State University, Mississippi State, 1979.)

Fans and Airstreams

The purpose of the fan is to create an airstream that will remove the light trash, dust, and light seeds from the crop seed. On machines with one airstream, the seed will pass through the airstream either at the bottom of the feed hopper or below the end of the last bottom screen. On machines with two airstreams, the seed will pass through an airstream at both of these locations (Figure 4). The airstream just below the bottom of the feed hopper is known as the top or upper air whereas the one below the last bottom screen is known as the bottom or lower air.

The top airstream removes the light trash and dust from the seed before it reaches the first screen. The bottom airstream removes more of the light seed and heavier trash not removed by the top airstream and screens. The strength of each airstream is controlled by an adjustable door in the air passage or by regulating the speed of the fan.

For efficient cleaning, both airstreams should be strong enough to remove an occasionally good seed along with the other materials being lifted out. The bottom air should be relied on most for removing light and immature seeds since the particles are more uniform in size at this point and more uniformity of separations is possible.

Screens

The screens in an air-screen machine consist of a flat piece of thin, perforated metal or woven wire mesh fastened on all sides to a rectangular wooden frame. The screens are placed flat in the machine but slope downward at an angle of approximately 10°.

Most screens have a solid metal section approximately 9 in. wide across one end. This is called a spreader section. The screen should be placed in the machine so that this solid section is elevated above the other end of it. Then, as the seed discharges from the feed hopper or a preceding screen, it falls onto the spreader section and has an opportunity to spread into a thin layer, the long pieces of trash lying flat before moving onto the perforated section.

The function of the screens is to remove the material that is greater or smaller in width and thickness than the crop seed.

Screens can be divided into two general classes according to their use and position in the machine. These two classes are top screens, which are sometimes referred to as scalpers, and bottom screens, which are often called graders. For simplicity and clarification, the terms top screen and bottom screen will be used throughout this chapter. The terms second top screen and second bottom screen, etc., will be introduced when describing machines with more than two screens.

In some of the literature, reference is made to the position of the screens in the cleaner. The locations or positions of the screens in a cleaner are counted from the top of the machine to the bottom. The top screen is in the first position, the second screen is in the second position, and so on down the line.

Perforated metal screens may have perforations that are round, oblong, or triangular in shape. The round and oblong perforations are most commonly used. Screens with triangular perforations have limited application and are seldom used. Screens constructed of woven wire mesh may have openings that are either rectangular or square.

There are many screens with different sizes of perforations available to the seed conditioner (Table 1). Each screen is identified by a number that indicates the size and shape of the perforation. The size of the perforations in screens manufactured in foreign countries is measured in millimeters, but those manufactured in the U.S. are still designated in fractions of an inch.

The overall length and width of the screen frame must also be indicated when obtaining screens for a particular machine. The different types of screen openings used for cleaning seed are shown in Figure 5 and the methods of measurement are described below.

Round perforations — The sizes of round perforations in perforated metal screens are specified by the diameter of the perforations. The unit of measurement is either millimeters or fractions of an inch. In expressing the size in fractions of an inch as in the U.S., two systems are used: (1) the diameter is measured in 64ths of an inch in sizes from 5/64 to 80/64. These screen sizes are usually designated by using only the numerator of the fraction, i.e., 5, 6, ... 78, 80, the denominator being understood to be 64. (2) The second system used is for the perforations smaller than 5/64 in. In this system the diameters are measured in fractions of an inch taken in harmonic progression from 1/25 to 1/12 in., i.e., 1/25, 1/24, 1/23, 1/13, 1/12, etc.

Oblong perforations — The oblong perforations in perforated metal screens are designated by both the width and the length of the perforations. The width of an oblong perforation is measured in the same manner as the diameter of a round perforation described above. The length of the perforations is usually 1/4, 5/16, 1/2, or 3/4 in.

The first number in the size listing of an oblong perforation is the width of perforation; the second number is the length of the perforation. For example, a screen designated as 7 × 3/4 has oblong perforations 7/64 in. wide and 3/4 in. long.

The long axis of the oblong perforations is usually parallel with the direction of the flow of the seed on the screen. However, screens with large oblong perforations are available in which the long axis of the opening is perpendicular to the flow of seed on the screen. This type of screen is known as a cross-slot. Cross-slot screens are useful in separating split beans (half beans) from whole beans. Screens are also available in which the oblong perforations are at an oblique angle to the direction of flow.

Triangular perforations — Two methods are used to designate the size of triangular

Table 1
SCREEN PERFORATON AND MESH SIZES AVAILABLE FOR AIR-SCREEN CLEANERS

Perforated metal sheet									Wire cloth			
Round holes			Oblong holes			Finished screens made only in 9 in. and 8 in. model widths; sheet sizes 26 in. × 61½ in. and 16 in. × 53¾ in.						
Fractions	64ths		Fractions	64ths	Triangles 64ths	Oblong cross slot	Round hole half sizes	Oblong half sizes	Square openings	Oblong openings		
1/25	5½	24	1/24×½	5×¾	5	6×¾	6½	8½×¾	3×3	2×8	4×8½	6×14
1/24	6	25	1/22×½	5½×¾	8	7×¾	7½	9½×¾	4×4	2×9	4×15	6×15
1/23	7	26	1/22×½ Diag.	6×¾	9	8×¾	8½	10½×¾	5×5	2×10	4×16	6×16
1/22	8	27	3/64×5/16	6½×¾	10	9×¾	9½	11½×¾	7×7	2×11	4×18	6×18
1/21	9	28	1/20×½	7×¾	11	10×¾	10½	12½×¾	8×8	2×12	4×19	6×19
1/20	10	29	1/18×¼	8×¾-D		11×¾	11½	13½×¾	9×9	3×14	4×20	6×20
1/19	11	30	1/18×¾	9×¾		12×¾	12½	14½×¾	10×10	3×16	4×22	6×21
1/18	12	31	1/16×¼-A	10×¾-E		13×¾	13½		12×12	3×16 SP.	4×24	6×22
1/17	13	32	1/16×½	11×¾-F		14×¾	14½		14×14	3×18	4×24 SP.	6×23
1/16	14	34	1/15×½	12×¾-G		15×¾	15½		15×15	3×20	4×26	6×24
1/15	15	36	1/14×¼-B	13×¾-H		16×¾	16½		16×16	3×21	4×28	6×25
1/14	16	38	1/14×½	14×¾-I		18×¾	17½		17×17		4×30	6×26
1/13	17	40	1/13×½	15×¾-J		10½×¾	18½		18×18		4×32	6×28
1/12	18	42	1/12×½-C	16×¾-K		11½×¾	19½		20×20		4×34	6×30
	19	44		17×¾		12½×¾	20½		22×22		4×36	6×32
	20	48		18×¾			21½		24×24			6×34
	21	56		19×¾			22½		26×26			6×36
	22	64		20×¾					28×28			6×38
	23	72		21×¾					30×30			6×40
		80		22×¾					32×32			6×42
				24×¾-L					34×34			6×50
				32×1					36×36			6×60
									38×38			18×20
									40×40			20×22
									45×45			
									50×50			
									60×60			

From Ferrell-Ross Company, Saginaw, Mich. With permission.

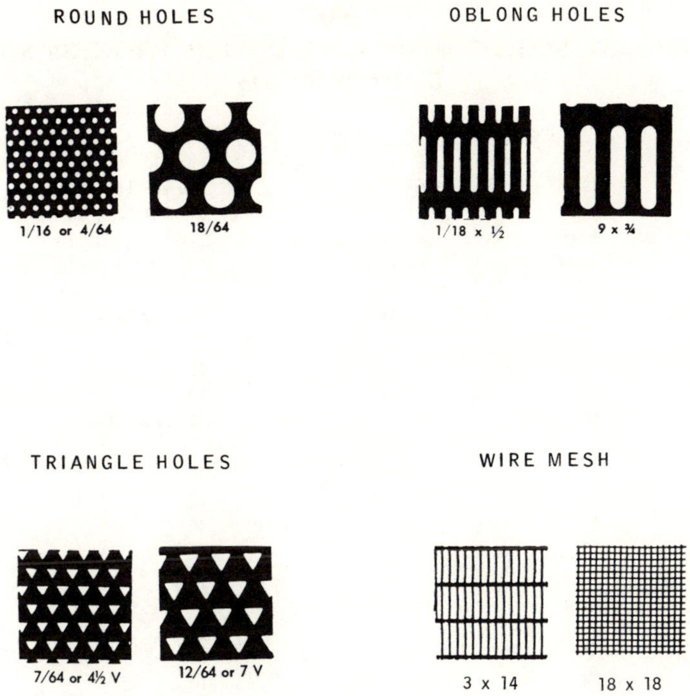

FIGURE 5. Types of perforated and woven wire mesh screens used for cleaning seeds. (Courtesy of the Crippen Manufacturing Company, Alma, Mich., 1978.)

perforations. In one case, the size is measured by the length of each side of the equilateral triangle formed by the perforation. In the second method, the size of the triangular opening is indicated by the diameter of the largest circle that can be inscribed within the triangle. The number indicating the size of a triangular perforation is ordinarily followed by the symbols "V" or "Tri" to distinguish it from a round perforation, i.e., 8 Tri and 5V.

The unit of measurement is 64ths of an inch. It should be noticed that two triangular screens having the same numerical size designation may not have the same size perforations. For example, triangular perforations designated as 7 Tri, 8 Tri, 9 Tri, 10 Tri, 11 Tri, and 12 Tri, are measured by the length of the sides of the triangle. The same perforations measured by the circle diameter method would be designated as 4 1/2V, 5V, 5 1/2V, 6V, 6 1/2V, and 7V, respectively.

Wire mesh —The size of a wire mesh screen is designated by the number of openings per inch in each direction. Both rectangular and square openings are available. A 6 × 20 wire mesh has six openings per inch in one direction and 20 openings per inch in the other. The long axis of the rectangle is parallel to the direction of flow of the seed on the screen. A square mesh such as 10 × 10 would have 10 openings per inch across and down the screen. These screens are generally used for grasses and small seeded legumes.

The selection of proper screens is probably the most important factor affecting the performance of the air-screen cleaner, (Table 2). The seed must either go over or through the screen to make a separation. Thus, one can easily see that better separations can be obtained when the size of the screen perforations is as near the thickness and width of the crop seed as possible without allowing too much of the crop seed to go over or through the screens.

Table 2
SCREEN OPENING SIZES COMMONLY USED FOR PROCESSING VARIOUS KIND OF SEEDS

Seed	Col. 1	Col. 2	Col. 3	Col. 4
Alfalfa	1/14 or 1/13	6×26	1/16 or 1/14	6×24
Sorrel from	1/18	6×34	6×20 6×21, 6×22	6×32
Mustard from	1/18	6×34	3/64×6/16, 6/20, 6/21	6/32
Yellow trefoil from	1/18	6×34	6×22, 1/21, 1/22	6×32
Night flowering catchfly from	1/18	6×34	6×20 or 6×21	6×32
Allspice, whole	21	6	20	8/64×¾
Anise	14	1/20	12	1/18
Barley	11/64×¾ or 24	1/14×½	22, 21 or 20	1/13×½
Beans				
Cranberry	32	14/64×¾	30	16/64×¾
Yellow eye	24	11/64×¾	22	12/64×¾
Lima	56	16	48	20
Baby Lima	32	17	30	19
Red kidney	80	13/64×¾	28	14/64×¾
Navy or pea	22	10/64×¾	20	11/64×¾
Pinto	26	9/64×¾	24	10/64×¾
Great Northern	26	10/64×¾	24	11/64×¾
Mung	14	7	13	8/64×¾
Beet	24	7	22 or 20	8
Bent grass	26×26	60×60	28×28	50×50
Bermudagrass	6×28	6×42	6×30	6×40
Bramegrass	10	9 tri or 6×26	1/13×½	6×24
Blue grass				
Canada	1/18	6×42	1/20	6×40
Kentucky	6×28	6×42	26×26	6×40
Bulbous	10	4×24	7	4×22
Buckwheat	16	7	14	6/64×¾
Cabbage	10	1/18	8	8/64×5/16
Canary grass	9	6×26	3×16	6×24
Cantaloupe	20	7	16	9
Caraway	1/14×¼	6×24	1/16×½	6×22
Carrot	7	6×28	1/12	6×26
Carpet grass	12	1/14×½	10 6×28	1/13×½ 6×38
Celery	1/14	30×30	1/16	28×28
Clover				
Hop	1/20	6×36	1/25	6×34
Ladine	1/18	6×34	1/20 or 1/21	6×32
Alsike	1/18	6×34	1/19	6×32
Crimson	6	6×24	1/13	6×22
Bur hulled	1/12	6×24	1/13	6×22
Dalea or woods	6	6×23	1/13	6×22
Red	1/15	6×24 or 6×26	8/64×5/16	6×22 or 6×24
Red, hare's ear mustard from	3/64×5/16	6×24	3/64×5/16	6×24
White sweet	1/12	6×26	1¼	6×24
White sweet unhulled	10	6×26	8	6×24
White	1/16–1/18	6×34	1/18, 1/19, 1/20	6×32
White, dock from	1/16	6×34	6×22 or 6×23	6×32
Cumin	14	1/25	12	1/22
Coriander	15	1/13×1/2	13	1/12×1/2
Corn, cleaning only	32	14	28, 30	16

Table 2 (continued)
SCREEN OPENING SIZES COMMONLY USED FOR PROCESSING VARIOUS KIND OF SEEDS

Seed	Col. 1	Col. 2	Col. 3	Col. 4
Cotton seed				
Delinted	20	9/64×¾	18	10/64×¾
Not delinted	40	12/64×¾	36	13/64×¾
Cowpea	22	12 or 10/64×¾	21	11/64×¾
Crested wheat grass	1/16×½	6×30	1/18×¼	5 tri.
Crotalaria	12	6	1/12×½	7
Cucumber	18	8	17 or 16	9
Dallas grass	8	1/15	3×14	1/14
Fescue				
Meadow	1/13×1½	6×21	1/14×¼	4×22 or 5 tri.
Chewings	3/64×5/16	6×32	1/22×½	6×30 or 5 tri.
Ky. 31	3/64×5/16	6×32	1/22×½	5 tri. or 6×30
Fennel	14	1/16	12	1/14
Flax				
Bison	1/16×½	1/14	1/18×¾	1/13
Golden	1/13×½	1/12	1/14×¼	6
Small	3×16	1/14	3/64×5/16	1/12
Walsh & viking same as bison				
Garbanzo (chick pea)	80	11/64×¾	26	12/64×¾
Hemp	14	1/13×½	12	1/12×½
Johnson grass	10	1/18	8	1/16
Lentil	18	12	7/64×¾	12 or 13
Lespedeza				
Korean	6	1/17, 1/18	6×15 or 1/18×¼	1/16
Kobe	8	1/16	6×14 or 1/18×¾	1/14
Sericea, unhulled	7	1/15	1/14×¼	6×26
Sericea, hulled	1/15	6×28	1/16 or 3/64×5/16	6×26
Lettuce	4×18	24×24	6×20	20×20
Milo, maize	14	1/13×½	13	1/12×½
Millet				
Siberian	7	1/20	1/12	1/15
Proso	9	3/64×5/16	7	1/18×¼
Mustard				
White	9	1/22×½	7	3/64×5/16
Brown	7	1/22	1/12	1/20
Oats	24	1/16×½ or 11 tri.	11/64×¾	1/14×½
Oats, corn from	10/64×¾ or 9/64×¾	1/16×½	9/64×¾ or 8/64×¾	1/14×½
Oat grass, tall	1/12×½	6×30	1/14×½	6×28
Okra	16	3/64×5/16	14	1/18×¾
Onion	11	1/15	9.8	1/14
Onion sets	1/¼	3×3	1 in.	8×3
Orchard grass	3/64×5/16	6×32	1/22×½ or 1/24×½	6x32 or 5 tri.
Peas				
Austrian winter	20	12 or 9/64×¾	17	10/64×¾
Field	22	10 or 9/64×¾	20	10/64×¾
Pepper, red	14	1/12	12	7
Pepper, black,	18	6/64×¾	16	7/64×¾
Pinion nuts	32	12/64×¾	30	14/64×¾
	10	4×24	7	4×22
Popcorn, pearl	21	10	20	11
Psyllium	4×18	6×38	6×20	26×26
Pumpkin	36	18	82	20
Radish	10	1/14	9 or 8	1/12

Table 2 (continued)
SCREEN OPENING SIZES COMMONLY USED FOR PROCESSING VARIOUS KIND OF SEEDS

Seed	Col. 1	Col. 2	Col. 3	Col. 4
Rape, dwarf essex	7	1/23	1/12	1/22
Rape, German	9	1/18	7	1/16
Redtop	26×26	60×60	28×28	50×50
Redtop, timothy from	26×26	60×60	6×34	50×50
Rice, paddy	22	6 or 7	20 or 21	1/14×½, 1/18×½
Rice, hulled	14	1/12	12	14×14
Ryegrass	3/64×5/16	6×32	1/22×½	6×32 or 5 tri.
Rye	14	1/16×½	12	1/14×½
Rye, vetch from (100% Impossible)	7/64×¾	8	7/64×¾	9
Sesbania	10	1/18×¼	8 or 9	1/18×¾
Sesame	6	4×26	1/18×¼	1/16
Sorghum				
Grain	12	1/14×½	10	1/13×½
Sweet	8	1/22×½	6	3/64×5/16
Soybeans				
Large	22	10/64×¾	20	11/64×¾
Small	20	6/64×¼	18	7/64×¾
Black flat	18	7/64×¾	16	8/64×¾
Morninglory from	18	7/64×¾	16	11 or 12
Spinach	14	1/14	11 or 12	1/12
Squash	34	20	32	23
Sudangrass	12	1/22×½	10	3/64×5/16
Sunflower	24 to 32	10	24 to 32	11
Timothy	1/18	6×36	1/19 or 1/20	6×34 or 6×32
Timothy				
Sorrel from	6×20 or 6×21	6×36	6×20 or 6×21	6×34 or 6×32
Red clover from	6×24	6×36	6×26	6×34
Peppergrass from	1/19	6×34	1/20	6×32
Black plantain from	1″19	6×32	1/20	6×30
Redtop from	1/19	6×36	1/20	6/34
Buckhorn plantain from	1/20	6×30	1/22	6×30 or 6×28
Alsike clover from	1/20	6×34	1/25 or 6×26	6×32
Tobacco	30×30	50×50	32×32	40×40, 50×50
Tomato	12	1/14	10	1/12
Trefoil, birdsfoot	1/14	6×26	1/16	6×24
Turnip	1/12	1/22	1/14	1/20
Vetch	16	6/64×¾	14	6/64×¾ or 7/64×¾
Watermelon	28	14	24	16
Wheat	16	1/18×½ or 8 or 9	14	1/18×½ or 10 tri.
Wheat oats from	12	1/18×½	11	1/18×½
Wheatgrass, western	1/18×½	4×24	1/14×½	4×22

Note: Four-screen cleaners — Use Column 1 for top screen in top shoe, Column 2 for bottom screen in top shoe, Column 3 for top screen in bottom shoe, and Column 4 for bottom screen in bottom shoe. Three-screen cleaners — Use Column 1 for top screen, Column 3 for middle screen, and Column 4 for bottom screen.

From Ferrell-Ross Co., Catalogue of Seed Cleaning Equipment, Saginaw, Michigan, 1978. With permission.

When cleaning round-shaped seed, select a screen with round perforations for the top screen and a screen with oblong perforations for the bottom screen. Using a round perforation slightly larger than the crop seed will allow the crop seed to pass through the screen, but any material larger than the crop seed will flow over the screen. The bottom screen with oblong perforations will hold up the crop seed but will allow material such as broken seed, other small seed, and small inert material to pass through more easily. The crop seed will flow over the bottom screen. Thus, the smallest possible perforation is used for the top screen, while the largest possible perforation is used for the bottom screen.

When cleaning elongated seed, screens with oblong perforations are ordinarily used in both the top and bottom screen positions. However, in some cases when a machine has two or more top screens, a screen with round perforations is used in the first position to take out the largest trash.

In selecting an oblong perforation, be sure to select one long enough. The oblong perforations in a top screen should be slightly longer than the seed so that it can pass through the perforations. The perforations in the bottom screen should also be long enough to allow the elongated, undesirable material to pass through easily.

Shoes and Slope of Screens

The vibrating box-like section of the machine into which the screens are placed is called a shoe. The shoe is sloped to assist the seed to flow down the screen. The slope of the screen may be fixed or adjustable within the shoe so that the screen can be set in a steeper or flatter position. Increasing the slope of a screen causes material to move over the screen more rapidly. The top screens are usually set at a greater slope to facilitate more rapid removal of the trash which would tend to pass through the screen with the seed or remain on the screens and interfere with the seed passing through.

The bottom screens should have less slope so as to allow ample time for the small material to pass through the screen and not be carried along with the crop seed.

Eccentric and Rate of Vibration

The vibrating or shaking action of the shoes is developed by an off-center drive bearing called an eccentric. In machines with two shoes, the eccentrics are set so that the shoes travel in opposite directions all the time and the vibration of the entire machine is held to a minimum. On machines having only one shoe, a counterbalance is sometimes used to counteract the shoe vibration.

The vibration can be regulated on some cleaners to control the movement of the seed on the screens. A fast vibration causes the seed to bounce rapidly and turn all the sides to the openings in the screens. If the vibration is too slow, the seed will not bounce along the screen enough to affect a good separation. The vibration should be fast enough to move the seed along the screens with sufficient time to allow all or most of the small material to pass through the screens.

Principles of Operation

The four-screen air-screen cleaner with two airstreams is probably the most widely used cleaner in seed processing plants. The cleaning sequence is ordinarily arranged to take place in the following order and will be used in this discussion: top air separation, first top screen, first bottom screen, second top screen, second bottom screen, and bottom air separation (Figure 4).

The seed to be cleaned is first deposited in the feed hopper. As it is metered from the bottom of the feed hopper, it passes through the top airstream, which removes light trash and dust, before it contacts the first top screen.

The first top screen is used for removing the largest material such as stems, straw, and unthreshed seed pods from the seed mass. This material slides over the screen and is discarded. All the other material including the crop seed pass through the first top screen onto the first bottom screen.

The first bottom screen has perforations large enough to allow the small trash and small weed seed to pass through. The crop seed passes over the screen and is deposited onto the second top screen.

The second top screen has perforations smaller than the first top screen. The large material which was small enough to pass through the first screen passes over the second top screen, while the crop seed and remaining small material passes through onto the second bottom screen.

The perforations in the second bottom screen are slightly larger than the ones in the first bottom screen. This allows small seed and split seed to pass through while the crop seed passes over the screen.

The crop seed discharges over the second bottom screen and passes through the bottom airstream which removes light seed, hulls, and heavier trash not removed by upper air.

In some seed lots that contain only a small amount of undesirable material, the crop seed can be cleaned to a sufficient purity with only an air-screen cleaner. However, in most cases there is still some broken seed, immature seed, short seed, and particles of dirt remaining in the crop seed after it has been cleaned with the air-screen cleaner. This is because this material had the same width and thickness as the crop seed and could not be separated from the crop seed with a machine utilizing flat, perforated screens. This should not be considered a fault of the machine. It is only a limitation that is normal to this type of machine.

When there is some undesirable material remaining in the crop seed after cleaning with an air-screen cleaner, or any other machine, the operator should examine the seed and look for another difference in physical characteristics that could be used for separating the undesirable material. This difference might be found in length, surface texture, degree of roundness, or weight. Once it has been determined that there is sufficient difference in some other characteristic for making a separation, the conditioning should proceed to the machine utilizing this difference for making a separation.

Procedure for Setting Up and Operating

The following steps are given as a guide for setting up, starting, and adjusting an air-screen cleaner. As an operator becomes better acquainted with a machine, he will be able to make initial settings that will be close to what is needed for a particular crop.

1. When changing to a different crop or variety, remove the screens and throughly clean the machine.
2. Select the screens by making a preliminary test with small test screens. If these are not available, place a regular size screen on canvas on the floor and then place about 1 lb of seed on the screen. Bounce the screen several times and observe whether the undesirable material is removed without losing many of the crop seed.
3. After selecting the screens, place them in the machine.
4. If the machine has screen brushes that can be raised and lowered, put the brushes in raised position, so that the brushes press firmly against the underside of the screens.
5. Close adjustable doors in upper and lower airstream passageways. If speed of fan is adjustable, set it near minimum speed.

6. Fill seed hopper with seed.
7. Start machine.
8. If screen vibration is adjustable, set it at a slow to medium speed, not fast.
9. Open adjustable door at bottom of feed hopper to allow a small stream of seed to flow onto the screen.
10. Increase screen vibration if necessary to cause seed to move down screens.
11. Open gate in upper airstream until light trash is being lifted out.
12. Open gate in lower airstream until light seed and hulls are being lifted out.
13. Readjust steps 8, 9, 10, 11, and 12 to obtain maximum capacity and good cleaning.

Separating and Upgrading

To remove undesirable materials left by the air-screen cleaner, the operator has to utilize other machines which have other differences in physical characteristics for making a finer separation. These can be a length grader, thickness and width grader, gravity table, spiral separator, etc. depending upon the particular problem involved. But to effect a separation, the crop seed must differ from the undesirable material sufficiently in some physical characteristic, so that the machine can differentiate between the crop seed and the material to be removed.

Length Separators

In many cases, the seed of different crops are so nearly the same size in width and thickness that they cannot be separated with a screen. Also, some of the undesirable materials such as broken seed, soil particles, and weed seed may have the same width and thickness as the crop seed. However, if there is a difference between the length of the seed and the undesirable material, a precise separation can be made with either an indented cylinder or indented disc.

Indented Cylinder Separator

The indented cylinder separator consists of a rotating, horizontal cylinder and an adjustable horizontal separating trough inside the cylinder. The inside surface of the cylinder has thousands of small, closely spaced, hemispherical indentations. In operation, the seed enters the cylinder at one end and discharges at the other. A layer of the seed mass to be separated lies approximately 3 in. deep on the bottom of the cylinder. As the cylinder rotates, the short seed and other short material fit into the indents and are lifted out of the seed mass. The rotation of the cylinder lifts the short seed to the point where the indents are inverted enough for the seed to fall out of the indent and drop into the trough. The trough extends the length of the cylinder and has a small auger inside that conveys the lifted material to the discharge spout at the end of the trough. The long seed which is not lifted by the indents gradually moves through the cylinder and discharges through a separated spout at the end of the cylinder (Figure 6).

All the indents in a single cylinder are the same size. Cylinders with different indents are available. The cylinder in some models of cylinder separators can be removed and exchanged for one with a different indent. Thus, by having several cylinders with different indent sizes, the same machine can be used for making different separations (Table 3).

The size of the indent is indicated by a numerical notation which is the diameter that is measured near the top of the indent. For cylinders manufactured in Brazil and the European countries, the indents are measured in millimeters, while the indents in cylinders manufactured in the U.S. are measured in 64ths of an inch. For example, a No. 10 indent cylinder made in Brazil has indents 10 mm in diameter, while a No. 10 indent cylinder from the U.S. has indents 10/64 of an inch in diameter.

Table 3
INDENT CYLINDER SIZES USED FOR SEPARATING GRASSES, GRAINS, AND SEEDS

Indent number	Will lift	Will reject
2	Small pigweed, Alsike clover, small dodder, mullenweed, sand, etc.	Buckhorn, timothy, black medic clover, bluegrass (all varieties), alfalfa, crimson clover, lespedeza, etc.
3	Small sweet clover, pigweed, dodder, white clover, alsike, etc.	Thistles, buckhorn, sticks, alfalfa, red clover, etc.
4	Timothy, small clover (red and white), dodder, hulled water grass, mustard, sheep sorrel	Canada thistle, quackgrass, sticks, alfalfa, bluegrass, etc.
5	Red clover, alfalfa, small flax, water grass, mustard, bluegrass, etc.	Meadow fescue, wild brome, large buckhorn, quackgrass, cheat, chess, sticks, etc.
6 1/2	Small broken grain, small wild buckwheat, small vetch and cockle, wild mustard	Fescue, wheat, ryegrass, wheat grass, hulled orchard grass, flax, etc.
8 1/2	Buckwheat, cockle, vetch, sudangrass, small sugar beet seed, etc.	Wheat, rye, fescue, ryegrass, orchard grass, etc.
11	Broken grain, vetch, small onion and garlic, wild peas, coffee weed, etc.	Spring wheat, rye, rice, alta fescue
13	Spring wheat, small or broken durum, pearled and broken barley, flax	Durum, large spring wheat, barley, pin oats
19	Spring wheat, small durum	Oats, wild oats, barley, etc.
22	Wheat, winter wheat, hulled oats, rye, etc.	Oats, wild oats
24	Barley	Oats, wild oats, barley with tails, etc.
26		
28	Used primarily in length grading of seed corn and similar sized material	
32		
S-3	Equal to #22 Indent but has flat bottom; used primarily on corn	

From Carter-Day Company, Minneapolis, Minn. With permission.

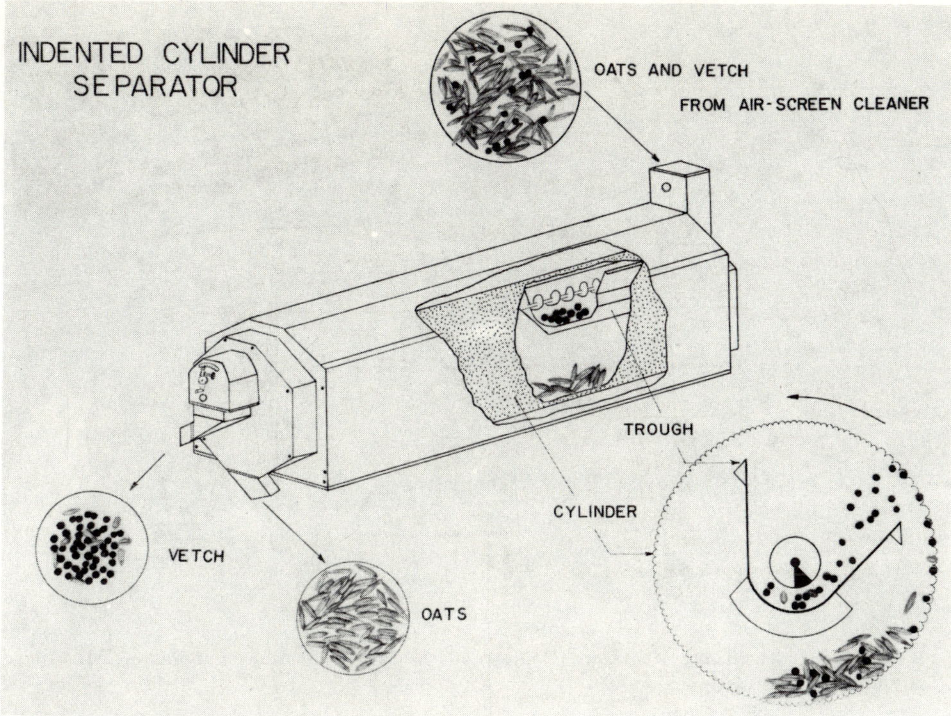

FIGURE 6. Indented cylinder separator. (Courtesy of the Seed Technology Laboratory, Mississippi State University, Mississippi State, 1979.)

Assuming there is sufficient difference in the length of the seed and the undesirable material and that the correct size of indent has been selected, the efficiency of separation depends very much on the position of the trough and the speed of the cylinder. Sometimes there may be a problem with the cylinder separator lifting out some long seed with the short seed or not getting all the short seed from the long seed. The probable causes of these problems are listed below.

If some of the long seed are falling into the trough with the short seed, this could be caused by:

1. Indents too large
2. Separating edge of trough set too low
3. Cylinder speed too fast
4. Rate of feed too fast, which causes the seed layer to be too deep at the feed end of the cylinder

If some of the short seed is not being separated from the long seed, this could be caused by:

1. Indents too small
2. Separating edge of trough set too high
3. Cylinder speed too slow

Indented Disc Separator

This consists of a series of indented discs that revolve together on a horizontal shaft. Each disc contains hundreds of recesses or pockets on each side. As the disc revolves,

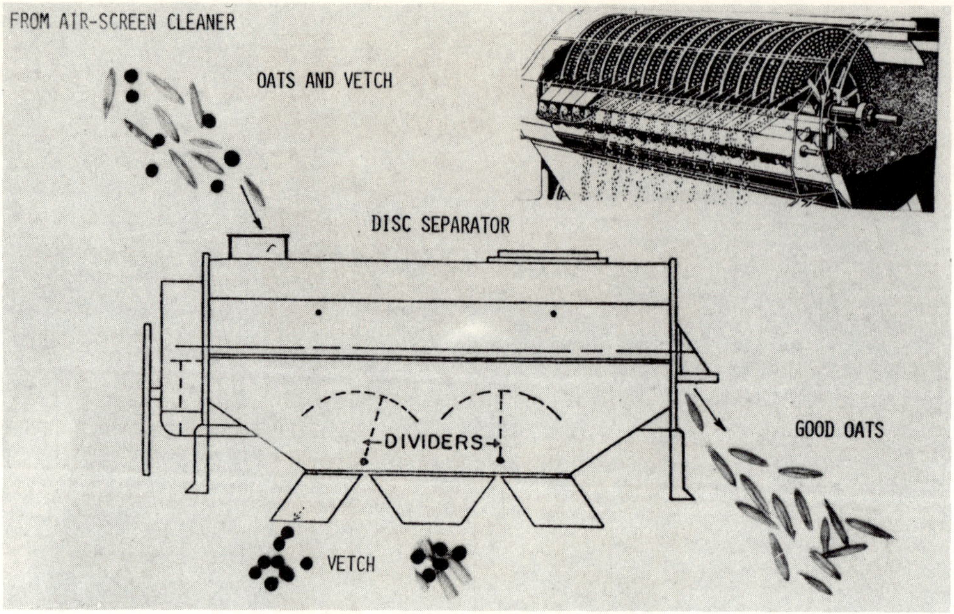

FIGURE 7. Indented disc separator. (Courtesy of the Seed Technology Laboratory, Mississippi State University, Mississippi State, 1979.)

the seed enters the intake end of the separator and moves through the open centers of the discs toward the opposite end. During this process, the materials shorter than the crop seed fit into the disc pockets and are lifted out into a trough at the side of the machine. The long seed that does not fit into the pockets is rejected and moves on through the discs to the discharge end, where it goes out through the tailings gate (Figure 7).

The seed is conveyed through the open portions of the disc by conveyor blades attached to spokes of the discs. These conveyor blades are small, rectangular pieces of metal bent at an angle so that they act as a screw conveyor as the discs revolve and push seed through the discs. The movement of the spokes and conveyor blades keep the seed mass thoroughly mixed so that the discs have a greater chance to lift out the short material.

The bottom section of the machine housing is shaped so as to fit closely to the lower edge of the revolving discs. This forces practically all of the seed to move through the center of the disc.

Types of Disc Pockets

Disc pockets are made in three shapes to solve specific separation problems. Each shape is available in several different sizes (Figure 8). The pocket size is indicated by its width in millimeters. The length or height of the pocket is about the same as the width, and the depth is about one half of the width.

V pocket — This type of pocket derives its name from vetch. It was designed to remove round vetch seed from wheat. The top edge of the pocket is straight and the bottom edge is round.

R pocket — This type of pocket derives its name from rice, since it was originally designed to remove cross-broken rice seed from whole seed. The top of the pocket is round and the bottom edge is straight. The R and the V types of pockets are made only in the smaller sizes, usually from 2 1/2 to 6 mm. The letters R and V are always followed by a number, which indicates the size in millimeters. For example, a V6 pocket is 6 mm wide and has a straight edge at the top and a round edge at the bottom.

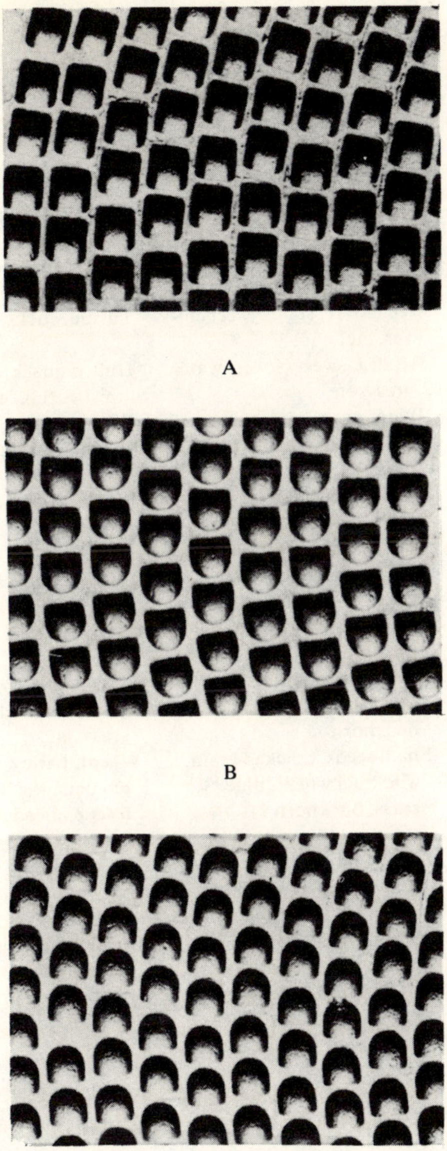

FIGURE 8. Types of disc indents. (A) Square pocket; (B) V pocket; (C) R pocket. (Courtesy of the Carter-Day Company, Minneapolis, 1979.)

Other pocket designations — Some discs are made with pockets larger than the R and V pockets. These pockets are square-faced and are designated only by letters, such as A, BB, M, and J. They are usually made only in sizes greater than 6 mm in width.

The general shape of the disc pocket selected should conform as closely as possible to the shape of the material to be lifted. This allows the smallest size possible to be used and results in a more precise and accurate separation to be made (Table 4).

Arrangement of Disc-Pocket Type and Size

A disc separator can be set up to have a combination of different pocket sizes or all the pockets of the same size.

Table 4
DISC POCKET SIZES USED FOR SEPARATING GRASSES, GRAINS, AND SEEDS

Type of pocket	Will lift	Will reject
V2 1/2	Alsike, timothy, ladino, white dutch, etc.	Buckhorn, Canada thistle
V3	Red clover, small sweet clover, etc.	Thistle, sticks, etc.
V3 1/2 V3 3/4	Alfalfa, sweet clover, water grass, etc.	Hulled quack, large thistle, flax, etc.
V4	Water grass, mustard, small cockle, small cracked grain, etc.	Wheat, barley, pin oats, large flax, etc.
V4 1/2 V5	Wild buckwheat, large cockle, vetch, cracked grain, onion bulblets	Wheat, oats, barley, rye, etc.
V5 1/2 V5 3/4 V6 V6 1/2	Flax, extra large cockle, vetch, small or broken wheat, large wild buckwheat, etc.	Unhulled quack, pin oats, wheat, barley, oats, rye, etc.
R3 1/2 R3 3/4 R4	Water grass, mustard, smartweed, bentgrass, buckhorn	Flax, fescue, ryegrass, orchard grass, etc.
R4 1/2 R5 R5 1/2 R6	Small seeds, cracked grain, wild buckwheat, bluegrass, buckhorn	Wheat, barley, oats, pin oats, rye, alta fescue, meadow brome, etc.
K L	Large wild buckwheat, broken wheat, broken barley, etc.	Wheat, barley, oats, pin oats, rye, etc.
M AC EE	Spring wheat, small or broken durum, pearled or broken barley, fescue, orchardgrass (Kentucky 31, Alta fescue — may use J& A)	Durum, large spring wheat, barley, oats, pin oats, etc.
J	Spring wheat, small durum, Kentucky 31 fescue, alta fescue, etc.	Oats, wild oats, barley, etc.
A MM	Wheat, winter wheat, durum, small barley, hulled oats, rye, safflower	Oats, wild oats, ragged barley, etc.
B	Barley	Oats, wild oats, sticks, etc.
RR-SS	Tailless barley	Sticks, stems, etc.
DD	Oats and all shorter grains	
AE-AD SS-DD	Peanuts	Sticks, stems, etc.

From Carter-Day Company, Minneapolis, Minn. With permission.

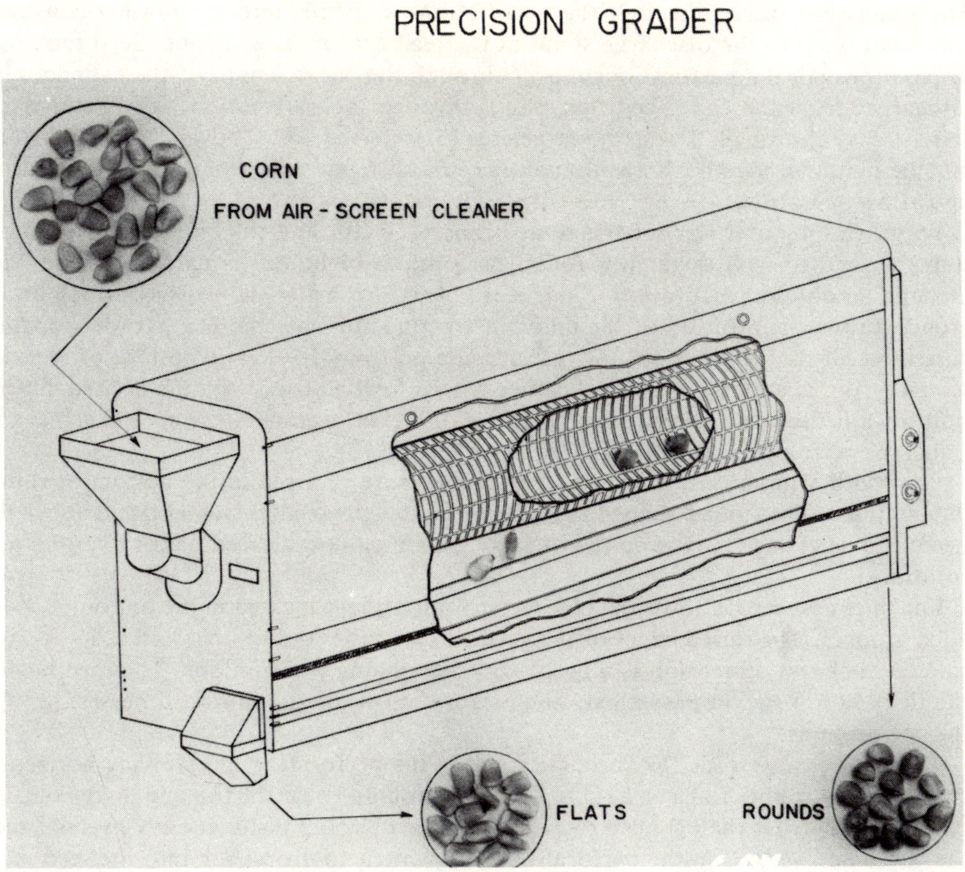

FIGURE 9. Width and thickness grader. (Courtesy of the Seed Technology Laboratory, Mississippi State University, Mississippi State, 1979.)

In a combination type of arrangement, the discs are grouped into sections, with the smaller size pocket disc installed near the seed intake and progressively larger pockets toward the discharge end. This arrangement provides a wider range in the size of materials that can be separated, but the capacity of the machine is usually reduced because of the fewer discs in each pocket-size group. If only one size of pocket is needed for making the required separation, it would be better to have all of the discs with the same type and size of pocket. This arrangement has a larger capacity than the combination arrangement.

The disc separator does not effectively length grade corn because the corn seed tends to lodge in the disc pockets.

Width and Thickness Separator

The width and thickness separator is widely used for size grading seeds such as corn, peanuts, and beans (Figure 9). They are also used effectively in removing contaminating weed seed from crop seed. The width and thickness separator is capable of making a very precise separation of seed according to either width and thickness dimensions.

The heart of the separator is a horizontal perforated metal cylinder that normally revolves at about 65 rpm. The seed enters the cylinder at one end just as in a cylinder length separator. The rotation of the cylinder causes a stirring of the seed so that each individual seed has an opportunity to align its proper dimension with the perforations.

The small seed passes through the perforations and falls into a vibrating conveyor which carries it to the discharge spout at the feed end of the machine. Seed too large to pass through the perforations moves through the entire length of the cylinder and discharges from the end. Seed that passes through the perforations is sometimes referred to as "throughs" and the larger seed is referred to as "overs". Several units may be mounted together so as to make several different size separations in a continuous-flow operation.

To grade long and flat seed such as beans by width and thickness, the seed must stand on end to pass through a round perforation or lie on its narrow side to pass through an oblong perforation. Since it is normal for a flat-shaped seed to lay on its broadest side, the position of the individual kernel must be changed in order for it to pass through the perforations. By placing the perforations in the bottom of recesses such as grooves and indents a seed can be forced to align itself with the desired dimension toward the perforation. Most width and thickness graders use special screens in which the perforations are placed in recesses (Figure 10).

For width separations, the round perforations are recessed in the bottom of round indentations. The funnel-shaped indentation causes the seed to stand on end above the perforations. The narrow seed passes through while the wide seed passes over the perforations.

For thickness separations, the oblong perforations are located in the bottom of spiralled grooves. The raised edges of the grooves cause the seed to turn so that its narrow side or thickness dimension is aligned with the oblong perforations. The seed having the thickest dimension passes over the perforation while the thin seed passes through the perforations.

To dislodge any seeds that become stuck in the perforation, a revolving horizontal shaft holding rubber flaps or a rubber roller is mounted near the top and on the outside of the cylinder. As the cylinder rotates, the flaps or roller dislodges any oversize seed that becomes wedged in the perforations, allowing it to drop back into the seed mass inside of the cylinder.

A large number of perforation sizes are available. They are available in four types:

1. Oblong perforations located in the bottom of spiralled grooves.
2. Oblong perforations in wire screen cylinders.
3. Large round perforations in individual recesses.
4. Small round perforations in ribbed cylinders. The ribs are pressed lengthwise in the cylinder wall and cause the seed to turn and tumble and align its proper dimension with the perforations.

All the perforations in a single cylinder are of the same size. Therefore each cylinder will make a single width or thickness separation. Changing cylinders for making a different separation is relatively easy.

Gravity Separators

Crop seed often contains immature or insect-damaged seed or inert material such as soil particles or small stones that cannot be separated from the crop seed because of their similarity in size, shape, etc. However, if there is a difference in the weights of the materials, a separation may be made with a gravity table (Figure 11). The gravity table employs a flotation process to separate light material from heavy material. The light material is actually floated away from the heavy material (Figure 11).

A gravity separator consists of two principal components — air chest and deck. The air chest is the bottom box-like part in which the fan and motor are located. The deck is mounted above the chest. The deck is a rectangular or triangular shaped table that

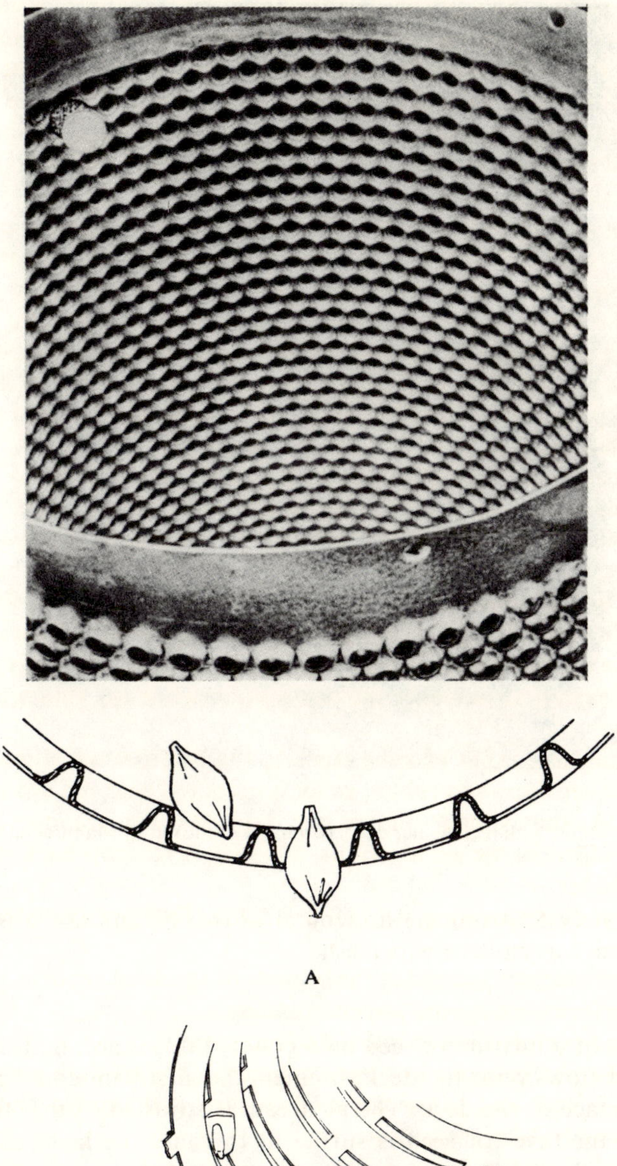

A

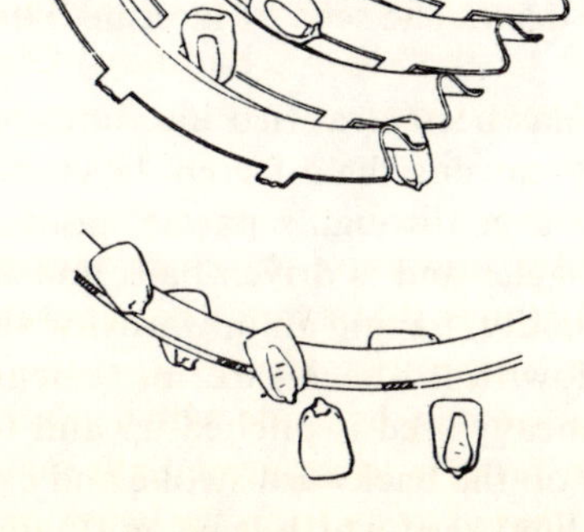

B

FIGURE 10. Sections of cylinders used in width and thickness graders, showing how perforations are placed in recesses to position seed to pass through more easily. (A) Round perforation; (B) Oblong perforation. (Courtesy of the Carter-Day Company, Minneapolis, 1979.)

FIGURE 11. Gravity table with rectangular deck. (Courtesy of the Oliver Manufacturing Company, Rocky Ford, Colo., 1978.)

slopes approximately 5° from the horizontal in two directions. It is covered with a porous material such as cloth or wire mesh.

Operation

The separation of a mixture of seed on a gravity table is accomplished in two steps. First, as the seed flows onto the deck, it enters the airstream coming up through the entire porous surface of the deck. The airstream is adjusted so that the volume of air passing through the layer of seed is sufficient to cause the layer of seed to stratify vertically. That is, the lightest seeds are on top and the heaviest ones are on the bottom next to the surface of the deck. The seed of intermediate weight is in between the light and heavy layers.

Second, after the seed has been stratified into horizontal layers, the layers must be separated. This is done by causing the different layers to move laterally to the edge of the deck, where they discharge through separate spouts. To accomplish this, the deck is mounted on inclined toggles and is driven back and forth by an eccentric drive system. The deck vibrates rapidly, having an upward-forward movement toward the high part of the deck and a downward-backward movement toward the low part. As the deck moves forward the heavy seed is pitched up and forward. The deck then moves backward and downward on the backward stroke and catches the heavy seed at a point farther up the deck. The light seed which is being floated on top of the heavy layer of seed by the air currents will flow by gravity to the lowest part of the deck.

A gravity table does not make a distinct separation between light and heavy seed. There is a zone of seed of intermediate weight between the light and heavy seed. This is commonly called the middling product. This middling product usually contains too much good seed to be discarded. It is a general practice to collect the middling product until enough has been collected to rerun and recover the good seed or recirculate the middling product back to the feed of the machine.

Uses

The gravity table can be thought of as an upgrading machine. It is capable of removing low quality seeds that cannot be separated by differences in dimensions or shape. Thus, the gravity table is often used in conditioning corn and beans to remove the bad seed from a seed lot and thereby improve the germination of the finished lot. The gravity table is used in processing alfalfa and clover seed to remove sand and small soil particles before the seed is processed through a roll mill. This eliminates wearing out the coverings on the rollers of the roll mill as rapidly as they would be if the sand and soil particles were not removed.

Seeds to be conditioned with a gravity table should first be cleaned with an air-screen cleaner. In some cases, such as with corn or beans, it may be advantageous to further grade the seed into uniform sizes before conditioning over a gravity table.

Deck Covering

Gravity table decks that are used in the seed industry are covered with a material like oxford cloth or closely woven wire. The cloth-covered deck is used in the separation of small seeds such as clovers. The wire deck is used when separating large seeds such as beans and corn. It is important that the right deck is used for a given seed species.

Some wire decks have coarse mesh wire about 1/2 × 1/2 in. on top of the closely woven wire. The purpose of the coarse mesh wire is to give more traction for smooth seeds. Some wire decks also have metal strips about 1/2 in. high and 3 in. apart on top of the wire. These strips are known as riffles. They serve as dams to prevent the smooth, heavy seed from moving toward the discharge edge before it is carried to the highest part of the deck.

Adjustments

There are five main adjustments provided on a gravity table:

1. Amount of air — The air should be adjusted so as to cause the layer of seed to become fluid and stratified. If excessive air is used, the strong air currents will keep the seed mixed and prevent it from becoming stratified, and the heavy seed will flow toward the low end of the deck with the light material. If insufficient air is used, the seed will not be stratified and the lighter seed will travel up the deck with the heavy seed.
2. End raise (elevation of the deck in the direction of the upward-forward motion) — The purpose of the end slope adjustment is to elevate the end of the deck so the heavy seed and the light seed can be separated after it becomes stratified.
3. Back raise (elevation of back side) — The purpose of the side slope adjustment is to regulate the rate at which the seed travels toward the discharge edge of the deck. An increase in the amount of side slope will increase the rate of travel. Seed mixtures with only a slight difference in weight require a relatively flat slope in order to give the seed more time to become stratified. A steep side slope can be used for mixtures that have a wide difference in specific gravities and stratify more quickly.
4. Speed of the eccentric — The eccentric speed determines the rate at which the seed will flow across the deck. An increase in the eccentric speed will cause the

seed to travel faster up the deck, while a decrease in speed will result in a slower rate of travel.

5. Rate of feed — It is important that a gravity table be fed uniformly and to its capacity. Feeding the machine too fast will cause an excess of good seed to be discharged with the light material. At no time should any portion of the deck be bare as this causes an uneven distribution of air through the deck.

All of the adjustments on a gravity table are very sensitive and must be coordinated. Only one adjustment should be made at a time after which the machine should be allowed to run several minutes for the adjustment to show its effect on the flow of seed across the deck. Other adjustments are then made as needed.

When a gravity table is out of adjustment all material on the deck will tend to move toward the highest or lowest part of the deck (Figure 12).

If all the material tends to move toward the highest part of the deck, it could be caused by one or more of the following.

1. Insufficient air currents
2. Not enough back raise
3. Not enough end raise
4. Speed of vibration too fast

If all of the material tends to move toward the lowest part of the deck, it could be caused by one or more of the following.

1. Excessive air currents
2. End raise too high
3. Back raise too high
4. Speed of vibration too slow

Steps in Setting Up a Gravity Table

1. Select the right type of deck. A cloth deck is used for small seeds such as the clovers, while a wire mesh deck is used for larger seeds.
2. Close the air valve.
3. Adjust the side and end tilt for a moderate amount of slope.
4. Start the machine.
5. Start the feed and adjust the vibration so that the seed will collect against the back rail until the deck is about half-covered.
6. Increase the air until the seed becomes stratified and the light material begins to flow downhill.
7. Adjust the feed again and then adjust the air to make the light material flow downhill. This operation is repeated until the deck is completely covered with a layer of seed.
8. The vibration and slope can now be adjusted to increase capacity.

Causes of Poor Operation

To obtain the best performance with a gravity table, it is important that it is properly installed. Some of the mistakes often made in the installation of a gravity table are listed below.

1. Weak foundation — The gravity table is a vibrating machine and should be installed on a firm foundation. Vibrations in the air chest caused by a weak foun-

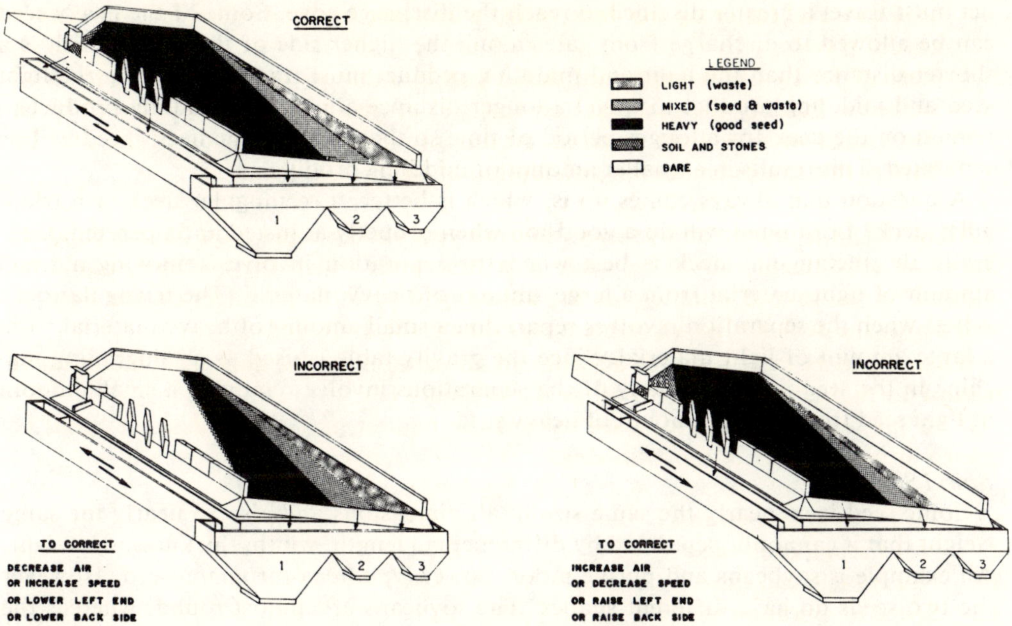

FIGURE 12. Patterns of materials on gravity table during correct and incorrect operation. (Courtesy of the Seed Technology Laboratory, Mississippi State University, Mississippi State, 1979.)

dation will reduce the effectiveness of the vibration of the deck in separating the seed.

2. Unclean air — A large amount of air is blown through the deck. If the air is laden with dust, the deck soon becomes clogged, which will prevent proper flotation of the seed. A clean filtered source should be provided. Often clean air is brought in from the outside of the building.
3. Fan running backward — Proper air delivery cannot be obtained if the fan is running backward. If a directional arrow cannot be located, remove the deck and start the motor. The fan blades should rotate toward the discharge opening in the fan housing.

   This appears to be an unlikely problem but it is actually quite common, especially on new installations or after electrical repairs in the processing plant.
4. Loose belts — The machine will not operate at a uniform speed if the drive belts are slack. Irregular speed will cause the layer of seed to surge across the deck.
5. Loose deck clamps — Loose deck clamps will allow the deck to bounce and the vibration will not be effective in making a separation of the seed.

Shape of Gravity Deck

Gravity separators are built with either a rectangular or triangular deck. Both operate by the same principles. The major difference between the two decks is the distance the heavy and light seed travel across the deck before they discharge.

On the triangular deck, the light seed travels a short distance down the lower side from the stratifying zone to the discharge side. The middling product (seed having intermediate weight) also travels a short distance but slightly farther than the lightest seed. The heavy seed travels a longer distance to discharge from the higher section of the deck.

The rectangular deck has a longer lower side, and the light seed and middling product must travel a greater distance to reach the discharge edge. Some of the heavy seed can be allowed to discharge from gates along the higher side of the deck, which is a shorter distance than the light and middling product must travel. Requiring the light seed and middling product to travel a longer distance allows the middling product to remain on the deck for a longer period of time so that more of the heavy seed will be separated. This results in a smaller amount of middling product.

A question that always comes up is, which is better, a rectangular deck or a triangular deck? Both types will do a good job when properly adjusted and operated. Generally the rectangular deck is best when the separation involves removing a small amount of light material from a large amount of heavy material. The triangular deck is best when the separation involves separating a small amount of heavy material from a large amount of light material. Since the gravity table is used as an upgrading machine in the seed industry, many of the separations involve removing a small amount of light seed from a large amount of heavy seed.

Spiral Separator

Some seed is so nearly the same size in all dimensions and also so nearly the same weight that it cannot be separated by differences in length, width, thickness, or weight. An example is soybeans and purple moon flower (*Ipomoea turbinata*) seed. However the two seeds do have different shapes. The soybeans are almost round, whereas the purple moon flower seeds have two flat sides and are shaped like a quarter section of an orange. If these two kinds of seed are placed on an inclined surface, the soybeans will roll much faster than the purple moon flower. Thus by taking advantage of the ability of one seed to roll faster than the other, the two may be separated with a spiral separator (Figure 13). In addition to its use in removal of certain weed seed, it also removes immature and damaged soybeans which are malformed and sometimes effects an upgrading of germination.

The spiral separator consists of a vertical column surrounded by one to five sheet metal spirals. The spirals are so constructed that they slope downward toward the center column.

The spiral separator is a finishing machine and should follow the air-screen cleaner in the conditioning sequence. In operation the seed is fed onto the spirals from a hopper at the top. As the seed rolls down the spirals, the round seed attains a higher velocity than the irregularly shaped seed. The momentum of the round seed increases until it runs over the edges of the inner spirals, drops into the outer housing and discharges through a spout under the machine. The other seeds or particles continue sliding down the inner spirals to the bottom and discharge through another spout.

This is a very simple machine. It operates completely by gravity and has no motor or moving parts. Once adjusted, it requires little supervision except for periodic checking to see that the space between the spirals has not become plugged with trash or dust. This can be a problem if the seed has not been well cleaned before passing through the spiral separator.

Treating and Bagging

After all or enough of the undesirable material has been removed from the seed to obtain an acceptable product, the crop seed is ready for treatment with a fungicide or insecticide. The treatment may be in the form of a powder, liquid, or slurry. Special machines are available for applying each type of treatment. If it is desired not to treat the seed, it goes directly to the bagging machine and is placed in a paper, plastic, or burlap bag.

Because of the rapid changes in seed treatments being used and changes in government regulations, no specific chemical treatment information is presented. Examples of commonly used seed treatment equipment are shown in Figures 14 and 15.

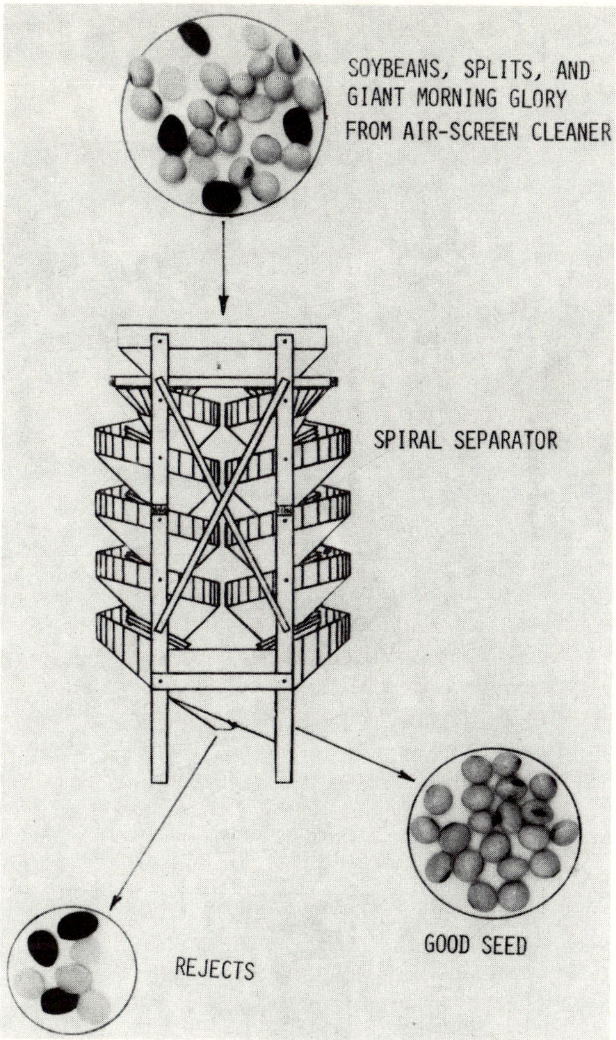

FIGURE 13. View of Spiral separator. (Courtesy of the Seed Technology Laboratory, Mississippi State University, Mississippi State, 1979.)

Shipping and Storage

After the seed is placed in bags, it is ready for shipment into the market channels or to be placed in storage to await shipment at a later date. The storage area should be of ample size so the seed lots can be kept separated. It should also be well-constructed, with a good roof and dry floor. The bags should be placed on pallets or on boards to keep them from being in contact with the concrete floor. A good loading area should be provided so the seed can be easily loaded into trucks. Warehousing and inventory control is a discipline within itself and is outside the scope of this chapter. However it is a very important part of the system and is worthy of further study by the serious seed conditioner.

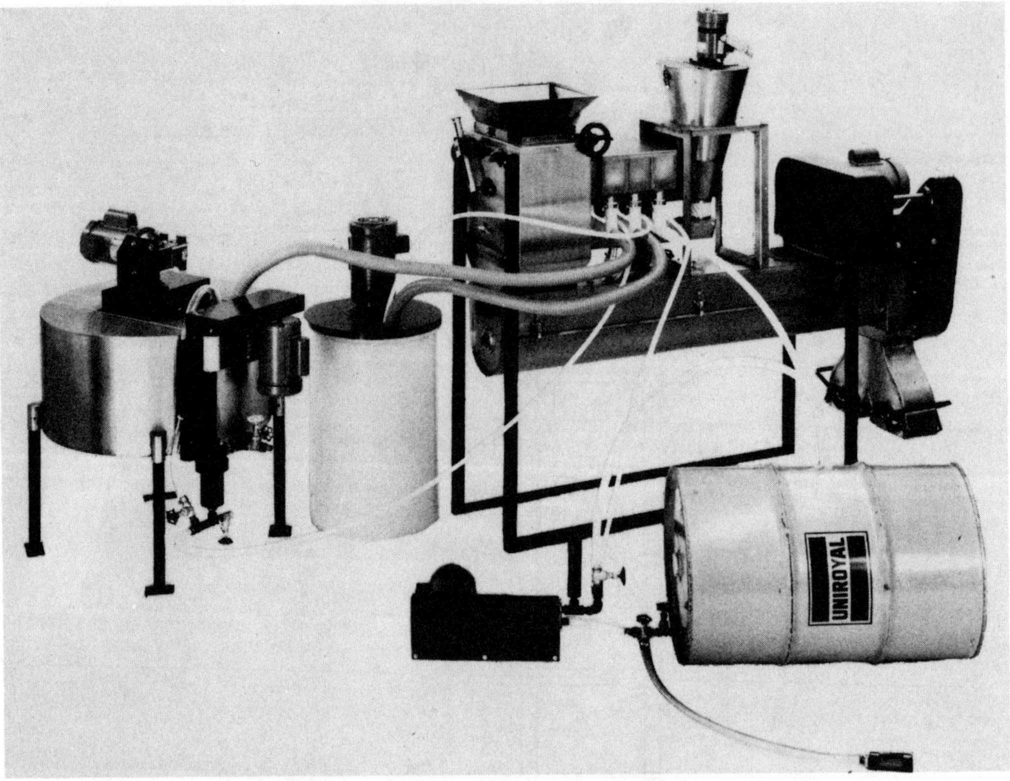

FIGURE 14. Seed treater with mixing auger for coating seeds. (Courtesy of Gustafson, Inc., Dallas, 1979.)

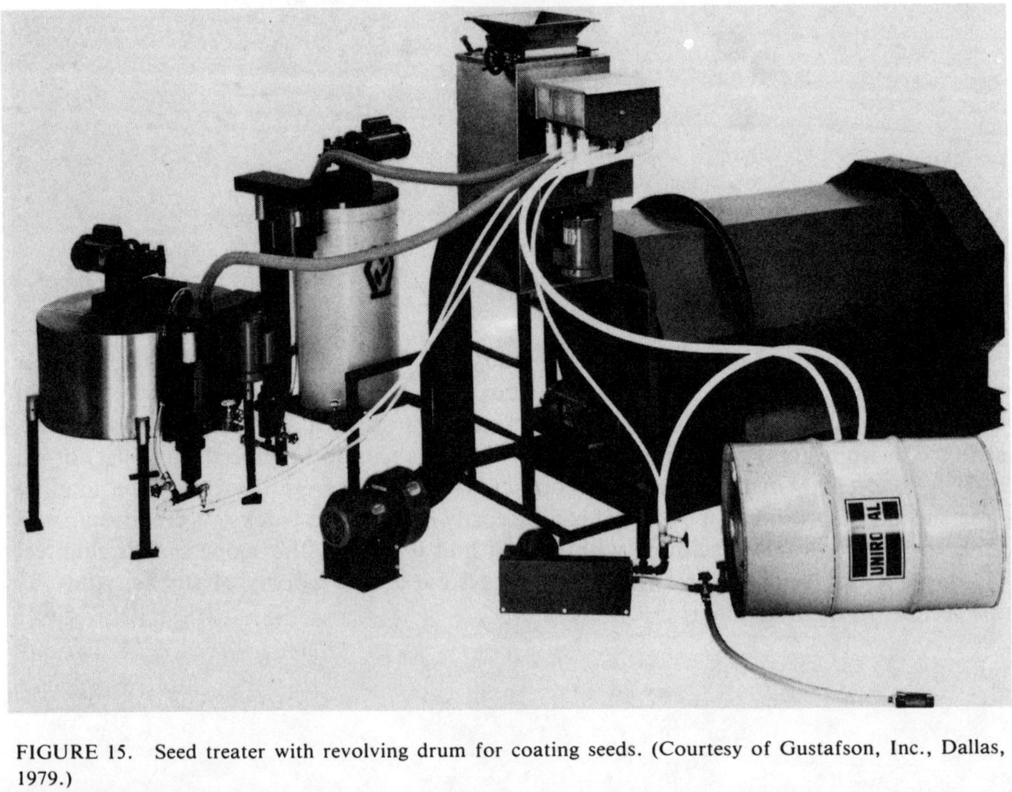

FIGURE 15. Seed treater with revolving drum for coating seeds. (Courtesy of Gustafson, Inc., Dallas, 1979.)

REFERENCES

1. **Boyd, A. H., Dougherty, G. M., Matthes, R. K., and Rushing, K. W.,** Seed Drying and Processing, in *Cereal Seed Technology,* Food and Agriculture Organization, Rome, 1975.
2. CEA, Carter-Day Company, Brochures Pertaining to Carter Equipment, Minneapolis, 1979.
3. Crippen Manufacturing Company, Catalog of Seed Cleaning Equipment, Alma, Mich., 1978.
4. Ferrell-Ross Company, Catalog of Seed Cleaning Equipment, Saginaw, Mich., 1978.
5. Gustafson Company, Brochures on Seed Treating Equipment, Dallas, 1978.
6. Oliver Manufacturing Company, Brochures on Gravity Tables, Rocky Ford, Colo., 1978.
7. Triple/S Dynamics, Brochures on Gravity Tables, Dallas, 1978.
8. **Vaughan, C. E., Gregg, B. R., and Delouche, J. C.,** *Seed Processing and Handling Handbook No. 1,* Seed Technology Laboratory, Mississippi State University, Mississippi State, 1968.
9. **Welch, G. B,** Seed Processing Equipment, *Miss. Agric. Exp. Stn. Bull. 520,* Mississippi State, 1954.
10. **Welch, G. B.,** *Beneficiamento De Sementes No Brasil,* Universidade do Estado de Mississippi/Ministerio da Agricultura do Brasil, Brasilia, 1973.

SEED TESTING: PURITY, GERMINATION, VIGOR, AND VARIETY

L. W. Woodstock

Seed testing is the science of evaluating seed quality for agricultural purposes.[1] By seed quality we mean those factors such as germination percentage, vigor, freedom from fungal and weed seed contamination, varietal purity, etc. which determine the overall planting value of the seed. Each of the above factors may be determined by special and appropriate tests. However, the objective of most routine seed testing is limited to determining the germination percentage, the purity composition, and the occurrence of noxious-weed seeds. Because seed is both an item of commerce and an input in agricultural production, information from seed tests is used for labeling purposes, establishing prices, and seed control work as well as an aid to the farmer in planting.[2]

Most seed tests are performed by analysts in laboratories which are members of one or more seed testing associations. The Association of Official Seed Analysts (AOSA), founded in 1908, has contributed in the U.S. and Canada to the development of seed testing rules and the standardization of their interpretation, influenced seed legislation at the federal and state levels, and encouraged the use of referee testing to evaluate testing procedures.[1] The International Seed Testing Association (ISTA) has played an active role in the development of seed testing rules and standardization of methods at the international level. It has encouraged and aided developing countries to establish seed testing laboratories and has organized training courses for seed analysts. The Society of Commercial Seed Technologists (SCST) has established high standards for commercial seed analysts by means of a thorough examination for membership leading to the title of Registered Seed Analyst. The Commercial Seed Analysts Association of Canada was formed to help analysts in technical problems and in keeping up with new testing methods. A list of seed testing associations is given in Table 1.

Seed is usually sampled for testing during storage or while being offered for sale. A representative sample drawn from a bulk seed lot and sent to a seed testing laboratory for analysis is referred to as the *submitted sample.* This is reduced at the laboratory to obtain a *working sample* the weight of which is specified by the AOSA[3] and ISTA[4] Rules and by the Federal Seed Act[5] for each kind of seed. Weights for agricultural and vegetable seeds are given in Table 2. It is essential that the submitted sample accurately represent the seed lot from which it is drawn and that the dividing procedures used for the working sample be precise and unbiased.[1,2] For sampling seed in bags and bins the use of double-tubed sleeve-type triers of various lengths is preferred.[2] Subsampling in the laboratory is generally done with mechanical dividers such as the Gamet Precision divider, the Boerner divider, or the Ottawa divider.[6] For separations such as nonfree-flowing chaffy grasses a Hay-Bates or similar divider is used.

The object of purity analysis is to determine the percent by weight of each component: (1) pure seed, the crop species being tested; (2) other crop seed, crop seeds other than the crop being tested; (3) weed seed, seeds from plants classified as weeds; and (4) inert matter such as chaff, stems, small stones, etc.[6] Purity testing is a meticulous hand and eye process in which the analyst identifies the various components by visual observation. In purity testing the analyst often uses only the simplest equipment such as spatula and forceps, magnifier, wide-field stereoscopic microscopes, and balances. Analysis of grasses is aided by use of a seedblower to remove lighter material and by diaphanoscope in which a strong beam of light coming from below the sample helps the analyst determine whether certain kinds of florets contain caryopses.[6] Some weed seeds are regarded as noxious or particularly objectionable and each state as well as the federal government has established a list of noxious-weed seeds.[2,6] A list of noxious-weed seeds arranged by common name is given in Table 3.[7,8] In contrast to the other purity components, the rate of occurrence of noxious weed seed is expressed as the number of seed of each kind per ounce or per pound.

In seed laboratory practice, germination is defined as the emergence and development from the seed embryo of those essential structures which, for the kind of seed in question, are indicative of the seed's capacity to produce a normal plant under favorable conditions.[6] The environmental conditions of moisture, substrate, temperature, and light should therefore not only favor maximum germination, but should permit seedling development to a stage where normal and abnormal seedlings can be distinguished. The requirements often include special treatments to overcome dormancy which would otherwise interfere with interpretation of the germination test.

A type of dormancy especially common in legumes is hardseededness where an impermeable seed coat prevents water absorption. Hard seeds are reported as part of the total germination percentage. The viability of dormant tree seeds may be determined by germinating excised embryos on blotters or by the use of Tetrazolium compounds where dehydrogenase enzymes in living cells reduce the Tetrazolium to a red water-insoluble formazan. The TZ Test is also used to detect weakness in a number of other crop seeds.

Germination requirements for indicated seed kinds are given in Table 4. In seed laboratories, most germination tests are on sand, soil, paper towels, blotters, or creped cellulose paper. The substrate should be free of toxic chemicals. The seed plus substrate are placed in germination chambers which provide the specified conditions of temperature and light, usually at a high relative humidity. First and final counts of germinated seedlings are made at times indicated by the rules (Table 4) with only "normal" seedlings contributing to the germination percentage. Criteria for distinguishing between "normal" and "abnormal" seedlings are specific for each seed kind and successful evaluation depends upon experience of the analyst. Photographs like those in Figures 1 and 2 for spinach and *Crotalaria* aid the analyst in making evaluations. Generally, a normal seedling is one having well-balanced growth in all essential parts without lesions or defects, whereas abnormal seedlings are those exhibiting stunting or weakness, broken parts, malformations, or other abnormalities.[6] Causes of seedling abnormality may include declining vitality, infection with pathogenic organisms, mechanical injury, insect injury, chemical treatments, use of metal germination trays, substrate toxicity, frost damage, and mineral deficiency.[6]

Purity, noxious weed, and germination are the standard tests routinely performed by seed testing laboratories.[1,2] However, many additional tests, usually performed on request, and lacking the "official" status of tests in the rules reveal important aspects of seed quality. These include tests for vigor, variety, pathological quality of "seed health", moisture, and effectiveness of seed treatments and inoculation of legume seed.

Seed vigor has been tentatively defined as the sum total of all those properties in seeds which, upon planting, result in rapid and uniform production of healthy seedlings under a wide range of environment including both favorable and stress conditions.[9] Information from vigor tests is used to reveal deterioration or injury during harvesting, processing, transporting and storage; aid in the detection and removal of causes of injury; supply essential information needed by the farmer or home gardener relative to planting rates and dates; and aid in the better marketing, distribution, and use of seeds. Some vigor tests are listed in Table 5.

Seeds are tested for variety to determine whether they have the characteristics of the variety represented on the label.[6] Variety tests include: (1) visual observations of seed characteristics such as color, pubescence, shape of the seed and its parts, etc.; (2) seedling characteristics such as leaf shape, color and venation patterns, length of internodes, color of the crown area; (3) grow-out tests in the field or greenhouse; (4) chemical tests such as the phenol test for wheat and bluegrass; (5) use of ultraviolet light to distinguish between annual and perennial ryegrass where the roots of the former fluoresce under UV; (6) chromosome counts for tetraploid grass and legume varieties; (7) disease resistance; and (8) electrophoretic and chromatographic methods for separation of plant constituents such as proteins.[1]

The sanitary condition of seeds involves the seed-borne fungi, bacteria, viruses, eelworms, and insects which cause disease and injury.[2] Although pathological testing is routine in some European countries, this is not the case in North America. Methods for pathological testing are visual examination, agar media tests, blotter methods, serological techniques, bacteriophage, and plant injection tests.[1] The identification and control of seed-borne diseases is described in another section of this Handbook.

Seed moisture is determined by: (1) the air oven method of 130°C for 1 hr for most seed kinds, (2) the air oven method of 105°C for 16 hr or (3) the toluene distillation method depending on seed kind and as specified in the ISTA Rules.[4]

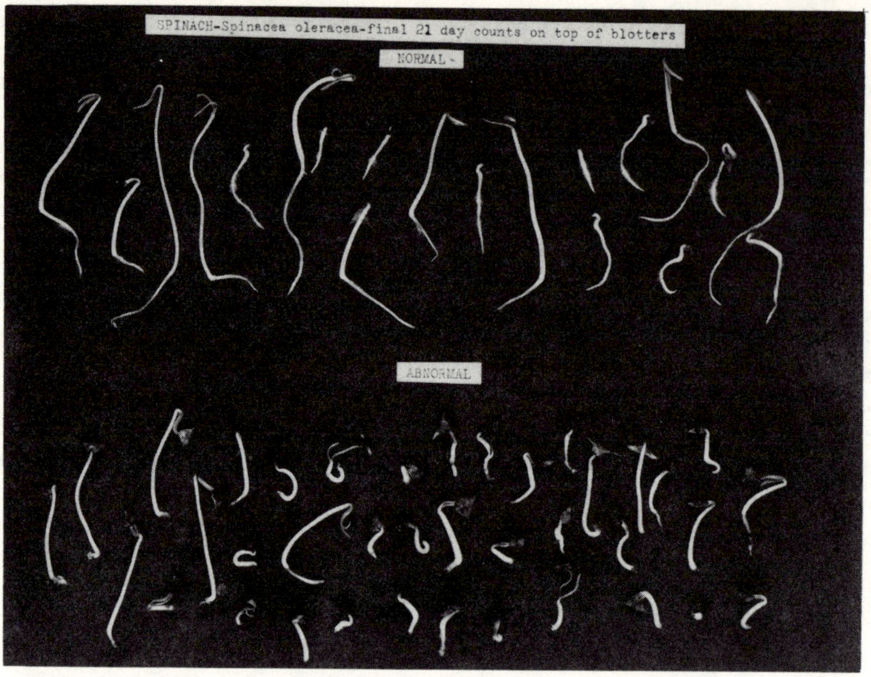

FIGURE 1. Normal and abnormal seedlings of spinach. Photo No. 2509. (Courtesy of Agricultural Marketing Service, Seed Standardization Branch, U.S. Department of Agriculture, Washington, D.C.)

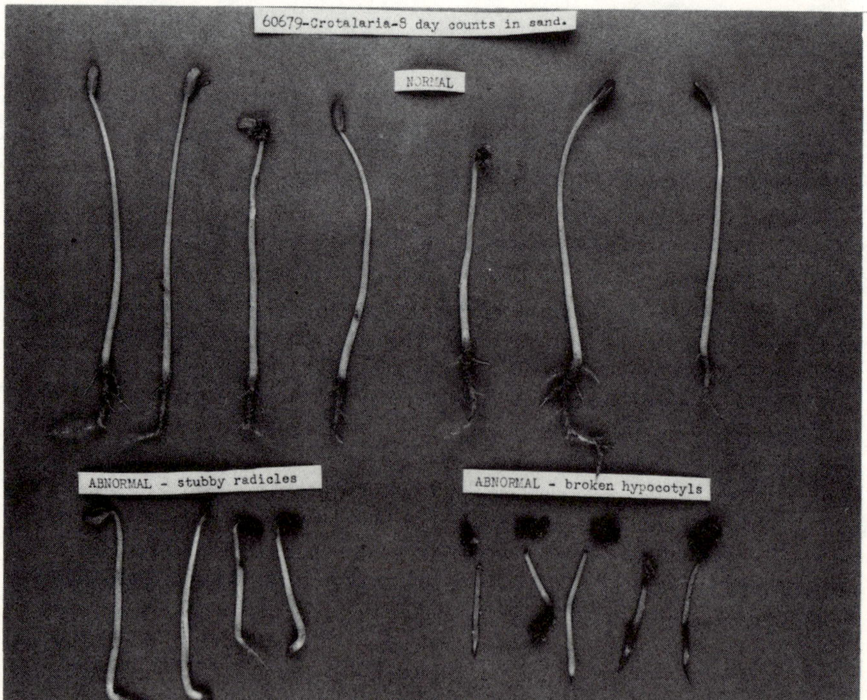

FIGURE 2. Normal and abnormal seedlings of *Crotalaria*. Photo No. 60679. (Courtesy of Agricultural Marketing Service, Seed Standardization Branch, U.S. Department of Agriculture, Washington, D.C.)

Table 1
SEED TESTING ASSOCIATIONS

Organization	Year founded	Membership	Publications	Secretariat/ editorial office
Association of Official Seed Analysts (AOSA)	1908	Official state and federal laboratories, U.S. and Canada	*J. Seed Technol.* Newsletter, Handbooks	L. O. Copeland, Ed., Dept. of Crop and Soil Sciences, Michigan State University, East Lansing, 48824
International Seed Testing Association (ISTA)	1921—1924	By country	*Seed Sci. Technol.*	ISTA-Secretariat, Reckenholz, P.O. Box 412, CH-8046 Zurich, Switzerland
Society of Commercial Seed Technologists (SCST)	1922	Registered seed technologists, U.S. and Canada		Mary K. Bristol, Sec./ Treas., AMM Seed Testing, Box 1771, Fresno, Calif., 93717
Commercial Seed Analysts Association of Canada (CSAAC)	1944—1945	Canadian seed analysts	*Maple Leaf*	Gail Livingstone, Sec., 22 Langfield Crescent, Roxdale, Ontario, M9V 3L7, Canada

Table 2
WEIGHT OF WORKING SAMPLE[a]

Name of seed	Minimum weight for purity analysis (g)	Minimum weight for noxious-weed seed examination (g)	Approximate[b] number of seeds per gram (no.)
Agricultural Seed			
Alfalfa — *Medicago sativa*	5	50	500
Alfilaria — *Erodium cicutarium*	5	50	440
Alyceclover — *Alysicarpus vaginalis*	5	50	665
Bahiagrass — *Paspalum notatum*			
Var. Pensacola	5	50	600
All other vars	7	50	365
Barley — *Hordeum vulgare*	100	500	30
Barrelclover — *Medicago tribuloides*	50	300	—
Bean			
Adzuki — *Phaseolus angularis*	200	500	11
Field — *Phaseolus vulgaris*	500	500	4
Mung — *Phaseolus aureus*	100	500	24
Beet, field — *Beta vulgaris*	50	500	55
Beet, sugar — *Beta vulgaris*	50	500	55
Beggarweed, Florida — *Desmodium tortuosum*	5	50	440
Bentgrass			
Colonial (incl. Astoria and Highland) — *Agrostis tenuis*	¼	2.5	13,000
Creeping — *Agrostis palustris*	¼	2.5	13,515
Velvet — *Agrostis canina*	¼	2.5	18,180
Bermudagrass — *Cynodon dactylon*	1	10	3,930
Bermudagrass, giant — *Cynodon dactylon* var. *aridus*	1	10	2,950
Bluegrass			
Bulbous — *Poa bulbosa*	4	40	585
Canada — *Poa compressa*	½	5	5,050
Glaucantha — *Poa glaucantha*	1	10	—
Kentucky (all vars.) — *Poa pratensis*	1	10	3,060
Nevada — *Poa nevadensis*	1	10	2,305
Rough — *Poa trivialis*	½	5	4,610
Texas — *Poa arachnifera*	1	10	2,500
Wood — *Poa nemoralis*	½	5	4,330
Bluestem			
Big — *Andropogon gerardi*	7	70	320
Little — *Andropogon scoparius*	5	50	525
Sand — *Andropogon hallii*	10	100	215
Yellow — *Andropogon ischaemum*	1	10	1,945

Table 2 (continued)
WEIGHT OF WORKING SAMPLE[a]

Name of seed	Minimum weight for purity analysis (g)	Minimum weight for noxious-weed seed examination (g)	Approximate[b] number of seeds per gram (no.)
Brome			
Field — *Bromus arvensis*	5	50	465
Mountain — *Bromus marginatus*	20	200	140
Smooth — *Bromus inermis*	7	70	315
Broomcorn — *Sorghum vulgare* var. *technicum*	40	400	60
Buckwheat — *Fagopyrum esculentum*	50	500	45
Buffalograss — *Buchloe dactyloides*			
(Burs)	20	200	110
(Caryopses)	3	30	740
Buffelgrass — *Pennisetum ciliare*			
(Fascicles)	6	60	355
(Caryopses)	2	20	1,940
Burclover, California — *Medicago hispida* (in bur)	50	500	—
Burclover, California — *Medicago hispida* (out of bur)	7	70	375
Burclover, spotted — *Medicago arabica* (in bur)	50	500	50
Burclover, spotted — *Medicago arabica* (out of bur)	5	50	550
Burnet, little — *Sanguisorba minor*	25	250	110
Buttonclover — *Medicago orbicularis*	7	70	335
Canarygrass — *Phalaris canariensis*	20	200	150
Canarygrass, reed — *Phalaris arundinacea*	2	20	1,185
Carpetgrass — *Axonopus affinis*	1	10	2,230
Castorbean — *Ricinus communis*	500	500	5
Chess, soft — *Bromus mollis*	5	50	555
Chickpea — *Cicer arietinum*	500	500	2
Clover			
Alsike — *Trifolium hybridum*	2	20	1,500
Berseem — *Trifolium alexandrinum*	5	50	455
Cluster — *Trifolium glomeratium*	1	10	2,925
Crimson — *Trifolium incarnatum*	10	100	330
Kenya — *Trifolium semipilosum*	2	20	—
Ladino — *Trifolium repens*	2	20	1,935
Lappa — *Trifolium lappaceum*	2	20	1,500
Large hop — *Trifolium procumbens (T. campestre)*	1	10	5,435

Table 2 (continued)
WEIGHT OF WORKING SAMPLE[a]

Name of seed	Minimum weight for purity analysis (g)	Minimum weight for noxious-weed seed examination (g)	Approximate[b] number of seed per gram (no.)
Persian — *Trifolium resupinatum*	2	20	1,415
Red — *Trifolium pratense*	5	50	600
Rose — *Trifolium hirtum*	7	70	360
Small hop (Suckling) — *Trifolium dubium*	2	20	1,950
Strawberry — *Trifolium fragiferum*	5	50	635
Sub — *Trifolium subterraneum*	25	250	120
White — *Trifolium repens*	2	20	1,500
Corn			
Field — *Zea mays*	500	500	3
Pop — *Zea mays* var. *everta*	500	500	—
Cotton — *Gossypium* spp.	300	500	8
Cowpea — *Vigna sinensis*	300	500	8
Crambe — *Crambe abyssinica*	25	250	—
Crested dogtail — *Cynosurus cristatus*	2	20	1,900
Crotalaria			
Lance — *Crotalaria lanceolata*	7	70	375
Showy — *Crotalaria spectabilis*	25	250	80
Slenderleaf — *Crotalaria intermedia*	10	100	205
Striped — *Crotalaria mucronata*	10	100	215
Sunn — *Crotalaria juncea*	75	500	35
Crownvetch — *Coronilla varia*	10	100	305
Dallisgrass — *Paspalum dilatatum*	4	40	620
Dichondra — *Dichondra repens*	5	50	470
Dropseed, sand — *Sporobolus cryptandrus*	¼	2.5	12,345
Emmer — *Triticum dicoccum*	100	500	25
Fescue			
Chewings — *Festuca rubra* var. *commutata*	3	30	900
Hair — *F. capillata*	1	10	—
Hard — *F. ovina* var. *duriuscula*	2	20	1,305
Meadow — *F. elatior*	5	50	495
Red — *F. rubra*	3	30	900
Sheep — *F. ovina*	2	20	1,165
Tall — *F. arundinacea*	5	50	455
Flax — *Linium usitatissimum*	15	150	180
Grama			
Blue — *Bouteloua gracilis*	2	20	1,595
Side-oats — *B. curtipendula*			
(Other than caryopses)	6	60	350
(Caryopses)	2	20	1,605

Table 2 (continued) WEIGHT OF WORKING SAMPLE[a]

Name of seed	Minimum weight for purity analysis (g)	Minimum weight for noxious-weed seed examination (g)	Approximate[b] number of seeds per gram (no.)
Guat — *Cyamopsis tetragonoloba*	75	500	35
Guineagrass — *Panicum maximum*	2	20	2,205
Hardinggrass — *Phalaris stenoptera*	3	30	750
Hemp — *Cannabis sativa*	50	500	45
Indiangrass, yellow — *Sorghastrum nutans*	7	70	395
Indigo, hairy — *Indigofera hirsuta*	7	70	435
Japanese lawngrass — *Zoysia japonica*	2	20	1,325
Johnsongrass — *Sorghum halepense*	10	100	265
Kudzu — *Pueraria thunbergiana*	25	250	80
Lentil — *Lens culinaris*	120	500	14—23
Lespedeza			
Korean — *Lespedeza stipulacea*	5	50	525
Sericea or Chinese — *L. cunceta* (L. sericea)	3	30	820
Siberian — *L. hedysaroides*	3	30	820
Striate (Common, Kobe, Tenn. 76) *L. striata*	5	50	750
Lovegrass, sand — *Eragrostis trichodes*	1	10	3,585
Lovegrass, weeping — *E. curvula*	1	10	3,270
Lupine			
Blue — *Lupinus angustifolius*	500	500	7
White — *L. albus*	500	500	7
Yellow — *L. luteus*	300	500	9
Manilagrass — *Z. matrella*	2	20	—
Meadow foxtail — *Alopecurus pratensis*	3	30	895
Medick, black — *Medicago lupulina*	5	50	585
Millet			
Browntop — *Panicum ramosum*	8	80	315
Foxtail, Such as Common, White Wonder, German, Hungarian, Siberian, or Golden — *Setaria italica*	5	50	480
Japanese — *Echinochloa crusgalli* var. *frumentacea*	9	90	315
Pearl — *Pennisetum glaucum*	15	150	180
Proso — *Panicum miliaceum*	15	150	185
Molassesgrass — *Melinis minutiflora*	½	5	7,750

Table 2 (continued)
WEIGHT OF WORKING SAMPLE[a]

Name of seed	Minimum weight for purity analysis (g)	Minimum weight for noxious-weed seed examination (g)	Approximate[b] number of seeds per gram (no.)
Mustard	2	20	1,255
Black — *Brassica nigra*			
India — *B. juncea*	5	50	624
White — *B. hirta*	15	150	160
Napiergrass — *Pennisetum purpureum*	5	50	—
Oat — *Avena* spp.	75	500	35—50
Oatgrass, tall — *Arrhenatherum elatius*	6	60	415
Orchardgrass — *Dactylis glomerata*	3	30	945
Panicgrass, blue — *Panicum antidotale*	2	20	1,370
Panicgrass, green — *P. maximum* var. *trichoglume*	2	20	1,305
Peanut — *Arachis hypogaea*	500	500	1—3
Pea, field — *Pisum sativum* var. *arvense*	500	500	4
Rape			
Annual — *Brassica napus* var. *annua*	7	70	345
Bird — *B. campestris*	7	70	425
Turnip — *B. campestris* vars	5	50	535
Winter — *B. napus* var. *biennis*	10	100	230
Redtop — *Agrostis alba*	¼	2.5	10,695
Rescuegrass — *Bromus catharticus*	20	200	115
Rhodesgrass — *Chloris gayana*	1	10	4,725
Rice — *Oryza sativa*	50	500	65
Ricegrass, Indian — *Oryzopsis hymenoides*	7	70	355
Roughpea — *Lathyrus hirsutus*	75	500	40
Rye — *Secale cerale*	75	500	40
Ryegrass			
Annual (Italian) — *Lolium multiflorum*	5	50	420
Perennial — *L. perenne*	5	50	530
Wimmera — *L. rigidum*	5	50	—
Safflower — *Carthamus tinctorius*	100	500	30
Sainfoin— *Onobrychis viciaefolia*	50	500	50
Sesame — *Sesamum indicum*	7	70	360
Sesbania — *Sesbania exaltata*	25	250	105
Smilo — *Oryzopsis miliacea*	2	20	2,010
Sorghum — *Sorghum vulgare*	50	500	55
Sorghum almum — *S. almum*	15	150	150
Sorghum — sudangrass hybrid, *S. vulgare* × *S. sudanense*	50	500	55

Table 2 (continued)
WEIGHT OF WORKING SAMPLE[a]

Name of seed	Minimum weight for purity analysis (g)	Minimum weight for noxious-weed seed examination (g)	Approximate[b] number of seed per gram (no.)
Sorgrass[c]	15	150	135
Sourclover — *Melilotus indica*	5	50	660
Soybean — *Glycine max*	500	500	6—13
Spelt — *Triticum spelta*	100	500	25
Sudangrass — *Sorghum sudanense*	25	250	100
Sunflower (Cult.) — *Helianthus annuus*	100	500	—
Sweetclover			
White — *Melilotus alba*	5	50	570
Yellow — *M. officinalis*	5	50	570
Sweet vernalgrass — *Anthoxanthum odoratum*	2	20	1,600
Switchgrass — *Panicum virgatum*	4	40	570
Timothy — *Phleum pratense*	1	10	2,565
Tobacco — *Nicotiana tabacum*	½	5	15,625
Trefoil			
Big — *Lotus uliginosus* (L. major)	2	20	1,945
Birdsfoot — *L. corniculatus*	3	30	815
Triticale — *Triticosecale*	100	500	—
Vaseygrass — *Paspalum urvillei*	3	30	970
Veldtgrass — *Ehrharta calycina*	4	40	655
Velvetbean — *Stizolobium deeringianum*	500	500	2
Velvetgrass — *Holcus lanatus*	1	10	3,360
Vetch			
Common — *Vicia sativa*	150	500	19
Hairy — *V. villosa*	75	500	35
Hungarian — *V. pannonica*	100	500	24
Monantha — *V. articulata (V. monantha)*	100	500	—
Narrowleaf — *V. angustifolia*	50	500	60
Purple — *V. benghalensis*	100	500	22
Woolypod — *V. dasycarpa*	100	500	25
Wheat			
Common — *Triticum aestivum*	100	500	25
Club — *T. compactum*	100	500	25
Durum — *T. durum*	100	500	25
Polish — *T. polonicum*	100	500	25
Poulard — *T. turgidum*	100	500	25
Wheatgrass			
Beardless — *Agropyron inerme*	8	80	275
Fairway crested — *A. cristatum*	4	40	685
Standard crested — *A. desertorum*	5	50	425

Table 2 (continued)
WEIGHT OF WORKING SAMPLE[a]

Name of seed	Minimum weight for purity analysis (g)	Minimum weight for noxious-weed seed examination (g)	Approximate[b] number of seeds per gram (no.)
Intermediate — *A. intermedium*	15	150	175
Pubescent — *A. tricophorum*	15	150	180
Siberian — *A. sibiricum*	10	50	—
Slender — *A. trachycaulum*	7	70	295
Streambank — *A. riparium*	10	50	370
Tall — *A. elongatum*	15	150	165
Western — *A. smithii*	10	100	250
Wildrye			
Canada — *Elymus canadensis*	11	110	190
Russian — *E. junceus*	6	60	360
Vegetable Seed			
Artichoke — *Cynara scolymus*	100	500	24
Asparagus — *Asparagus officinalis*	100	500	25
Asparagusbean — *Vigna sesquipedalis*	300	500	8
Beans			
Garden — *Phaseolus vulgaris*	500	500	4
Lima — *P. lunatus* var. *macrocarpus*	500	500	2
Runner — *P. coccineus*	500	500	1
Beet — *Beta vulgaris*	50	300	60
Broadbean — *Vicia faba*	500	500	—
Broccoli — *Brassica oleracea* var. *botrytis*	10	50	315
Brussels sprouts — *B. oleracea* var. *gemmifera*	10	50	315
Burdock, great — *Arctium lappa*	15	150	—
Cabbage — *B. oleracea* var. *capitata*	10	50	315
Cabbage, Chinese (Petsai) — *B. pekinensis*	5	50	635
Cabbage, tronchuda — *B. oleracea* var. *tronchuda*	10	100	—
Cardoon — *Cynara cardunculus*	100	500	—
Carrot — *Daucus carota*	3	50	825
Cauliflower — *Brassica oleracea* var. *botrytis*	10	50	315
Celeriac — *Apium graveolens* var. *rapaceum*	1	25	2,520
Celery — *A. graveolens* var. *dulce*	1	25	2,520
Chard, Swiss — *Beta vulgaris* var. *cicla*	50	300	60
Chicory — *Cichorium intybus*	3	50	940
Chives — *Allium schoenophrasum*	5	50	—

Table 2 (continued)
WEIGHT OF WORKING SAMPLE[a]

Name of seed	Minimum weight for purity analysis (g)	Minimum weight for noxious-weed seed examination (g)	Approximate[b] number of seed per gram (no.)
Citron — *Citrulius vulgaris*	200	500	11
Collards — *Brassica oleracea* var. *acephala*	10	50	315
Corn, sweet — *Zea mays*	500	500	—
Cornsalad — *Valerianella locusta* var. *olitoria*			
Vars. Fullhearted and Dark Green Fullhearted	5	50	—
All other varieties	10	50	380
Cowpea — *Vigna sinensis*	300	500	8
Cress			
Garden — *Lepidium sativum*	5	50	425
Upland — *Barbarea verna*	2	35	1,160
Water — *Rorippa nasturtium aquaticum*	1	25	5,170
Cucumber — *Cucumis sativus*	75	500	40
Dandelion — *Taraxacum officinale*	2	35	1,240
Eggplant — *Solanum melongena* var. *esculentum*	10	50	230
Endive — *Cichorium endivia*	3	50	940
Kale — *Brassica oleracea* var. *acephala*	10	50	315
Kale, Chinese — *Brassica oleracea* var. *alboglabra*	10	50	—
Siberian — *B. napus* var. *pabularia*	10	50	—
Kohlrabi — *B. oleracea* var. *gongylodes*	10	50	315
Leek — *Allium porrum*	7	50	395
Lettuce — *Lactuca sativa*	3	50	890
Muskmelon (cantaloupe) — *Cucumis melo*	50	500	45
Mustard, India — *Brassica juncea*	5	50	625
Mustard, spinach — *B. perviridis*	5	50	535
Okra — *Hibiscus esculentus*	100	500	19
Onion — *Allium cepa*	7	50	340
Onion, Welsh — *A. fistulosum*	10	50	—
Pak-choi — *Brassica chinensis*	5	50	635
Parsley — *Petroselinum hortense (P. crispum)*	5	50	650
Parsnip — *Pastinaca sativa*	5	50	430
Pea — *Pisum sativum*	500	500	3
Pepper — *Capsicum* spp.	15	150	165
Pumpkin — *Cucurbita pepo*	500	500	5
Radish — *Raphanus sativus*	30	300	75
Rhubarb — *Rheum rhaponticum*	50	300	60
Rutabaga — *Brassica napus* var. *napobrassica*	5	50	430

Table 2 (continued)
WEIGHT OF WORKING SAMPLE[a]

Name of seed	Minimum weight for purity analysis (g)	Minimum weight for noxious-weed seed examination (g)	Approximate[b] number of seeds per gram (no.)
Salsify — *Tragopogon porrifolius*	50	300	65
Sorrel — *Rumex acetosa*	2	35	1,080
Soybean — *Glycine max*	500	500	6—13
Spinach — *Spinacia oleracea*	25	150	100
Spinach, New Zealand — *Tetragonia expansa*	200	500	13
Squash — *Cucurbita moschata* and *C. maxima*	200	500	14
Tomato — *Lycopersicon esculentum*	5	50	405
Tomato, husk — *Physalis pubescens*	2	35	1,240
Turnip — *Brassica rapa*	5	50	535
Watermelon — *Citrullus vulgaris*	200	500	11

[a] From Rules and Regulations under the Federal Seed Act, Table 1, 1975.[5] Weights for herb, tree, and shrub seeds may be found in Table 1, AOSA Rules for Testing Seeds, 1978.[3]

[b] The number of seeds per gram or the weight of 1000 seeds can be calculated in the laboratory for individual lots from the sample submitted.

[c] Rhizomatous derivatives of a johnsongrass X sorghum cross or a johnsongrass X sudangrass cross.

Table 3
NOXIOUS-WEED SEEDS

English or common name[a]	Latin name[b]	State or states in which noxious[c]
Acuate	*Lagascea mollis*	Hawaii
Alyssum, hoary	*Berteroa incana*	Mich., Minn., Mont., Wis.
Anoda, spurred	*Anoda cristata*	N.C.
Aramina Caesarweed	*Urena lobata*	Hawaii
Ballcress (see hoary cress)		
Balloonvine	*Cardiospermum halicacabum*	Ala., Ark., Miss.
Barbwiregrass	*Cymbopogon refractus*	Hawaii
Barley, wild	*Hordeum* spp.	Hawaii
Beancaper, Syrian	*Zygophyllum fabago*	Calif., Idaho
Bedstraw	*Galium* spp.	Conn., Mass., N.Y., Pa., Vt.
Bedstraw	*G. asprellum*	Conn., N.Y., Mass., Pa., Vt.
Bellflower, creeping	*Campanula rapunculoides*	Mont.
Bermudagrass	*Cynodon* spp.	Ala., Fla., Md., Miss., Tex., Va.
Bermudagrass	*C. dactylon*	Ala., Ark., Washington, D.C., Fla., Ga., Idaho, La., Md., Miss., N.J., N.C., S.C., Tex., Utah, Va., Wash.
Bermudagrass, giant	*C. dactylon* var. *Aridus*	Ala., Washington, D.C., Fla., Md., Miss., Tex., Va.
Betony, marsh (see woundwort)		
Biddy biddy (see piripiri)		
Bindweed Wild morningglory	*Convolvulus* spp.	Pa., N.C., Utah
Bindweed, field Wild morningglory European morningglory Creeping Jennie (see bindweed)	*C. arvensis*	In Washington, D.C. and all States *except* Ky.
Bindweed, hedge Wild morningglory (see bindweed)	*C. sepium*	Ala., Ark., Ga., Kans., La., Miss., Hawaii, Mo., N.C., N.D., N.J., Okla., Pa., S.C., S.D., Tenn., Tex., Wash., Utah
Blackeyesusan	*Rudbeckia hirta*	Hawaii
Bluegrass, annual	*Poa annua*	Conn., Del. Fla., Ky., Md., Mass., Washington, D.C., N.Y., S.D., Va.
Blueweed, Texas Blue sunflower	*Helianthus ciliaris*	Ariz., Ark., Calif., Ga., Hawaii, Kans., La., N.M., Okla., S.C., Tex.
Bromes (see cheat or chess)		

Table 3 (Continued)
NOXIOUS-WEED SEEDS

English or common name[a]	Latin name[b]	State or states in which noxious[c]
Brome, downy	*Bromus tectorum*	Wis.
Broom, French	*Cytisus monspessulanus*	Hawaii
Broom, Scotch	*C. scoparius*	Hawaii
Broom, Spanish	*Spartium junceum*	Hawaii
Broomrape, hemp Branched broomrape	*Orobanche ramosa*	Calif.
Broomrape, Coopers	*O. ludoviciana*	Calif.
Buckhorn (see plantain, buckhorn)		
Buffalobur	*Solanum rostratum*	Hawaii
Bullnettle (see horsenettle)		
Bur bush	*Triumfetta bartramia*	Hawaii
Bur-grass (see sandbur)		
Bur, Sacramento	*T. semitriloba*	Hawaii
Bursage (see povertyweed)	*Franseria* spp.	Neb.
Bursage, slimleaf	*F. conferti flora, F. tenuifolia*	Neb.
Butterprint Velvetleaf	*Abutilon theophrasti*	Iowa, N.C.
Caesarweed (see Aramina)		
Camelthorn	*Alhagi pseudalhagi, Alhagi camelorum*	Ariz., Calif., Hawaii, Idaho, Nev., N.M., Ore., Wash., Wyo.
Canegrass Japanese silvergrass	*Miscanthus japonicus*	Hawaii
Carrot, wild	*Daucus carota*	Idaho, Ill., Iowa, Kans., Mich., Mo., S.D., Tex., W.Va.
Charlock (see mustard, wild)		
Cheat or chess Hairy chess	*B. commutatus*	Ala., Ark., Fla., Ga., Hawaii, La., Miss., S.C., Tenn., Tex.
Cheat or chess	*Bromus secalinus*	Ala., Ark., Del., Fla., Ga., Hawaii, Kan., La., Miss., N.J., Okla., S.C., Tenn., Tex.
Chess, hairy (see cheat or chess)		
Chickweed	*Stellaria* spp.	Mont.
Cocklebur	*Xanthium* spp.	Ala., Ark., Ga., Fla., Hawaii, La., Miss., N.C., S.C., Tenn., Tex.
Cocklebur	*X. pensylvanicum, Xanthium commune*	Ala., Ark., Ga., Fla., Hawaii, Iowa, La., Miss., N.C., S.C., Tenn., Tex.
Cockle, corn corncockle	*Agrostemma githago*	Ala., Ark., Del., Fla., Ga., Hawaii, Ind., Ky., La., Maine, Md., Mass., Miss., N.C., N.J., N.Y., Ohio, Okla, Ore., Pa., R.I., S.C., Tenn., Tex., Vt., W. Va.

Table 3 (Continued)
NOXIOUS-WEED SEEDS

English or common name[a]	Latin name[b]	State or states in which noxious[c]
Cockle, white	*Lychnis alba*	Conn., Wis.
Coffee bean (see indigo, tall)		
Cogongrass (see Kogon)		
Cornflower (see Ragged Robin)		
Crabgrass		
Large	*Digitaria sanguinalis*	N.H., Nev.
Smooth	*D. ischaemum*	Nev.
Crazyweed (see loco weeds)		
Creeping Jenny (see bindweed, field)		
Cress, hoary (see hoary cress)		
Crotalaria	*Crotalaria* spp.	Ala., Ark., Miss., N.C.
Crotalaria Showy crotalaria	*C. spectabilis*	Ala., Ark., Fla., Hawaii, Miss., N.C., Tenn.
Crotalaria, longbeaked	*C. longisrostrata*	Hawaii, N.C.
Crotalaria, smooth	*C. mucronata*	Fla., N.C.
Curse	*Clidemia hirta*	Hawaii
Daisy, oxeye	*Chrysanthemum leucanthemum* (var. pinnatifidum)	Hawaii, Ill., Ind., Kan., Ky., Mo., Mont., Tenn., W.Va., Wyo., Wis.
Darnel	*Lolium temulentum*	Ala., Ark., Fla., Ga., Hawaii, La., Miss., Okla., S.C., Tex., Tenn.
Deathweed (see povertyweed)		
Ding ding Indian jointvetch	*Aeschynomene indica*	Hawaii
Dock (see sorrel)	*Rumex* spp. (except *R. haustatulus* in Ala.)	Ala., Ark., Conn., Ga., Kan., La., Miss., Neb., Okla., S.C., Tex., Wash., Ore., Tenn.
Dock, broadleaf	*R. obtusifolius*	Ala., Ark., Conn., Fla., Ga., Kan., La., Miss., Neb., N.C., Okla., Ore., S.C., Tenn., Tex., Wash.
Dock, cluster	*R. conglomeratus*	Ala., Ark., Conn., Fla., Ga., Kan., La., Miss., Neb., Okla., Ore., S.C., Tenn., Tex., Wash.
Dock, curled (see dock, curly)		

Table 3 (Continued)
NOXIOUS-WEED SEEDS

English or common name[a]	Latin name[b]	State or states in which noxious[c]
Dock, curly (Curled dock, Sour dock)	*R. crispus*	Ala., Ariz., Ark., Colo., Conn., Fla., Ga., Hawaii, Ill., Ind., Iowa, Kan., La., Miss., Mo., Mont., Neb., Nev., N.C., Ohio, Okla., Ore., S.C., Tenn., Tex., Wash.
Dock, smooth	*R. altissimus*	Ala., Ark., Conn., Ga., Iowa, Kan., La., Miss., Neb., Okla., Ore., S.C., Tenn., Tex., Wash.
Dock, sour (see dock, curly)		
Dock, veiny Winged dock	*R. venosus*	Ala., Ark., Conn., Ga., Kan., La., Miss., Neb., Okla., S.C., Tenn., Tex., Ore., Wash.
Dodder	*Cuscuta* spp.	In all states except Alaska, any species except *C. coryli* in N.D.
Elephantopus Elephant-foot	*Elephantopus* spp.	Hawaii
Emex	*Emex australis*	Hawaii
Spiny emex	*E. spinosa*	Hawaii
Fanweed (see pennycress)		
Fieldcress, Austrian	*Rorippa austriaca*, *Radicula austriaca*	Alaska, Calif., Hawaii, Idaho, Nev., Ore., Wash., Wyo.
Fountaingrass	*Pennisetum setaceum*	Hawaii
Foxtail, giant	*Setaria faberi*	Ala., Ark., Del., Ga., Ill., Ind., Iowa, Kan., Ky., Mich., Minn., Miss., Mo., N.C., Okla., S.C., Tex., Va., Tenn., Wis.

Table 3 (Continued)
NOXIOUS-WEED SEEDS

English or common name[a]	Latin name[b]	State or states in which noxious[c]
Foxtailgrass, West Indian	*Andropogon bicornis*	Hawaii
Franseria, white-leaved (see povertyweed)		
Frenchweed (see penny-cress)		
Garlic, wild (Onion, wild)	*Allium* spp.	Ala., Ark., Del., Fla., Ga., Ind., Kan., Ky., La., Mass., Miss., N.C., Okla., S.C., Tex., Vt., Va., Tenn.
	A. vineale	Ala., Ark., Del., Calif., D.C., Fla., Ga., Hawaii, Ill., Ind., Kan., Ky., La., Maine, Md., Mass., Miss., Mo., N.J., N.Y., N.C., Ohio, Okla., Ore., Pa., R.I., S.C., Tenn., Tex., Va., Vt., W.Va.
	A. canadense	Ala., Ark., Washington, D.C., Del., Fla., Ga., Ill., Ind., Kan., Ky., La., Maine, Md., Mass., Mich., Miss., Mo., N.J., N.Y., N.C., Ohio, Okla., Pa., R.I., S.C., Tenn., Tex., Vt., Va.
Gaura	*Gaura* spp.	Hawaii
Goatgrass, barb (Goatgrass)	*Aegilops* spp.	Tex.
Goatgrass, jointed (Joint goatgrass)	*A. cylindrica*	Colo., Tex.
Goatgrass	*A. triuncialis*	Hawaii, Nev., Tex.
Goathead (see puncturevine)		
Goatweed (see St. Johnswort)		
Gorse	*Ulex europaeus*	Hawaii
Gromwell Corn gromwell	*Lithospermum arvense*	Wash.
Groundcherry, smooth Perenial ground cherry	*Physalis subgrabrata*	Calif., Idaho
Groundcherry, purple-flower	*p. lobata*, *Quincula lobata*	Colo.
Halogeton	*Halogeton glomeratus*	Calif., Colo., Hawaii, Idaho, Mont., N.M., Nev., Ore., Utah, Wash.
Hamakua pamakani River eupatorium	*Eupatorium riparium*	Hawaii
Hawkweed	*Hieracium* spp.	W. Va.
Hoary cress Whitetop	*Cardaria* spp.	Utah

Table 3 (continued) NOXIOUS-WEED SEEDS

English or common name[a]	Latin name[b]	State or states in which noxious[c]
Hoary cress Perennial peppergrass	*C. repens*	N.D.
Hoary cress Whitetop, Whiteweed, Perennial peppergrass, Heartpodded hoary cress, Lenspodded hoary cress	*C. draba* *Lepidium draba* *L. repens* *L. draba* var. *repens* *C. chalepensis*	Alaska, Ariz., Calif., Colo., Hawaii, Idaho, Ill., Ind., Iowa, Kan., Mich., Minn., Mo., Mont., Neb., N.M., N.D., Ohio, Okla., Ore., S.D., Utah, Wash., Wyo.
Hoary cress Whitetop, Whiteweed, Hairy whitetop, Ballcress, Perennial peppergrass, Globepodded hoary cress	*C. pubescens* *Hymenophysa pubescens*	Alaska, Ariz., Calif., Colo., Hawaii, Idaho, Neb., Nev., N.M., N.D., Ohio, Ore., Utah, Wash., Wyo.
Hogpotato (see pignut)		
Horsenettle Silverleaf nightshade, Bullnettle, Purple nightshade, White horsenettle, Nightshade	*Solanum elaeagnifolium*	Ala., Alaska, Ariz., Ark., Calif., Colo., Fla., Ga., Hawaii, Idaho, Kan., La., Miss., Nev., N.M., Okla., S.C., Tex., Wash., Wyo.
Horsenettle, Carolina Bullnettle	*S. carolinense*	Ala., Alaska, Ariz., Ark., Calif., Colo., Conn., Del., Fla., Ga., Hawaii, Ill., Ind., Iowa, Kan., La., Maine, Mass., Mich., Minn., Miss., Neb., Nev., N.J., N.Y., N.C., Ohio, Okla., Pa., R.I., S.C., S.D., Tex., Vt., W.Va., Wyo.
Horsenettle, white (see horsenettle)		
Indigo, curly Northern jointvetch	*Aeschynomene virginica*	Ark.
Indigo, tall Coffee bean Hemp sesbania	*Sesbania exaltata*	Ark.
Ivy, German	*Senecio mikanioides*	Hawaii
Jaraguagrass	*Hyparrhenia rufa*	Hawaii
Johnsongrass (see sorghum)	*Sorghum halepense* *Holcus halepensis* *Andropogon halepensis*	Ala., Ariz., Ark., Calif., Colo., Del., Fla., Ga., Idaho, Hawaii, Ill., Ind., Kan., Ky., La., Md., Miss., Mo., Neb., N.M., N.C., Nev., Ohio, Okla., Pa., S.C., Tenn., Tex., Utah, Va., Wash., W.Va.

Table 3 (Continued)
NOXIOUS-WEED SEEDS

English or common name[a]	Latin name[b]	State or states in which noxious[c]
Jointvetch, Indian (see ding ding)		
Jointvetch, northern (see curly indigo)		
Kahili flower Banksgrevillea	*Grevillea banksii*	Hawaii
Khakiweed	*Alternanthera repens*	Hawaii
Klamathweed (see St. Johnswort)		
Knapweed, diffuse	*Centaurea diffusa*	Calif., Idaho, Wash.
Knapweed, Russian	*C. repens* *C. picris*	Alaska, Ariz., Calif., Colo., Conn., Ga., Hawaii, Idaho, Ill., Iowa, Kan., La., Mich., Minn., Mont., Neb., Nev., N.M., N.Y., N.D., Ohio, Okla., Ore., S.C., S.D., Tex., Utah, Wash., Wyo.
Knapweed, spotted	*C. maculosa*	Calif., Idaho, Mont., Wash.
Knapweed, squarrose	*C. squarrosa*	Calif.
Kogon Cogongrass	*Imperata cylindrica*	Hawaii
Lettuce, blue	*Lactuca pulchella*	Alaska, Colo., Hawaii, Idaho, Mont., Ore., Wash., Wyo.
Lettuce, blue flowering (see lettuce, blue)		
Loco weeds (Loco; crazyweed)	*Astragalus* spp. *Oxytropsis* spp.	Hawaii
Malachra	*Malachra capitata* *M. alceaefolia*	Hawaii
Mallow, alkali Alkali sida	*Sida hederacea*	Ariz., Calif., Ore.
Marabu	*Dichrostachys nutans*	Hawaii
Marigold, wild	*Tagetes minuta*	Calif., Hawaii
Maui pamakani Pamakani	*Eupatorium adenophorum*	Hawaii
Meadow-foxtail Pacific	*Alopecurus myosuroides*	Wash.
Medusahead (Medusahead rye, Medusahead wildrye)	*Taeniatherum asperum* *Elymus caput-medusae* *Taeniatherum caput-medusae*	Calif., Hawaii, Idaho, Mont., Nev., Wash.
Melastoma (Banks melastoma)	*Melastoma malabathricum*	Hawaii
Melon, dudaim	*Cucumis melo* var. *dudaim*	Calif.
Mile-a-minute vine	*Mikania micrantha*	Hawaii
Milkweed	*Asclepies* spp.	Hawaii
Millet, Texas Texas panicum	*Panicum texanum*	Fla., N.C.
Moonflower Purple moonflower, small moonflower, giant morningglory	*Calonyction muricatum*	Ala., Ark., La., Miss., N.C., Okla., S.C.

Table 3 (Continued)
NOXIOUS-WEED SEEDS

English or common name[a]	Latin name[b]	State or states in which noxious[c]
Morningglory (see bindweed, field)	*Ipomoea* spp.	Ariz., Ark., Kan., Neb., N.M., N.C., Okla., Tex.
Morningglory European (see bindweed, field)		
Morningglory, giant (see moonflower)		
Morningglory, wild (see morningglory)		
Morningglory, ivyleaf	*I. hederacea*	Ariz., Ark., Kan., Neb., N.M., N.C., Okla., S.D., Tex.
Morningglory, Wooly morningglory	*I. hirsutula*	Ariz., Ark., Kan., Neb., N.M., N.C., Okla., Tex.
Morningglory, wild (see bindweed, field and bindweed, hedge) Tall morningglory	*I. purpurea*	Ariz., Ark., Kan., Neb., N. M., N. C., Okla., Tex., W. Va.
Mullein Common mullein	*Verbascum thapsus*	Hawaii
Mustard Mustard, wild; turnip, wild	*Brassica* spp.	Ala., Ariz., Conn., Fla., Ga., Ill., Kan., Maine, Miss., N.C., N.D., R.I., S.C., Tenn., Tex., Vt., Va.
Mustard, black mustard, wild	*B. nigra*	Ala., Ariz., Colo., Conn., Fla., Ga., Ill., Iowa, Kan., Maine, Mass., Mich., Miss., Neb., N.C., N.D., Pa., R.I., S.C., S.D., Tenn., Tex., Vt., Va.
Mustard blue Tenella, purple	*Chorispora tenella*	Calif., Wyo.
Mustard, Indian mustard, wild	*B. juncea*	Ala., Ariz., Colo., Conn., Fla., Ga., Ill., Iowa, Kan., Maine, Mich., Miss., Mass., Neb., N.C., N.D., Pa., R.I., S.C., S.D., Tenn., Tex., Va., Vt., W. Va., Wis.
Mustard, white mustard, yellow	*B. hirta*	Ala., Ariz., Conn., Fla., Ga., Ill., Kan., Maine, Miss., N.C., N.D., R.I., S.C., Tenn., Tex., Vt., Va.

Table 3 (Continued)
NOXIOUS-WEED SEEDS

English or common name[a]	Latin name[b]	State or states in which noxious[c]
Mustard, wild (charlock (see mustard, black and Indian) wild turnip)	*B. kaber* *B. arvensis* *B. sinapistrum* *Sinapis arvensis*	Ala., Ariz., Colo., Conn., Fla., Ga., Hawaii, Ill., Ind., Iowa, Kan., Maine, Mass., Miss., Mich., Minn., Neb., Nev., N.C., N.D., Pa., R.I., S.C., S. D., Tenn., Tex., Vt., Va., Wash., W. Va., Wis.
Mustard, yellow (see mustard, white)		
Needlegrass (see punagrass)	*Stipa* spp.	Hawaii
Nettle, stinging stinging weed	*Hesperocnide sandwicensis*	Hawaii
Nightshade (see Horsenettle)		
Nightshade, purple (see horsenettle)		
Nightshade, silverleaf (see horsenettle)		
Nutgrass (see nutsedge)	*Cyperus* spp.	R.I., N.C.
Nutgrass yellow nutgrass, yellow nutsedge	*C. esculentus*	Alaska, Ariz., Calif., Conn., Hawaii, Maine, Miss., N.C., N.M., R.I., Tex., Wyo.,
Nutgrass (Perennial nutgrass, purple nutgrass, purple nutsedge)	*C. rotundus*	Ariz., Ark., Calif., Fla., Ga., Hawaii, La., Miss, N.M., N.C., Okla., R.I., S.C., Tenn., Tex., Wyo.
Nutgrass, perennial (see nutgrass)		
Nutgrass, yellow (see nutgrass)		
Nutgrass tubers	*C.* spp.	N.C., Tex.
Nutsedge (see nutgrass and nutgrass tubers)		
Oat, wild Wild oats	*Avena fatua*	Ariz., Colo., Hawaii, Idaho, Mont., N.M., N.D., Okla., S.C., S.D., Utah, Wash., Wis.
Oat, wild	*A. sterilis*	Okla.
Onion, wild (see garlic, wild)		
Panicum, Texas (see millet, Texas)		
Passionflower, wingleaf	*Passiflora pulchella*	Hawaii
Peaweed, Austrian	*Swainsona salsula* *Sphaerophysa salsula*	Calif., Idaho, Nev., Wyo.

Table 3 (Continued)
NOXIOUS-WEED SEEDS

English or common name[a]	Latin name[b]	State or states in which noxious[c]
Pennycress fanweed, Frenchweed, field penny-cress	*Thlaspi arvense*	Ind., Kan., Mich., Minn., Neb., Nev., N.D., Ohio, S.D., Wash.
Pepper, hooked spiked	*Piper aduncum*	Hawaii
Peppercress, field Field peppergrass	*Lepidium campestre*	Ind., W. Va.
Peppercress, perennial Broad-leaved peppergrass tall whitetop, perennial pepperweed, whitetop	*Lepidium latifolium*	Alaska, Calif., Colo., Hawaii, Utah, Wash.
Peppergrass, field (see peppercress, field)		
Peppergrass, perennial (see hoary cress and pepper-cress, perennial)		
Pepperweed (see pepper-cress, field and perennial)		
Pignut Indian rush pea, hog-potato	*Hoffmanseggia densiflora*	Kan.
Piripiri biddy biddy	*Acaena sanguisorbae* *A. anserinifolia*	Calif., Hawaii
Plantain	*Plantago* spp.	Conn., Wash.
Plantain, bracted	*P. aristata*	Ala., Conn., La., Miss., N.C., Okla., S.C., Tex., Wash.
Plantain, common or broadleaf	*P. major*	Conn., Wash.
Plantain, buckhorn	*P. lanceolata*	All states *except* Alaska, Ariz., Calif., Del., Md., Neb., Nev., N.H., N.J., N.M., N.Y., N.D, S.D., Utah, Vt., Va.
Povertygrass	*Aristida* spp.	Hawaii
Povertyweed Bursage	*Franseria* spp.	Hawaii, Neb.
Povertyweed Mouse-ear povertyweed, sumpweed, deathweed	*Iva axillaris*	Calif., Colo., Idaho, Oreg., Utah, Wash., Wyo.
Povertyweed, mouse-ear (see povertyweed)		
Povertyweed, silverleaf (bur ragweed, white-leaved franseria, creeping ragweed, skeletonleaf bursage)	*Franseria discolor*	Colo., Hawaii, Idaho, Neb., N.M., Wyo.
Povertyweed, woolly-leaved Bur ragweed, woollyleaf bursage	*F. tomentosa*	Colo., Hawaii, Kan., Neb.
Punagrass	*Stipa brachychaeta*	Calif.
Puncturevine Goathead	*Tribulus terrestris*	Ariz., Calif., Colo., Hawaii, Idaho, Neb., Nev., Ore., Tex., Wash., Wyo.

Table 3 (Continued)
NOXIOUS-WEED SEEDS

English or common name[a]	Latin name[b]	State or states in which noxious[c]
Quackgrass	*Agropyron repens*	In Washington, D.C., and all states *except* Ariz., Ark.
Radish	*Raphanus* spp.	Ala.
Radish, wild	*R. raphanistrum*	Ala., Conn., Fla., Ga., Maine, Mass., Miss., Minn., N.C., R.I., S.C., Tex., Vt., Va., Wis.
Ragged Robin cornflower	*Centaurea cyanus*	N.C.
Ragweed (see povertyweed)		
Ragweed, bur (see povertyweed)		
Ragweed, creeping (see povertyweed)		
Ragweed, perennial western ragweed	*Ambrosia psilostachya*	Hawaii, Idaho, Wash., Wyo.
Ragwort, tansy	*Senecio jacobaea*	Oreg., Wash.
Rape, bird Rape, wild turnip	*Brassica campestris*	Neb., Ill., (also see states listed after Mustard, *Brassica* spp.)
Rhodomyrtus Downy rosemyrtle	*Rhodomyrtus tomentosa*	Hawaii
Rice, red	*Oryza sativa*	Ark., Fla., Ga., La., S.C., Miss., Tex.
Rocket, yellow (see wintercress, bitter)		
Rosemyrtle, downy (see Rhodomyrtus)		
Russian-thistle (see thistle, Russian)		
Rye, common rye	*Secale cereale*	Wash.
Rye, Medusahead (see medusahead)		
Ryegrass, Persian (see darnel)	*Lolium persicum*	Tex.
Sage, Mediterranean	*Salvia aethiopis*	Calif.
Sandbur sandbur grass	*Cenchrus* spp.	Hawaii
Sandbur bur-grass, sandbur-grass, field sandbur	*C. pauciflorus*	Ariz., Ga., Hawaii, Nev., N.C., S.C., Wash.
Sandbur grass, coast (field sandbur)	*C. incertus*	Calif.
Sandbur grass, mat Longspine sandbur	*C. longispinus*	Calif.
Sandbur grass, southern Southern sandbur	*C. echinatus*	Calif.
Sesbania, hemp (see tall indigo)		
Sensitiveplant, giant	*Mimosa invisa*	Hawaii
Sicklepod	(*Cassia* spp.)	N.C.
Sida, alkali (see alkali mallow)		

Table 3 (continued)
NOXIOUS-WEED SEEDS

English or common name[a]	Latin name[b]	State or states in which noxious[c]
Silvergrass, Japanese (see canegrass)		
Skeletonweed, yellow-flowering, Rush skeleton-weed	*Chondrilla juncea*	Calif., Idaho, Nev., Wash.
Smartweed Ladysthumb, knotweed	*Polygonum* spp.	Hawaii
Smartweed, swamp	*P. coccineum*	Kan.
Snakeroot, white	*Eupatorium rugosum*	Hawaii
Sorghum almum (see sorghum)	*Sorghum almum*	Colo., Fla., Ga., Hawaii, S.C., Tenn., Tex.
Sorghum, perennial (such as, but not limited to johnsongrass, sorghum almum, and perennial sweet sudangrass)	*S.* spp.	Ariz., Calif., Del., Kan., Md., Nev., N.J., Ind., Okla., Va., Wash., Pa., Tex.
Sorgrass (see sorghum)		S.C.
Sorrel, red (see sorrel, sheep)		
Sorrel, sheep Red sorrel, dock	*Rumex acetosella*	Ala., Ark., Conn., Fla., Ga., Iowa, Kan., Ky., La., Miss., Mo., Neb., Okla., Ore., S.C., Tenn., Tex., Wash.
Sowthistle Perennial sowthistle, creeping sowthistle	*Sonchus arvensis*	Alaska, Ariz., Calif., Colo., Conn., Hawaii, Idaho, Ill., Ind., Iowa, Kan., Maine, Mass., Mich., Minn., Mont., Nev., N.D., Ohio, Ore., Pa., R.I., S.D., Vt., Wash., Wis., Wyo., W.Va.
Spearhead	*Rhynchospora* spp.	La.
Spikenard, wild	*Hyptis suaveolens*	Hawaii
Spurge, leafy	*Euphorbia esula*	Alaska, Ariz., Calif., Colo., Hawaii, Idaho, Ill., Iowa, Kan., Mich., Minn., Mo., Mont., Neb., Nev., N.Y., N.D., Utah, Ohio, Ore., S.D., Wash., Wis., Wyo,.
Spurge, leafy (shining spurge)	*E. lucida*	S.D.
Spurge, shining (see leafy spurge)		
Spurge, toothed	*E. serrata*	Hawaii
St. Johnswort Klamath-weed Goatweed	*Hypericum perforatum*	Alaska, Calif., Colo., Hawaii, Idaho, Mont., Nev., Ore., Wash., Wyo.

Table 3 (Continued)
NOXIOUS-WEED SEEDS

English or common name[a]	Latin name[b]	State or states in which noxious[c]
Starbur Bristly starbur	*Acanthospermum hispidum*	Hawaii
Stargrass, Australian	*Chloris devaricata*	Hawaii
Starthistle, Iberian	*Centaurea iberica*	Nev., Ore., Calif.
Starthistle, Malta	*C. melitensis*	Ore.
Starthistle, purple	*C. calcitrapa*	Nev., Ore.
Starthistle, yellow	*C. solstitialis*	Calif., Hawaii, Idaho, Mo., Nev., Ore., Wash., W.Va., Wyo.
Sudangrass, perennial sweet type		S.C., Wash.
Sumpweed (see poverty-weed)		
Swainsonpea (see Austrian peaweed)		
Terongan	*Solanum torvum*	Hawaii
Thistle, blessed	*Cnicus benedictus*	Ala., Fla., Ga., Miss., N.C., S.C., Tex.
Thistle, Canada	*Cirsium arvense*	In Washington, D.C., and all states *except* Ark.
Thistle, golden	*Scolymus hispanicus*	Calif.
Thistle, Italian	*Carduus pycnocephalus* *C. tenuiflorus*	Calif., Va.
Thistle, musk Nodding thistle	*C. nutans*	Calif., Colo., Kan., Neb., Utah
Thistles, onopordum	*Onopordum* spp.	Calif.
Thistle, plumeless including musk and curled thistle	*C.* spp.	Hawaii, Va.
Thistle, plumeless	*C. acanthoides*	Calif., Neb., Hawaii, Va.
Thistle, Russian	*Salsola kali* var. *tenuifolia* *S. pestifer*	Ariz., Hawaii, Mo.
Thistle, Scotch Cotton thistle	*Onopordum acanthium*	Calif., Neb., Utah
Thistle, smooth distaff	*Carthamus baeticus*	Calif.
Thistle, wavyleaf	*Cirsium undulatum*	Calif.
Thistle, whitestem distaff	*Carthamus leucocaulos*	Calif.
Thistle, woolly distaff	*C. lanatus*	Calif.
Thistle, yellowspine	*Cirsium ochrocentrum*	Calif.
Threeawn (see poverty-grass)		
Toadflax, Dalmatian	*Linaria dalmatica*	Calif., Colo., Idaho, Mont., Nev., Wash.
Toadflax, yellow	*L. vulgaris*	Alaska, Colo., Mont., Wash.
Torpedograss	*Panicum repens*	Hawaii
Turnip, wild (see mustard)		
Velvetleaf (see butterprint)		
White top (see hoary cress)		
White top, hairy (see hoary cress)		
White top (tall) (see pep-pergrass, perennial)		
Whiteweed (see hoary cress)		

Table 3 (continued)
NOXIOUS-WEED SEEDS

English or common name[a]	Latin name[b]	State or states in which noxious[c]
Wildrye, medusahead (see Medusahead)		
Wintercress, bitter Yellow rocket	*Barbarea vulgaris* *Campe barbarea*	Conn., Ind., Maine, Mich., Wis.
Witchweed	*Striga* spp.	Hawaii
Witchweed	*S. lutea* *S. asiatica*	Hawaii, N.C., S.C.
Woundwort	*Stachys palustris*	Maine

[a] A weed may have more than one common name. There is a list of common names in *Weed Science,* 19(No. 4), 435, 1971, *(Weed Science* names).[8] Different common names are sometimes used in different state laws for the same weed (duplicate names). *Weed Science* names are usually, but not always, one of the duplicate names. The common name entries on this list are arranged alphabetically using either a *Weed Science* name or a duplicate name. Differing duplicate and *Weed Science* names, if any, are listed after each entry. Cross references appear. The cross reference can be to a duplicate name or a *Weed Science* name.

[b] Latin names appear with, and without, underlining. The underlined names are synonyms.

[c] The state or states in which a weed is noxious are listed with *Weed Science* names only.

From Livestock, Poultry, Grain and Seed Division, State Noxious-Weed Seed Requirements Recognized in the Administration of the Federal Seed Act, U.S. Department of Agriculture, Washington, D.C., 1978.

Table 4
GERMINATION REQUIREMENTS FOR INDICATED KINDS

Name of seed	Substrata[a]	Temp (°C)	First count (days)	Final[b] count (days)	Additional directions[c]: Specific requirements and photo numbers	Additional directions[c]: Fresh and dormant seed
			Agricultural Seed			
Alfalfa — *Medicago sativa*	B, T, S	20	4	[1]7	Photos 2481, 2486; see par. (b)(10)	
Alfilaria — *Erodium cicutarium*	B, T	20—30	3	14	Clip seeds	
Alyceclover — *Alysicarpus vaginalis*	B, T	35	4	[1]21	See par. (b)(1) for swollen seeds	
Bahiagrass — *Paspalum notatum*						
Var. *Pensacola*	P, S	20—35	7	[2]28	Light; see par. (b)(2)	
All other vars	P	30—35	3	21	Light; remove glumes; see par. (b)(2)	Scratch caryopses; KNO_3
Barley — *Hordeum vulgare*	B, T, S	20; 15	4	7	—	Prechill 5 days at 5 or 10°C or predry
Barrelclover — *Medicago tribuloides*	B, T	20	4	[1]14	Remove seeds from bur; see par. (b)(11)	
Bean						
Adzuki — *Phaseolus angularis*	B, T, S	20—30	4	[1]10	—	
Field — *P. vulgaris*	B, T, S	20—30; 25	5	[1]8	—	
Mung — *P. aureus*	B, T, S	20—30	3	[1]7	—	
Beet, field — *Beta vulgaris*	B, T, S	20—30	3	14	Photos 19557, 19558; see par. (b)(3)	
Beet, sugar — *B. vulgaris*	B, T, S	20—30; 20	3	10	Do	
Beggarweed, Florida — *Desmodium tortuosum*	B, T	30	5	28	—	

Table 4 (continued)
GERMINATION REQUIREMENTS FOR INDICATED KINDS

Name of seed	Substrata[a]	Temp (°C)	First count (days)	Final[b] count (days)	Additional directions[c]: Specific requirements and photo numbers	Additional directions[c]: Fresh and dormant seed
			Agricultural Seed			
Bentgrass						
Colonial (including Astoria and Highland) — *Agrostis tenuis*	P	15—30; 10—30; 15—25	7	28	Light; KNO_3	Prechill at 5 or 10°C for 7 days; see par. (a)(2)
Creeping — *A. palustris*	P	15—30; 10—30; 15—25	7	28	Do	Prechill at 5 or 10°C for 7 days
Velvet — *A. canina*	P	20—30	7	21	Do	
Bermudagrass — *Cynodon dactylon*	P	20—35	7	21	Light; KNO_3; photo 2518; see par. (a)(9)	
Bermudagrass, giant — *C. dactylon* var. *aridus*	P	20—35	7	21	Do	Prechill at 10°C for 7 days and then test at 20—35°; continue tests of hulled seed for 14 days and of unhulled seed for 21 days
Bluegrass						
Bulbous — *Poa bulbosa*	P, S	10	10	35	KNO_3 or soil	Prechill all samples at 5°C for 7 days
Canada — *P. compressa*	P	15—25; 15—30	10	28	Light; KNO_3; see par. (a)(2)	10—30°C
Glaucantha — *P. glaucantha*	P	15—25; 15—30	10	28	Light; KNO_3	
Kentucky (all vars.) — *P. pratensis*	P	15—25	10	28	Do	Prechill at 10°C for 5 days

Nevada — *P. nevadensis*	P	20—30	7	21	Do	
Rough — *P. trivialis*	P	20—30	7	21	Light	
Texas — *P. arachnifera*	P	20—30	7	28	Light; KNO_3	Prechill at 5°C for 2 weeks
Wood — *P. nemoralis*	P	20—30	7	28	Light	
Bluestem						
Big — *Andropogon gerardi*	P, TS	20—30	7	28	Light; KNO_3	Prechill at 5°C for 2 weeks
Little — *A. scoparius*	P, TS	20—30	7	28	Do	Do
Sand — *A. hallii*	P, TS	20—30	7	28	Do	Do
Yellow — *A. ischaemum*	P, TS	20—30	5	21	Do	Do
Brome						
Field — *Bromus arvensis*	P, TB	15—25; 20—30	6	14	Light	Prechill at 10°C for 5 days
Mountain — *B marginatus*	P	20—30	6	14	Do	
Smooth — *B. inermis*	P, B, TB	20—30	6	14	Light optional	Prechill at 5 or 10°C for 5 days, then test at 30°C for 9 additional days
Broomcorn — *Sorghum vulgare* var. *technicum*	B, T, S	20—30	3	10	—	
Buckwheat — *Fagopyrum esculentum*	B, T	20—30	3	6	—	
Buffalograss — *Buchloe dactyloides*						
(Burs)	P, TB, TS	20—35	7	28	Light; KNO_3	Prechill at 5°C for 6 weeks; test 14 additional days
(Caryopses)	P	20—35	5	14	Light; KNO_3	
Buffelgrass — *Pennisetum ciliare*	S	30	7	28	Light; press fascicles into well-packed soil and prechill at 5°C for 7 days	See par. (b)(4)
Burclover, California — *Medicago hispida*	B, T	20	4	[1]14	Remove seeds from bur; see par. (b)(10)	
Burclover, spotted — *M. arabica*	B, T	20	4	[1]14	Do	
Burnet, little — *Sanguisorba minor*	B, T	15	5	14		

Table 4 (continued)
GERMINATION REQUIREMENTS FOR INDICATED KINDS

Name of seed	Substrata[a]	Temp (°C)	First count (days)	Final[b] count (days)	Additional directions[c]: Specific requirements and photo numbers	Additional directions[c]: Fresh and dormant seed
Buttonclover — *M. orbicularis*	B, T	20	4	[1]10	See par. (b)(10)	15°C
Canarygrass — *Phalaris canariensis*	B, T	20—30	3	7		
Canarygrass, reed — *P. arundinacea*	P	20—30	5	21	Light; KNO_3	
Carpetgrass — *Axonopus affinis*	P	20—35	10	21	Light	KNO_3
Castorbean — *Ricinus communis*	T, S	20—30	7	14	Remove caruncle if mold interferes with test	
Chess, soft — *Bromus mollis*	P	20—30	7	14	Light	Prechill at 5 or 10°C for 7 days
Chickpea — *Cicer arietinum*	T, S	20—30	3	7		
Clover						
Alsike — *Trifolium hybridum*	B, T, S	20	3	[1]7	See par. (b)(10)	15°C
Berseem — *T. alexandrinum*	B, T, S	20	3	[1]7	Do	Do
Cluster — *T. glomeratum*	B, T	20	4	[1]10	Do	Do
Crimson — *T. incarnatum*	B, T, S	20	4	[1]7	See par. (b)(10); photos 2479, 2482	Do
Kenya — *T. semipilosum*	B, T, S	20	3	[1]7		
Ladino — *T. repens*	B, T, S	20	3	[1]7	See par. (b)(10)	15°C
Lappa — *T. lappaceum*	B, T	20	3	[1]7	Do	Do
Large hop — *T. procumbens (T. campestre)*	B, T	20	4	[1]14	Do	Do
Persian — *T. resupinatum*	B, T	20	3	[1]7	Do	Do
Red — *T. pratense*	B, T, S	20	4	[1]7	See par. (b)(10); photos 2483, 2484	Do
Rose — *T. hirtum*	B, T	20	4	[1]10	See par. (b)(10)	Do

Small hop (Suckling) — *T. dubium*	B, T	20	4	[1]14	Do	Do
Strawberry — *T. fragiferum*	B, T	20	3	[1]7	Do	Do
Sub — *T. subterraneum*	B, T	20	4	[1]14	Do	Do
White — *T. repens*	B, T, S	20	3	[1]7	Do	Do
Corn						
Field — *Zea mays*	B, T, S	20—30; 25	4	7	Photos 2510, 2511, 2512,	
Pop — *Z. mays* var. *everta*	B, T, S	20—30; 25	4	7		
Cotton — *Gossypium* spp.	B, T, S	20—30; 30	4	[1]12	Photos 19553, 19554	Test by alternate method; see par. (b)(5)
Cowpea — *Vigna sinensis*	B, T, S	20—30	5	[1]8	Photos 1989, 1990, 2377	
Crambe — *Crambe abyssinica*	T	25	3	7		
Crested dogtail — *Cynosurus cristatus*	P	20—30	10	21	Light	Prechill for 3 days at 5 or 10°C
Crotalaria						
Lance — *Crotalaria lanceolata*	B, T, S	20—30	4	[1]10		
Showy — *C. spectabilis*	B, T, S	20—30	4	[1]10	Photos 2496, 2497	
Slenderleaf — *C. intermedia*	B, T, S	20—30	4	[1]10		
Striped — *C. mucronata*	B, T, S	20—30	4	[1]10		
Sunn — *C. juncea*	B, T, S	20—30	4	[1]10		
Crownvetch — *Coronilla varia*	B, T, S	20	7	[1]14		
Dallisgrass — *Paspalum dilatatum*	P	20—35	7	21	Light; KNO_3	
Dichondra — *Dichondra repens*	B, T	20—30	7	[1]28		
Dropseed, sand — *Sporobolus cryptandrus*	P	5—35; 15—35	5	28	Do	Prechill at 5°C for 4 weeks
Emmer — *Tricticum dicoccum*	B, T, S	20; 15	4	7	Photos 2507, 2520—2522	Prechill at 5 or 10°C for 5 days or predry
Fescue						
Chewing — *Festuca rubra* var. *commutata*	P	15—25	7	21	Light and KNO_3 optional	Prechill at 5 or 10°C for 5 days
Hard — *F. ovina* var. *duriuscula*	P	15—25	7	21	Light and KNO_3 optional	

Table 4 (continued)
GERMINATION REQUIREMENTS FOR INDICATED KINDS

Name of seed	Substrata[a]	Temp (°C)	First count (days)	Final[b] count (days)	Additional directions[c]: Specific requirements and photo numbers	Additional directions[c]: Fresh and dormant seed
Hair — *F. capillata*	P	10—25	10	28	Light; KNO_3	
Meadow — *F. elatior*	P	15—25; 20—30	5	14	Light and KNO_3 optional	
Red — *F. rubra*	P	15—25	7	21	Light and KNO_3 optional	
Sheep — *F. ovina*	P	15—25	7	21	Light and KNO_3 optional	
Tall — *F. arundinacea*	P	15—25; 20—30	5	14	Light and KNO_3 optional	Prechill at 5 or 10°C for 5 days and test for 21 days
Flax — *Linum usitatissimum*	B, T, S	20—30	3	7	Photos 2003, 2008, 2485, 2487	
Grama						
Blue — *Bouteloua gracilis*	P, TB	20—30	7	28	Light	KNO_3
Side-oats — *B. curripendula*	P	15—30	7	28	Light, KNO_3	
Guar — *Cyamopsis tetragonoloba*	B, T	30; 20—30	5	'14		
Guineagrass — *Panicum maximum*	P	15—35	10	28	Light; KNO_3 optional	
Hardinggrass — *Phalaris tuberosa* var. *stenoptera*	P	10—30	7	28	Light	Do
Alternate method	P	15—25	7	14	Light; KNO_3 presoak at 15°C for 24 hr	
Hemp — *Cannabis sativa*	B, T	20—30	3	7		
Indiangrass, yellow — *Sorghastrum nutans*	P, TS	20—30	7	28	Light; KNO_3	Prechill at 5°C for 2 weeks
Indigo, hairy — *Indigofera hirsuta*	B, T	20—30	5	'14		
Japanese lawngrass — *Zoysia japonica*	P	35—20	10	28	Light; KNO_3	
Johnsongrass — *Sorghum halepense*	P	20—35	7	35	Light	KNO_3
Kudzu — *Pueraria thunbergiana*	B, T	20—30	5	'14		

Lentil — *Lens culinaris*	B, T	20	5	[1]10		
Lespedeza						
Korean — *Lespedeza stipulacea*	B, T, S	20—35	5	[1]14		
Sericea or Chinese — *L. cuneata (L. sericea)*	B, T, S	20—35	7	[1]21	Photo 2494	
Siberian — *L. hedysaroides*	B, T, S	20—35	7	[1]21		
Striate (Common, Kobe, Tenn. 76) — *L. striata*	B, T, S	20—35	7	[1]14		
Lovegrass, sand — *Eragrostis trichodes*	P	20—30	5	14	Light; KNO_3	Prechill at 5 or 10°C for 6 weeks
Lovegrass, weeping — *E. curvula*	P	20—35	5	14	Light	KNO_3
Lupine						
Blue — *Lupinus angustifolius*	B, T, S	20	4	[1]10	Photos 14535-14542	
White — *L. albus*	B, T	20	3	[1]10		
Yellow — *L. luteus*	B, T	20	7	[1]10		
Manilagrass — *Zoysia matrella*	P	35—20	10	28	Light; KNO_3	
Meadow foxtail — *Alopecurus pratensis*	P	20—30	7	14	Light	
Medick, black — *Medicago lupulina*	B, T, S	20	4	[1]7	See par. (b)(10)	
Millet						
Browntop — *Panicum ramosum*	B, P, T	20—30; 30	4	14	Light and KNO_3 optional	Predry at 35 or 40°C for 7 days; and test at 30°C
Alternate method	B, P, T	5—35	4	14	Light; KNO_3	
Foxtail — Such as Common, White Wonder, German, Hungarian, Siberian, or Golden — *Setaria italica*	B, T	15—30; 20—30	4	10		
Japanese — *Echinochloa crusgalli* var. *frumentacea*	B, T	20—30	4	10		
Pearl — *Pennisetum glaucum*	B, T	20—30	3	7		

Table 4 (continued)
GERMINATION REQUIREMENTS FOR INDICATED KINDS

Name of seed	Substrata[a]	Temp (°C)	First count (days)	Final[b] count (days)	Additional directions[c]: Specific requirements and photo numbers	Additional directions[c]: Fresh and dormant seed
Proso — *Panicum miliaceum*	B, T	20—30	3	7		
Molassesgrass — *Melinis minutiflora*	P	20—30	7	21	Light	
Mustard						
Black — *Brassica nigra*	P	20—30	3	7	Do	KNO_3 and prechill at 10°C for 3 days
India — *B. juncea*	P	20—30	3	7	Light	Prechill at 10°C for 7 days and test for 5 days; KNO_3
White — *B. hirta*	P	20—30	3	5	Do	
Napiergrass — *Pennisetum purpureum*	B, T	20—30	3	10		
Oat — *Avena* spp.	B, T, S	20; 15	5	10	Photos 2407, 2408, 2524-2527, 19545, 19546	Prechill at 5 or 10°C for 5 days and test for 7 days or predry and test for 10 days
Oatgrass tall — *Arrhenatherum elatius*	P	20—30	6	14	Light	
Orchardgrass — *Dactylis glomerata*	P, TS	15—25	7	21	Light; germination more rapid on soil	Prechill at 5 or 10°C for 7 days and test at 15—25°C
Panicgrass, blue — *Panicum antidotale*	P, TS	20—30	7	28	Light	
Panicgrass, green — *P. maximum* var. *trichoglume*	P	15—35	10	28	Light; KNO_3 optional	
Peanut — *Arachis hypogaeo*	B, T, S	20—30; 25	5	10	Remove shells; photos 19541, 19542	Predry up to 14 days at 40°C
Pea, field — *Pisum sativum* var. *arvense*	B, T, S	20	3	[1]8	Photos 2503, 2506, 14543-14547	

Rape						
Annual — *Brassica napus* var. *annua*	B, T	20—30	3	7		
Bird — *B. campestris*	P	20—30	3	10	Light	KNO_3
Turnip — *B. campestris* vars.	B, T	20—30	3	7		
Winter — *B. napus* var. *biennis*	B, T	20—30	3	7		
Redtop — *Agrostis alba*	P, TB	20—30	5	10	Light	Do
Rescuegrass — *Bromus catharticus*	P, S	10—30	7	28	Light, see par. (b)(7) for alternate method	In soil at 15°C
Rhodesgrass — *Chloris gayana*	P	20—30	6	14	Light; KNO_3	Presoak, see par. (b)(9)
Rice — *Oryza sativa*	T, S	20—30; 30	5	14	Photos 19549, 19550; see par. (b)(8) for alternate method	
Ricegrass, Indian — *Oryzopsis hymenoides*	S	5—15; 15; 15—25	7	28		Dark; prechill in soil at 5°C for 4 weeks
Roughpea — *Lathyrus hirsutus*	B, T	20	7	[1]14		
Rye — *Secale cereale*	B, T, S	20; 15	4	7	Photos 2403, 2406, 2528-2531	Prechill at 5 or 10°C for 5 days or predry
Ryegrass						
Annual (Italian) — *Lolium multiflorum*	P, TB	15—25	5	14	Light optional; see par. (b)(9) for fluorescence test	Light; KNO_3; prechill at 5 or 10°C for 5 days and test at 15—25°C; if the seeds are still dormant rechill for 3 days and continue the test at 15—25°C an additional 4 days
Perennial — *L. perenne*	P, TB	15—25	5	14	Do	Do
Wimmera — *L. rigidum*	P, TB	15—25; 20—30	5	14	Do	Do
Safflower — *Carthamus tinctorius*	P, B, T, S	15; 20	4	14	Light at 15°C	

Table 4 (continued)
GERMINATION REQUIREMENTS FOR INDICATED KINDS

Name of seed	Substrata[a]	Temp (°C)	First count (days)	Final[b] count (days)	Additional directions[c]: Specific requirements and photo numbers	Additional directions[c]: Fresh and dormant seed
Sainfoin — *Onobrychis viciaefolia*	B, T	20—30	4	'14		
Sesame — *Sesamum indicum*	B, T, TB	20—30	3	6		
Sesbania — *Sesbania exaltata*	B, T	20—30	5	'7		
Smilo — *Oryzopsis miliacea*	P	20—30	7	42	Light	Prechill at 5°C for 2 weeks
Sorghum						
Grain and Sweet — *Sorghum vulgare*	B, T, S	20—30	4	10	Photos 2413-2416	Prechill grain varieties at 5 or 10°C for 5 days; test sweet varieties at 30—45°C for 2—4 hr/day
Sorghum almum — *S. almum*	T, S	20—35; 15—35	5	21		Prechill at 5°C for 5 days. Upon the 10th day of test, clip or pierce the distal end of ungerminated seeds
Sorghum-sudangrass hybrid — *S. vulgare × S. sudanense*	B, T, S	20—30	4	10		Prechill at 5 or 10°C for 5 days
Sorgrass[d]	B, T, S	15—35; 20—35	5	21	Photos 2413—2416	Prechill at 5 or 10°C for 7 days
Sourclover — *Melilotus indica*	B, T	20	3	'14	See paragraph (b)(10)	
Soybean — *Glycine max*	B, T, S	20—30; 25	5	'8	Photos 2371, 2372, 2378	
Spelt — *Triticum spelta*	B, T, S	20; 15	4	7	Photos 2507, 2520-2522	Prechill at 5 or 10°C for 5 days or predry
Sudangrass — *Sorghum sudanense*	B,T, S	20—30; 15—30	4	10	Photos 2449-2452	Prechill at 10°C for 5 days
Sunflower (Cult.) — *Helianthus annuus*	T, B	20—30	3	7		

Sweetclover						
White — *Melilotus alba*	B, T, S	20	4	[1]7	Photos 2374, 2375, 2376, 2381; see par. (b)(10)	
Yellow — *M. officinalis*	B, T, S	20	4	[1]7	Do	
Sweet vernalgrass — *Anthoxanthum odoratum*	P	20—30	6	14	Light	
Switchgrass — *Panicum virgatum*	P, TS	15—30	7	28	Light, KNO_3	Prechill at 5°C for 2 weeks
Timothy — *Phleum pratense*	P, TB	15—25; 20—30	5	10	Light; photo 2399; see par. (a)(9)	KNO_3 and prechill at 5 or 10°C for 5 days
Tobacco — *Nicotiana tabacum*	P, TB	20—30	7	14	Light	
Trefoil						
Big — *Lotus uliginosus (L. major)*	B, T	20	5	12		
Birdsfoot — *L. corniculatus*	B, T	20	5	12	Photos 19531, 19532	
Triticale — *Tritico secale*	B, T, S	20; 15	4	7		Prechill 5 days at 5 or 10°C or predry
Vaseygrass — *Paspalum urvillei*	P	20—35	7	21	Light	KNO_3
Veldtgrass — *Ehrharta calycina*	P	10—30	7	28	Do	
Velvetbean — *Stizolobium deeringianum*	B, T, S, C	20—30	3	[1]14	Photos 19539, 19540	
Velvetgrass — *Holcus lanatus*	P	20—30	6	14	Light	
Vetch						
Common — *Vicia sativa*	B, T	20	5	[1]10		
Hairy — *V. villosa*	B, T	20	5	[1]14		
Hungarian — *V. pannonica*	B, T	20	5	[1]10		
Monantha — *V. articulata*	B, T	20	5	[1]10		
V. monantha						
Narrowleaf — *V. angustifolia*	B, T	20	5	[1]14		
Purple — *V. benghalensis*	B, T	20	5	[1]10		
Woolypod — *V. dasycarpa*	B, T	20	5	[1]14		Prechill at 10°C for 5 days and test at 15°C

Table 4 (continued)
GERMINATION REQUIREMENTS FOR INDICATED KINDS

Name of seed	Substrata[a]	Temp (°C)	First count (days)	Final[b] count (days)	Additional directions[c]: Specific requirements and photo numbers	Additional directions[c]: Fresh and dormant seed
Wheat						
Common — *Triticum aestivum*	B, T, S	20; 15	4	7	Photos 2507, 2520-2522	Prechill at 5 or 10°C for 5 days, or predry
Club — *T. compactum*	B, T, S	20; 15	4	7	Do	Do
Durum — *T. durum*	B, T, S	20; 15	4	10	Do	Do
Polish — *T. polonicum*	B, T, S	20; 15	4	7	Do	Do
Poulard — *T. turgidum*	B, T, S	20; 15	4	7	Do	Do
Wheatgrass						
Beardless — *Agropyron inerme*	P, TB	15—25	7	14	Light and KNO_3 optional	KNO_3 and prechill at 5 or 10°C for 7 days
Fairway crested — *A. cristatum*	P, TB	15—25; 20—30	5	14	Do	Do
Standard crested — *A. desertorum*	P, TB	15—25; 20—30	5	14	Do	Do
Intermediate — *A. intermedium*	P	15—25	5	28	Do	Do
Pubescent — *A. tricophorum*	P	15—25	5	28	Do	Do
Alternate method	P	20—30	5	28	Light	Do
Siberian — *A. sibiricum*	P, TB	15—25	7	14	Light and KNO_3 optional	Do
Slender — *A. trachycaulum*	P, TB	15—25; 10—30	5	14	Do	Prechill at 5 or 10°C for 5 days. If still dormant on 10th day rechill 2 days
Streambank — *A. riparium*	P, TB	15—25	5	14	Do	Prechill at 5 or 10°C for 5 days
Tall — *A. elongatum*	P	15—25	5	21	Do	Do
Alternate method	P	20—30	5	21	Light	Do

Western — *A. smithii*	P, B, T	15—25	7	28	Light and KNO_3 optional	KNO_3 or soil and prechill at 5 or 10°C for 7 days
Wildrye						
Canada — *Elymus canadensis*	P	15—30	7	21	Light	Prechill at 5°C for 2 weeks
Russian — *E. junceus*	P	20—30	5	14	Do	Prechill at 5 or 10°C for 5 days
			Vegetable Seed			
Artichoke — *Cynara scolymus*	B, T	20—30	7	21	Photos 19533, 19534	
Asparagus — *Asparagus officinalis*	B, T, S	20—30	7	21		
Asparagusbean — *Vigna sesquipedalis*	B, T, S	20—30	5	[1]8		
Beans						
Garden — *Phaseolus vulgaris*	B, T, S	20—30; 25	5	[1]8	Photos 1834, 1835, 1846, 1854, 1855	
Lima — *P. lunatus* var. *macrocarpus*	B, T, C, S	20—30	5	[1]9	Photos 2380, 2400, 2401	
Runner — *P. coccineus*	B, T, S	20—30	5	[1]9		
Beet — *Beta vulgaris*	B, T, S	20—30	3	14	See par. (b)(3); photos 19557, 19558	
Broadbean — *Vicia faba*	S, C	20	4	14	See par. (b)(10)	Prechill at 10°C for 3 days
Broccoli — *Brassica oleracea* var. *botrytis*	B, P, T	20—30	3	10		Prechill at 5 or 10°C for 3 days; KNO_3 and light
Burdock, great — *Arctium lappa*	B, T	20—30	7	14		
Brussels sprouts — *B. oleracea* var. *gemmifera*	B, P, T	20—30	3	10		Prechill at 5 or 10°C for 3 days; KNO_3 and light
Cabbage — *B. oleracea* var. *capitata*	B, P, T	20—30	3	10	Photos 19551, 19552	Do
Cabbage, Chinese — *B. pekinensis*	B, T	20—30	3	7		

Table 4 (continued)
GERMINATION REQUIREMENTS FOR INDICATED KINDS

Name of seed	Substrata[a]	Temp (°C)	First count (days)	Final[b] count (days)	Additional directions[c]: Specific requirements and photo numbers	Additional directions[c]: Fresh and dormant seed
Cabbage, tronchuda — *B. oleracea* var. *tronchuda*	B, P	20—30	3	10	Photos 19551, 19552	Do
Cardoon — *Cynara cardunculus*	B, T	20—30	7	21	Photos 19547, 19548	
Carrot — *Daucus carota*	B, T	20—30	6	21	Photo 19561	
Cauliflower — *B. oleracea* var. *botrytis*	B, P, T	20—30	3	10		Prechill at 5 or 10°C for 3 days; KNO_3 and light
Celeriac — *Apium graveolens* var. *rapaceum*	P	15—25; 20	10	21	Light; see par. (a)(9)	
Celery — *A. graveolens* var. *dulce*	P	15—25; 20	10	21	Do	
Chard, Swiss — *Beta vulgaris* var. *cicia*	B, T, S	20—30	3	14	See par. (b)(13)	
Chicory — *Cichorium intybus*	P, TS	20—30	5	14	Light; KNO_3 or soil; photo 2504; see par. (a)(9)	
Chives — *Allium schoenophrasum*	B, T	20	6	14		
Citron — *Citrullus vulgaris*	B, T	20—30	7	14	Soak seeds 6 hr	Test at 30°C
Collards — *Brassica oleracea* var. *acephala*	B, P, T	20—30	3	10		Prechill at 5 or 10°C for 3 days; KNO_3 and light
Corn, sweet — *Zea mays*	B, T, S	20—30; 25	4	7	Photos 2510—2512, 2514	
Cornsalad (Fetticus) — *Valerianella locusta* var. *olitoria*	B, T	15	7	28		Test at 10°C
Cowpea — *Vigna sinensis*	B, T, S	20—30	5	'8	Photos 1989, 1990, 2377	
Cress						
Garden — *Lepidium sativum*	B, P, T	15	4	10		Light
Upland — *Barbarea verna*	P, TB	20—35	4	7	Light; KNO_3	

Water — *Rorippa nasturtium-aquaticum*	P	20—30	4	14	Light	
Cucumber — *Cucumis sativus*	B, T, S	20—30	3	7	Keep substratum on dry side; see par. (a)(3); photos 19535, 19536	
Dandelion — *Taraxacum officinale*	P, TB	20—30	7	21	Light; see par. (a)(9)	
Eggplant — *Solanum melongena* var. *esculentum*	P, TB, RB	20—30	7	14		Light; KNO_3
Endive — *Cichorium endivia*	P, TS	20—30	5	14	Light; KNO_3 or soil	See par. (b)(6)
Kale — *Brassica oleracea* var. *acephala*	B, P, T	20—30	3	10		Prechill at 5 or 10°C for 3 days; KNO_3 and light
Kale, Chinese — *Brassica oleracea* var. *alboglabra*	B, P, T	20—30	3	10		Light; KNO_3; prechill at 5 or 10°C for 3 days
Kale, Siberian — *B. napus* var. *pabularia*	B	20—30	3	7		
Kohlrabi — *B. oleracea* var. *gongylodes*	B, P, T	20—30	3	10		Prechill at 5 or 10°C for 3 days; KNO_3 and light
Leek — *Allium porrum*	B, T	20	6	14		
Lettuce — *Lactuca sativa*	P	20	None	7		Prechill at 10°C for 3 days or test at 15°C
Muskmelon (cantaloupe) — *Cucumis melo*	B, T, S	20—30	4	10	Keep substratum on dry side [par. (a)(3)]	
Mustard, India — *Brassica juncea*	P	20—30	3	7	Light	Prechill at 10°C for 7 days and test for 5 additional days KNO_3
Mustard, spinach — *B. perviridis*	B, T	20—30	3	7		
Okra — *Hibiscus esculentus*	B, T	20—30	4	'14	Photos 19543, 19544	
Onion — *Allium cepa*	B, T	20	6	10	Photos 1962, 2253, 2254, 2328, 2330, 2340, 2341, 2469	

Table 4 (continued)
GERMINATION REQUIREMENTS FOR INDICATED KINDS

Name of seed	Substrata[a]	Temp (°C)	First count (days)	Final[b] count (days)	Additional directions[c]: Specific requirements and photo numbers	Additional directions[c]: Fresh and dormant seed
Alternate method	S	20	6	12		
Onion, Welsh — *A. fistulosum*	B, T	20	6	10		
Pak-choi — *B. chinensis*	B, T	20—30	3	7		
Parsley — *Petroselinum hortense (P. crispum)*	B, T, TS	20—30	11	28		
Parsnip — *Pastinaca sativa*	B, T, TS	20—30	6	28		
Pea — *Pisum sativum*	B, T, S	20	5	'8	Photos 2492, 2498—2500	
Pepper — *Capsicum* spp.	T, TB, RB	20—30	6	14		Light and KNO_3
Pumpkin — *Cucurbita pepo*	B, T, S	20—30	4	7	Keep substratum on dry side [par. (a)(3)]	
Radish — *Raphanus sativus*	B, T	20	4	6	Photos 2554, 19555, 19556	
Rhubarb — *Rheum rhaponticum*	TB, TS	20—30	7	21	Light	
Rutabaga — *Brassica napus* var. *napobrassica*	B, T	20—30	3	14		
Salsify — *Tragopogon porrifolius*	B, T	15	5	10		Prechill at 10°C for 3 days
Sorrel — *Rumex acetosa*	P, TB, TS	20—30	3	14	Light	Test at 15°C
Soybean — *Glycine max*	B, T, S	20—30; 25	5	'8	Photos 2371, 2372, 2378	
Spinach — *Spinacia oleracea*	TB, T	15; 10	7	21	Keep substratum on dry side [par. (a)(3)]	
Spinach, New Zealand — *Tetragonia expansa*	TS	10—30	5	28	Do	
Alternate method	B, T	15	5	21	Remove pulp from "seeds"	
Squash — *Cucurbita moschata* and *C. maxima*	B, T, S	20—30	4	7	Keep substratum on dry side [par. (a)(3)]; photos 19537, 19538	
Tomato — *Lycopersicon esculentum*	B, P, RB, T	20—30	5	14	Photo 2513	Light; KNO_3

Tomato, husk — *Physalis pubescens*	P, TB	20—30	7	28	Light; KNO_3	
Turnip — *Brassica rapa*	B, T	20—30	3	7		
Watermelon — *Citrullus vulgaris*	B, T, S	20—30; 25	4	14	Keep substratum on dry side [par. (a)(3)]	Test at 30°C

[a] The symbols used for substrate are: B, between blotters; TB, top of blotters; T, paper toweling; S, sand; TS, top of sand or soil; P, covered Petri dishes; C, creped cellulose paper; and RB, blotters with raised covers. Conditions as specified in Rules and Regulations, p. 24.[5]

[b] Hard seeds often present (1) or firm ungerminated seeds frequently present (2).

[c] Specific requirements given in Rules and Regulations.[5] Information concerning the availability of photographs of seedlings may be obtained from the U.S. Department of Agriculture, Agricultural Marketing Service, Livestock, Poultry, Grain and Seed Div., Seed Standardization Branch, Building 306, Beltsville Agric. Res. Center — East, Beltsville, Md., 20705. Specific photos or paragraphs referred to in this column may be found in the source listed for this table.

[d] Rhizomatous derivatives of a johnsongrass × sorghum cross or a johnsongrass × sudangrass cross.

From Agricultural Marketing Service, Livestock, Poultry, Grain and Seed Division, Rules and Regulations under the Federal Seed Act, U.S. Department of Agriculture, Washington, D.C., 1975.

Table 5
SOME TESTS INTENDED TO MEASURE "VIGOR" OF SEEDS

Category of test		Kind of test	Characteristics indicated (and their likely effects)
I.	Physical	Seed size (weight) X-ray	Food reserves (early growth) physical integrity
II.	Biochemical	Triphenyl tetrazolium chloride (TTC)	Dehydrogenase activity, indicating respiration potential
		Acid fuchsin	Intactness of cell membranes (physiological fitness)
		Glutamic acid decarboxylase (GADA)	Physiological activity (yield capacity)
III.	Physiological	Conductivity test	Electric conductivity of leachate from seed, leakiness of cell membranes indicating physiological aging
		Respiration test	Oxygen uptake and respiratory quotient (RQ), indicating seedling vigor
IV.	Modified laboratory germination tests	Time to germination	Weighted mean time to germination (1) from sowing; "germination speed" (2) from first sign of germination within the population: = "uniformity" of germination
		Exhaustion test	Food reserves utilized for growth in the dark
		Seedling evaluation	Seedling vigor: sturdiness; seedlings grown in light after emergence from favourable medium
		Temperature stress	(1) survival and (2) germination over a wide range of temperatures
		Moisture stress and reduced aeration	(1) survival and (2) germination over a range of moisture potentials
		Penetration test	Ability of seedling to emerge through mechanical obstruction
V.	Emergence tests	In soil including different soil types, ranges of moisture, temperature, bulk density, pathogenic potential	Ability to survive, germinate and emerge under a wide range of potential field conditions
		Cold test	Ability to survive in cold soil of high pathogenic potential and to germinate in warmth later
		Self-protection test	Comparison of establishment with and without fungicidal dressing

From Heydecker, W., The "vigor" of seeds — a review, *Proc. Int. Seed Test. Assoc.*, 34, 201, 1969. With permission.

REFERENCES

1. **Copeland, L. O.**, *Principles of Seed Science and Technology,* Burgess Publishing, Minneapolis, 1976, chap. 12.
2. **Justice, O. L.**, Testing Agricultural and Vegetable Seeds, Handbook No. 30 U.S. Department of Agriculture, Washington, D. C., 1952.
3. Association of Official Seed Analysts, Rules for testing seeds, *J. Seed Technol.*, 3, 1, 1978.
4. International Seed Testing Association, International rules for seed testing, *Seed Sci. Technol.*, 4, 1, 1976.
5. Agricultural Marketing Service, Livestock, Poultry, Grain and Seed Division, Rules and Regulations Under the Federal Seed Act, U.S. Department of Agriculture, Washington, D. C., 1975.
6. **Justice, O. L.**, Essentials of seed testing, in *Seed Biology,* Vol. 3, Kozlowski, T. T., Ed., Academic Press, New York, 1972, chap. 5.
7. Agriculture Marketing Service, Livestock, Poultry, Grain and Seed Division, State Noxious-Weed Seed Requirements Recognized in the Administration of the Federal Seed Act, U.S. Department of Agriculture, Washington, D.C., 1978.
8. Report, Subcommittee on Standardization of Common and Botanical Names of Weeds, Composite list of weeds, *Weed Sci.*, 19, 435, 1971.
9. **Anon.**, Progress report on the seed vigor testing handbook, *Assoc. Off. Seed Anal. Newsletter,* 50, 1, 1976.
10. **Heydecker, W.**, The "vigour" of seeds — a review, *Proc. Int. Seed Test. Assoc.*, 34, 201, 1969.

IDENTIFICATION AND CONTROL OF SEED-BORNE PATHOGENIC MICROORGANISMS

M. M. Kulik

The microorganisms responsible for a large number of plant diseases, some of which are of major economic importance, may be transmitted by seeds. Some of the more important pathogens are of common occurrence in seeds but others are infrequently encountered. Seeds that carry pathogenic microorganisms are of agricultural importance because:

1. They may be low in vigor or nongerminable, leading to reduced seedling stands and a possible lowering of yields.
2. They can supply inoculum that may spread disease in the developing crop, perhaps leading to yield reduction.
3. They can introduce exotic pathogens into areas that are free of them.
4. Preharvest infection of seeds by pathogenic microorganisms may reduce seed quality and the yield of viable seed.[25]

Seed-borne pathogenic microorganisms consist of fungi, bacteria, viruses, and nematodes (listed in descending order of occurrence).

IDENTIFICATION OF SEED-BORNE PATHOGENIC MICROORGANISMS

Seed-borne fungi are most often detected by placing seeds on a suitable substrate such as a nutrient agar or a moist blotter in a petri dish, followed by incubation to allow the fungi to grow out from the seeds. The fungi are then identified using a microscope to observe their fruiting structures and conidia. There are special techniques for certain seed-borne fungi. See Neergaard[27] for a detailed treatment of seed-borne fungi and their identification.

Seed-borne bacteria are usually detected by culturing seeds on a nutrient agar, followed by incubation to allow the bacteria to develop around the seeds. The bacteria are then transferred to special diagnostic media for identification. Certain seed-borne bacteria may be identified by the phage-plaque method, by serological methods, or by symptom expression in seedlings that develop from infected seeds. See Gorlenko[21] for a discussion of bacteria that cause plant disease, and Neergaard[27] for seed-borne bacteria.

Seed-borne viruses may be detected by sowing the seeds and observing the symptoms that develop in the resulting plants. It may also be necessary for positive identification to determine certain viral properties such as the thermal inactivation temperature or the dilution end point. Serological tests and examination of the virus under an electron microscope may also be necessary to make a positive identification. Refer to Gibbs and Harrison[20] for a general treatment of plant viruses, and to Phatak[33] for detailed information on the identification of seed-borne viruses.

Seed-borne nematodes are detected by maceration of infected seeds to liberate the nematodes. Identification of nematodes is based on their morphology as observed under a compound microscope. See Jenkins and Taylor[20] for a detailed discussion of plant nematodes.

CONTROL OF PATHOGENIC SEED-BORNE MICROORGANISMS

Methods that have been recommended for the control of diseases incited by seed-borne bacteria and fungi include:

1. Sowing of noninfected seeds through production of seeds in disease-free areas; eradication of seed-borne inoculum by chemical, hot water, or hot air treatment; and use of seed that has been tested for seed-borne bacterial and fungal pathogens and found to be free of them
2. Fumigation, chemical, or heat treatment of infested soil
3. Use of resistant or less susceptible cultivars
4. Use of certain cultural practices — deep plowing to bury infected crop debris, infected weeds, and fungal bodies; crop rotation using nonsusceptible cultivars, or allowing the land to remain fallow; burning of crop residues; flooding; good drainage; use of furrow irrigation in place of sprinkler irrigation; and adjusted date of seed sowing
5. Disinfestation of tools and machinery
6. Adjustment of soil pH
7. Application of fertilizers
8. Seed treatment with protectant chemicals
9. Spraying with fungicides and insecticides
10. Quarantine measures
11. Removal of diseased host plants (i.e., rogueing) or weeds

Seed-borne viruses may be controlled by:

1. Sowing of noninfected seeds through production of seeds in disease-free areas, eradication of virus through hot water treatment, and use of seeds from lots that have been indexed (i.e., grown out and found to be free of virus symptoms)
2. Spraying with insecticides to control insect vectors
3. Quarantine measures
4. Eradication of diseased plants (i.e., rogueing) and weed hosts
5. Use of resistant cultivars

Methods used to control nematodes include:

1. Crop rotation using nonsuceptible plants
2. Disinfestation of tools and machinery
3. Fumigation, chemical, or heat treatment of infested soil
4. Use of resistant cultivars
5. Soaking seeds in hot water to kill nematodes

TABULAR INFORMATION

In Tables 1 to 4 are listed seed-borne diseases of economic importance, according to Noble and Richardson,[28] incited by bacteria, fungi, nematodes, and viruses that affect commonly grown crops. The scientific names of the fungi are those most commonly used, and may not always be the latest acceptable name, according to the rules of botanical nomenclature. The scientific names of the seed-borne viruses are from Phatak.[33] In selecting the references for the identification and control of the approximately 400 seed-borne diseases listed in Tables 1 to 4, it was the author's intent to:

1. Cite as few references as possible, for the convenience of the reader
2. Cite references that would be available in the library of an agricultural college or university

To accomplish this, works of a general nature were cited in most instances. Any reader who wishes to refer to an original description of a pathogen or method of control will find the pertinent literature citations given in these general works. The scientific names of the hosts are according to Terrell.[42]

Table 1
SEED-BORNE BACTERIA

Host: Common name	Host: Scientific name	Disease	Bacterium	Identification ref.	Method of control[a]	Control ref.
Cereals						
Barley	*Hordeum vulgare*	Basal glume rot	*Pseudomonas atrofaciens*	6, 18	ST burning	28
		Bacterial blight	*Xanthomonas translucens* f. sp. *hordei*	6, 18	ST	18
Corn	*Zea mays*	Holcus spot	*P. syringae*	6, 18	RC, CR	1
		Bacterial leaf bight	*Erwinia stewartii*	6, 18	RC, insect	1
		Nebraska bacterial wilt and leaf freckles	*Corynebacterium nebraskense*	37	CR, CP	1
		Bacterial stalk rot	*E. Carotovora* f. sp *zeae*	6	RC, CP	1
Oat	*Avena sativa*	Bacterial stripe blight	*P. striafaciens*	12, 18	—	
		Halo blight	*P. coronafaciens*	12, 18	ST, CR, CP	17
Rice	*Oryza sativa*	Bacterial blight	*X. oryzae*	12, 31	RC, CP, spray	31
		Bacterial sheath rot	*P. oryzicola*	12	—	
Wheat	*Triticum aestivum*	Basal glume rot	*P. atrofaciens*	18	ST	49
		Black chaff	*X. translucens* f. sp. *undulosa*	6, 18	RC, CP	49
		Spike blight	*C. tritici*	6, 12	CR, good drainage	49
Forage grasses and legumes						
Alfalfa	*Medicago sativa*	Bacterial wilt	*C. insidiosum*	6, 12	RC	17
Cowpea	*Vigna unguiculata* subsp. *unguiculata*	Bacterial spot	*P. syringae*	18	—	
		Bacterial blight	*X. vignicola*	6,12,18	—	
Lespedeza, Korean	*Lespedeza stipulacea*	Bacterial wilt	*X. lespedezae*	18	—	

Orchardgrass	*Dactylis glomerata*	Yellow slime disease	*C. rathayi*	6, 12, 18	ST	17
Sorghum	*Sorghum bicolor*	Leaf streak	*X. holcicola*	18	ST, CR, CP	41
Sudangrass	*Sorgum sudanense*	Leaf streak	*X. holcicola*	18	ST, CR, CP	41
Wheatgrass beardless	*Agropyron spicatum* f. *inerme*	Bacterial blight	*P. coronafaciens* var *atropurpurea*	18	—	
Wheatgrass, western	*Agropyron smithii*	Head blight	*C. agropyri*	6, 18	ST	17
Fiber and oilseeds						
Castorbean	*Ricinus communis*	Bacterial leaf spot	*X. ricini*	12	—	
Cotton	*Gossypium* spp.	Angular leaf spot	*X. malvacearum*	12	CR, RC, ST, CP	17
Sesame	*Sesamum indicum*	Bacterial leaf spot	*P. sesame*	12	ST	28
		Bacterial leaf spot	*X. sesame*	12	—	
Soybean	*Glycine max*	Bacterial pustule	*X. phaseoli* var. *sojense*	18	CR, RC, CP	2
		Wildfire	P. tabaci	18	CR, RC, CP	2
		Bacterial wilt	*C. flaccumfaciens*	6, 12, 18	CP	2
		Bacterial wilt	*P. solanacearum*	6, 12	CR, CP	2
		Bacterial blight	*P. glycinea*	18	CR, CP	2
Miscellaneous						
Poppy, Opium	*Papaver somniferum*	Bacterial blight	*X. papavericola*	18	CP, spray	18
Tobacco	*Nicotiana tabacum*	Angular leaf spot	*P. angulata*	18	CR, RC, CP, spray	17
		Philippine leaf spot	*P. polycolor*	18	ST	18
		Bacterial leaf spot	*P. tabaci*	12, 18	CR, RC, CP, spray	17
		Wisconsin leaf spot	*P. mellea*	18	—	

Table 1 (continued)
SEED-BORNE BACTERIA

Host						
Common name	Scientific name	Disease	Bacterium	Identification ref.	Method of control[a]	Control ref.
Vegetables						
Bean, green	*Phaseolus vulgaris*	Brown spot blight	*P. syringae*	6, 12, 18	CP, spray	9
		Bacterial wilt	*C. flaccumfaciens*	6, 18	CR, RC, ST	9
		Bacterial wilt	*C. flaccumfaciens* var *aurianticum*	6	—	
		Common blight	*X. phaseoli*	12, 18	CR, RC, ST	9
		Fuscous blight	*X. phaseoli* var. *fuscans*	12, 18	CR, RC, ST	9
		Halo blight	*P. phaseolicola*	12, 18	CR, RC, ST	9
Bean, lima	*Phaseolus lunatus*	Brown spot blight	*P. syringae*	6, 12, 18	CP, spray	9
Cabbage	*Brassica oleracea* var. *capitata*	Bacterial leaf spot	*P. maculicola*	18	CR, HW, SOT	9
		Black rot	*X. campestris,*	6,12,18	CR, HW, ST, spray	9
Carrot	*Daucus carota*	Bacterial blight	*X. carotae*	18	CR, HW	9
Cauliflower	*Brassica oleracea* var. *botrytis*	Bacterial leaf spot	*P. maculicola*	18	CR, HW, SoT	9
Celery	*Apium graveolens*	Bacterial blight	*P. apii*	18:55	CR, spray	9
		Soft rot	*E. carotovora*	6, 18	HW, ST	28
Cucumber	*Cucumis sativus*	Angular leaf spot	*P. lachrymans*	12, 18	CR, ST, spray	9
Lettuce	*Lactuca sativa*	Bacterial leaf blight	*P. cichori*	6, 18	CR, CP	9
Pea, garden	*Pisum sativum*	Bacterial blight	*P. pisi*	12, 18	CR, ST, CP	9
Pepper	*Capsicum annuum*	Bacterial leaf spot	*X. vesicatoria*	12, 18	CR, ST, CP, spray	9
Radish	*Raphanus sativus*	Bacterial leaf spot	*X. vesicatoria* var. *raphani*	18	CR, ST	9
Squash	*Cucurbita* spp.	Bacterial leaf spot	*X. cucurbitae*	18	CR, ST, spray	9

Tomato	*Lycopersicon esculentum*	Bacterial canker	*C. michiganense*	6, 12, 18	CR, ST, CP, spray	9
		Bacterial spot	*X. vesicatoria*	12, 18	CR, ST, CP, spray	9
		Bacterial speck	*P. tomato* (syn *P. punctulans*)	18	CR, ST, CP, spray	9

[a] Abbreviations for methods of control: CP — cultural practices; CR — crop rotation; EVR — eliminate virus reservoirs; HW — soak seed in hot water; Insect — spray with insecticides to eliminate vectors; RC — resistant cultivars; SI — seed indexing; SOT — fumigation, chemical, or heat treatment of infested soil; Spray — spray with fungicides or antibiotics; ST — seed treatment with chemicals.

Table 2
SEED-BORNE FUNGI

Host			Fungus				
Common name	Scientific name	Disease	Imperfect state	Perfect state	Identification Ref.	Method of control[a]	Control Ref.
Cereals							
Barley	*Hordeum vulgare*	Black point	*Helminthosporium sativum*	*Cochliobolus sativus*	39	CR, RC, ST, CP	17
		Black smut		*Ustilago arenae*	12, 17	ST	17
		Cephalosporium stripe disease	*Cephalosporium gramineum*		12, 39	see "wheat"	
		Covered smut		*Ustilago hordei*	17	ST	17
		Ergot		*Claviceps purpurea*	39, 49	See "rye"	
		Glume blotch	*Septoria nodorum*	*Leptosphaeria nodorum*	12, 49	See "wheat"	
		Loose smut		*Ustilago nuda*	12, 17	HW	17
		Net blotch	*Helminthosporium teres*	*Pyrenophora teres*	12, 39	CR, ST, CP	17
		Scab	*Fusarium graminearum*	*Gibberella zeae*	12, 31	see "wheat"	
		Scald	*Rhynchosporium secalis*		12, 39	CR, RC	17
		Septoria leaf blotch	*Septoria passerinii*		12, 39	—	
		Snow mold	*Fusarium nivale*	*Griphosphaeria nivalis*	12, 49	See "wheat"	
		Stripe disease	*Helminthosporium gramineum*	*Pyrenophora graminea*	12, 39	RC, ST	17
Corn	*Zea mays*	Anthracnose	*Colletotrichum graminicola*		39	CR, RC, CP	1
		Brown spot	*Physoderma maydis*		1	RC	1
		Charred ear mold	*Helminthosporium carbonum*	*Cochliobolus carbonum*	1, 12	RC	1
		Crazy spot	*Sclerophthora macrospora*		1	Soil drainage	1

		Diplodia ear rot	*Diplodia maydis*		1	RC, CP, less plants	1
		Dry rot	*Diplodia macrospora*		39	—	
		Ear rot	*Fusarium moniliforme*	*Gibberella fujikuroi*	1, 12	RC, CP, Less plants	1
		False smut		*Ustilaginoidea virens*	1, 12	—	
		Gray ear rot	*Macrophoma zeae*	*Physalospora zeae*	1	Adapted cultivars	1
		Kernel rot	*Cephalosporium acremonium*		1	—	1
		Late wilt	*Cephalosporium maydis*	*Gaeumannomyces graminis*	1	CR, RC	1
		Nigrospora cob and stalk rot	*Nigrospora oryzae*	*Khuskia oryzae*	1, 12	RC, CP	1
		Pink ear rot	*Fusarium graminearum*	*Gibberella zeae*	12, 31	RC, CP, less plants	1
		Smut		*Ustilago maydis*	1, 12	CP, avoid susceptible cultivars	1
		Southern leaf blight	*Helminthosporium maydis*	*Cochliobolus heterostrophus*	1, 12	RC, spray	1
		Stalk and ear rot	*Diplodia frumenti*	*Physalospora zeicola*	39	ST	28
Millet	*Panicum miliaceum*	Head smut		*Sphacelotheca destruens*	12, 35	ST	35
Oat	*Avena sativa*	Black loose smut		*Ustilago avenae*	12, 17	RC, ST	17
		Cephalosporium stripe disease	*Cephalosporium gramineum*		12, 39	See "wheat"	
		Covered smut		*Ustilago hordei*	17	RC	17
		Downy mildew	*Sclerophthora macrospora*		1	—	
		Helminthosporium blight	*Helminthosporium victoriae*	*Cochliobolus victoriae*	39	CR, RC, ST	17
		Helminthosporium leaf blotch	*Helminthosporium avenae*	*Pyrenophora avenae*	12, 39	ST	28
		Scab	*Fusarium graminearum*	*Gibberella zeae*	12, 31	See "wheat"	

Table 2 (Continued)
SEED-BORNE FUNGI

Host			Fungus				
Common name	Scientific name	Disease	Imperfect state	Perfect state	Identification Ref.	Method of control[a]	Control Ref.
		Seedling blight	*Helminthosporium sativum*	*Cochliobolus sativus*	39	See "wheat"	
		Septoria leaf blotch	*Septoria avenae f. sp. avenae*	*Leptosphaeria avenaria f. sp. avenaria*	12, 39	RC	17
		Snow mold	*Fusarium nivale*	*Griphosphaeria nivalis*	12, 49	See "wheat"	
Rice	*Oryza sativa*	Fusarium blight; bakanae disease	*Fusarium moniliforme*	*Gibberella fujikuroi*	12, 31	ST	31
		Black kernel	*Curvularia* spp.		31	ST	28
		Blast	*Pyricularia oryzae*		12, 31	Spray	31
		Brown spot	*Helminthosporium oryzae*	*Cochliobolus miyabeanus*	12, 31	CP	31
		False smut		*Ustilaginoidea virens*	12, 31	Spray	31
		Glume spot	*Cercospora oryzae*	*Sphaerulina oryzina*	12, 31	RC	31
		Kernel spot		*Tilletia barclayana*	12, 31	RC	31
		Scab	*Fusarium graminearum*	*Gibberella zeae*	12, 31	See "wheat"	
		Stackburn	*Trichoconis padwickii*		12, 31	HW	31
		Udbatta disease	*Ephelis oryzae*	*Balansia oryzae*	31	HW	31
Rye	*Secale cereale*	Blind seed disease		*Gloeotinia temulenta*	39	See "bromegrasses"	
		Bunt		*Tilletia contraversa*	17	ST	28
		Ergot		*Claviceps purpurea*	39, 49	CR, clean seed	17
		Scab	*Fusarium graminearum*	*Gibberella zeae*	12, 31	See "wheat"	
		Snow mold	*Fusarium nivale*	*Griphosphaeria nivalis*	12, 49	See "wheat"	
		Stem smut		*Urocystis occulta*	17	ST	17

Wheat	*Triticum aestivum*	Cephalosporium stripe disease	*Cephalosporium gramineum*		12, 49	CR, CP	49
		Common bunt; rough-spored bunt		*Tilletia caries*	49	RC, ST, clean seed	49
		Common bunt; smooth-spored bunt		*Tilletia foetida*	49	RC, ST, clean seed	49
		Crown rot	*Helminthosporium sativum*	*Cochliobolus sativus*	49	CR, ST, CP	49
		Downy mildew	*Sclerophthora macrospora*		49	CR, good drainage	49
		Dwarf bunt		*Tilletia contraversa*	17	ST, clean seed	49
		Ergot		*Claviceps purpurea*	39, 49	CR, clean seed	49
		Flag smut		*Urocystis agropyri*	49	RC, ST	49
		Karnal bunt		*Neovossia indica*	49	RC, ST	49
		Leaf spot	*Helminthosporium cyclops*	*Pyrenophora semeniperda*	39	—	
		Leaf spot	*Helminthosporium tritici-repentis*	*Pyrenophora tritici-repentis*	12, 49	RC, ST, CP, spray	49
		Loose smut		*Ustilago tritici*	49	HW, RC, ST	49
		Scab	*Fusarium graminearum*	*Gibberella zeae*	12, 31	CR, ST, CP	49
		Septoria leaf and glume blotch	*Septoria nodorum*	*Leptosphaeria nodorum*	12, 49	CR, RC, CP,spray	49
		Snow mold	*Fusarium nivale*	*Griphosphaeria nivalis*	12, 49	ST	49
Forage grasses and legumes							
Alfalfa	*Medicago sativa*	Leaf, stem, and pod spot	*Stemphylium botryosum*	*Pleospora herbarum*	12, 17	—	
		Spring black stem	*Phoma medicaginis*		17, 36	CR, RC, ST	17
		Wilt	*Verticillium albo-atrum*		12, 17	—	
Bluegrasses	*Poa* spp.	Dwarf bunt		*Tilletia contraversa*	17	See "rye"	
		Ergot		*Claviceps purpurea*	14, 49	See "rye"	
		Flag smut		*Urocystis agropyri*	14, 49	RC, ST, spray	14
		Melting out	*Helminthosporium vagans*		14, 39	—	

Table 2 (Continued) SEED-BORNE FUNGI

Host			Fungus				
Common name	Scientific name	Disease	Imperfect state	Perfect state	Identification Ref.	Method of control[a]	Control Ref.
Bromegrasses	*Bromus* spp.	Blind seed disease		*Gloeotinia temulenta*	14, 39	ST, CP	14
		Head smut		*Ustilago bullata*	17	RC, ST	28
		Leaf spot	*Helminthosporium bromei*	*Pyrenophora bromi*	39	—	
		Leaf spot	*Selenophoma bromigena*		39	—	
		Helminthosporium leaf spot	*Helminthosporium sativum*	*Cochliobolus sativus*	14, 39	See "wheat"	
Clovers	*Trifolium* spp.	Anther mold	*Botrytis anthophila*		32, 36	—	
		Black patch disease	*Rhizoctonia leguminicola*		48	—	
		Leaf spot	*Stemphylium botryosum*	*Pleospora herbarum*	12, 17	—	
		Leaf spot	*Stemphylium sarciniforme*		17, 36	—	
		Northern anthracnose	*Kabatiella caulivora*		17, 36	RC	17
		Sclerotinia root rot and crown rot		*Sclerotinia sclerotiorum*	2, 12	RC	17
		Sclerotinia root rot and crown rot		*Sclerotinia trifoliorum*	36	RC	17
		Southern anthracnose	*Collelotrichum trifolii*		17	RC	17
		Summer black stem and leaf spot	*Cercospora zebrina*		8, 17	RC, burning	17
		Wilt	*Verticillium albo-atrum*		12, 17	—	
Cowpea	*Vigna unguiculata* sub sp. *unguiculata*	Leaf and pod spot	*Ascochyta phaseolorum* (syn. *Phoma exigua)*		12, 5	—	

Fescues	*Festuca* spp.	Blind seed disease		*Gloeotinia temulenta*	14, 39	See "bromegrasses"	
		Choke		*Epichloe typhina*	14,39	—	
		Ergot		*Claviceps purpurea*	14, 49	See "rye"	
		Head smut		*Ustilago bullata*	17	—	
		Helminthosporium leaf spot	*Helminthosporium sativum*	*Cochliobolus sativus*	39	See "wheat"	
		Net blotch	*Helminthosporium dictyoides*	*Pyrenophora dictyoides*	12, 39	—	
Lupines	*Lupinus* spp.	Anthracnose		*Glomerella cingulata*	12	Hot air	28
		Brown spot	*Pleiochaeta setosa*		12	HW	28
Orchardgrass	*Dactylis glomerata*	Crown rot	*Helminthosporium sativum*	*Cochliobolus sativus*	39	See "wheat"	
		Dwarf bunt		*Tilletia contraversa*	17	See "rye"	
		Ergot		*Claviceps purpurea*	14, 49	See "rye"	
Ryegrasses	*Lolium* spp.	Blind seed disease		*Gloeotinia temulenta*	14, 39	See "bromegrasses"	
		Ergot		*Claviceps purpurea*	14, 49	See "rye"	
Sorghum	*Sorghum bicolor*	Anthracnose	*Colletotrichum graminicola*		12, 39	CR, RC, CP	41
		Covered smut		*Sphacelolotheca sorghi*	12, 41	RC, ST	41
		Gibberella seedling blight	*Fusarium moniliforme*	*Gibberella fujikuroi*	12, 41	See "rice"	
		Gloeocercospora leaf spot	*Gloeocercospora sorghi*		12, 41	CR, ST, CP	41
		Helminthosporium leaf blight	*Helminthosporium turcicum*	*Tricometasphaeria turcica*	12, 41	CR, ST, CP, spray	41
		Leaf spot	*Phoma insidiosa*		12, 41	CR, ST, CP	41
		Long smut		*Tolyposporum ehrenbergii*	12, 41	CR, Clean seed	41
		Loose smut		*Spacelotheca cruenta*	12, 41	ST	41
		Melo root rot	*Periconia circinata*		12, 41	RC	41
		Red rot	*Colletotrichum falcatum*	*Glomerella tucumensis*	12, 41	—	

Table 2 (Continued)
SEED-BORNE FUNGI

Host			Fungus				
Common name	Scientific name	Disease	Imperfect state	Perfect state	Identification Ref.	Method of control[a]	Control Ref.
		Sooty stripe	*Ramulispora sorghi*		39, 41	CR, ST, CP, Clean seed	41
Sudangrass	*Sorghum sudanense*	Anthracnose	*Colletotrichum graminicola*		12, 39	See "sorghum"	
		Covered smut		*Sphacelotheca sorghi*	12, 41	See "sorghum"	
		Helminthosporium leaf blight	*Helminthosporium turcicum*	*Trichometasphaeria turcica*	12, 41	See "sorghum"	
		Long smut		*Tolysporium ehrenbergii*	12, 41	See "sorghum"	
Sweetclovers	*Melilotus* spp.	Stem canker	*Ascochyta caulicola*		17	CR, ST, burning	17
		Summer black stem and leaf spot	*Cercospora davisii*	*Mycosphaerella davisii*	17	RC, burning	17
Timothy	*Phleum pratense*	Anthracnose	*Colletotrichum graminicola*		12, 39	—	
		Ergot		*Claviceps purpurea*	14, 49	See "rye"	
		Leaf streak	*Helminthosporium dictyoides* var. *phlei*		14, 39	—	
		Stripe smut		*Ustilao striiformis*	14, 17	ST	28
Vetches	*Vicia* spp.	Chocolate spot	*Botrytis fabae*		12	HW	28
		Downy mildew	*Peronospora viciae*		12, 36	—	
		Rust		*Uromyces viciae-fabae*	12	—	
Wheatgrasses	*Agropyron* spp.	Common bunt		*Tilletia caries*	17	See "wheat"	
		Dwarf bunt		*Tilletia contraversa*	17	See "rye"	
		Head smut		*Ustilago bullata*	17	—	

Fiber and oilseeds							
Castorbean	*Ricinus communis*	Gray mold	*Botrytis ricini*	*Sclerotinia ricini*	12	—	
		Seedling blight	*Alternaria ricini*		19	—	
Cotton	*Gossypium* spp.	Anthracnose	*Colletotrichum gossypii*	*Glomerella gossypii*	17	CR, ST	17
		Ascochyta blight	*Ascochyta gossypii*		17	CR, ST, CP	17
		Boll rot		*Nematospora coryli*	12	See "soybean"	
		Diplodia rot	*Diplodia gossypina*	*Physalospora rhodina*	39	—	
		Fusarium wilt	*Fusarium oxysporum* f. sp. *vasinfectum*		12, 17	RC, CP	17
		Soreshin	*Rhizoctonia solani*		12, 17	CR, ST	17
		Stem blight	*Macrophomina phaseolina*		12, 39	—	
		Verticillium wilt	*Verticillium alboatrum*		12, 17	CR, RC	17
		Verticillium wilt	*Verticillium dahliae*		12, 17	CR, RC	17
Flax	*Linum usitatissimum*	Alternaria blight	*Alternaria linicola*		26	ST	26
		Anthracnose	*Colletotrichum lini*		17, 26	CR, RC, ST	17
		Browning and stem break	*Aureobasidium lini*	*Guignardia fulvida*	17, 26	RC, ST	17
		Foot rot	*Phoma exigua* var. *linicola*		26	ST	26
		Fusarium wilt	*Fusarium oxysporum* f. sp. *lini*		17, 26	RC	17
		Gray mold	*Botrytis cinerea*	*Sclerotinia fuckeliana*	12,26	ST,CP	26
		Pasmo	*Septoria linicola*	*Mycosphaerella linorum*	17, 26	CR, RC, ST, CP	17
Peanut	*Arachis hypogaea*	Cercospora leaf spot	*Cercospora arachidicola*	*Mycospharella arachidis*	8, ,12	CR, CP, spray	17
		Cercospora leaf spot	*Cercospora personata*	*Mycosphaerella berkeleyi*	8, 12	CR, CP, spray	17
		Crown rot	*Aspergillus niger*		12, 45	ST	28

Table 2 (Continued)
SEED-BORNE FUNGI

Host: Common name	Host: Scientific name	Disease	Fungus: Imperfect state	Fungus: Perfect state	Identification Ref.	Method of control[a]	Control Ref.
		Dry rot	*Diplodia gossypina*	*Physalospora rhodina*	15, 39	—	
		Root and stem rot	*Macrophomina phaseolina*		12, 39	See "soybean"	
		Seedling disease	*Rhizoctia solani*		12, 17	See "bean, green"	
Safflower	*Carthamus tinctorius*	Leaf spot	*Cercospora carthami*		8	—	
		Rust		*Puccinia carthami*	7	ST	28
		Wilt	*Fusarium oxysporum* f. sp. *carthami*		24	—	
Sesame	*Sesamum indicum*	Blight	*Alternaria sesami*		12, 19	ST	43
		Charcoal rot	*Macrophomina phaseolina*		12, 39	See "soybean"	
		Leaf blotch	*Helminthosporium sesami*		34	CP	28
		Leaf spot	*Cercospora sesami*		8	HW	30
Soybean	*Glycine max*	Anthracnose	*Colletotrichum dematium* var *truncata*		2, 17	ST, CP, spray	2
		Anthracnose	*Colletotrichum glycines*	*Glomerella glycines*	2, 17	ST, CP, spray	2
		Black patch	*Rhizoctonia leguminicola*		48	—	
		Brown spot	*Septoria glycines*		2, 12	CR, CP	2
		Charcoal rot	*Macrophomina phaseolina*		2, 12	CR, CP, irrigation	2
		Damping off	*Rhizoctonia solani*		2, 12	See "bean green"	
		Downy mildew	*Peronospora manshurica*		2, 17	CR, RC, ST, CP	2

		Phyllosticta leaf spot	*Phyllosticta sojaecola*	*Pleosphaerulina sojaecola*	2	CR, CP	2
		Pod and collar rot	*Fusarium semitectum*		2, 12	RC, CP	2
		Pod and stem blight	*Phomopsis sojae*	*Diaporthe phaseolorum* var. *sojae*	2, 12	CR, CP, spray	2
		Purple stain	*Cercospora kikuchii*		8, 12	RC, ST, spray, later sowing	2
		Root and stem rot	*Phytophthora megasperma* var. *sojae*		2, 12	RC, CP	2
		Stem canker		*Diaporthe phaseolorum* var. *caulivora*	2, 12	CR, RC, CP, spray	2
		Target spot	*Corynespora cassiicola*		2, 12	RV	2
		Yeast spot		*Nematospora coryli*	2, 12	Insect	2
Sunflower	*Helianthus annuus*	Downy mildew	*Plasmoplara halstedii*		44	—	
		Gray mold	*Botrytis cinerea*	*Sclerotinia fuckeliana*	12, 45	ST	28
		Wilt		*Sclerotinia sclerotiorum*	2, 12	ST	28
		Wilt	*Verticillium alboatrum*		12, 17	—	
Miscellaneous							
Poppy, opium	*Papaver somniferum*	Downy mildew	*Peronospora arborescens*		3	—	
		Leaf and spot	*Helminthosporium papaveris*	*Pleospora papaveracea*	19	—	
		Seedling blight	*Fusarium acuminatum*		39	—	
Tobacco	*Nicotiana tabacum*	Brown spot	*Alternaria longipes*		12	ST	28
		Downy mildew, blue mold	*Peronospora tabacina*		17	SoT, spray	17
		Frog-eye leaf spot	*Cercospora nicotianae*		8, 12	ST, CP, spray	17

Table 2 (Continued)
SEED-BORNE FUNGI

Host			Fungus				
Common name	Scientific name	Disease	Imperfect state	Perfect state	Identification Ref.	Method of control[a]	Control Ref.
Vegetables							
Bean broad	*Vicia faba*	Chocolate spot	*Botrytis fabae*		12	HW	28
		Leaf and pod spot	*Ascochyta fabae*		12	—	
		Rust	*Uromyces viciae-fabae*		12	—	
green	*Phaseolus vulgaris*	Anthracnose	*Colletotrichum lindemuthianum*		12, 45	CR, RC, CP, spray	9
		Asochyta leaf spot	*Ascochyta boltshauseri*		45, 50	CR, CP	9
		Ashy stem blight	*Macrophomina phaseolina*		12, 45	ST, CP	9
		Damping off	*Rhizoctonia solani*		12, 45	CP	9
		Isariopsis leaf spot	*Phaeoisariopsis griseola*		45, 50	CR, RC, CP	9
		Soft rot		*Sclerotinia sclerotiorum*	12, 45	CR, CP, spray	9
		Yellows	*Fusarium oxysporum* f. sp. *phaseoli*		45	CR, CP	9
lima	*Phaseolus lunatus*	Anthracnose	*Colletotrichum lindemuthianum*		12, 45	CR, RC, CP, spray	9
		Ascochyta leaf spot	*Ascochyta boltshauseri*		45, 50	CR, CP	9
		Ashy stem blight	*Macrophomina phaseolina*		12, 45	ST, CP	9
		Damping off	*Rhizoctonia solani*		12, 45	CR, RC, CP	9
		Downy mildew	*Phytophthora phaseoli*		45	CR, CP, spray	9
		Pod blight		*Diaporthe phaseolorum*	12, 45	CR, spray	9
		Scab		*Elsinoe phaseoli*	12, 45	CR, spray	9

		Stem anthracnose	*Colletotrichum dematium* var. *truncata*		2, 45	CR, CP, spray	9
		soft rot		*Sclerotinia sclerotiorum*	12, 45	CR, CP, spray	9
Beet	*Beta vulgaris*	black leg	*Phoma betae*	*Pleospora bjoerlingii*	12, 45	CR, CP	9
		downy mildew	*Peronospora farinosa* (syn. *Peronospora schachtii)*		45	CR, HW, CP, spray	9
		leaf spot	*Cercospora beticola*		8, 45	CR, RC, ST, CP	9
		rust		*Uromyces betae*	12, 45	spray	9
Cabbage and	*Brassica oleracea* var *capitata* and	Alternaria leaf spot and blight	*Alternaria brassicae*		12, 45	CR, HW, CP, spray	9
cauliflower	*Brassica oleracea* var. *botrytis*	Black leg	*Phoma lingam*	*Leptosphaeria maculans*	12, 45	CR, HW, ST, CP, SoT	9
		Black spot	*Alternaria brassicicola*		12, 45	CR, HW, CP, spray	9
		Ring spot		*Mycosphaerella brassicicola*	12, 45	CR, HW, CP, spray	9
		Watery soft rot		*Sclerotinia sclerotiorum*	12, 45	CR, CP, spray	9
Carrot	*Daucus carota*	Alternaria blight	*Alternaria porri* f. sp. *dauci*		45	CR, HW, ST, CP	9
		Stem root rot	*Stemphylium radicinum*		45	CR, HW, ST, CP	9
Celery	*Apium graveolens*	Late blight	*Septoria apiicola* (syn *Septoria apii)*		45	HW, spray	9
		Phoma root and crown rot	*Phoma apiicola*		45	HW, spray	9
Chickpea	*Cicer arietinum*	Anthracnose	*Ascochyta rabei*	*Mycosphaerella rabei*	12, 47	—	
		Wilt	*Fusarium lateritum* f. sp. *ciceri*	*Gibberella baccata*	12, 47	—	

Table 2 (Continued)
SEED-BORNE FUNGI

Host			Fungus				
Common name	Scientific name	Disease	Imperfect state	Perfect state	Identification Ref.	Method of control[a]	Control Ref.
Cucumber	*Cucumis sativis*	Anthracnose	*Colletotrichum orbiculare* (syn. *Colletotrichum lagenarium)*		45	CR, RC, ST, CP, spray	9
		Black rot	*Phoma cucurbitacearum*	*Didymella bryoniae* (syn. *Mycosphaerella melonis)*	12, 45	CR, ST, CP, spray	9
		Leaf blight	*Alternaria cucumerina*		12, 19	CR, ST, CP	9
Eggplant	*Solanum melongena*	Anthracnose	*Colletotrichum acutatum*		38	—	
		Phomopsis blight	*Phomopsis vexans*	*Diaporthe vexans*	45	CP, CR, HW, RC, ST, spray	9
		Wilt	*Verticillium alboatrum*		12, 45	CP, CR	9
Lettuce	*Lactuca sativa*	Downy mildew	*Bremia lactucae*		45	CP, CR, RC, spray	9
		Gray mold rot	*Botrytis cinerea*	*Sclerotinia fuckeliana*	12, 45	CP, ST, spray, SoT	9
Okra	*Abelmoschus esculentus*	Ascochyta blight	*Aschochyta abelmoschi*		47	CP, CR	9
Onion	*Allium cepa*	Gray mold neck rot	*Botrytis allii*		12, 45	Spray	9
		Purple blotch	*Alternaria porri*		12, 45	CP, CR, ST, spray	9
		Rust		*Puccinia allii*	12, 45	CP, CR, spray	9
Pea	*Pisum sativum*	Ascochyta foot rot	*Ascochyta pinodella*		45	CP, CR	9
		Ascochyta leaf and pod rot	*Ascochyta pisi*		12, 45	CP, CR	9
		Downy mildew	*Peronospora viciae*		12, 45	CP, CR, RC	9
		Mycosphaerella blight	*Ascochyta pinodes*	*Mycosphaerella pinodes*	12, 45	CP, CR	9

		Near wilt	*Fusarium oxysporum* f. sp. *pisi*		45	CP, CR, RC	9
		Powdery mildew		*Erysiphe pisi* (syn. *Erysiphe polygoni*)	12, 45	HW	9
Pepper	*Capsicum annuum*	Frog-eye leaf spot	*Cercospora capsici*		8, 45	CP, CR, spray	9
Radish	*Raphanus sativus*	Anthracnose	*Colletotrichum higginsianum*		45	CP, CR, HW	9
		Black leg	*Phoma lingam*	*Leptosphaeria maculans*	12, 45	CP, CR, HW, ST, SoT	9
		Leaf spot	*Alternaria raphani*		19, 45	CR, HW	9
Tomato	*Lycopersicon esculentum*	Early blight	*Alternaria porri* f. sp. *solani*		45	CP, CR, HW, spray	9
		Fusarium wilt	*Fusarium oxysporum* f. sp. *lycopersici*		12, 45	RC	9
		Late blight	*Phytophthora infestans*		45	Spray	9
		Leaf mold	*Cladosporium fulvum*		12, 45	RC, ST, spray	9
		Stem rot	*Phoma lycopersici*	*Didymella lycopersici*	12, 45	CP, CR, spray, SoT	9
		Verticillium wilt	*Verticillium dahliae*		12, 45	CP, CR	9
Watermelon	*Citrullus lanatus*	Anthracnose	*Colletotrichum orbiculare* (syn *Colletotrichum lagenarium*		45	CP, CR, RC, ST, spray	9
		Gummy stem blight	*Phoma cucurbitacearum*	*Didymella bryoniae* (syn. *Mycosphaerella melonis*)	12, 45	CP, CR, ST, spray	9
		Wilt	*Fusarium oxysporum* f. sp. *niveum*		12, 45	CR, RC	45

[a] Abbreviations for methods of control: CP — cultural practices; CR — crop rotation; EVR — eliminate virus reservoirs; HW — soak seed in hot water; Insect — spray with insecticides to eliminate vectors; RC — resistant cultivars; SI — seed indexing; SoT — fumigation, chemical, or heat treatment of infested soil; Spray — spray with fungicides or antibiotics; ST — seed treatment with chemicals.

Table 3
SEED-BORNE NEMATODES

Host: Common name	Host: Scientific name	Nematode: Common name	Nematode: Scientific name	Identification ref.	Method of control[a]	Control ref.
Cereals						
Oat	*Avena sativa*	Seed gall nematode	*Anguina tritici*	10	See "wheat"	
Rice	*Oryza sativa*	White tip nematode	*Aphelenchoides besseyi* (syn. *A. oryzae*	10	HW, ST, CP	31
Rye	*Secale cereale*	Seed gall nematode	*Anguina tritici*	10	See "wheat"	
Wheat	*Triticum aestivum*	Seed gall nematode	*Anguina tritici*	10	CR, HW, SC	49
Forage grasses and legumes						
Alfalfa	*Medicago sativa*	Stem eelworm	*Ditylenchus dipsaci*	10	SoT	28
Bentgrasses	*Agrostis* spp.	Grass seed nematode	*Anguina agrostis*	10	SoT	28
Bluegrasses	*Poa* spp.	Grass seed nematode	*Anguina agrostis*	10	see "bentgrasses"	
Clover, red	*Trifolium pratense*	Stem eelworm	*Ditylenchus dipsaci*	10	SoT	28
Fescues	*Festuca* spp.	Grass seed nematode	*Anguina agrostis*	10	See "bentgrasses"	
Orchardgrass	*Dactylis glomerata*	Grass seed nematode	*Anguina agrostis*	10	See "bentgrasses"	
Timothy	*Phleum pratense*	Grass seed nematode	*Anguina agrostis*	10	See "bentgrasses"	
Wheatgrass, western	*Agropyron smithii*	Seed gall nematode	*Anguina agropyronifloris*	29		

Oilseeds						
Soybean	*Glycine max*	Soybean cyst nematode	*Heterodera glycines*	2	CR, RC, SoT	2
Vegetables						
Bean, broad	*Vicia faba*	Stem eelworm	*Ditylenchus dipsaci*	10	CR, SoT, HW, SF, CP	9
Beet	*Beta vulgaris*	Canker eelworm	*Ditylenchus dipsaci*	10	See "bean, broad"	
Onion	*Allium cepa*	Bloat nematode	*Ditylenchus dipsaci*	10	See "bean, broad"	

[a] Abbreviations for methods of control: CP — cultural practices; CR — crop rotation; EVR — eliminate virus reservoirs; HW — soak seed in hot water; Insect — spray with insecticides to eliminate vectors; RC — resistant cultivars; SI — see indexing; SoT — fumigation, chemical, or heat treatment of infested soil; Spray — spray with fungicides or antibiotics; and ST — seed treatment with chemicals.

Table 4
SEED-BORNE VIRUSES

Host		Virus		Identification ref.	Method of control[a]	Control ref.
Common name	Scientific name	Common name	Scientific name			
Cereals						
Barley	*Hordeum vulgare*	Barley mottle mosaic virus	—	16	—	
		Barley stripe mosaic virus	*Minpachus hordei*	11	SI, HW, CP, RC	17
Corn	*Zea mays*	Sugarcane mosaic virus	*Maphiflexus sacchari*	11	EVR	1
Oat	*Avena sativa*	Barley stripe mosaic virus	*Minpachus hordei*	11	See "barley"	
Wheat	*Triticum aestivum*	Barley stripe mosaic virus	*Minpachus hordei*	11	CP	49
Forage grasses and legumes						
Alfalfa	*Medicago sativa*	Alfalfa mosaic virus	*Maphibrevia medicaginis*	11	—	
Bluegrasses	*Poa* spp.	Arabis mosaic virus	*Genemaglobus arabidis*	11	—	
		Tomato black ring virus	*Genemaglobus betae*	11	—	
Bromegrasses	*Bromus* spp.	Barley stripe mosaic virus	*Minpachus hordei*	11	See "barley"	
Clovers	*Trifolium* spp.	White clover mosaic virus	*Minflexus trifolii*	11	—	
		Red clover mosaic virus	—	11	—	
		Red clover vein mosaic virus	*Maphilancea trifolii*	11	—	
Cowpea	*Vigna unguiculata* subsp. *unguiculata*	Cowpea mosaic beetle-transmitted virus	*Minglobus vignae*	11	—	
		Cowpea ringspot virus	*Minglobus vignae-orbis*	33	—	

		Cucumber mosaic virus	*Maphiglobus cucumeris*	11	See "cucumber"
		Tobacco ringspot virus	*Genemaglobus nicotianae*	11	—
		Tomato black ring virus	*Genemaglobus betae*	11	—
		Bean common mosaic virus	*Maphiflexus phaseoli*	11	See "bean, green"
		Cowpea aphid-borne mosaic virus	*Maphiflexus vignae*	11	—
		Cowpea mild mottle virus	—	11	—
		Cowpea mosaic virus	*Maphivirus* sp.	11	—
		Soybean stunt virus	*Maphivirus sojae-nanae*	25	—
Lupines	*Lupinus* spp.	Cucumber mosaic virus	*Maphiglobus cucumeris*	11	See "cucumber"
		Bean yellow mosaic virus	*Maphiflexus phaseoli-flavi*	11	—
Ryegrasses	*Lolium* spp.	Barley stripe mosaic virus	*Minpachus hordei*	11	See "barley"
Sweet clovers	*Melilotus* spp.	Bean yellow mosaic virus	*Maphiflexus phaseoli-flavi*	11	See "bean, green"
Wheatgrasses	*Agropyron* spp.	Barley stripe mosaic virus	*Minpachus hordei*	11	See "barley"
Oilseeds					
Peanut	*Arachis hypogaea*	Peanut stunt virus	*Maphiglobus arachidis*	11	—
		Peanut mottle virus	*Maphiflexus arachidis*	11	—
Soybean	*Glycine max*	Tobacco streak virus	*Minglobus nico-lineatae*	11	—
		Soybean mild mosaic virus	—	40	

Table 4 (continued) SEED-BORNE VIRUSES

Host: Common name	Host: Scientific name	Virus: Common name	Virus: Scientific name	Identification ref.	Method of control[a]	Control ref.
		Arabis mosaic virus	*Genemaglobus arabidis*	11	—	
		Cherry leaf roll virus	*Genemaglobus pruni*	11	—	
		Tobacco ringspot virus	*Genemaglobus nicotianae*	11	—	
		Tomato black ring virus	*Genemaglobus betae*	11	—	
		Tomato ringspot virus	*Genemaglobus lycopersici*	11	—	
		Raspberry ringspot virus	*Genemaglobus rubi*	11	—	
		Grapevine yellow vein virus	—	13	—	
		Soybean mosaic virus	*Maphiflexus sojae*	11	SI, EVR, Insect	2
		Cowpea mild mottle virus	—	11	—	
Miscellaneous						
Tobacco	*Nicotiana tabacum*	Cherry leaf roll virus	*Genemaglobus pruni*	11	—	
		Tobacco ringspot virus	*Genemaglbus nicotianae*	11	—	
		Tomato ringspot virus	*Genemaglobus lycopersici*	11	—	
Vegetables						
Bean						
Broad	*Vicia faba*	Broadbean stain virus	*Minglobus viciae-necrospermae*	11	—	
		Broadbean true mosaic virus	*Minglobus viciae-varians*	11	—	
		Bean yellow mosaic virus	*Maphiflexus phaseoli-flavi*	11	See "bean, green"	

		Red clover mosaic virus	—	11	—	
		Red clover vein mosaic virus	*Maphilancea trifolii*	11	—	
Green	*Phaseolus vulgaris*	Bean southern mosaic virus 2	—	11	RC, 1-year-old seed	9
		Tobacco streak virus	*Minglobus nico-lineatae*	11	—	
		Cucumber mosaic virus	*Maphiglobus cucumeris*	11	See "cucumber"	
		Cherry leaf roll virus	*Genemaglobus pruni*	11	—	
		Cocoa necrosis virus	—	11	—	
		Bean common mosaic virus	*Maphiflexus phaseoli*	11	RC	9
		Cowpea mild mottle virus	—	11	—	
Mung	*Vigna radiata* var. *radiata*	Broad bean mottle virus	*Minglobus viciae-maculans*	11	—	
Beet	*Beta vulgaris*	Arabis mosaic virus	*Genemaglobus arabidis*	11	—	
		Raspberry ringspot virus	*Genemaglobus rubi*	11	—	
		Tomato black ring virus	*Genemaglobus betae*	11	—	
		Lychnis ringspot virus	*Minpachus lychnidis*	4	—	
Carrot	*Daucus carota*	Carrot red leaf virus	*Daphivirus dauci*	46	—	
Celery	*Apium graveolens*	Strawberry latent ringspot virus	*Genemaglobus fragariae*	11	—	
Cucumber	*Cucumis sativus*	Cucumber green mottle mosaic virus	*Minchorda cucumeris*	11	—	
		Cucumber mosaic virus	*Maphiglobus cucumeris*	11	EVR	9

Table 4 (continued)
SEED-BORNE VIRUSES

Host		Virus		Identification ref.	Method of control[a]	Control ref.
Common name	Scientific name	Common name	Scientific name			
Lettuce	*Lactuca sativa*	Arabis mosaic virus	*Genemaglobus arabidis*	11	—	
		Tobacco ringspot virus	*Genemaglobus nicotianae*	11	—	
		Lettuce mosaic virus	*Maphiflexus lactucae*	11	CR, SI, EVR	9
		Onion yellow dwarf virus	*Maphiflexus allii*	11	SI, RC, Insect	9
Pea	*Pisum sativum*	Bean yellow mosaic virus	*Maphiflexus phaseoli-flavi*	11	—	
		Pea leafrolling mosaic virus	*Maphivirus pisi-tubulans*	23	—	
		Pea seed-borne mosaic virus	*Maphiflexus pisi*	11	—	
		Pea early browning virus	*Genemapachus pisi*	11	—	
Pepper	*Capsicum annuum*	Tobacco mosaic virus	*Minchorda nicotianae*	11	—	
		Alfalfa mosaic virus	*Maphibrevia medicaginis*	11	Isolate from alfalfa	9
Pumpkin and squash	*Cucurbita* spp.	Squash mosaic virus	*Minglobus cucurbitae*	11	SI	9
		Cucumber mosaic virus	*Maphiglobus cucumeris*	11	See "cucumber"	
		Tobacco ringspot virus	*Genemaglobus nicotianae*	11	—	
Tomato	*Lycopersicon esculentum*	Arabis mosaic virus	*Genemaglobus arabidis*	11	—	
		Tomato black ring virus	*Genemaglobus betae*	11	—	

		Tomato ringspot virus	*Genemaglobus lycopersici*	11	—	
		Tobacco mosaic virus	*Minchorda nicotianae*	11	HW, EVR	9
		Potato spindle tuber virus	*Mexenudum solani*	11	—	
Watermelon	*Citrullus lanatus*	Squash mosaic virus	*Minglobus cucurbitae*	11	See "pumpkin and squash"	

[a] Abbreviations for methods of control: CP — cultural practices; CR — crop rotation; EVR — eliminated virus reservoirs; HW — soak seed in hot water; Insect — spray with insecticides to eliminate vectors; RC — resistant cultivars; SI — seed indexing; SoT — fumigation, chemical, or heat treatment of infested soil; Spray — spray with fungicides or antibiotics; ST — seed treatment with chemicals.

REFERENCES

1. **Anon.,** *A Compendium of Corn Diseases,* American Phytopathological Society, St. Paul, Minn., 1973.
2. **Anon.,** *Compendium of Soybean Diseases,* American Phytopathological Society, St. Paul, Minn., 1975.
3. **Behr, L.,** The false mildew of the poppy [*Peronospora arborescens* (Berk.) de By.]. Investigation of the biology of the causal agent, *Phytopathol. Z.,* 27, 289, 1956.
4. **Bennett, C. W.,** Lychnis ringspot, *Phytopathology,* 49, 706, 1959.
5. **Boerema, G. H. and Howeler, L. H.,** *Phoma exigua* Desm. and its varieties, *Persoonia,* 5, 15, 1967.
6. **Buchanan, R. E. and Gibbons, N. E.,** *Bergey's Manual of Determinative Bacteriology,* 8th ed., Williams & Wilkins, Baltimore, 1974.
7. **Calvert, O. H. and Thomas, C. A.,** Some factors affecting seed transmission of safflower rust, *Phytopathology,* 44, 609, 1954.
8. **Chupp, C.,** A Monograph of the fungus genus *Cercospora,* C. Chupp, Ithaca, N.Y., 1953.
9. **Chupp, C. and Sherf, A. F.,** *Vegetable Diseases and Their Control,* Ronald Press, New York, 1960.
10. C.I.H., Descriptions of Plant-Parasitic Nematodes, Nos. 1—90 issued 1972—1975, Commonwealth Institute of Helminthology, St. Albans, Herts, England.
11. C.M.I./A.A.B., Descriptions of Plant Viruses, Nos. 1—185 issued 1970—77, Commonwealth Mycological Institute, Kew, Surrey, England.
12. C.M.I., Description of Pathogenic Fungi and Bacteria, Nos. 1—580 issued 1964—78, Commonwealth Mycological Institute, Kew, Surrey, England.
13. **Cory, L. and Hewett, W. B.,** Some grapevine viruses in pollen and seeds, *Phytopathology,* 58, 1316, 1968.
14. **Couch, H. B.,** *Diseases of Turfgrasses,* Reinhold, New York, 1962.
15. **Crawford, R. F.,** Fungi isolated from the interior of cotton seed, *Phytopathology,* 13, 501, 1923.
16. **Dhanraj, K. S. and Raychaudhuri, S. P.,** A note on barley mosaic in India, *Plant Dis. Rep.,* 53, 766, 1969.
17. **Dickson, J. G.,** *Diseases of Field Crops,* McGraw-Hill, New York, 1956.
18. **Elliott, C.,** *Manual of Bacterial Plant Pathogens,* Chronica Botanica, Waltham, Mass., 1951.
19. **Ellis, M. B.,** Demataceous Hyphomycetes, Commonwealth Mycological Institute, Kew, Surrey, England, 1971.
20. **Gibbs, A. and Harrison, B.,** *Plant Virology: The Principles,* Edward Arnold, London, 1976.
21. **Gorlenko, M. V.,** Bacterial Diseases of Plants, Israel Program for Scientific Translations, Jerusalem, 1965.
22. **Jenkins, W. R. and Taylor, D. P.,** *Plant Nematology,* Reinhold, New York, 1967.
23. **Kaiser, W. J.,** Diseases of food legumes caused by pea leaf roll virus in Iran, *F.A.O. Plant Prot. Bull.,* 20, 127, 1972.
24. **Klisiewicz, J. M.,** Wilt incitant *Fusarium oxysporum* f. *carthami* present in seed from infected safflower, *Phytopathology,* 53, 146, 1963.
25. **Kulik, M. M. and Schoen, J. F.,** Procedures for the routine detection of seed-borne pathogenic fungi in the seed-testing laboratory, *J. Seed Technol.* 2, 29, 1977.
26. **Muskett, A. E. and Colhoun, J.,** *The Diseases of the Flax Plant,* W. & G. Baird, Belfast, 1947.
27. **Neergaard, P.,** *Seed Pathology,* 2 vols., Macmillan, London, 1977.
28. **Noble, M. and Richardson, M. J.,** An Annotated List of Seed-Borne Diseases, 2nd ed., International Seed Testing Association, Wageningen, The Netherlands, 1968.
29. **Norton, D. C. and Sass, J. E.,** Pathological changes in *Agropyron smithii* induced by *Anguina agropyronifloris, Phytopathology,* 56, 769, 1966.
30. **Nusbaum, C. J.,** Studies of the control of Cercospora blight of benne, *S. C. Agric. Exp. Stn.* 54; 173, 1941.
31. **Ou, S. H.,** Rice Diseases, Commonwealth Mycological Institute, Kew, Surrey, England, 1972.
32. **Paulech, C.,** Occurrence of *Botrytis anthophila* Bond on red clover (*Trifolium pratense* L.) in the CSSR, *Biologia Bratislava,* 20, 491, 1965.
33. **Phatak, H. C.,** Seed-borne plant viruses. Identification and diagnosis in seed health testing, *Seed Sci. Technol.,* 2, 3, 1974.
34. **Poole, D. D.,** Aerial stem rot of sesame caused by *Helminthosporium sesami* in Texas, *Plant Dis. Rep.,* 40, 235, 1956.
35. **Ramakrishnan, T. S.,** *Diseases of Millets,* Indian Council of Agricultural Research, New Delhi, 1971.
36. **Sampson, K. and Western, J. H.,** *Diseases of British Grasses and Herbage Legumes,* Cambridge University Press, New York, 1941.
37. **Schuster, M. L., Hoff, B., Mandel, M., and Lazar, I.,** Leaf freckles and wilt, a new corn disease, *Proc. Ann. Corn-Sorghum Res Conf.,* 27, 176, 1973.

38. **Simmonds, J. H.,** A study of the species of *Colletotrichum* causing ripe fruit rots in Queensland, *Queensl. J. Agric. Sci.*, 22, 437, 1965.
39. **Sprague, R.,** *Diseases of Cereals and Grasses in North America,* Ronald Press, New York, 1950.
40. **Takahashi, K., Tanaka, T., and Tsuda, Y.,** Soybean mild mosaic virus, *Ann. Phytopathol. Soc. Jpn.*, 40, 103, 1974.
41. **Tarr, S. A. J.,** Diseases of Sorghum, Sudan Grass and Broom Corn, Commonwealth Mycological Institute, Kew, Surrey, England, 1962.
42. **Terrell, E. E.,** A checklist of names for 3,000 vascular plants of economic importance, *U.S. Dep. Agric. Agric. Handb.*, 505, 1977.
43. **Thomas, C. A.,** Control of pre-emergence damping-off and two leaf-spot diseases of sesame by seed treatment, *Phytopathology,* 49, 461, 1959.
44. **Tircomnicu, M. and Iliescu, H.,** Laboratory methods for the identification of latent infections in apparently healthy sunflower seeds and seedlings, Analele Inst. de Certetsri pentru Cereale si Plante Tehnice. — Fundulea, C., 39, 267, 1973; abstracted in *Rev. Plant Pathol.*, 54, 212, 1975.
45. **Walker, J. C.,** *Diseases of Vegetable Crops,* McGraw-Hill, New York, 1952.
46. **Watson, M. A. and Serjeant, E. P.,** Carrot motley dwarf, *Rothamsted Exp. Stn. Rep.*, 1961, 106, 1962.
47. **Weber, G. F.,** *Bacterial and Fungal Diseases of Plants in the Tropics,* University of Florida Press, Gainesville, 1973.
48. **Weimer, J. L.,** Blackpatch of soybean and other forage legumes, *Phytopathology,* 40, 782, 1950.
49. **Wiese, M. V.,** *Compendium of Wheat Diseases,* American Phytopathological Society, St. Paul, Minn., 1973.
50. **Zaumeyer, W. J. and Thomas, H. R.,** A monographic study of bean diseases and methods for their control, *U.S. Dep. Agric. Tech. Bull.*, 868, 1957.

STORAGE CONDITIONS FOR MAINTAINING SEED QUALITY

L. N. Bass

Storage conditions for maintaining seed quality vary with the kind of seeds and the storage period. Orthodox seeds, those which can be artificially dried without damage, store well under a variety of conditions. "Recalcitrant seeds", those which cannot be dried by conventional methods without loss of viability, do not store well under normal storage conditions.

For storage from harvest to planting time, orthodox seeds require no special storage conditions in temperate or cold climates, but in warm, humid regions low-temperature, dehumidified storage is usually required.

For extended periods of storage, seed moisture content and storage temperature are the more important factors to control, with seed moisture content being most critical.

Seed moisture content can be controlled either by drying seeds to the desired moisture content, followed by storage in sealed, moistureproof containers, or by storage in porous containers under controlled relative humidity conditions. For seeds in sealed containers, it is important that they be dry enough when sealed for safe storage at the highest temperature to which they might be subjected. For many orthodox seeds, the optimum seed moisture content for sealed storage is between 3 and 7%. For controlled storage, a relative humidity of 50% or lower is recommended. At a controlled relative humidity, seed moisture content equilibrates with the surrounding atmosphere. Consequently, moisture content of the seeds when they are put into storage is not as critical as it is for seeds placed in sealed containers. For long-term storage in sealed containers or under controlled relative humidity, temperature is also critical. In general, orthodox seeds, that are air-dry or drier, can be stored safely at temperatures below as well as above freezing. Temperatures of −18 to −20°C are preferrred to those near zero and higher. Research in progress in the National Seed Storage Laboratory has shown that many kinds of orthodox seeds can be stored air dry in liquid nitrogen (−196°C) for at least a year or two with little or no loss of viability. Studies are in progress to determine the long-term effects of storage in liquid nitrogen on seed viability. Theoretically, viability should be maintained at a high level longer at −196°C than at −18 to −20° C. It is possible that controlled drying rates in combination with controlled cooling rates down to −196°C may provide conditions suitable for long-term storage of recalcitrant seeds.

Although much research has been conducted on various aspects of seed storage, few studies have covered a wide variety of conditions on more than one or two kinds of seeds. Because of the large volume of literature on seed storage, it was not possible to include data in the tables from every published paper. A compilation of data for seeds stored in open or porous containers under a variety of conditions is given in Table 1.

Table 2 contains data for seeds with various moisture contents sealed in air in metal or glass containers stored under a range of temperatures.

Heated air drying is widely used by seedsmen to reduce seed moisture content to a safe level for controlled humidity or sealed storage. In research, drying with a desiccant sealed in the container with the seeds has received some attention (Table 3).

With the development of flexible, moisture-barrier, heat-sealable, packaging materials came the need to evaluate the efficacy of such materials for protective seed packages. Results of several such studies are summarized in Table 4.

Theoretically, reducing the oxygen available to seeds in a sealed container should increase their storage life. Results of studies in which various gases or a partial vacuum were substituted for air in sealed containers (Table 5) indicate that for short-term storage, the time and expense of substituting atmospheres are not worthwhile. For long-term storage, such substitutions may or may not be worthwhile, depending upon individual circumstances.

In recent years, application of coating materials to seeds to enhance precision planting has become popular. With any new seed processing procedure comes a need to know the effect of that procedure on seed longevity. The data in Table 6 show the effects of several seed coatings on the viability of seeds stored under various conditions. The data do not cover all the kinds of seeds which are processed with a coating material, but do show that under good storage conditions, coated seeds will store as well as uncoated seeds.

In general, to maintain seed quality for a long time in storage, most kinds of seeds should be dried to 3 to 7% moisture content, then sealed in a moistureproof container and stored at a temperature of −10°C or colder, preferably colder, for most orthodox seeds.

Suitable long-term storage procedures have not been established for most recalcitrant seeds.

Table 1
LONGEVITY OF SEEDS STORED UNDER VARIOUS CONDITIONS IN OPEN AND POROUS CONTAINERS

Species	Moisture (% when stored)	Storage			Germination (%)		Ref.
		RH (%)	Time (years)	Temperature (°C)	When stored	After storage	
African daisy							
Dimorphotheca sinuata de Candolle	Not given	40	16	5	64	76	19
D. pluvialis (L.) Moench	Not given	40	16	5	62	73	19
Ageratum							
Ageratum mexicanum Sims now *A. houstonianum* Mill.	Not given	35—40	16	5	89	89	21
Alfalfa, lucerne							
Medicago sativa L.	Not given	Ambient	16	18	99	1	44
	9.4	77—87	2	5—15	95	47	67
	9.4	60—70	4	0	95	93	67
	Not given	Ambient	20	Ambient	99	83	91
	Not given	Ambient	31	Ambient	Not given	63	55
	Not given	Ambient	32	Ambient	Not given	75	55
	Not given	Not given	4	Ambient	94	79	57
	Not given	Not given	4	−7 to −11	94	89	57
	Not given	Not given	18	Ambient	94	31	74
	Not given	Not given	18	−7 to −11	94	80	74
	13.9	15	6.25	22—27	100	98	1
	13.9	30	6.25	22—27	100	96	1
	13.9	45	6.25	22—27	100	94	1
	13.9	60	6.25	22—27	100	12	1
	13.9	75	2.25	22—27	100	3	1
	13.9	90	1.25	22—27	100	5	1
	Not given	30	6	7	87	94	1
	Not given	60	6	7	87	89	1
	Not given	Uncontrolled	6	7	87	94	1
	Not given	30	6	10	87	97	1
	Not given	60	6	10	87	91	1
	Not given	Uncontrolled	6	10	87	95	1

Table 1 (continued)
LONGEVITY OF SEEDS STORED UNDER VARIOUS CONDITIONS IN OPEN AND POROUS CONTAINERS

Species	Moisture (% when stored)	Storage: RH (%)	Storage: Time (years)	Storage: Temperature (°C)	Germination (%): When stored	Germination (%): After storage	Ref.
	Not given	30	6	15	87	94	1
	Not given	60	6	15	87	88	1
	Not given	Uncontrolled	6	15	87	97	1
	Not given	30	6	20	87	98	1
	Not given	60	6	20	87	78	1
	Not given	Uncontrolled	6	20	87	96	1
	Not given	35—40	18	5	60	60	21
	9.4	77—87	2	5—15	95	47	67
	9.4	60—70	4	0	95	93	67
Scarified (dry heat)	Not given	Ambient	13	Ambient	Not given	17	26
Unscarified	Not given	Ambient	13	Ambient	Not given	23	26
Amaranth, feathered							
Celosia cristata L. (Plumosa Group)	Not given	35—40	16	5	98	99	21
Apricot							
Prunus armeniaca L.	10—11	50—65	2.50	20—25	100	39	90
	10—11	80—90	2.50	2—10	100	0	90
Apple							
Malus sylvestris Mill.	10—11	50—65	2.50	20—25	95	33	90
	10—11	80—90	0.83	2—10	95	4	90
Artichoke							
Cynara scolymus L.	Not given	35—40	17	5	69	68	21
Ash, European							
Fraxinus excelsior L.	15.4	Not given	2	5	53	22	6
Asparagus							
Asparagus officinalis L.	Not given	35—40	17	5	87	78	21
Aster							
Callistephus chinensis (L.) Nees	Not given	Not given	16.33	−4	91	21	9
	Not given	35—40	14	5	82	75	21

Bahiagrass							
Paspalum notatum Fluegge	Not given	35—40	17	5	66	60	21
Bamboo							
Bamboo arundinacea Retz	18	Not given	0.6	21—32	Not given	0	97
Barley							
Hordeum vulgare L.	Not given	Ambient	5	18	100	5	44
	12.5	77—87	4	5—15	95	68	67
	12.5	60—70	4	0	95	94	67
	Not given	Ambient	32	Ambient	Not given	83	55
	Not given	35—40	17	5	99	99	21
Bean, garden							
Phaseolus vulgaris L.	10	Not given	6	Ambient	89	29	13
	10	Not given	6	30	89	12	13
	10	Not given	4	20	89	0	13
	10	Not given	15	10	89	98	13
	10	Not given	4	5	89	0	13
	10	Not given	15	−2	89	100	13
	10	Not given	15	−18	89	100	13
	7	Not given	8	Ambient	97	9	13
	7	Not given	8	30	97	6	13
	7	Not given	4	20	97	18	13
	7	Not given	15	10	97	86	13
	7	Not given	4	5	97	0	13
	7	Not given	15	−2	97	97	13
	7	Not given	15	−18	97	99	13
	10	Not given	15	−18	89	100	13
	10	Not given	15	−2	89	100	13
	Not given	35—40	18	5	99	99	21
	Not given	75—80	3.50	4—7	97	93	81
	Not given	Ambient	3.50	19—31	97	5	81
	Not given	90	0.25	32	93	20	61
	Not given	70	1	32	93	8	61
	Not given	70	1.25	21	93	82	61
	Not given	50	5	10	93	88	61
	Not given	35—40	18	5	87	84	21
	Not given	15	5.25	22—27	92	23	1
	Not given	30	6.25	22—27	92	20	1

Table 1 (continued)
LONGEVITY OF SEEDS STORED UNDER VARIOUS CONDITIONS IN OPEN AND POROUS CONTAINERS

		Storage			Germination (%)		
Species	Moisture (% when stored)	RH (%)	Time (years)	Temperature (°C)	When stored	After storage	Ref.
	Not given	45	6.25	22—27	92	31	1
	Not given	60	2.75	22—27	92	15	1
	Not given	75	1	22—27	92	29	1
	Not given	90	0.50	22—27	92	0	1
	Not given	Ambient	2.25	22—27	92	3	1
	Not given	30	6	7	95	61	1
	Not given	60	6	7	95	87	1
	Not given	Ambient	6	7	95	75	1
	Not given	30	6	10	95	83	1
	Not given	60	6	10	95	76	1
	Not given	Ambient	6	10	95	78	1
	Not given	30	6	15	95	48	1
	Not given	60	6	15	95	87	1
	Not given	Ambient	6	15	95	67	1
	Not given	30	6	20	95	37	1
	Not given	60	6	20	95	71	1
	Not given	Ambient	6	20	95	42	1
Bean, lima (green seed) *Phaseolus vulgaris* L.	Not given	35—40	18	5	90	99	21
	Not given	90	0.50	21	97	52	23
	Not given	70	1.50	21	97	94	23
	Not given	50	3	21	97	99	23
Bean, lima (bleached seed)	Not given	90	0.50	21	76	21	21
	Not given	70	1.50	21	76	30	21
	Not given	50	3	21	76	70	21
Bean, scarlet runner *Phaseolus coccineus* L.	Not given	35—40	16	5	92	95	21

Beet, garden							
Beta vulgaris L.	Not given	Ambient	13	18	84	1	44
	11.1	85—95	6	−20	83	88	96
	11.1	85—95	6	0—2	83	70	96
	11.1	70—80	6	5—20	83	78	96
	Not given	35—40	17	5	87	88	21
Beet, Mangel							
Beta vulgaris L.	Not given	35—40	17	5	87	96	21
Beetsugar							
Beta vulgaris L.	Not given	35—40	17	5	91	90	21
	Not given	Ambient	22	10 to −10	84	75	79
Whole seed	Not given	Ambient	20	Ambient	67	65	84
Segmented	Not given	Ambient	20	Ambient	73	64	84
Swiss chard							
Beta vulgaris L.	Not given	35—40	15	5	90	94	21
Bentgrass, Colonial							
Agrostis tenuis Sibth.	11.0	77—87	4	5—15	72	2	67
	11.0	60—70	4	0	82	63	67
Bindweed							
Convolvulus arvensis L.	Not given	Ambient	50	Ambient	Not given	62	33
Bluegrass, Kentucky							
Poa pratensis L.	8.7	80	2.50	1	86	80	15
	8.7	80	2	10	86	5	15
	8.7	20—40	2	21	86	3	15
	8.7	10—15	2.50	32	86	81	15
	8.7	4—6	2.50	43	86	68	15
	8.7	Ambient	2.50	Ambient	86	75	15
	3.8	50—60	2	1	75	64	54
	3.8	80—95	2	10	75	64	54
	3.8	25—30	2	27	75	57	54
	3.8	Ambient	2	Ambient	75	55	54
	3.8	15	2	43	75	60	54
	Not given	35—40	17	5	96	93	21
	Not given	Ambient	9	18	78	1	44
Bluegrass, Wood							
P. nemoralis L.	Not given	Ambient	5	18	84	2	44
	Not given	35—40	10	5	99	96	21

Table 1 (continued)
LONGEVITY OF SEEDS STORED UNDER VARIOUS CONDITIONS IN OPEN AND POROUS CONTAINERS

Species	Moisture (% when stored)	Storage: RH (%)	Storage: Time (years)	Storage: Temperature (°C)	Germination (%): When stored	Germination (%): After storage	Ref.
Bluegrass, roughstalk							
P. trivialis L.	Not given	35—40	6	18	98	9	44
Bluestem, big							
Andropogon gerardii Vitm.	Not given	35—40	12	5	73	80	21
Bluestem, sand							
A. hallii Hack.	Not given	35—40	7	5	64	66	21
Bluestem, little							
A. scoparius (Michx.) Nash now *Schizachyrium scoparium* (Michx.) Nash	Not given	35—40	15	5	79	32	21
Broad bean, fava bean							
Vicia faba L.	Not given	35—40	17	5	91	88	21
Brome, field							
Bromus arvensis L.	Not given	Ambient	4	18	95	8	44
	Not given	35—40	15	5	75	58	21
Brome, mountain							
B. marginatus Nees ex Steud.	7—9	Not given	5–	15—23	96	93	68
	Not given	35—40	10	5	89	90	21
B. polyanthus Scribn.	7—9	Not given	5–	15—23	94	93	68
Brome, smooth	7—9	Not given	5–	15—23	95	78	68
B. inermis Leyss.	8	80	2+	10	91	39	53
	8	80	2+	20	91	0	53
	8	80	0.50	30	91	1	53
	8	50	2+	10	91	91	53
	8	50	2+	20	91	87	53
	8	50	1	30	91	87	53
	8	20	2+	10	91	91	53
	8	20	1	20	91	90	53
	8	20	1	30	91	87	53

	8	Ambient	2+	Ambient	91	82	53
	Not given	40	10	5	93	77	35
	Not given	30—40	10	21	93	3	35
	Not given	50—60	10	10—15	93	6	35
	Not given	35—40	13	5	91	57	21
Brussels sprouts							
Brassica oleracea L. var. *gemminfera*	6—8	58—83	6.50	11—19	71	13	71
	Not given	35—40	19	5	92	94	21
Buckwheat							
Fagopyrum esculentum Moench	Not given	35—40	15	5	97	92	21
Buffel grass							
Pennisetum ciliare (L.)	Not given	15	6.25	22—27	46	26	1
Link now *Cenchrus ciliaris* L.	Not given	30	6.25	22—27	46	19	1
	Not given	45	6.25	22—27	46	29	1
	Not given	60	4.25	22—27	46	1	1
	Not given	75	0.83	22—27	46	3	1
	Not given	90	0.66	22—27	46	1	1
	Not given	Ambient	1.75	22—27	46	5	1
	Not given	35—40	15	5	85	15	21
Buttercup, creeping							
Ranunculus repens L.	11.2	77—87	4	5—15	96	100	67
	11.2	60—70	4	0	96	100	67
Cabbage							
Brassica L. var. *capitata*	3.1	50—60	2	1	96	95	54
	3.1	80—95	1.50	10	96	10	54
	3.1	25—30	2	27	96	95	54
	3.1	20—80	2	16—35	96	86	54
	3.1	15	2	43	96	85	54
	2.9	50—60	2	1	94	93	54
	2.9	80—95	1.50	10	94	0	54
	2.9	25—30	2	27	94	89	54
	2.9	20—80	2	16—35	94	75	54
	2.9	15	2	43	94	67	54
	6—8	58—83	4.66	11—19	95	95	71
	Not given	35—40	18	5	84	82	21
B. oleracea L. capitata alba	Not given	Ambient	8	18	94	11	44
B. oleracea L. var. *quercifolia*	Not given	Ambient	9	18	92	18	44

Table 1 (continued)
LONGEVITY OF SEEDS STORED UNDER VARIOUS CONDITIONS IN OPEN AND POROUS CONTAINERS

Species	Moisture (% when stored)	Storage			Germination (%)		Ref.
		RH (%)	Time (years)	Temperature (°C)	When stored	After storage	
Campion, white							
Lychnis alba Mill.	11.5	77—87	4	5—15	95	100	67
now *Silene alba* (Mill.) Krause	11.5	60—70	4	0	95	92	67
Canarygrass							
Phalaris canariensis L.	Not given	Ambient	15	18	99	3	44
		35—40	10	5	75	95	21
Canarygrass, reed							
Phalaris arundinacea L.	Not given	Ambient	3	18	49	2	44
	8.2	70	12	−12	89	88	16, 21
	8.2	40	12	−1	89	90	16, 21
	8.2	35	12	5	89	83	16, 21
	8.2	50	12	10	89	22	16, 21
	8.2	70	8	10	89	4	16, 21
	8.2	90	4	10	89	13	16, 21
	8.2	50	5	21	89	7	16, 21
	8.2	70	1	21	89	33	16, 21
	8.2	90	0.75	21	89	20	16, 21
	8.2	50	1.50	32	89	8	16, 21
	8.2	70	0.50	32	89	27	16, 21
	8.2	90	0.50	32	89	10	16, 21
	Not given	35—40	17	5	68	65	21
Cantaloupe; muskmelon							
Cucumis melo L.	Not given	15	12	32	91	64	17, 21
	Not given	30	12	21	91	70	17, 21
	Not given	60	12	10	91	84	17, 21
	Not given	35—40	18	5	90	88	21
Caraway							
Carum carvi L.	Not given	Ambient	2	18	77	13	44
	Not given	35—40	17	5	66	65	21

Carrot							
Daucus carota L.	Not given	Not given	10	Ambient	67	22	9
	Not given	Not given	20	−4	67	63	9
	Not given	Ambient	7	18	77	6	44
	10.2	85—95	6	−20	89	76	96
	10.2	85—95	6	0—2	89	41	96
	10.2	70—80	6	5—20	89	47	96
	Not given	35—40	17	5	91	88	21
Castorbean							
Ricinus communis L.	Not given	35—40	17	5	96	90	21
Cedar, Port-Orford							
Chamaecyparis lawsoniana (A. Murr.) Parl.	8.0	Not given	7	−18	56	43	2
	8.0	Not given	7	0	56	43	2
Cedar, western red	12	Not given	3	−18	60	76	10
	7.9	Not given	7	−18	68	34	2
	7.9	Not given	3	0	68	6	2
	7.9	Not given	3	Ambient	68	0	2
	17	Not given	2	−4	60	27	10
Celery							
Apium graveolens L. var. *dulce* (Mill.) Pers.	Not given	Ambient	7	18	67	1	44
	8.3	85—95	6	−20	79	76	96
	8.3	85—95	6	0—2	79	51	96
	8.3	70—80	6	5—20	79	61	96
	Not given	35—40	17	5	77	80	21
Chess, soft							
Bromus mollis L.	12.5	77—87	2	5—15	88	1	67
	12.5	60—70	4	0	88	82	67
	Not given	35—40	4	18	73	12	44
	Not given	35—40	10	5	89	90	21
Chives	11	85—95	6	−20	98	86	96
Allium schoenoprasum L.	11	85—95	6	0—2	98	1	96
	11	70—80	6	5—20	98	0	96
	Not given	35—40	16	5	92	83	21
Chrysanthemum							
Chrysanthemum coronarium L.	Not given	Ambient	8	18	48	4	44
Clover, alsike							
Trifolium hybridum L.	Not given	Ambient	25	18	99	3	44

Table 1 (continued)
LONGEVITY OF SEEDS STORED UNDER VARIOUS CONDITIONS IN OPEN AND POROUS CONTAINERS

		Storage			Germination (%)		
Species	Moisture (% when stored)	RH (%)	Time (years)	Temperature (°C)	When stored	After storage	Ref.
	Not given	Not given	17	Ambient	93	19	88
	Not given	Not given	15	Ambient	93	22	88
	Not given	Not given	10	Ambient	93	45	88
	Not given	35—40	11	5	90	90	21
Clover, crimson							
Trifolium incarnatum L.	Not given	Ambient	5	18	98	1	44
	Not given	Not given	4	Ambient	80	16	57
	Not given	Not given	4	−7 to −11	80	80	57
	Not given	Not given	18	−7 to −11	80	20	74
	Not given	35—40	15	5	86	52	21
	4	70	16	−12	81	86	20
	4	40	16	−1	81	61	20
	4	60	16	10	81	9	20
	4	30	16	21	81	17	20
	4	15	16	32	81	22	20
	7	70	16	−12	91	81	20
	7	40	16	−1	91	64	20
	7	60	16	10	91	20	20
	7	30	16	21	91	17	20
	7	15	16	32	91	26	20
	10	70	16	−12	91	78	20
	10	40	16	−1	91	74	20
	10	60	16	10	91	12	20
	10	30	16	21	91	25	20
	10	15	16	32	91	25	20
Clover, ladino							
Trifolium repens L.	Not given	Ambient	26	18	98	8	44

Clover, red							
Trifolium repens L.	10.8	77—87	4	5—15	80	2	67
	10.8	60—70	4	0	80	76	67
	12.7	Ambient	16	0	93	43	41
	12.7	Ambient	13	Ambient	93	0	41
	Not given	Ambient	12	18	100	3	44
	Not given	Ambient	20	Ambient	97	33	91
	Not given	Not given	17	Ambient	96	9	88
	Not given	Not given	15	Ambient	96	15	88
	Not given	Not given	10	Ambient	96	44	88
	9—11	58—83	5.50	11—19	92	60	70
Continuously frozen	Not given	60—90	13	−5 to −15	96	97	83
Thawed weekly	Not given	60—90	13	−5 to −15	96	96	83
	Not given	60—90	10	Ambient	97	0	48
	Not given	60—90	23	0—5	97	34	48
	Not given	35—40	17	5	95	67	21
Clover, small hop							
Trifolium dubium Sibth.	10.2	77—87	4	5—15	90	10	67
	10.2	60—70	4	0	90	93	67
Clover, white	9.8	Ambient	16	0	92	92	41, 42
Trifolium repens L.	9.8	Ambient	16	Ambient	92	2	41, 42
	Not given	Not given	4	Ambient	96	80	57
	Not given	Not given	4	−7 to −11	96	92	57
	Not given	Not given	18	Ambient	96	22	74
	Not given	Not given	18	−7 to −11	96	70	74
	Not given	35—40	15	5	99	88	21
	Not given	35—40	14	5	95	60	21
	10	77—87	4	5—15	98	5	67
	10	60—70	4	0	98	84	67
Cocoa (pods)							
Theobroma cacao L.	In moist steam-sterilized charcoal	Not given	0.09	Ambient	100	100	49
Corn, field inbred lines							
Zea mays L.	10	Ambient	7	13—21	97	55	76
	10	Ambient	7	4	97	90	76
	Not given	35—40	15	5	94	93	21
Open pollinated	Not given	35—40	17	5	97	98	21
	Not given	15	6	22—27	95	83	1

Table 1 (continued)
LONGEVITY OF SEEDS STORED UNDER VARIOUS CONDITIONS IN OPEN AND POROUS CONTAINERS

		Storage			Germination (%)		
Species	Moisture (% when stored)	RH (%)	Time (years)	Temperature (°C)	When stored	After storage	Ref.
	Not given	30	6.25	22—27	95	80	1
	Not given	45	6.25	22—27	95	70	1
	Not given	60	4.75	22—27	95	21	1
	Not given	75	1.75	22—27	95	24	1
	Not given	90	0.83	22—27	95	3	1
	Not given	Ambient	3.75	22—27	95	15	1
	Not given	Ambient	32	Ambient	Not given	29	55
	Not given	35—40	17	5	98	99	21
	Not given	90	0.25	32	89	1	61
	Not given	70	1	32	89	1	61
	Not given	70	1.25	21	89	81	61
	Not given	50	4	21	89	85	61
	Not given	50	—	10	89	91	61
	Not given	35—40	17	5	91	94	21
Cotton, upland *Gossypium hirsutum* L.	8—11	Ambient	11	1—32	96	82	37
	Not given	35—40	15	5	82	84	21
Cotton, sea island *G. barbadense* L.	Not given	35—40	13	5	80	80	21
Cowpea (southern pea) *Vigna unguiculata* (L.) Walp subsp. *unguiculata*	Not given	Not given	4	5	84	84	46
	Not given	Humid	4.50	Warm	84	45	46
	Not given	35—40	17	5	92	96	21
Crambe *Crambe abyssinica* Hochst. ex R. E. Fries	Not given	50	4	32	93	51	38
	Not given	70	1.50	21	93	33	38
	Not given	50	8	21	93	80	38
	Not given	70	8	10	93	67	38

	Not given	50	8	10	93	82	38
	Not given	35	8	4	93	88	38
	Not given	40	8	−1	93	88	38
	Not given	35—40	5	5	88	84	21
Cress, garden							
Lepidium sativum	Not given	35—40	14	5	74	82	21
	Not given	35—40	17	5	94	31	21
Crotalaria							
Crotalaria juncea L.	Not given	35—40	14	5	72	70	21
Cucumber							
Cucumis sativus L.	Not given	90	0.25	32	96	19	61
	Not given	70	1	32	96	25	61
	Not given	70	1.25	21	96	83	61
	Not given	50	5	10	96	94	61
	Not given	35—40	17	5	97	98	21
Datura							
Datura ferox L.	Not given	35—40	10	5	99	96	21
Datura							
D. innoxia Mill.	Not given	35—40	10	5	84	98	21
Datura							
D. metal L.	Not given	35—40	10	5	90	74	21
Datura, jimson weed							
D. stramonium L.	Not given	35—40	10	5	98	98	21
Delphinium, annual							
Delphinium spp.	Not given	Not given	5	5	57	0	9
	Not given	Not given	10	−4	57	6	9
Delphinium, perennial							
D. spp.	Not given	Not given	5	5	43	0	9
	Not given	Not given	10	−4	43	6	9
	Not given	35—40	16	5	87	88	21
Dill							
Anethum graveolens L.	Not given	35—40	18	5	92	64	21
Dock, curly							
Rumex crispus L.	9.8	77—87	4	5—15	63	98	67
	9.8	60—70	4	0	63	88	67
Dogtail, crested							
Cynosurus cristatus L.	Not given	Ambient	4	18	63	5	67

Table 1 (continued)
LONGEVITY OF SEEDS STORED UNDER VARIOUS CONDITIONS IN OPEN AND POROUS CONTAINERS

		Storage			Germination (%)		
Species	Moisture (% when stored)	RH (%)	Time (years)	Temperature (°C)	When stored	After storage	Ref.
Eggplant							
Solanum melongena L.	Not given	Ambient	2.16	Ambient	73	25	30
	Not given	Not given	10	Ambient	86	27	9
	Not given	Not given	10	−4	86	76	9
	Not given	35—40	17	5	99	97	21
Elm, American							
Ulmus americana L.	Not given	Not given	10	−4	96	14	9
Evening primrose							
Oenothera biennis L.	Not given	35—40	11	5	99	99	21
	Not given	35—40	11	5	99	14	21
Fescue, creeping red							
Festuca rubra L. subsp. *rubra*	3.6	50—60	2	1	95	91	54
	3.6	80—95	2	10	95	0	54
	3.6	25—30	2	27	95	83	54
	3.6	20—80	2	16—35	95	4	54
	3.6	15	2	43	95	83	54
	Not given	35—40	17	5	90	72	21
	11.4	80	2.50	1	95	7	15
	11.4	80	1.50	10	95	7	15
	11.4	20—40	0.50	21	95	7	15
	11.4	10—15	2	32	95	12	15
	11.4	4—6	1.50	43	95	4	15
	11.4	Ambient	2.50	Ambient	95	3	15
	13.6	77—87	4	5—15	87	1	67
	13.6	60—70	4	0	87	80	67
Fescue, hard							
Festuca ovina var. duriuscula (L.) Koch	Not given	Ambient	5	18	89	1	1
now *F. longifolia* Thuill.	Not given	35—40	13	5	94	85	21

Fescue, meadow							
Festuca pratensis Huds.	Not given	35—40	5	18	98	6	44
	Not given	35—40	10	5	95	85	21
	12.5	77—87	2	5—15	90	29	67
	12.5	60—70	4	0	90	89	67
Fescue, tall							
Festuca arundinacea Schreb.	12.8	77—87	2	5—15	76	15	67
	12.8	60—70	4	0	76	63	67
	Not given	35—40	5	18	95	14	44
	Not given	Not given	4	Ambient	91	3	57
	Not given	Not given	4	−7 to −11	91	93	57
	Not given	Not given	18	−7 to −11	91	66	74
	Not given	35—40	17	5	89	66	21
Fir, Balsam							
Abies balsamea (L.) Mill.	Not given	Ambient	5	18	29	2	44
Fir, Douglas							
Pseudotsuga menziesii	15	Not given	3	−4	88	7	10
(Mirbel) Franco	10	Not given	3	−18	88	78	10
	8.3	Not given	7	−18	43	27	2
	8.3	Not given	7	0	43	14	2
	8.3	Not given	7	Ambient	43	0	2
Fir, noble							
Abies nobilis Dougl. ex D. Don) Lindl.	9.0	Not given	7	−18	43	33	2
Now *A. procena* Rehd.	9.0	Not given	7	0	43	32	2
	9.0	Not given	7	Ambient	43	2	2
Fir, white							
A. concolor (Gord. & Glend.)	6.3	Not given	7	−18	72	62	2
Lindl. ex Hildebr.	6.3	Not given	7	0	72	55	2
	6.3	Not given	3	Ambient	72	20	2
Flax							
Linum usitatissimum L.	Not given	Ambient	10	18	72	1	44
	Not given	35—40	17	5	99	95	21
Foxtail, black grass							
Alopercurus myosuroides Huds.	13.6	77—87	4	5—15	44	7	67
	13.6	60—70	4	0	44	40	67
Foxtail, creeping							
A. arundinaceus Poir.	Not given	35—40	14	5	87	88	21

Table 1 (continued)
LONGEVITY OF SEEDS STORED UNDER VARIOUS CONDITIONS IN OPEN AND POROUS CONTAINERS

		Storage			Germination (%)		
Species	Moisture (% when stored)	RH (%)	Time (years)	Temperature (°C)	When stored	After storage	Ref.
Foxtail, meadow							
A. pratensis L.	Not given	Ambient	9	18	89	1	44
	Not given	40	10	5	84	87	35
	Not given	30—40	10	21	84	82	35
	Not given	50—60	10	10—15	84	80	35
	Not given	35—40	11	5	75	70	21
Geranium, Cutleaf							
Geranium dissectum L.	9.1	77—87	4	5—15	99	90	67
	9.1	60—70	4	0	99	99	67
Geranium, Dovefoot							
G. molle L.	9.1	77—87	4	5—15	99	83	67
	9.1	60—70	4	0	99	98	67
Gherkin, West Indian							
Cucumis anguria L.	Not given	35—40	17	5	85	90	21
Gladiolus							
Gladiolus spp.	Not given	Not given	5	5	82	0	9
	Not given	Not given	10	−4	82	4	9
Grapefruit							
Citrus paradisi Macfad.	60	Not given	1	5	94	88	5
	78	Not given	1	5	99	83	5
	100	Not given	1	5	96	88	5
	123	Not given	1	5	32	16	5
	131	Not given	1	5	94	89	5
Guar							
Cyamopsis tetragonoloba (L.) Taub.	Not given	35—40	16	5	96	96	21
Guayule							
Parthenium argentatum A. Gray	Not given	35—40	16	5	66	62	21

Hemlock, western							
Tsuga heterophylla (Raf.)	8	Not given	1	−4	12	5	10
Sarg.	8	Not given	3	−18	12	13	10
	7	Not given	7	−18	70	70	2
	7	Not given	7	0	70	56	2
	7	Not given	7	Ambient	70	0	2
Hemp							
Cannabis sativa L.	5.7	50	8.25	10	98	97	94
	5.7	Not given	8.25	−10	98	93	94
	5.7	50	15	10	98	86	40
	5.7	Ambient	15	−10	98	87	40
	8.3	50	15	10	100	96	40
	8.3	Ambient	15	−10	100	97	40
Indiangrass							
Sorghastrum nutans (L.) Nash	Not given	35—40	16	5	66	41	21
Juniper, alligator							
Juniperus deppeana Steud.	Not given	Ambient	9	Ambient	Not known	16	62
Juniper, one-seed							
J. monosperma (Engelm.) Sarg.	Not given	Ambient	21	Ambient	Not known	54	62
J. osteosperma (Torr.) Little	Not given	Ambient	45	Ambient	Not known	17	62
Kenaf							
Hibiscus cannabinus L.	Not given	35—40	13	5	78	77	21
Ladysthumb							
Polygonum persicaria L.	10.5	77—87	4	5—15	95	73	67
	10.5	60—70	4	0	95	99	67
Lambsquarters							
Chenopodium album L.	8.6	77—87	4	5—15	86	98	67
	8.6	60—70	4	0	86	98	67
Larch							
Larix spp.	Not given	Ambient	3	18	32	4	44
Lemon, Rough							
Citrus limon L.) Burm. f.	56	Ambient	1.50	5	84	14	5
Leek							
Allium ampeloprasum L.	Not given	35—40	17	5	97	99	21
Lentil							
Lens culinaris Medik.	Not given	35—40	5	5	99	99	21
Lespedeza, Kobe	Not given	Not given	4	Ambient	87	39	57

Table 1 (continued)
LONGEVITY OF SEEDS STORED UNDER VARIOUS CONDITIONS IN OPEN AND POROUS CONTAINERS

Species	Moisture (% when stored)	Storage: RH (%)	Storage: Time (years)	Storage: Temperature (°C)	Germination (%): When stored	Germination (%): After storage	Ref.
Lespedeza striata (Thunb. ex Murr.) Hook & Arn.	Not given	Not given	4	−7 to −11	87	90	57
	Not given	Not given	18	Ambient	87	12	74
	Not given	Not given	18	−7 to −11	87	64	74
	Not given	35—40	10	5	95	96	21
Lespedeza, Korean *L. stipulacea* Maxim.	Not given	Not given	4	Ambient	97	54	57
	Not given	Not given	4	−7 to −11	97	95	57
	Not given	Not given	18	Ambient	97	4	74
	Not given	Not given	18	−7 to −11	97	77	74
	Not given	35—40	16	5	95	78	21
Lespedeza, Sericea *L. cuneata* (Dum. G. Don)	Not given	Not given	4	Ambient	89	46	57
	Not given	Not given	4	−7 to −11	89	89	57
	Not given	Not given	18	Ambient	89	4	74
	Not given	Not given	18	−7 to −11	89	72	21
	Not given	35—40	15	5	93	74	21
Lettuce *Lactuca sativa* L.	Not given	15	6.25	22—27	96	88	1
	Not given	30	6.25	22—27	96	10	1
	Not given	45	4.75	22—27	96	2	1
	Not given	60	2.75	22—27	96	4	1
	Not given	75	1.25	22—27	96	2	1
	Not given	90	0.50	22—27	96	0	1
	Not given	Ambient	2.25	22—27	96	0	1
	Not given	30	6	7	95	90	1
	Not given	60	6	7	95	89	1
	Not given	Ambient	6	7	95	88	1
	Not given	30	6	10	95	92	1

Not given	60	6	10	95	86	1
Not given	Ambient	6	10	95	84	1
Not given	30	6	15	95	79	1
Not given	60	6	15	95	88	1
Not given	Ambient	6	15	95	84	1
Not given	30	6	20	95	62	1
Not given	60	4.25	20	95	11	1
Not given	Ambient	6	20	95	81	1
Not given	Ambient	20	−4	98	88	9
Not given	Ambient	18	5	100	84	14
Not given	Ambient	18	−2	100	83	14
Not given	Ambient	18	−18	100	89	14
4	70	19	−12	97	94	21
4	40	19	−1	97	10	21
4	60	8	10	97	0	21
4	30	5	21	97	2	21
4	15	8	32	97	2	21
7	70	19	−12	97	93	21
7	40	19	−1	97	17	21
7	60	8	10	97	0	21
7	30	5	21	97	2	21
7	15	8	32	97	2	21
10	70	19	−12	95	94	21
10	40	19	−1	95	18	21
10	60	8	10	95	0	21
10	30	5	21	95	3	21
10	15	8	32	95	0	21
Not given	90	0.25	32	98	35	21
Not given	90	0.25	21	98	17	21
Not given	90	4	10	98	58	21
Not given	70	0.91	32	98	8	21
Not given	70	1.16	21	98	6	21
Not given	70	8.50	10	98	7	21
Not given	70	10	−12	98	95	21
Not given	60	4.50	10	98	31	21
Not given	50	1.25	32	98	7	21
Not given	50	2.25	21	98	55	21
Not given	50	4.50	10	98	18	21

Table 1 (continued)
LONGEVITY OF SEEDS STORED UNDER VARIOUS CONDITIONS IN OPEN AND POROUS CONTAINERS

Species	Moisture (% when stored)	Storage: RH (%)	Storage: Time (years)	Storage: Temperature (°C)	Germination (%): When stored	Germination (%): After storage	Ref.
	Not given	40	10	−1	98	9	21
	Not given	35—40	10	5	98	2	21
	Not given	30	3.50	21	98	6	21
	Not given	Ambient	3	Ambient	98	42	21
	Not given	15	2.25	32	98	55	21
	4.7	20	6	Ambient	99	36	77
Lily, Regal							
Lilium regale Wilson	Not given	Not given	5	5	90	0	9
	Not given	Not given	15	−4	90	72	9
	9.9	Not given	15	−5	86	74	8
	9.9	Not given	15	5	86	8	8
Lobelia, Cardinal flower							
Lobelia cardinalis L.	6.9	Not given	3	Ambient	63	13	11
	6.9	Not given	2	5	63	6	11
	6.2	Not given	16	−5	63	14	11
Lotus angustissimus L.	Not given	35—40	11	5	68	64	21
L. arabicus L.	Not given	35—40	11	5	96	95	21
L. arenarius Brot.	Not given	35—40	5	5	94	94	21
L. caucasicus Kuprian	Not given	35—40	5	5	92	89	21
L. collinus (Boiss.) Heldr.	Not given	35—40	6	5	97	88	21
L. conimbricensis Brot.	Not given	35—40	6	5	98	99	21
L. conjugatus L.	Not given	35—40	11	5	91	96	21
Trefoil, birdsfoot							
L. corniculatus L.	Not given	Ambient	25	18	93	14	44
	Not given	35—40	11	5	92	94	21
L. creticus L.	Not given	35—40	11	5	92	91	21
L. decumbens Poir.	Not given	35—40	4		61	77	21
L. divaricatus Boiss.	Not given	35—40	11	5	95	94	21

L. edulis L.	Not given	35—40	6	5	99	74	21
L. frondosus Freyn.	Not given	35—40	11	5	75	68	21
L. glinoides Del.	Not given	35—40	11	5	91	97	21
L. hispidus Desf.	Not given	35—40	11	5	91	89	21
L. major Scop.	Not given	35—40	6	5	96	87	21
L. mearnsii Britton	Not given	35—40	5	5	82	84	21
L. ornithopodioides L.	Not given	35—40	11	5	96	83	21
L. palustris Willd.	Not given	35—40	11	5	94	92	21
L. parviflorus Desf.	Not given	35—40	6	5	87	88	2
L. pedunculatus Cav.	Not given	35—40	11	5	90	98	21
L. peregrinus L.	Not given	35—40	6	5	93	93	21
L. purshianus (Benth.) Clements and Clements	Not given	35—40	5	5	91	81	21
L. pusillus Medic	Not given	35—40	5	5	91	96	21
L. strictus Fisch. & C. A. Mey	Not given	35—40	5	5	72	39	21
L. tenuis Waldst. & Kit. ex Willd.	Not given	35—40	11	5	96	93	21
L. tetragonolobus L. now *Tetrogonolobus purpureus* L.	Not given	35—40	6	5	90	97	21
L. weilleri Maire	Not given	35—40	5	5	87	90	21
Luffa, angled							
L. acutangula (L.)	Not given	35—40	12	5	90	86	21
Luffa, vegetable sponge							
L. aegyptiaca Mill.	Not given	35—40	11	5	69	42	21
Lupine, yellow							
Lupinus luteus L.	Not given	Ambient	8	18	85	10	44
	Not given	35—40	15	5	86	76	21
Macadamia Nut							
Macadamia spp.	Not given	Ambient	0.91	18—27	35	8	56
	Not given	40—45	6	5	52	53	21
Marigold							
Tagetes erecta L.	Not given	35—40	16	5	97	94	21
Marigold, French							
T. patula L.	Not given	35—40	16	5	91	82	21
Mayweed, scentless							
Matricaria inodora L.	12.2	77—87	4	5—15	91	96	67
	12.2	60—70	4	0	91	99	67
Medic, black							
Medicago lupulina L.	9.5	77—87	2	5—15	84	14	67

Table 1 (continued)
LONGEVITY OF SEEDS STORED UNDER VARIOUS CONDITIONS IN OPEN AND POROUS CONTAINERS

Species	Moisture (% when stored)	Storage RH (%)	Storage Time (years)	Storage Temperature (°C)	Germination (%) When stored	Germination (%) After storage	Ref.
	9.5	60—70	4	0	84	93	67
	Not given	Ambient	15	18	99	1	44
Mesquitegrass, vine							
Panicum obtusum H.B.K.	Not given	Ambient	25	Ambient	98	7	92
Mignonette, common							
Reseda odorata L.	Not given	Ambient	10	18	59	2	44
Mustard							
Brassica campestris L. now *B. rapa* L.	Not given	Ambient	12	18	98	3	44
Mustard, white							
Sinapis alba L.	Not given	Ambient	15	18	96	1	44
Needlegrass, green							
Stipa viridula Trin.	7 to 9	Not given	5	15—23	28 (dormant)	98	68
Oatgrass, tall							
Avena elatior L. now *Arrhenatherum elatius* (L.) Beauv. ex J. & C. Presl.	Not given	Ambient	6	18	98	6	44
	Not given	40	10	5	90	92	35
	Not given	30—40	10	21	90	36	35
	Not given	50—60	10	10—15	90	54	35
	Not given	35—40	11	5	98	95	21
Oats							
Avena sativa L.	Not given	Ambient	10	18	84	2	44
	Not given	Not given	19	Ambient	95	41	88
	Not given	Not given	15	Ambient	95	83	88
	Not given	Not given	10	Ambient	95	96	88
	Not given	Ambient	32	Ambient	Not given	53	55
	12	77—87	4	5—15	92	22	67
	12	60—70	4	0	92	95	67
	Not given	35—40	16	5	99	95	21

Oat, Bristle							
A. strigosa Schreb.	13.5	77—87	4	5—15	92	29	67
	13.5	60—70	4	0	92	88	67
Oat, red							
A. byzantina K. Koch	Not given	35—40	16	5	99	99	21
Oat, wild *A. fatua* L.	14	77—87	4	5—15	34	8	67
	14	60—70	4	0	34	42	67
Okra							
Abelmoschus esculentus (L.) Moench	Not given	35—40	17	5	82	92	21
Onion							
Allium cepa L.	Not given	Ambient	10	−4	98	69	9
	7.7	Ambient	5	Ambient	95	41	14
	7.7	Ambient	14	5	95	69	14
	7.7	Ambient	18	−2	95	15	14
	7.7	Ambient	18	−18	95	87	14
	8	Ambient	7	Ambient	86	0	27
	8	85	7	4	86	63	27
	8	25	7	−7	86	68	27
	10	Ambient	7	Ambient	86	0	27
	10	85	7	4	86	57	27
	10	25	7	−7	86	67	27
	8.6	35	1	20	96	81	5
	11.4	55	0.75	20	96	7	5
	14.9	70	0.50	20	96	1	5
	4.1	50—60	2	1	95	93	54
	4.1	80—95	1	10	95	13	54
	4.1	25—30	2	27	95	88	54
	4.1	20—80	2	16—35	95	88	54
	4.1	15	2	43	95	88	54
	3.8	50—60	2	1	93	92	54
	3.8	80—95	1	10	93	22	54
	3.8	25—30	2	27	93	84	54
	3.8	20—80	1.50	16—35	93	11	54
	3.8	15	2	43	93	76	54
	11.4	85—95	7	−20	93	89	96
	11.4	85—95	7	0—2	93	29	96
	11.4	70—80	6	5—20	93	0.5	96
	6.3	20	9	Ambient	96	87	7

Table 1 (continued)
LONGEVITY OF SEEDS STORED UNDER VARIOUS CONDITIONS IN OPEN AND POROUS CONTAINERS

Species	Moisture (% when stored)	Storage RH (%)	Storage Time (years)	Storage Temperature (°C)	Germination (%) When stored	Germination (%) After storage	Ref.
	Not given	35—40	17	5	93	93	21
Onion, beltsville bunching *Allium cepa* L. × *A. fistulosum* L.	Not given	35—40	13	5	93	64	21
Onion, Welsh *A. fistulosum* L.	Not given	35—40	17	5	97	78	21
Orange, sour *Citrus aurantium* L.	80	Ambient	1.50	5	90	70	5
Orange, sweet *C. sinensis* (L.) Osb.	82	Ambient	1.50	5	71	15	5
Orchardgrass, cocksfoot *Dactylis glomerata* L.	10.9	77—87	4	5—15	90	1	67
	10.9	60—70	4	0	90	88	67
	Not given	40	10	5	96	91	35
	Not given	30—40	10	21	96	61	35
	Not given	50—60	10	10—15	96	34	35
	Not given	Ambient	9	18	90	1	44
	Not given	Not given	3	Ambient	90	3	57
	Not given	Not given	4	−7 to −11	90	81	57
	Not given	Not given	18	−7 to −11	90	30	74
	Not given	35—40	15	5	86	85	21
Osteospermum *Osteospermum ecklonis* (DC.) Norl.	Not given	50	1	32	79	13	25
	Not given	50	1.50	21	79	80	25
	Not given	50	3.50	10	79	84	25
	Not given	50	6	10	79	77	21
	Not given	70	6	10	79	38	21
	Not given	70	3.50	10	79	60	25
Pansy *Viola tricolor* L.	Not given	35—40	16	5	90	80	21

Papaya							
Carica papaya L.	Not given	50	6	10	68	53	18
Parsley							
Petroselinum sativum Hoffm. now *P. crispum* (Mill.) Nym. ex A. W. Hill	Not given	Ambient	6	18	65	3	44
	Not given	35—40	18	5	74	84	21
Parsnip							
Pastinaca sativa L.	Not given	Ambient	4	18	92	3	44
	6.3	Ambient	3	Ambient	94	58	63
	6.1	Not given	3	Ice chest	95	84	63
	9.9	85—95	5	−20	93	1	96
	9.9	85—95	5	0—2	93	3	96
	9.9	70—80	5	5—20	93	0	96
	Not given	35—40	18	5	90	88	21
Pea, garden							
Pisum sativum L.	Not given	90	0.50	32	94	50	61
	Not given	70	1	32	94	29	61
	Not given	70	1.25	21	94	80	61
	Not given	50	5	10	94	87	61
	Not given	Ambient	10	18	87	7	44
	Not given	35—40	16	5	90	96	21
Pea, pigeon							
Cajanus cajan (L.) Huth	Not given	35—40	10	5	85	71	21
	Not given	35—40	15	5	85	20	21
Peanut, Spanish (Groundnut)							
Arachis hypogaea L.	Ambient	Not given	6	21—29	85	56	28
	Not given	35—40	14	5	98	92	21
Peanut, Valencia	Ambient	Not given	6	21—29	53	47	28
	Not given	35—40	17	5	88	94	21
Pear							
Pyrus communis L.	10—11	50—65	2.50	20—25	96	61	90
	10—11	80—90	0.83	2—10	96	36	90
Peony							
Paeonia suffruticosa Andr.	Not given	Not given	5	5	50	3	9
	Not given	Not given	10	−4	50	29	9
Pepper							
Capsicum annuum L.	Not given	Not given	20	−4	73	19	9
	Not given	35—40	18	4	87	87	21

Table 1 (continued)
LONGEVITY OF SEEDS STORED UNDER VARIOUS CONDITIONS IN OPEN AND POROUS CONTAINERS

Species	Moisture (% when stored)	Storage: RH (%)	Storage: Time (years)	Storage: Temperature (°C)	Germination (%): When stored	Germination (%): After storage	Ref.
Petunia							
Petunia × *hybrida* Vilm.	Not given	35—40	16	5	68	74	21
Pine, Austrian							
Pinus nigra Arnold var. *austriaca* (Hoess) Baddux Pine, loblolly	Not given	Ambient	8	18	66	2	44
P. taeda L.	Not given	Not given	12	5	82	15	9
	Not given	Not given	12	−4	82	80	9
Pine, lodgepole							
P. contorta Dougl. ex Loud.	8.8	Not given	7	0	32	32	9
Pine, mugo							
P. montana Mill. now *P. mugo* Turra Pine, Norway	Not given	Not given	14	18	96	2	44
P. resinosa Ait.	6.5	Not given	30	2—10	95	8	87
Pine, ponderosa							
P. ponderosa Dougl. ex P. & C. Lawson	8.7	Not given	35	10—27	Not given	16	58
	8.1	Not given	7	−18	66	57	2
	8.1	Not given	7	0	66	67	2
	8.1	Not given	7	Ambient	66	35	2
	15	Not given	3	−4	95	92	10
	10	Not given	3	−18	95	72	10
Pine, Scotch							
P. sylvestris L.	Not given	Ambient	11	18	69	5	44
Pine, shortleaf							
P. echinata Mill.	Not given	Not given	16	5	59	3	9
	Not given	Not given	16	−4	59	36	9
Plantain, buckhorn							
Plantago lanceolata L.	10.2	77—87	4	5—15	98	48	67
	10.2	60—70	4	0	98	98	67

Plum, American							
Prunus americana Marsh.	10—11	50—65	2.50	20—25	100	62	90
	10—11	80—90	2.50	2—10	100	23	90
Potato							
Solanum tuberosum L.	Not given	35—40	10	5	95	98	21
Pyrethrum							
Chrysanthemum cinerariifolium (Trevir.) Vis.	6.5	Not given	15	−18.5	57	38	12
	6.5	Not given	4	−4	57	5	12
	6.5	Not given	15	5	57	29	12
	6.5	Not given	5	10	57	5	12
	6.5	Not given	13	20	57	1	12
	6.5	Not given	1	30	57	4	12
	6.0	Not given	4	Ambient	74	7	7
Quinine							
Cinchona ledgerana Moens ex Trimen	6.0	Not given	1.50	5	74	3	7
now *C. calisaya* Wedd.	6.0	Not given	4	−4	74	43	7
Radish							
Raphanus sativus L.	Not given	Ambient	11	18	71	1	44
	Not given	35—40	13	5	99	99	21
Rice							
Oryza sativa L.	11.5—13.2	60	3.50	15	96	98	45
	Not given	15	6.25	22—27	91	18	1
	Not given	30	6.25	22—27	91	6	1
	Not given	45	6.25	22—27	91	5	1
	Not given	60	1.66	22—27	91	27	1
	Not given	75	0.83	22—27	91	16	1
	Not given	90	0.83	22—27	91	1	1
	Not given	Ambient	1.25	22—27	91	14	1
	Not given	30	6	7	92	42	1
	Not given	60	6	7	92	12	1
	Not given	Ambient	6	7	92	2	1
	Not given	30	6	10	92	31	1
	Not given	60	6	10	92	33	1
	Not given	Ambient	6	10	92	20	1
	Not given	30	6	15	92	14	1
	Not given	60	6	15	92	5	1

Table 1 (continued)
LONGEVITY OF SEEDS STORED UNDER VARIOUS CONDITIONS IN OPEN AND POROUS CONTAINERS

Species	Moisture (% when stored)	Storage: RH (%)	Storage: Time (years)	Storage: Temperature (°C)	Germination (%): When stored	Germination (%): After storage	Ref.
	Not given	Ambient	6	15	92	12	1
	Not given	30	6	20	92	3	1
	Not given	60	4.25	20	92	10	1
	Not given	Ambient	6	20	92	4	1
	Not given	35—40	18	5	99	90	21
Roselle							
Hibiscus sabdariffa L.	Not given	35—40	13	5	78	80	21
Rye							
Secale cereale L.	Not given	Ambient	4	18	93	3	44
	9	58—83	3 +	9—20	90	90	69
	Not given	Ambient	32	Ambient	Not given	1	55
	Not given	35—40	17	5	82	58	21
Ryegrass, Italian							
Lolium multiflorum Lam.	12.2	77—87	4	5—15	93	6	67
	12.2	60—70	4	0	93	84	67
	Not given	Ambient	9	18	83	2	44
	9	35	4	Ambient	96	92	52
	11	53	4	4—21	96	92	52
	13	70	4	4—21	96	67	52
	16	83	2	4—21	96	28	52
	19	89	1	4—21	96	10	52
	Not given	35—40	14	5	93	83	21
Ryegrass, perennial							
L. perenne L.	10—12	58—83	6.50	11—19	95	92	70
	Not given	Ambient	8	18	92	3	44
	Not given	35—40	13	5	98	96	21
	12.6	77—87	4	5—15	92	51	67
	12.6	60—70	4	0	92	85	67

Safflower							
Carthamus tinctorus L.	4	70	17	−12	95	74	21
	4	40	17	−1	95	88	21
	4	60	17	10	95	53	21
	4	30	17	21	95	7	21
	4	15	17	32	95	7	21
	7	70	17	−12	94	89	21
	7	40	17	−1	94	84	21
	7	60	17	10	94	55	21
	7	30	17	21	94	10	21
	7	15	17	32	94	7	2
	10	70	17	−12	95	91	21
	10	40	17	−1	95	91	21
	10	60	17	10	95	53	21
	10	30	17	21	95	10	21
	10	15	17	32	95	3	21
	Not given	35—40	16	5	74	72	21
Sage							
Salvia officinalis L.	Not given	Ambient	7	18	74	3	44
	Not given	35—40	15	5	75	92	21
Sainfoin							
Onobrychis sativa Lam.	Not given	Ambient	7	18	60	1	44
Sainfoin							
O. viciifolia Scop.	Not given	Ambient	14	Ambient	Not given	36	55
	Not given	35—40	5	5	94	84	21
Salsify							
Tragopogon porrifolius L.	Not given	35—40	14	5	98	99	21
Savory, summer							
Satureja hortensis L.	Not given	35—40	15	5	67	46	21
Scabiosa							
Scabiosa atropurpurea L. and *S. caucasica* Bieb.	Not given	35—40	16	5	77	68	21
Serradella							
Ornithopus sativus Brot.	Not given	Ambient	10	18	97	5	44
Sesame							
Sesamum indicum L.	4	70	17	−12	94	89	21
	4	40	17	−1	94	92	21
	4	60	17	10	94	83	21

Table 1 (continued)
LONGEVITY OF SEEDS STORED UNDER VARIOUS CONDITIONS IN OPEN AND POROUS CONTAINERS

		Storage			Germination (%)		
Species	Moisture (% when stored)	RH (%)	Time (years)	Temperature (°C)	When stored	After storage	Ref.
	4	30	17	21	94	82	21
	4	15	17	32	94	43	21
	7	70	17	−12	92	94	21
	7	40	17	−1	92	94	21
	7	60	17	10	92	61	21
	7	30	17	21	92	78	21
	7	15	17	32	92	37	21
	10	70	17	−12	88	96	21
	10	40	17	−1	88	95	21
	10	60	17	10	88	72	21
	10	30	17	21	88	84	21
	10	15	11	32	88	51	21
	Not given	35—40	17	5	96	92	21
Silver bells							
Browallia viscosa HBK.	Not given	35—40	14	5	99	99	21
Snapdragon							
Antirrhinum majus L.	Not given	35—40	11	5	90	80	21
	Not given	35—40	16	5	90	40	21
Sorghum							
Sorghum bicolor (L.) Moench	4	70	16	−12	92	95	24
	4	40	16	−1	92	92	24
	4	60	16	10	92	40	24
	4	30	16	21	92	32	24
	4	15	16	32	92	1	24
	7	70	16	−12	95	94	24
	7	40	16	−1	95	92	24
	7	60	16	10	95	42	24
	7	30	16	21	95	47	24

	7	15	11	32	95	42	21
	10	70	16	−12	91	94	24
	10	40	16	−1	91	83	24
	10	60	16	10	91	41	24
	10	30	16	21	91	36	24
	10	15	11	32	91	35	21
	Not given	75—80	3.50	4—7	95	85	81
	Not given	Ambient	3.50	19—31	95	1	81
	Not given	35—40	17	5	90	90	21
Soybean							
Gycine max (L.) Merr.	11.4	15	6.25	22—7	100	3	1
	11.4	30	6.25	22—27	100	4	1
	11.4	45	4.75	22—27	100	5	1
	11.4	60	2.66	22—27	100	5	1
	11.4	75	1.25	22—27	100	1	1
	11.4	90	0.66	22—27	100	0	1
	11.4	Ambient	1.66	22—27	100	3	1
	Not given	30	6	7	97	8	1
	Not given	60	6	7	97	7	1
	Not given	Ambient	6	7	97	6	1
	Not given	30	6	10	97	6	1
	Not given	60	6	10	97	6	1
	Not given	Ambient	6	10	97	12	1
	Not given	30	4.33	15	97	3	1
	Not given	60	3.33	15	97	4	1
	Not given	Ambient	6	15	97	2	1
	Not given	30	4.50	20	97	4	1
	Not given	60	3.25	20	97	2	1
	Not given	Ambient	4.25	20	97	4	1
	Not given	35—40	16	5	92	79	21
	Not given	Ambient	9	Ambient	Not given	83	55
Spinach, New Zealand							
Tetragonia tetragonioides (Pall.) Ktze.	Not given	Ambient	10	Ambient	1	95	59
	Not given	35—40	17	5	74	42	21
Spinach							
Spinacia oleracea L.	Not given	35—40	17	5	88	88	21

Table 1 (continued)
LONGEVITY OF SEEDS STORED UNDER VARIOUS CONDITIONS IN OPEN AND POROUS CONTAINERS

Species	Moisture (% when stored)	Storage: RH (%)	Storage: Time (years)	Storage: Temperature (°C)	Germination (%): When stored	Germination (%): After storage	Ref.
Spruce, Engelmann							
Picea engelmannii Parry ex	7.2	Not given	7	−18	71	46	2
Engelm.	7.2	Not given	7	0	71	54	2
	7.2	Not given	7	Ambient	71	0	2
Spruce, Norway							
Picea excelsa (Lam.) Link	Not given	Ambient	8	18	83	2	44
now *P. abies* (L.) K Arst. Spruce, Sitka	7.3	Not given	7	−18	80	81	2
P. sitchensis (Bong.) Carr.	7.3	Not given	7	0	80	78	2
	7.3	Not given	7	Ambient	80	1	2
	13	Not given	2	−4	56	11	10
	9	Not given	3	−18	56	47	10
Spruce, white							
Picea glauca (Moench) Voss	6.7	Not given	7	−18	52	46	2
	6.7	Not given	7	0	52	47	2
	6.7	Not given	3	Ambient	52	27	2
	Not given	Ambient	5	18	87	6	44
Spurry							
Spergula maxima Weike	Not given	Ambient	13	18	83	2	44
now *S. arvensis* L. Spurry *Spergula sativa* Boenn.	Not given	Ambient	14	18	86	1	44
now *S. arvensis* L. Stock *Matthiola incana* (L.) R. Br.	Not given	35—40	16	5	94	92	21
Strawberry							
Fragaria spp.	Not given	35—40	10	5	72	82	21
Sudangrass							
Sorghum sudanense (Piper) Stapf	Not given	35—40	17	5	88	82	21
Sunflower							
Helianthus annuus L.	Not given	35—40	16	5	90	90	21

Sweetclover, white							
Melilotus alba Medik.	Not given	Ambient	20	Ambient	95	70	91
	Not given	Ambient	15	Ambient	Not given	76	55
	Not given	35—40	13	5	81	82	21
Sweetclover, yellow							
Melilotus officinalis Lam.	Not given	Ambient	14	Ambient	Not given	24	55
	Not given	35—40	17	5	84	64	21
Sweet pea							
Lathyrus odoratus L.	Not given	35—40	16	5	96	96	21
Switchgrass							
Panicum vigatum L.	Not given	35—40	16	5	93	97	21
Tampala							
Amaranthus tricolor L.	Not given	35—40	17	5	88	85	21
Tea							
Thea sp.	Not given	100	0.09	0	70+	50—70	95
	Not given	70	0.09	0	70+	15—30	95
	Not given	55	0.09	0	70+	15	95
	Not given	40	0.09	0	70+	0	95
Thyme							
Thymus spp.	Not given	Ambient	5	18	62	3	44
Timothy							
Phleum nodosum L. now *P. pratense* L.	10.1	77—87	2	5—15	89	13	67
	10.1	60—70	4	0	89	76	67
Timothy							
Phleum pratense L.	10.6	77—87	4	5—15	97	1	67
	10.6	60—70	4	0	97	85	67
	Not given	Ambient	17	Ambient	97	0	88
	Not given	Ambient	15	Ambient	97	2	88
	Not given	Ambient	10	Ambient	97	54	88
	10—12	58—83	6.50	11—19	98	95	70
	Not given	Ambient	7	18	98	1	44
	Not given	40	10	5	98	99	35
	Not given	30—40	10	21	98	87	35
	Not given	50—60	10	10—15	98	79	35
	Not given	35—40	10	5	84	72	21
Unhulled	Not given	Not given	10	Not given	Not given	11	51
Hulled	Not given	Not given	10	Not given	Not given	4	51

Table 1 (continued)
LONGEVITY OF SEEDS STORED UNDER VARIOUS CONDITIONS IN OPEN AND POROUS CONTAINERS

Species	Moisture (% when stored)	Storage RH (%)	Storage Time (years)	Storage Temperature (°C)	Germination (%) When stored	Germination (%) After storage	Ref.
Tobacco							
Nicotiana tabacum L.	9	Not given	25	5	Not given	5	65,66
	Not given	35—40	18	5	91	92	21
Tomato							
Lycopersicon esculentum Mill.	Not given	90	0.25	32	90	43	61
	Not given	70	1	32	90	56	61
	Not given	70	1.25	21	90	84	61
	Not given	50	5	10	90	90	61
	Not given	Not given	4	5	81	72	46
	Not given	Humid	4	Warm	84	15	46
	Not given	Not given	12	Ambient	93	3	9
	Not given	Not given	12	−4	93	88	9
	Not given	Not given	18	Ambient	93	22	14
	Not given	Not given	18	5	93	88	14
	Not given	Not given	18	−18	93	87	14
	Not given	Ambient	32	Ambient	Not given	87	60
	Not given	Ambient	43	Ambient 32	Not given	76	21
		32					
		35—40		+ 5 for 11	—	—	—
		11					
Turnip							
Brassica rapa L. (Rapifera group)	6—8	58—83	6.50	11—19	98	98	71
	Not given	35—40	19	5	90	39	21
	Not given	35—40	19	5	99	98	21
Velvetgrass							
Holcus lanatus L.	12.9	77—87	4	5—15	88	6	67
	12.9	60—70	4	0	88	85	67
	Not given	Ambient	8	18	86	1	44

Verbena							
Verbena spp.	Not given	Not given	16.33	−4	45	5	9
	Not given	35—40	16	5	56	38	21
Vetch, Kidney							
Anthyllis vulneraria L.	Not given	Ambient	11	18	96	1	44
Watermelon							
Citrullus lanatus (Thunb.)	Not given	90	0.25	32	90	41	61
Matsum. and Nakai	Not given	70	1	32	90	40	61
	Not given	70	1.25	21	90	81	61
	Not given	50	5	10	90	90	61
	Not given	35—40	17	5	95	94	21
Wheat							
Triticum aestivum L.	Not given	Not given	19	Ambient	99	0	88
	Not given	Not given	15	Ambient	99	11	88
	Not given	Not given	10	Ambient	99	92	88
	12	58—83	6+	11—19	97	96	69
	Not given	Ambient	32	Ambient	Not given	16	55
	Not given	Ambient	31	Ambient	Not given	23	55
	Not given	Ambient	32	Ambient	97	1	50
	Not given	Ambient	30	Ambient	94	12	50
	Not given	Ambient	25	Ambient	93	14	50
	Not given	Ambient	7	18	92	3	44
	Not given	35—40	17	5	92	86	21
Wheat, spring	12.5	77—87	4	5—15	99	37	67
	12.5	60—70	4	0	99	92	67
Wheat, winter	12	77—87	4	5—15	92	19	67
	12	60—70	4	0	92	91	67
Wheatgrass, fairway crested							
Agropyron cristatum (L.) Gaertn.	7—9	Not given	5−	15—23	95	84	68
	Not given	35—40	15	5	85	93	21
Wheatgrass, intermediate							
A. intermedium. (Host) Beauv. var. *intermedium*	Not given	40	10	5	97	98	35
	Not given	30—40	10	21	97	55	35
	Not given	50—60	10	10—15	97	15	35
	Not given	35—40	16	5	86	78	21

Table 1 (continued)
LONGEVITY OF SEEDS STORED UNDER VARIOUS CONDITIONS IN OPEN AND POROUS CONTAINERS

Species	Moisture (% when stored)	Storage RH (%)	Storage Time (years)	Storage Temperature (°C)	Germination (%) When stored	Germination (%) After storage	Ref.
Wheatgrass, slender							
A. trachycaulum	7—9	Not given	5–	15—23	98	24	68
(Link) Malte ex H. F. Lewis	Not given	35—40	11	5	99	33	21
Wheatgrass, western							
A. smithii Rydb.	7—9	Not given	5–	15—23	93	63	68
	Not given	35—40	13	5	70	29	21
Wildrye, blue							
Elymus glaucus Buckl.	7—9	Not given	5–	15—23	92	2	68
Wildrye, Canada							
E. canadensis L.	Not given	35—40	12	5	92	76	21
Wildrye, Russian							
E. junceus Fisch.	Not given	35—40	14	5	95	86	21
now *Psathyrostachys juncea* (Fisch.) Nevski	Not given	35—40	14	5	95	86	21
Zinnia							
Zinnia elegans Jacq.	Not given	35—40	16	5	84	84	21

Table 2
EFFECT OF VARIOUS STORAGE CONDITIONS ON VIABILITY OF SEEDS IN AIR IN SEALED GLASS OR METAL CONTAINERS

Species	Moisture (% when stored)	Storage: Time (years)	Storage: Temperature (°C)	Germination (%): When stored	Germination (%): After storage	Ref.
Alfalfa, Lucerne						
Medicago sativa L.	Not given	4.75	22—27	100	16	1
	Not given	6	8	87	71	1
	Not given	6	10	87	82	1
	Not given	6	15	87	90	1
	Not given	6	20	87	65	1
	Not given	12 +	Ambient	90	50 +	72
Ash, European						
Fraxinum excelsior L.	10.8	7	5	53	55	6
Ash, green						
F. pennsylvanica Marsh.	13.3	2	5	68	0	6
Aster, China						
Callistephus chinensis (L.) Nees	7.9	12.33	5	91	14	9
	7.9	16.33	−4	91	89	9
	4.6	12.33	5	91	1	9
	4.6	16.33	−4	91	81	9
Barley						
Hordeum vulgare L.	Not given	8	Ambient	95 +	50−	72
	Not given	6	Ambient	95 +	80−	72
Bean, garden						
Phaseolus vulgaris L.	10	15	−18	89	100	13
	10	15	−2	89	100	13
	10	15	5	89	100	13
	10	15	10	89	85	13
	10	15	20	89	0	13
	10	6	30	89	5	13
	10	6	Ambient	89	6	13
		15	−18	97	99	13
	7	15	−2	97	100	13
	7	15	5	97	92	13
	7	15	10	97	92	13
	7	15	20	97	14	13
	7	8	30	97	3	13
	7	6	Ambient	97	4	13
	Not given	1.75	22—27	92	13	1
	Not given	6	7	95	80	1
	Not given	6	10	95	83	1
	Not given	6	15	95	62	1
	Not given	4.25	20	95	7	1
	Not given	3.50	4—7	97	95	81
Beech, European						
Fagus sylvatica L.	30.90	0.75	8	92	9	34
	30.90	1.75	2	92	1	34
	30.90	3.25	−10	92	3	34
	25.60	1.50	8	93	0	34
	25.60	1.75	2	93	8	34
	25.60	3.25	−10	93	5	34
	20.60	1.50	8	95	0	34
	20.60	1.75	2	95	5	34
	20.60	2.50	−10	95	3	34
	15.60	0.75	8	95	1	34
	15.60	1.50	2	95	0	34
	15.60	2.50	−10	95	4	34

Table 2 (continued)
EFFECT OF VARIOUS STORAGE CONDITIONS ON VIABILITY OF SEED IN AIR IN SEALED GLASS OR METAL CONTAINERS

Species	Moisture (% when stored)	Storage: Time (years)	Storage: Temperature (°C)	Germination (%): When stored	Germination (%): After storage	Ref.
Beet, garden *Beta vulgaris* L.	Not given	13	Ambient	90	50−	72
Beet, mangel *B. vulgaris* L.	Not given	13	Ambient	90	50−	72
Beet, sugar *B. vulgaris* L.	Not given	13	Ambient	90	50−	72
Bluegrass, Kentucky *Poa pratensis* L.	8.7	2.50	1	86	85	15
	8.7	2.50	10	86	86	15
	8.7	2.50	21	86	84	15
	8.7	2	32	86	1	15
	8.7	0.50	43	86	2	15
	8.7	2.50	Ambient	86	65	15
	6.2	2.50	1	84	84	15
	6.2	2.50	10	84	82	15
	6.2	2.50	21	84	84	15
	6.2	2.50	32	84	72	15
	6.2	1.50	43	84	1	15
	6.2	2.50	Ambient	84	86	15
	4.9	2.50	1	87	86	15
	4.9	2.50	10	87	85	15
	4.9	2.50	21	87	85	15
	4.9	2.50	32	87	85	15
	4.9	2.50	43	87	76	15
	4.9	2.50	Ambient	87	83	15
	4.1	2	1	74	66	54
	4.1	2	10	74	66	54
	4.1	2	27	74	65	54
	4.1	2	16—35	74	60	54
	4.1	2	43	74	63	54
	4.3	2	1	80	64	54
	4.3	2	10	80	63	54
	4.3	2	27	80	63	54
	4.3	2	16—35	80	64	54
	4.3	2	43	80	61	54
	7.1	2	1	78	62	54
	7.1	2	10	78	63	54
	7.1	2	27	78	60	54
	7.1	2	16—35	78	55	54
	7.1	0.75	43	78	1	54
	4.2	5	Ambient	86	82	78
	2.2	5	Ambient	84	77	78
	1.3	5	Ambient	82	45	78
	0.3	5	Ambient	75	6	78
	0.3	5	7	75	71	78
Smooth-stalked meadow grass	Not given	9	Ambient	90+	50−	72
Bluegrass, rough *Poa trivialis* L.	Not given	7	Ambient	90+	50−	72
Bluegrass, wood (wood meadow grass) *P. nemoralis* L.	Not given	7	Ambient	90+	50−	72

Table 2 (continued)
EFFECT OF VARIOUS STORAGE CONDITIONS ON VIABILITY OF SEED IN AIR IN SEALED GLASS OR METAL CONTAINERS

Species	Moisture (% when stored)	Storage: Time (years)	Storage: Temperature (°C)	Germination (%): When stored	Germination (%): After storage	Ref.
Cabbage						
Brassica oleracea L. var. capitata	3.2	2	1	95	94	54
	3.2	2	10	95	94	54
	3.2	2	27	95	94	54
	3.2	2	16—35	95	93	54
	3.2	2	43	95	88	54
	5.7	2	1	96	95	54
	5.7	2	10	96	96	54
	5.7	2	27	96	95	54
	5.7	2	16—35	96	95	54
	5.7	2	43	96	0	54
	6.6	2	1	94	93	54
	6.6	2	10	94	91	54
	6.6	2	27	94	93	54
	6.6	2	16—35	94	90	54
	6.6	1	43	94	0	54
	Not given	9	Ambient	90	50	72
Canarygrass, reed						
Phalaris arundinacea L.	3.8	12	−12	86	85	21
	3.8	12	−1	86	90	21
	3.8	12	5	86	84	21
	3.8	12	10	86	86	21
	3.8	12	21	86	57	21
	3.8	12	32	86	5	21
	6.9	12	−12	88	92	21
	6.9	12	−1	88	88	21
	6.9	12	5	88	82	21
	6.9	12	10	88	42	21
	6.9	12	21	88	7	21
	6.9	6	32	88	9	21
	10.3	12	−12	86	93	21
	10.3	12	−1	86	82	21
	10.3	12	5	86	75	21
	10.3	12	10	86	26	21
	10.3	6	21	86	5	21
	10.3	1	32	86	4	21
Carrot						
Daucus carota L.	7.2	20	Ambient	67	2	9
	7.2	20	−4	67	63	9
	10.7	20	−4	67	65	9
	Not given	9	Ambient	90	50−	72
Cashew						
Anacardium occidentale L.	Not given	1	Ambient	100	97.50	82
Clover						
Trifolium hybridum L.	Not given	7	Ambient	90	50−	72
Clover, crimson						
T. incarnatum L	4	16	−12	81	83	20
	4	16	−1	81	73	20
	4	16	10	81	81	20
	4	16	21	81	81	20
	4	16	32	81	77	20
	6	16	−12	91	87	20

Table 2 (continued)
EFFECT OF VARIOUS STORAGE CONDITIONS ON VIABILITY OF SEED IN AIR IN SEALED GLASS OR METAL CONTAINERS

Species	Moisture (% when stored)	Storage: Time (years)	Storage: Temperature (°C)	Germination (%): When stored	Germination (%): After storage	Ref.
	6	16	−1	91	84	20
	6	16	10	91	77	20
	6	16	21	91	85	20
	6	16	32	91	27	20
	11	16	−12	91	82	20
	11	16	−1	91	30	20
	11	16	10	91	22	20
	11	16	21	91	3	20
	11	2	32	91	5	20
	Not given	6	Ambient	90	50−	72
Clover, red						
T. pratense L.	Not given	6	Ambient	90	50−	72
	12.70	16	0	93	52	41, 42
	12.70	13	Ambient	93	15	41,42
Clover, white						
T. repens L.	9.8	16	0	92	91	41,42
	9.8	16	0	92	92	41, 42
	Not given	9	Ambient	90	50	72
Cotton, upland						
Gossypium hirsutum L.	7	19	1	87	92	89
	11	19	1	89	89	89
	13	19	1	90	51	89
	14	15	1	88	0	89
	7	19	21	93	51	89
	9	7	21	87	59	89
	7	25	1	87	91	80
	11	25	1	87	86	80
	13	25	1	87	16	80
	7	19	21	87	51	80
	11	5.50	21	87	1	80
	13	2	21	87	3	80
	7	7	Ambient	87	88	80
	11	3	Ambient	87	18	80
	7	5.50	32	87	0	80
	5	13	5	67	36	9
	5	13	−4	67	43	9
Crested dogtail						
Cynosurus cristatus L.	Not given	7	Ambient	90 +	50−	72
Cucumber						
Cucumis sativus L.	9.40	3	16	98	81	3
	8	3	21	98	86	3
	6.60	3	32	98	85	3
	5.50	3	32	98	95	3
	4.10	5	Ambient	98	95	78
	2.30	5	Ambient	98	93	78
	1.10	5	Ambient	95	89	78
	0.40	5	Ambient	96	84	78
Dandelion, common						
Taraxacum officinale Wiggers (formerly *T. officinale* Weber)	7.90	2	Ambient	91	0	9

Table 2 (continued)
EFFECT OF VARIOUS STORAGE CONDITIONS ON VIABILITY OF SEED IN AIR IN SEALED GLASS OR METAL CONTAINERS

Species	Moisture (% when stored)	Storage: Time (years)	Storage: Temperature (°C)	Germination (%): When stored	Germination (%): After storage	Ref.
	7.90	10	5	91	7	9
	7.90	16	−4	91	63	9
	3.90	6	Ambient	91	0	9
	3.90	10	5	91	17	9
	3.90	16	−4	91	79	9
	7.90	4	Ambient	91	12	9
	7.90	14	5	91	84	9
	7.90	12	−4	91	81	9
	3.90	8	Ambient	91	79	9
	3.90	14	5	91	88	9
	3.90	14	−4	91	82	9
Delphinium, annual *Delphinium* spp.	Not given	19	−4	43	41	9
Delphinium, perennial *D.* spp.	Not given	10	5	43	5	9
Eggplant *Solanum melongena* L.	10.40	10	Ambient	86	4	9
	10.40	14	−4	86	81	9
	7	14	Ambient	86	50	9
	7	12	−4	86	76	9
Elm, American *Ulmus americana* L.	7	15	−4	96	45	4, 9
	3	15	−4	96	75	4, 9
	7	10	5	96	2	4, 9
	3	10	5	96	25	4, 9
	3	4	Ambient	96	33	4, 9
Fescue, Chewings *Festuca rubra* subsp. *commutata* Gaud .	Not given	3	Ambient	90	50−	72
	15.20	2.75	−10	81	57	64
	15.20	2.75	2	81	28	64
	15.20	2.75	10	81	2	64
	9.80	2.75	2	81	47	64
	9.80	2.75	10	81	30	64
	9.80	2.75	20	81	14	64
Fescue, creeping red *Festuca rubra* L.	3.20	2	1	81	74	15
	3.20	2	10	81	67	15
	3.20	2	21	81	52	15
	3.20	2	32	81	40	15
	3.20	2	43	81	35	15
	3.20	2	Ambient	81	47	15
	11.40	2.50	1	95	60	15
	11.40	2.50	10	95	49	15
	11.40	2.50	21	95	1	15
	11.40	0.50	32	95	0	15
	11.40	0.25	43	95	0	15
	11.40	1	Ambient	95	0	15
	6.40	2.50	1	72	47	15
	6.40	2.50	10	72	30	15
	6.40	2.50	21	72	20	15

Table 2 (continued)
EFFECT OF VARIOUS STORAGE CONDITIONS ON VIABILITY OF SEED IN AIR IN SEALED GLASS OR METAL CONTAINERS

Species	Moisture (% when stored)	Storage: Time (years)	Storage: Temperature (°C)	Germination (%): When stored	Germination (%): After storage	Ref.
	6.40	2.50	32	72	7	15
	6.40	2.50	43	72	1	15
	6.40	2.50	Ambient	72	13	15
	3.60	2	1	95	84	54
	3.60	2	10	95	85	54
	3.60	2	27	95	84	54
	3.60	2	16—35	95	82	54
	3.60	2	43	95	79	54
	5.40	2	1	94	88	54
	5.40	2	10	94	86	54
	5.40	2	27	94	83	54
	5.40	2	16—35	94	85	54
	5.40	2	43	94	77	54
	5.60	2	1	93	93	54
	5.60	2	10	93	91	54
	5.60	2	27	93	87	54
	5.60	2	16—35	93	87	54
	5.60	2	43	93	5	54
Fescue, hard *Festuca longifolia* Thuill.	Not given	5	Ambient	90	50	72
Fescue, meadow *Festuca pratensis* Huds.	Not given	9	Ambient	90	50	72
Fescue, tall *Festuca arundinacea* Schreb.	Not given	6	Ambient	90 +	50−	72
Fescue, fine-leaved sheep (hair) *Festuca tenuifolia* Sibth.	Not given	4	Ambient	90	50	72
Fir, California red, shasta red *Abies magnifica* A. Murr. vars.	11	16	8 (9 years) then −4	38	12	9
	13	16	8 (9 years) then −4	32	19	9
	9	16	8 (9 years) then −4	58	17	9
Fir, grand *Abies grandis* (Dougl. ex D. Don) Lindl.	11	16	8 (9 years) then −4	98	26	9
Fir, noble *Abies nobilis* (Dough. ex D. Don) Lindl. (Now *A. procera* Rehd.)	8	16	8 (9 years) then −4	62	30	9
	11	16	8 (9 years) then −4	66	33	9
Flax *Linum usitatissimum* L.	7—8	18	−20 to 32	Not given	58	43
Foxtail, meadow *Alopercurus pratensis* L.	Not given	6	Ambient	90	50−	72
Gladiolus *Gladiolus* spp.	7.60	5	5	82	0	9
	7.60	12	−4	82	73	9

Table 2 (continued)
EFFECT OF VARIOUS STORAGE CONDITIONS ON VIABILITY OF SEED IN AIR IN SEALED GLASS OR METAL CONTAINERS

Species	Moisture (% when stored)	Storage: Time (years)	Storage: Temperature (°C)	Germination (%): When stored	Germination (%): After storage	Ref.
	5.80	10	5	82	0	9
	5.80	12	−4	82	70	9
Grapefruit						
Citrus paradisi Macfad.	78	1	5	99	66	5
	123	0.50	5	32	14	5
Hemp						
Cannabis sativa L.	5.70	8.25	10	98	98	94
	5.70	8.25	−10	98	99	94
	5.70	15	10	98	94	40
	5.70	15	−10 (10 years) −12 (5 years)	98	97	40
	8.30	15	10	100	93	40
	8.30	15	−10 (10 years) −12 (5 years)	100	98	40
	6.20	12	10	98	96	40
	6.20	12	0	98	98	40
	6.20	12	−10 (10 years) −12 (2 years)	98	98	40
	6.20	27	−10 (10 years) −12 (17 years)	98	95	21
	9.50	10	10	98	6	40
	9.50	12	0	98	95	40
	9.50	12	−10	98	97	40
Kale						
Brassica oleracea var. *acephala* DC	Not given	9	Ambient	90	50−	72
Kenaf						
Hibiscus cannabinus L.	8	5.50	−10	66	77	94
	8	5.50	0	66	71	94
	8	5.50	10	66	68	94
	8	5.50	21	66	26	94
	12	5.50	−10	73	70	94
	12	5.50	0	73	70	94
	12	5.50	10	73	33	94
	12	4	21	73	0	94
	8	12	10	66	66	40
	8	12	0	66	73	40
	8	12	−10 (10 years) −12 (2 years)	66	68	40
	8	27	−10 (10 years) −12 (14 years)	66	79	21
	9.20	10	10	91	88	40
	9.20	10	0	91	90	40
	9.20	10	−18 (8 years) −12 (2 years)	91	85	40
	10.40	10	10	88	73	40
	10.40	10	0	88	79	40
	10.40	10	−18 (8 years) −12 (2 years)	88	88	40
	10.40	25	−18 (8 years) −12 (17 years)	88	88	21

Table 2 (continued)
EFFECT OF VARIOUS STORAGE CONDITIONS ON VIABILITY OF SEED IN AIR IN SEALED GLASS OR METAL CONTAINERS

Species	Moisture (% when stored)	Storage: Time (years)	Storage: Temperature (°C)	Germination (%): When stored	Germination (%): After storage	Ref.
	12	12	10	91	10	40
	12	12	0	91	66	40
Lemon, rough						
Citrus limon (L.) Burm. f.	56	1.50	5	84	53	5
Lettuce						
Lactuca sativa L.	56	19	−12	97	93	21
	56	19	−1	97	93	21
	56	19	10	97	90	21
	4	19	21	97	85	21
	4	8	32	97	36	21
	7	19	−12	97	94	21
	7	11	−1	97	14	21
	7	8	10	97	1	21
	7	4	21	97	4	21
	7	1	32	97	0	21
	10	19	−12	95	96	21
	10	19	−1	95	18	21
	10	5	10	95	4	21
	10	1	21	95	11	21
	10	1	32	95	0	21
	8.20	20	−4	98	86	9
	5.70	18	Ambient	100	15	14
	9.70	14	Ambient	100	33	14
	5.70	18	5	100	88	14
	9.70	18	5	100	54	14
	5.70	18	−2	100	85	14
	9.70	18	−2	100	86	14
	5.70	18	−18	100	71	14
	9.70	18	−18	100	79	14
	15.30	18	−2	99	98	14
	19	18	−2	100	81	14
	15.30	18	5	99	72	14
	15.30	18	−18	99	64	14
	19	18	5	100	0	14
	19	18	−18	100	37	14
	Not given	2.25	22—27	96	3	1
	Not given	6	7	92	81	1
	Not given	6	10	92	84	1
	Not given	5.25	15	92	5	1
	Not given	1.50	20	92	14	1
	5.40	12	Ambient	98	67	9
	5.40	20	−4	98	80	9
Lily, Regal						
Lilium regale Wilson	9.90	15	5	90	20	8, 9
	9.90	15	−4	90	83	8, 9
	4.50	10	5	90	91	8, 9
	4.50	15	−4	90	85	8, 9
	4.50	7	Ambient	86	31	8, 9
	4.50	15	−5	86	93	8, 9
	9.90	15	−5	86	81	8, 9

Table 2 (continued)
EFFECT OF VARIOUS STORAGE CONDITIONS ON VIABILITY OF SEED IN AIR IN SEALED GLASS OR METAL CONTAINERS

Species	Moisture (% when stored)	Storage: Time (years)	Storage: Temperature (°C)	Germination (%): When stored	Germination (%): After storage	Ref.
Lobelia						
Lobelia cardinalis L.	6.90	3	Ambient	63	22	11
	6.90	25	5	63	58	11
	6.90	25	−5	63	45	11
	4.70	12.50	Ambient	63	24	11
	4.70	25	5	63	58	11
	4.70	25	−5	63	64	11
Oatgrass, tall						
Arrhenatherum elatius (L.) Beauv. ex J. & C. Presl	Not given	8	Ambient	90+	50−	72
Oats						
Avena sativa L.	Not given	11	Ambient	95+	80−	72
	Not given	15	Ambient	95+	50−	72
Okra						
Hibiscus esculentus L.	8.0	11	2—4	90	93	73
now *Abelmoschus esculentus* (L.) Moench	8.0	11	Ambient	90	55	73
Onion						
Allium cepa L.	8.3	10	Ambient	98	19	9
	8.3	10	−4	98	90	9
	12.5	10	−4	98	71	9
	5.6	10	Ambient	94	61	14
	5.6	18	5	94	71	14
	5.6	18	−18	94	93	14
	10.8	5	Ambient	95	55	14
	10.8	18	5	95	4	14
	10.8	18	−18	95	96	14
	8	7	Ambient	86	20	27
	8	7	4	86	81	27
	8	7	−7	86	84	27
	10	7	21—32	86	0	27
	10	7	4	86	74	27
	10	7	−7	86	82	27
	7.30	3	16	88	86	3
	7.30	3	32	88	70	3
	Not given	11	4—35	94	25	31
	Not given	13	1—4	94	6	31
	Not given	4	Ambient	90	50	72
	6.40	13	Ambient	94	90	32
	6.40	13	5—10	94	89	32
	4.10	2	1	95	92	54
	4.10	2	10	95	93	54
	4.10	2	27	95	93	54
	4.10	2	16—35	95	91	54
	4.10	2	43	95	73	54
	5.20	2	1	96	92	54
	5.20	2	10	96	93	54
	5.20	2	27	96	92	54
	5.20	2	16—35	96	93	54
	5.20	2	43	96	9	54
	6.70	2	1	95	93	54

Table 2 (continued)
EFFECT OF VARIOUS STORAGE CONDITIONS ON VIABILITY OF SEED IN AIR IN SEALED GLASS OR METAL CONTAINERS

Species	Moisture (% when stored)	Storage: Time (years)	Storage: Temperature (°C)	Germination (%): When stored	Germination (%): After storage	Ref.
	6.70	2	10	95	93	54
	6.70	2	27	95	94	54
	6.70	2	16—35	95	93	54
	6.70	1	43	95	12	54
Orange, sour						
Citrus aurantium L.	80	0.50	5	90	63	5
Orange, sweet						
C. sinensis (L.) Osb.	82	0.50	5	71	1	5
Orchardgrass (cocksfoot)						
Dactylis glomerata L.	Not given	9	Ambient	90	50−	72
Parsnip						
Pastinaca sativa L.	Not given	7	Ambient	Not given	50	29
	4	7	Ambient	Not given	47	29
	4	7	20	Not given	72	29
	8	7	20	Not given	52	29
	6.30	2	25	88	22	63
	6.10	3	25	95	48	63
	6.10	3	5	95	54	63
	6.80	3	25	95	10	63
	6.80	3	5	95	12	63
	Not given	4	Ambient	90	50−	72
Pea, garden						
Pisum sativum L.	Not given	12	Ambient	90	50−	72
Peanut (groundnut)						
Arachis hypogaea L.	Not given	6	21—25	69	52	28
Peony						
Paeonia suffruticosa Andr.	Not given	5	5	50	6	9
	Not given	5	−4	50	22	9
Pepper						
Capsicum annuum L. var. *annuum*	7	8	Ambient	73	1	9
	7	20	−4	73	54	9
	10.50	16	−4	73	64	9
	8.20	3	16	93	80	3
	6.70	3	16	93	89	3
	8.30	1.50	21	93	37	3
	6.70	1.50	21	93	81	3
	5.20	3	21	93	88	3
	8.10	3	32	93	1	3
	5.30	3	32	93	50	3
	4	3	32	93	84	3
Pine, loblolly						
Pinus taeda L.	Not given	12	5	82	13	9
	Not given	12	−4	82	73	9
Pine, shortleaf						
Pinus echinata Mill.	Not given	16	5	59	16	9
	Not given	16	−4	59	42	9
Pyrethrum						
Chrysanthemum spp.	6.50	15	−18.50	57	48	12
	6.50	15	−4	57	58	12
	6.50	15	5	57	47	12

Table 2 (continued)
EFFECT OF VARIOUS STORAGE CONDITIONS ON VIABILITY OF SEED IN AIR IN SEALED GLASS OR METAL CONTAINERS

Species	Moisture (% when stored)	Storage: Time (years)	Storage: Temperature (°C)	Germination (%): When stored	Germination (%): After storage	Ref.
	6.50	15	10	57	28	12
	6.50	15	10	57	6	12
	6.50	2	30	57	9	12
Quinine						
Cinchona ledgerana Moens ex Trimen. (now *C. calisaya* Wedd.)	4.80	4	5	75	62	7
	4.80	4	−4	75	68	7
	6	4	Ambient	74	1	7
	6	4	5	74	70	7
	6	4	−4	74	55	7
	9.40	3	Ambient	66	1	7
	9.40	4	5	66	60	7
	9.40	4	−4	66	67	7
Rape						
Brassica napus L.	Not given	9	Ambient	90	50−	72
Rice						
Oryza sativa L.	Not given	0.83	22—27	91	2	1
	Not given	4.25	7	92	15	1
	Not given	4.25	10	92	5	1
	Not given	4.25	15	92	1	1
	Not given	1	20	92	11	1
Rye						
Secale cereale L.	Not given	4	Ambient	95 +	80−	72
	Not given	6	Ambient	95 +	50−	72
Ryegrass, perennial						
Lolium perenne L.	Not given	7	Ambient	90	50−	72
Ryegrass, Italian						
L. multiflorum Lam.	Not given	11	Ambient	90 +	50−	72
Safflower						
Carthamus tinctorius L	4	17	−12	95	83	21
	4	17	−1	95	95	21
	4	17	10	95	97	21
	4	17	21	95	93	21
	4	11	32	95	89	21
	4	17	32	95	0	21
	7	17	−12	94	87	21
	7	17	−1	94	94	21
	7	11	10	94	81	21
	7	3	21	94	3	21
	7	1	32	94	0	21
	10	17	−12	95	78	21
	10	11	−1	95	47	21
	10	2	10	95	17	21
	10	1	21	95	0	21
	10	1	32	95	0	21
Sainfoin						
Onobrychis viciifolia Scop.	Not given	7	Ambient	90	50−	72
Sesame						
Sesamum indicum L.	4	17	−12	94	93	21

Table 2 (continued)
EFFECT OF VARIOUS STORAGE CONDITIONS ON VIABILITY OF SEED IN AIR IN SEALED GLASS OR METAL CONTAINERS

Species	Moisture (% when stored)	Storage Time (years)	Storage Temperature (°C)	Germination (%) When stored	Germination (%) After storage	Ref.
	4	17	−1	94	93	21
	4	17	10	94	93	21
	4	17	21	94	92	21
	4	17	32	94	2	21
	7	17	−12	92	91	21
	7	17	−1	92	23	21
	7	5	10	92	39	21
	7	2	21	92	0	21
	7	1	32	92	0	21
	10	1	−12	88	28	21
	10	1	−1	88	30	21
	10	2	10	88	6	21
	10	1	21	88	0	21
	10	1	32	88	0	21
Sorghum						
Sorghum bicolor (L.) Moench	4	16	−12	92	93	24
	4	16	−1	92	95	24
	4	16	10	92	94	24
	4	16	21	92	85	24
	4	16	32	92	69	24
	7	16	−12	95	95	24
	7	16	−1	95	90	24
	7	16	10	95	85	24
	7	16	21	95	83	24
	7	16	32	95	52	24
	10	16	−12	91	91	24
	10	16	−1	91	90	24
	10	16	10	91	77	24
	10	16	21	91	58	24
	10	5	32	91	43	21
	Not given	3.50	4—7	95	90	81
	Not given	3.50	19—31	95	35	81
Soybean						
Glycine max (L.) Merr.	18.10	10	−10	97	17	93
	13.90	10	−10	97	97	93
	13.90	10	2	97	97	93
	9.40	10	10	97	97	93
	11.40	1.75	22—27	100	12	1
	Not given	4.25	7	97	8	1
	Not given	5.25	10	97	2	1
	Not given	3.25	15	97	10	1
	Not given	3.25	20	97	1	1
Swede						
Brassica napus var. *napobrassica* (L.) Reichemb.	Not given	9	Ambient	90	50−	72
Timothy						
Phleum pratense L.	Not given	9	Ambient	90	50−	72
Tobacco						
Nicotiana tabacum L.	9	25	5	Not given	84	65, 66

Table 2 (continued)
EFFECT OF VARIOUS STORAGE CONDITIONS ON VIABILITY OF SEED IN AIR IN SEALED GLASS OR METAL CONTAINERS

Species	Moisture (% when stored)	Storage: Time (years)	Storage: Temperature (°C)	Germination (%): When stored	Germination (%): After storage	
Tomato *Lycopersicon esculentum* Mill.	6.60	10	Ambient	93	75	9
	6.60	10	−4	93	90	9
	4	5	Ambient	91	89	78
	2.30	5	Ambient	88	90	78
	1.20	5	Ambient	89	87	78
	0.40	5	Ambient	87	65	78
	0.40	5	7	87	81	78
	9.80	3	16	93	90	3
	9.80	1	32	93	18	3
	5.90	3	32	93	86	3
	9.70	3	21	93	82	3
	4.50	3	32	93	91	3
	10	8	Ambient	93	0	9
	10	12	−4	93	88	9
	10	18	Ambient	93	34	14
	10	18	5	93	89	14
	10	18	−18	93	93	14
Trefoil *Lotus* spp.	Not given	8	Ambient	90	50−	72
Turnip *Brassica rapa* L.	Not given	1.50	Ambient	90	50−	72
Verbena *Verbena* spp.	6.20	12.33	5	45	9	9
	6.20	16.33	−4	45	32	9
	4.20	12.33	5	45	4	9
	4.20	16.33	−4	45	20	9
Vetch, Common *Vicia sativa* L. subsp. *sativa*	Not given	9	Ambient	90	50−	72

Table 3
LONGEVITY OF SEEDS STORED AT VARIOUS TEMPERATURES IN SEALED CONTAINERS IN AN ATMOSPHERE OTHER THAN AIR

Species	Moisture (% when stored)	Storage: Time (years)	Storage: Temperature (°C)	Germination (%): When stored	Germination (%): After storage	Ref.
		Atmosphere: Argon				
Clover, crimson *Trifolium incarnatum* L.	4	16	−12	81	80	20
	4	16	−1	81	83	20
	4	16	10	81	81	20
	4	16	21	81	83	20
	4	16	32	81	83	20
	7	16	−12	91	78	20
	7	16	−1	91	86	20
	7	16	10	91	79	20

Table 3 (continued)
LONGEVITY OF SEEDS STORED AT VARIOUS TEMPERATURES IN SEALED CONTAINERS IN AN ATMOSPHERE OTHER THAN AIR

Species	Moisture (% when stored)	Storage		Germination (%)		Ref.
		Time (years)	Temperature (°C)	When stored	After storage	
	7	16	21	91	78	20
	7	16	32	91	27	20
	10	16	−12	91	83	20
	10	16	−1	91	78	20
	10	16	10	91	73	20
	10	16	21	91	4	20
	10	2	32	91	12	20
Lettuce	4	19	−12	97	94	21
Lactuca sativa L.	4	19	−1	97	93	21
	4	19	10	97	92	21
	4	19	21	97	90	21
	4	19	32	97	79	21
	7	19	−12	97	94	21
	7	19	−1	97	94	21
	7	19	10	97	86	21
	7	5	21	97	71	21
	7	1	32	97	8	21
	10	19	−12	95	94	21
	10	19	−1	95	15	21
	10	5	10	95	26	21
	10	1	21	95	32	21
	10	1	32	95	0	21
Safflower	4	17	−12	95	80	21
Carthamus tinctorius L.	4	17	−1	95	93	21
	4	17	10	95	98	21
	4	17	21	95	96	21
	4	17	32	95	5	21
	7	17	−12	94	67	21
	7	17	−1	94	94	21
	7	11	10	94	58	21
	7	3	21	94	15	21
	7	1	32	94	0	21
	10	17	−12	95	70	21
	10	11	−1	95	47	21
	10	3	10	95	5	21
	10	0	21	95	0	21
	10	0	32	95	0	21
Sesame	4	17	−12	94	94	21
Sesamum indicum L.	4	17	−1	94	93	21
	4	17	10	94	94	21
	4	17	21	94	96	21
	4	17	32	94	14	21
	7	17	−12	92	91	21
	7	17	−1	92	41	21
	7	5	10	92	62	21
	7	2	21	92	1	21
	7	1	32	92	0	21
	10	1	−12	88	23	21
	10	1	−1	88	24	21
	10	2	10	88	29	21
	10	1	21	88	0	21
	10	1	32	88	0	21

Table 3 (continued)
LONGEVITY OF SEEDS STORED AT VARIOUS TEMPERATURES IN SEALED CONTAINERS IN AN ATMOSPHERE OTHER THAN AIR

Species	Moisture (% when stored)	Storage: Time (years)	Storage: Temperature (°C)	Germination (%): When stored	Germination (%): After storage	Ref.
Sorghum	4	16	−12	92	92	24
Sorghum bicolor	4	16	−1	92	93	24
(L.) Moench	4	16	10	92	92	24
	4	16	21	92	70	24
	4	16	32	92	52	24
	7	16	−12	95	93	24
	7	16	−1	95	94	24
	7	16	10	95	84	24
	7	16	21	95	77	24
	7	16	32	95	50	24
	10	16	−12	91	94	24
	10	16	−1	91	91	24
	10	16	10	91	82	24
	10	16	21	91	59	24
	10	5	32	91	67	21
Atmosphere: Carbon Dioxide						
Clover, crimson	4	16	−12	81	86	20
Trifolium incarnatum L.	4	16	−1	81	78	20
	4	16	10	81	82	20
	4	16	21	81	78	20
	4	16	32	81	78	20
	7	16	−12	91	84	20
	7	16	−1	91	81	20
	7	16	10	91	81	20
	7	16	21	91	80	20
	7	16	32	91	48	20
	10	16	−12	91	84	20
	10	16	−1	91	78	20
	10	16	10	91	58	20
	10	16	21	91	3	20
	10	2	32	91	5	21
Clover, red	Not given	14	Ambient	97	2	48
T. pratense L.	Not given	23	0—5	97	55	48
Lettuce	4	19	−12	97	94	21
Lactuca sativa	4	19	−1	97	95	21
	4	19	10	97	93	21
	4	19	21	97	92	21
	4	19	32	97	70	21
	7	19	−12	97	97	21
	7	19	−1	97	94	21
	7	19	10	97	75	21
	7	8	21	97	0	21
	7	1	32	97	0	21
	10	19	−12	95	96	21
	10	19	−1	95	4	21
	10	5	10	95	18	21
	10	1	21	95	27	21
	10	1	32	95	0	21
Safflower	4	17	−12	95	89	21
Carthamus tinctorius L.	4	17	−1	95	92	21

Table 3 (continued)
LONGEVITY OF SEEDS STORED AT VARIOUS TEMPERATURES IN SEALED CONTAINERS IN AN ATMOSPHERE OTHER THAN AIR

Species	Moisture (% when stored)	Storage: Time (years)	Storage: Temperature (°C)	Germination (%): When stored	Germination (%): After storage	Ref.
	4	17	10	95	94	21
	4	17	21	95	97	21
	4	17	32	95	2	21
	7	17	−12	94	80	21
	7	17	−1	94	94	21
	7	11	10	94	52	21
	7	2	21	94	47	21
	7	1	32	94	0	21
	10	17	−12	95	76	21
	10	11	−1	95	65	21
	10	2	10	95	19	21
	10	1	21	95	0	21
	10	1	32	95	0	21
Sesame	4	17	−12	94	94	21
Sesamum indicum L.	4	17	−1	94	94	21
	4	17	10	94	90	21
	4	17	21	94	91	21
	4	17	32	94	6	21
	7	17	−12	92	93	21
	7	17	−1	92	65	21
	7	5	10	92	67	21
	7	2	21	92	3	21
	7	1	32	92	0	21
	10	2	−12	88	8	21
	10	2	−1	88	12	21
	10	2	10	88	11	21
	10	1	21	88	0	21
	10	1	32	88	0	21
Sorghum	4	16	−12	92	92	24
Sorghum bicolor	4	16	−1	92	91	24
(L.) Moench	4	16	10	92	92	24
	4	16	21	92	87	24
	4	16	32	92	69	24
	7	16	−12	95	95	24
	7	16	−1	95	92	24
	7	16	10	95	86	24
	7	16	21	95	76	24
	7	16	32	95	49	24
	10	16	−12	91	95	24
	10	16	−1	91	90	24
	10	16	10	91	76	24
	10	16	21	91	61	24
	10	5	32	91	58	21
Atmosphere: Helium						
Clover, crimson	4	16	−12	81	84	20
Trifolium incarnatum L.	4	16	−1	81	81	20
	4	16	10	81	77	20
	4	16	21	81	77	20
	4	16	32	81	71	20
	7	16	−12	91	82	20

Table 3 (continued)
LONGEVITY OF SEEDS STORED AT VARIOUS TEMPERATURES IN SEALED CONTAINERS IN AN ATMOSPHERE OTHER THAN AIR

Species	Moisture (% when stored)	Storage: Time (years)	Storage: Temperature (°C)	Germination (%): When stored	Germination (%): After storage	Ref.
	7	16	−1	91	78	20
	7	16	10	91	75	20
	7	16	21	91	78	20
	7	16	32	91	54	20
	10	16	−12	91	81	20
	10	16	−1	91	81	20
	10	16	10	91	63	20
	10	16	21	91	4	20
	10	2	32	91	7	21
Lettuce	4	19	−12	97	92	21
Lactuca sativa L.	4	19	−1	97	95	21
	4	19	10	97	96	21
	4	19	21	97	90	21
	4	19	32	97	70	21
	7	19	−12	97	96	21
	7	19	−1	97	92	21
	7	19	10	97	86	21
	7	8	21	97	0	21
	7	1	32	97	0	21
	10	19	−12	95	90	21
	10	19	−1	95	35	21
	10	5	10	95	22	21
	10	1	21	95	49	21
	10	1	32	95	0	21
Safflower	4	17	−12	95	93	21
Carthamus tinctorius L.	4	17	−1	95	94	21
	4	17	10	95	96	21
	4	11	21	95	94	21
	4	17	21	95	0	21
	4	17	32	95	4	21
	7	17	−12	94	83	21
	7	17	−1	94	96	21
	7	17	10	94	3	21
	7	3	21	94	24	21
	7	1	32	94	0	21
	10	17	−12	95	63	21
	10	11	−1	95	69	21
	10	2	10	95	47	21
	10	1	21	95	0	21
	10	1	32	95	0	21
Sesame	4	17	−12	96	95	21
Sesamum indicum L.	4	17	−1	94	94	21
	4	17	10	94	92	21
	4	17	21	94	93	21
	4	17	32	94	19	21
	7	17	−12	92	93	21
	7	17	−1	92	52	21
	7	5	10	92	57	21
	7	2	21	92	0	21
	7	1	32	92	0	21
	10	1	−12	88	26	21
	10	1	−1	88	28	21
	10	2	10	88	32	21

Table 3 (continued)
LONGEVITY OF SEEDS STORED AT VARIOUS TEMPERATURES IN SEALED CONTAINERS IN AN ATMOSPHERE OTHER THAN AIR

Species	Moisture (% when stored)	Storage: Time (years)	Storage: Temperature (°C)	Germination (%): When stored	Germination (%): After storage	Ref.
	10	1	21	88	0	21
	10	1	32	88	0	21
Sorghum	4	16	−12	92	91	24
Sorghum bicolor	4	16	−1	92	95	24
(L.) Moench	4	16	10	92	90	24
	4	16	21	92	85	24
	4	16	32	92	72	24
	7	16	−12	95	89	24
	7	16	−1	95	94	24
	7	16	10	95	90	24
	7	16	21	95	81	24
	7	16	32	95	45	24
	10	16	−12	91	94	24
	10	16	−1	91	86	24
	10	16	10	91	81	24
	10	16	21	91	65	24
	10	5	32	91	55	21
			Atmosphere: Nitrogen			
Bluegrass, Kentucky	3.50	2	1	75	47	54
Poa pratensis L.	3.50	2	10	75	47	54
	3.50	2	27	75	49	54
	3.50	2	16—35	75	46	54
	3.50	2	43	75	44	54
	4.60	2	1	73	49	54
	4.60	2	10	73	50	54
	4.60	2	27	73	47	54
	4.60	2	16—35	73	43	54
	4.60	1.50	43	73	2	54
	5.70	2	1	79	49	54
	5.70	2	10	79	50	54
	5.70	2	27	79	51	54
	5.70	2	16—35	79	49	54
	5.70	1	43	79	1	54
Cabbage	2.50	2	1	94	94	54
Brassica oleracea	2.50	2	10	94	91	54
var. *capitata* L.	2.50	2	27	94	93	54
	2.50	2	16—35	94	94	54
	2.50	2	43	94	82	54
	5.20	2	1	94	95	54
	5.20	2	10	94	93	54
	5.20	2	27	94	93	54
	5.20	2	16—35	94	94	54
	5.20	1	43	94	23	54
	7	2	1	93	93	54
	7	2	10	93	94	54
	7	2	27	93	65	54
	7	2	16—35	93	63	54
	7	0.25	43	93	0	54
Clover, crimson	4	16	−12	81	81	20
Trifoliom incarnatum L.	4	16	−1	81	82	20

Table 3 (continued)
LONGEVITY OF SEEDS STORED AT VARIOUS TEMPERATURES IN SEALED CONTAINERS IN AN ATMOSPHERE OTHER THAN AIR

Species	Moisture (% when stored)	Storage: Time (years)	Storage: Temperature (°C)	Germination (%): When stored	Germination (%): After storage	Ref.
	4	16	10	81	56	20
	4	16	21	81	81	20
	4	16	32	81	78	20
	7	16	−12	91	85	20
	7	16	−1	91	79	20
	7	16	10	91	74	20
	7	16	21	91	37	20
	7	16	32	91	31	20
	10	16	−12	91	87	20
	10	16	−1	91	75	20
	10	16	10	91	43	20
	10	16	21	91	3	20
	10	2	32	91	6	20
Clover, red	12.70	16	0	93	93	41, 42
Trifolium pratense L.	12.70	5	Ambient	93	0	41, 42
Clover, white	9.80	16	0	92	88	41, 42
Trifolium repens L.	9.80	16	Ambient	92	89	41, 42
Fescue, creeping red	3.10	2	1	88	84	54
Festuca rubra L.	3.10	2	10	88	83	54
	3.10	2	27	88	76	54
	3.10	2	16—35	88	79	54
	3.10	2	43	88	71	54
	4	2	1	88	89	54
	4	2	10	88	86	54
	4	2	27	88	82	54
	4	2	16—35	88	80	54
	4	2	43	88	62	54
	4.90	2	1	86	88	54
	4.90	2	10	86	85	54
	4.90	2	27	86	72	54
	4.90	2	16—35	86	75	54
	4.90	2	43	86	22	54
Lettuce	4	19	−12	97	95	21
Lactuca sativa L.	4	19	−1	97	91	21
	4	19	10	97	91	21
	4	19	21	97	91	21
	4	19	32	97	69	21
	7	19	−12	97	97	21
	7	19	−1	97	89	21
	7	19	10	97	80	21
	7	8	21	97	0	21
	7	1	32	97	0	21
	10	19	−12	95	93	21
	10	19	−1	95	1	21
	10	5	10	95	22	21
	10	1	21	95	11	21
	10	1	32	95	0	21
Onion	3.80	2	1	93	94	54
Allium cepa L.	3.80	2	10	93	93	54
	3.80	2	27	93	93	54
	3.80	2	16—35	93	92	54
	3.80	2	43	93	78	54

Table 3 (continued)
LONGEVITY OF SEEDS STORED AT VARIOUS TEMPERATURES IN SEALED CONTAINERS IN AN ATMOSPHERE OTHER THAN AIR

Species	Moisture (% when stored)	Storage: Time (years)	Storage: Temperature (°C)	Germination (%): When stored	Germination (%): After storage	Ref.
	4.40	2	1	94	92	54
	4.40	2	10	94	94	54
	4.40	2	27	94	92	54
	4.40	2	16—35	94	92	54
	4.40	2	43	94	26	54
	6.50	2	1	96	94	54
	6.50	2	10	96	94	54
	6.50	2	27	96	94	54
	6.50	2	16—35	96	92	54
	6.50	0.75	43	96	0	54
Ryegrass, perennial *Lolium perenne* L.	5.60	15	15 (8 years) 0 (7 years)	96	71	47
	3.80	15	15 (8 years) 0 (7 years)	96	41	47
Sealed glass bottles, air replaced with nitrogen after 7 years	2.60	15	15 (8 years) 0 (7 years)	90	45	47
	1.80	15	15 (8 years) 0 (7 years)	92	71	47
	0.70	15	15 (8 years) 0 (7 years)	93	57	47
Safflower *Carthamus tinctorius* L.	4	17	−12	95	84	21
	4	17	−1	95	97	21
	4	17	10	95	96	21
	4	17	21	95	95	21
	4	17	32	95	4	21
	7	17	−12	94	87	21
	7	17	−1	94	95	21
	7	17	10	94	1	21
	7	3	21	94	2	21
	7	1	32	94	0	21
	10	17	−12	95	59	21
	10	11	−1	95	67	21
	10	2	10	95	19	21
	10	1	21	95	0	21
	10	1	32	95	0	21
Sesame	4	17	−12	94	93	21
Sesamum indicum L.	4	17	−1	94	90	21
	4	17	10	94	95	21
	4	17	21	94	88	21
	4	17	32	94	14	21
	7	17	−12	92	88	21
	7	17	−1	92	55	21
	7	5	10	92	65	21
	7	2	21	92	1	21
	7	1	32	92	0	21
	10	1	−12	88	23	21
	10	1	−1	88	25	21
	10	2	10	88	20	21
	10	1	21	88	0	21
	10	1	32	88	0	21

Table 3 (continued)
LONGEVITY OF SEEDS STORED AT VARIOUS TEMPERATURES IN SEALED CONTAINERS IN AN ATMOSPHERE OTHER THAN AIR

Species	Moisture (% when stored)	Storage		Germination (%)		Ref.
		Time (years)	Temperature (°C)	When stored	After storage	
Sorghum	4	16	−12	92	93	24
Sorghum bicolor	4	16	−1	92	90	24
(L.) Moench	4	16	10	92	93	24
	4	16	21	92	82	24
	4	16	32	92	75	24
	7	16	−12	95	93	24
	7	16	−1	95	88	24
	7	16	10	95	86	24
	7	16	21	95	77	24
	7	16	32	95	46	24
	10	16	−12	91	94	24
	10	16	−1	91	90	24
	10	16	10	91	82	24
	10	16	21	91	57	24
	10	5	32	91	48	21
Atmosphere: Partial Vacuum						
Aster, China	Not given	2	10—15	68	70	36
Callistephus chinensis (L.)						
Nees						
Bluegrass, Kentucky	3.5	2	1	75	50	54
Poa pratensis L.						
	3.5	2	10	75	47	54
	3.5	2	27	75	46	54
	3.5	2	16—35	75	44	54
	3.5	2	43	75	42	54
	4.6	2	1	73	47	54
	4.6	2	10	73	44	54
	4.6	2	27	73	45	54
	4.6	2	16—35	73	42	54
	4.6	2	43	73	0	54
	5.7	2	1	79	54	54
	5.7	2	10	79	47	54
	5.7	2	27	79	50	54
	5.7	2	16—35	79	52	54
	5.7	1	43	79	3	54
Cabbage	2.5	2	1	94	94	54
Brassica oleracea	2.5	2	10	94	91	54
var. *capitata* L.	2.5	2	27	94	93	54
	2.5	2	16—35	94	94	54
	2.5	2	43	94	82	54
	5.2	2	1	94	95	54
	5.2	2	10	94	93	54
	5.2	2	27	94	93	54
	5.2	2	16	94	94	54
	5.2	1	43	94	23	54
	7.0	2	1	93	93	54
	7.0	2	10	93	94	54
	7.0	2	27	93	65	54
	7.0	2	16—35	93	63	54
	7.0	0.25	43	93	0	54

Table 3 (continued)
LONGEVITY OF SEEDS STORED AT VARIOUS TEMPERATURES IN SEALED CONTAINERS IN AN ATMOSPHERE OTHER THAN AIR

Species	Moisture (% when stored)	Storage: Time (years)	Storage: Temperature (°C)	Germination (%): When stored	Germination (%): After storage	Ref.
Clover, crimson	4	16	−12	81	81	20
Trifolium incarnatum L.	4	16	−1	81	78	20
	4	16	10	81	77	20
	4	16	21	81	73	20
	4	16	32	81	29	20
	6	16	−12	91	87	20
	6	16	−1	91	84	20
	6	16	10	91	77	20
	6	16	21	91	86	20
	6	16	32	91	52	20
	11	16	−12	91	83	20
	11	16	−1	91	80	20
	11	16	10	91	66	20
	11	16	21	91	5	20
	11	2	32	91	12	20
Clover, red *T. pratense* L.	12.7	16	0	93	95	41, 42
	12.7	15	Ambient	93	0	41, 42
Clover, white *T. repens* L.	9.8	16	0	92	94	41, 42
	9.8	16	Ambient	92	60	41, 42
Dandelion, common	7.9	3	Ambient	91	3	9
Taraxacum officinale Wiggers	7.9	3	5	91	30	9
	7.9	3	−5	91	93	9
	6.2	3	Ambient	91	60	9
	6.2	3	5	91	80	9
	6.2	3	−5	91	91	9
	3.9	3	Ambient	91	88	9
	3.9	3	5	91	92	9
	3.9	3	−5	91	86	9
Fescue, creeping red	3.1	2	1	88	89	54
Festuca rubra L.	3.1	2	10	88	86	54
	3.1	2	27	88	87	54
	3.1	2	16	88	81	54
	3.1	2	43	88	79	54
	4.0	2	1	88	86	54
	4.0	2	10	88	87	54
	4.0	2	27	88	88	54
	4.0	2	16	88	86	54
	4.0	2	43	88	72	54
	4.9	2	1	87	88	54
	4.9	2	10	87	88	54
	4.9	2	27	87	69	54
	4.9	2	16	87	74	54
	4.9	2	43	87	47	54
Lettuce	4	19	−12	97	97	21
Lactuca sativa L.	4	19	−1	97	91	21
	4	11	10	97	83	21
	4	19	21	97	90	21
	4	19	32	97	73	21
	7	19	−12	97	95	21
	7	19	−1	97	96	21

Table 3 (continued)
LONGEVITY OF SEEDS STORED AT VARIOUS TEMPERATURES IN SEALED CONTAINERS IN AN ATMOSPHERE OTHER THAN AIR

Species	Moisture (% when stored)	Storage: Time (years)	Storage: Temperature (°C)	Germination (%): When stored	Germination (%): After storage	Ref.
	7	19	10	97	72	21
	7	5	21	97	81	21
	7	1	32	97	0	21
	10	19	−12	95	94	21
	10	19	−1	95	45	21
	10	5	10	95	15	21
	10	1	21	95	21	21
	10	1	32	95	0	21
Onion	3.8	2	1	93	94	54
Allium cepa L.	3.8	2	10	93	94	54
	3.8	2	27	93	94	54
	3.8	2	16—35	93	94	54
	3.8	2	43	93	85	54
	4.4	2	1	94	93	54
	4.4	2	10	94	93	54
	4.4	2	27	94	94	54
	4.4	2	16—35	94	93	54
	4.4	2	43	94	25	54
	6.5	2	1	96	94	54
	6.5	2	10	96	94	54
	6.5	2	27	96	94	54
	6.5	2	16—35	96	95	54
	6.5	0.75	43	96	0	54
Pine, shortleaf	Not given	16	Ambient	84	6	9
Pinus echinata Mill.	Not given	16	5	84	48	9
	Not given	16	−4	84	68	9
Pine, lobolly	Not given	15	Ambient	84	22	9
P. taeda L.	—	15	5	84	71	9
	—	15	−4	84	89	9
Safflower	4	17	−12	95	86	21
Carthamus tinctorius L.	4	17	−1	95	93	21
	4	17	10	95	96	21
	4	17	21	95	96	21
	4	11	32	95	83	21
	4	17	32	95	5	21
	7	17	−12	94	86	21
	7	17	−1	94	92	21
	7	11	10	94	19	21
	7	3	21	94	20	21
	7	1	32	94	0	21
	10	17	−12	95	72	21
	10	11	−1	95	57	21
	10	2	10	95	18	21
	10	1	21	95	0	21
	10	1	32	95	0	21
Sage, scarlet *Salvia splendens* F. Sellow ex Roem. & Schult.	Not given	1.50	10—15	42	76	36
Sesame	4	17	−12	94	95	21
Sesamum indicum L.	4	17	−1	94	95	21
	4	17	10	94	95	21
	4	17	21	94	94	21

Table 3 (continued)
LONGEVITY OF SEEDS STORED AT VARIOUS TEMPERATURES IN SEALED CONTAINERS IN AN ATMOSPHERE OTHER THAN AIR

Species	Moisture (% when stored)	Storage: Time (years)	Storage: Temperature (°C)	Germination (%): When stored	Germination (%): After storage	Ref.
	4	17	32	94	29	21
	7	17	−12	92	93	21
	7	17	−1	92	60	21
	7	5	10	92	52	21
	7	3	21	92	0	21
	7	1	32	92	0	21
	10	3	−12	88	9	21
	10	5	−1	88	5	21
	10	2	10	88	7	21
	10	1	21	88	0	21
	10	1	32	88	0	21
Sorghum	4	16	−12	92	90	24
Sorghum bicolor	4	16	−1	92	92	24
(L.) Moench	4	16	10	92	94	24
	4	16	21	92	85	24
	4	16	32	92	74	24
	7	16	−12	95	92	24
	7	16	−1	95	91	24
	7	16	10	95	90	24
	7	16	21	95	82	24
	7	16	32	95	44	24
	10	16	−12	91	94	24
	10	16	−1	91	89	24
	10	16	10	91	78	24
	10	16	21	91	59	24
	10	5	32	91	53	21

Table 4
LONGEVITY OF SEEDS SEALED WITH A DESICCANT AND STORED AT VARIOUS TEMPERATURES

Species	Moisture (% when stored)	Storage: Time (years)	Storage: Temperature (°C)	Germination (%): When stored	Germination (%): After storage	Ref.
Desiccant: Calcium Chloride						
Amaranth, Globe *Gomphrena globosa* L.	Not given	6	Ambient	78	45	75
Apple *Malus sylvestris* Mill.	10—11	2.50	2—10	96	89	90
Apricot *Prunus armeniaca* L.	10—11	2.50	2—10	100	100	90
Aster, China *Callistephus chinensis* (L.) Nees	Not given	2	10—15	68	69	36
Baby's-breath *Gypsophila elegans* Bieb.	Not given	6	Ambient	98	95	75
Bamboo	18	0.60	21—32	Not given	76	97
Bambusa arundinacea Retz.	18	0.06	21—27	Not given	61	97
	18	0.60	16—21	Not given	76	97
	18	0.60	10	Not given	80	97
	23	0.60	21—32	49	81	97
	12	0.60	21—32	88	87	97
Bean, fava (broadbean) *Vicia faba* L.	Not given	5	Ambient	65	76	75
Bean, garden *Phaseolus vulgaris* L.	9.1	15	Ambient	97	88	75
Bluegrass, Kentucky	Not given	13	Ambient	91	64	75
Poa pratensis L.	Not given	13	0	91	75	75
Bright-eyes	Not given	6	5	95	73	75
Vinca rosea L. = *Catharanthus roseus* (L.) G. Don	Not given	6	Ambient	95	77	75

Table 4 (continued)
LONGEVITY OF SEEDS SEALED WITH A DESICCANT AND STORED AT VARIOUS TEMPERATURES

Species	Moisture (% when stored)	Storage: Time (years)	Storage: Temperature (°C)	Germination (%): When stored	Germination (%): After storage	Ref.
Brome, mountain *Bromus marginatus* Nees ex Steud.	Not given	5	Ambient	99	9	75
Brome, smooth *B. inermis* Leyss.	Not given	5	Ambient	89	6	75
Burdock, edible	6.3	15	Ambient	90	6	75
Arctium lappa L.	5.8	13	Ambient	92	40	75
Cabbage, Chinese *Brassica pekinensis* (Lour.) Rupr.	5.2	21	Ambient	93	3	75
Calendula *Calendula officinalis* L.	Not given	1	10—15	86	70	36
Canterbury bells *Campanula medium* L.	Not given	6	Ambient	97	86	75
Carnation *Dianthus caryophyllus* L.	Not given	6	Ambient	89	62	75
Carrot *Daucus carota* L.	7.2	21	Ambient	70	35	75
Clover, red	Not given	23	Ambient	97	64	48
Trifolium pratense L.	Not given	23	0—5	97	63	48
	Not given	13	Ambient	93	49	75
	Not given	13	0	93	89	75
Cockscomb *Celosia cristata* L.	Not given	6	Ambient	91	84	75
Cowpea (southern pea) *Vigna unguiculata* (L.) Walp. subsp. *unguiculata*	Not given	3.70	Warm	95	80	46
Cucumber *Cucumis sativus* L.	6.5	21	Ambient	93	87	75
Delphinium *Delphinium elatum* L.	Not given	1	10—15	57	58	36

Eggplant	6.7	21	Ambient	90	79	75
Solanum melongena L.						
Impatiens	Not given	1	10—15	35	35	36
Impatiens holstii Engl. & Warb. ex Engl. *I. walleriana* Hooker f. ex Oliv.						
Lettuce	6.3	15	Ambient	96	3	75
Lactuca sativa L.						
Moss rose	Not given	6	Ambient	77	60	75
Portulaca grandiflora Hook.						
Oatgrass, tall	Not given	13	0	95	70	75
Arrhenatherum elatius (L.) Beauv. ex J. and C. Presl	Not given	13	Ambient	95	35	75
Onion	Not given	1.50	10—15	73	64	36
Allium cepa L.						
	1.2	9	21—32	90	83	3
	7.9	15	Ambient	83	3	75
	7.1	15	Ambient	94	1	75
Onion, Welsh	5.9	15	Ambient	82	1	75
Allium fistulosum L.	7.9	15	Ambient	81	2	75
Orchardgrass (cocksfoot)	Not given	13	Ambient	94	40	75
Dactylis glomerata L.	Not given	13	0	94	62	75
Pansy						
Viola tricolor L.	Not given	1	10—15	68	46	36
	Not given	6	Ambient	73	55	75
Pea, garden	7.9	9	Ambient	62	44	75
Pisum sativum L.						
Pear	10—11	2.50	2—10	96	88	90
Pyrus communis L.						
Pepper	Not given	3.70	Warm	Not given	41	46
Capsicum annuum L.	6.1	15	Ambient	63	36	75
Perilla	4.5	3	Ambient	86	19	75
Perilla ocymoides L. = *P. frutescens* (L.) Britt.	4.5	8	Ambient	86	0	75
	Not given	14	0	86	28	75
Phlox	Not given	1	10—15	95	96	36
Phlox drummondii Hooker						
Plum	10—11	2.50	2—10	100	92	90
Prunus americana Marsh.						

Table 4 (continued)
LONGEVITY OF SEEDS SEALED WITH A DESICCANT AND STORED AT VARIOUS TEMPERATURES

Species	Moisture (% when stored)	Storage: Time (years)	Storage: Temperature (°C)	Germination (%): When stored	Germination (%): After storage	Ref.
Poppy, Californica *Eschscholzia californica* Cham.	Not given	6	Ambient	81	65	75
Primrose, Chinese *Primula sinensis* Sab. ex. Lindl.	Not given	1	10—15	74	52	36
Primrose, German *P. obconica* Hance.	Not given	1	10—15	9	16	36
Pussy-foot *Ageratum conyzoides* L.	Not given	6	Ambient	83	65	75
Radish *Raphanus sativus* L.	5.9	15	Ambient	85	57	75
Rescue Grass *Bromus uioloides* H.B.K.	Not given	10	Ambient	99	6	75
Ryegrass, Italian	Not given	13	Ambient	88	39	75
Lolium multiflorum Lam.	Not given	13	0	88	93	75
Sage, scarlet	Not given	1.50	10—15	42	65	36
Salvia splendens F. Sellow ex Roem.	Not given	6	8	89	41	75
and Schult	Not given	6	5	89	60	75
Salsify *Tragopogon porrifolius* L.	Not given	1.50	10—15	95	91	36
Smartweed *Polygonum hydropiper* L.	Not given	4	Ambient	95	2	75
Snapdragon *Antirrhinum majus* L.	Not given	6	Ambient	81	65	75
Spinach, common *Spinacia oleracea* L.	9.3	21	Ambient	83	56	75
Spurge, painted *Euphorbia heterophylla* L.	Not given	6	Ambient	94	90	75
Squash *Cucurbita* spp.	6.5	15	Ambient	82	71	75

Timothy	Not given	13	Ambient	99	43	75
Phleum pratense L.	Not given	13	0	99	95	75
Tobacco	9	25	5	Not given	81	65,66
Nicotiana tabacum L.	9	25	Ambient (basement)	Not given	56	65,66
	9	25	Ambient (laboratory)	Not given	12	65,66
	9	25	Ambient (attic)	Not given	25	65,66
Tomato *Lycopersicon esculentum* Mill.	Not given	3.7	Warm	96	50	46
Wallflower	Not given	6	5	72	45	75
Cheiranthus cheiri L.	Not given	6	Ambient	72	43	75
Watermelon *Citrullus lanatus* (Thunb.) Matsum. and Nakai)	Not given	3.70	Warm	87	77	46
Zinnia *Zinnia elegans* Jacq.	Not given	1	10—15	43	44	36
Desiccant: Calcium Oxide						
Amaranth, globe *Gomphrena globosa* L.	Not given	6	Ambient	78	20	75
Baby's-breath *Gypsophila elegans* Bieb.	Not given	6	Ambient	98	3	75
Bean, fava (broadbean) *Vicia faba* L.	Not given	2	Ambient	65	16	75
Bean, garden *Phaseolus vulgaris* L.	2.8	15	Ambient	97	32	75
Bluegrass, Kentucky *Poa pratensis* L.	Not given	13	Ambient	91	27	75
Bright-eyes *Vinca rosea* L. = *Catharanthus roseus* (L.) G. Don	Not given	6	Ambient	95	9	75

Table 4 (continued)
LONGEVITY OF SEEDS SEALED WITH A DESICCANT AND STORED AT VARIOUS TEMPERATURES

Species	Moisture (% when stored)	Storage: Time (years)	Storage: Temperature (°C)	Germination (%): When stored	Germination (%): After storage	Ref.
Brome, mountain *Bromus marginatus* Nees ex Steud.	Not given	10	Ambient	99	97	75
Brome, smooth *Bromus inermis* Leyss.	Not given	10	Ambient	89	48	75
Burdock, edible *Arctium lappa* L.	2.6	20	Ambient	90	9	75
	2.5	13	Ambient	92	71	75
Cabbage, Chinese *Brassica pekinensis* (Lour.) Rupr.	2.2	10	Ambient	93	11	75
Canterbury bells *Campanula medium* L.	Not given	6	Ambient	97	25	75
Carnation *Dianthus caryophyllus* L.	Not given	6	Ambient	89	20	75
Carrot *Daucus carota* L.	7.2	20	Ambient	67	1	9
	7.2	20	−4	67	65	9
	5.4	20	Ambient	67	6	9
	5.4	20	−4	67	69	9
	3.8	5	Ambient	70	29	75
Clover, red *Trifolium pratense* L.	Not given	13	Ambient	93	30	75
Cockscomb *Celosia cristata* L.	Not given	6	Ambient	91	60	75
Cucumber *Cucumis sativus* L.	2.3	21	Ambient	93	8	75
Eggplant *Solanum melongena* L.	7	14	Ambient	86	50	9
	7	14	−4	76	76	9
	5.2	20	Ambient	86	42	9
	5.2	20	−4	86	86	9
	3.2	21	Ambient	90	9	75

Lettuce	5.4	12	Ambient	98	1	9
Lactuca sativa L.	5.4	20	−4	98	76	9
	4.1	14	Ambient	98	3	9
	4.1	20	−4	98	84	9
	2.5	20	Ambient	96	41	75
Moss rose	Not given	6	Ambient	77	11	75
Portulaca grandiflora Hook.						
Oatgrass, tall	Not given	13	Ambient	95	24	75
Arrhenatherum elatius (L.) Beauv. ex J. & C. Presl						
Onion	8.3	8	Ambient	98	33	9
Allium cepa L.						
	8.3	18	−4	98	79	9
	6.3	14	Ambient	98	32	9
	6.3	20	−4	98	93	9
	2.5	15	Ambient	83	57	75
	2.7	15	Ambient	94	76	75
Onion, Welsh	2.6	15	Ambient	82	54	75
A. fistulosum L.	3.1	15	Ambient	81	52	75
Orchardgrass (cocksfoot)	Not given	13	Ambient	94	22	75
Dactylis glomerata L.						
Pansy	Not given	6	Ambient	73	4	75
Viola tricolor L.						
Pea, garden	2.8	9	Ambient	62	12	75
Pisum sativum L.						
Pepper	7	8	Ambient	73	4	9
Capsicum annuum L.	7	20	−4	73	67	9
	5.2	20	−4	73	61	9
	5.2	10	Ambient	73	20	9
	2.6	15	Ambient	63	7	75
Perilla	2.1	1	Ambient	86	9	75
Perilla ocymoides L. = *P. frutescens* (L.) Britt.						
Poppy, California	Not given	6	Ambient	81	12	75
Eschscholzia californica Cham.						
Pussy-foot	Not given	6	Ambient	83	25	75
Ageratum conyzoides L.						

Table 4 (continued)
LONGEVITY OF SEEDS SEALED WITH A DESICCANT AND STORED AT VARIOUS TEMPERATURES

Species	Moisture (% when stored)	Storage: Time (years)	Storage: Temperature (°C)	Germination (%): When stored	Germination (%): After storage	Ref.
Radish *Raphanus sativus* L.	2.6	5	Ambient	85	27	75
Rescue grass *Bromus unioloides* H.B.K.	Not given	10	Ambient	99	91	75
Ryegrass, Italian *Lolium multiflorum* Lam.	Not given	10	Ambient	88	9	75
Sage, scarlet *Salvia splendens* F. Sellow ex Roem. &Schult.	Not given	2	Ambient	89	4	75
Smartweed *Polygonum hydropiper* L.	Not given	3	Ambient	95	3	75
Snapdragon *Antirrhinum majus* L.	Not given	6	Ambient	81	39	75
Spinach, common *Spinacia oleracea* L.	4.1	17	Ambient	83	4	75
Spurge, painted *Euphorbia heterophylla* L.	Not given	6	Ambient	94	9	75
Squash *Cucurbita* spp.	3.1	15	Ambient	82	5	75
Timothy *Phleum pratense* L.	Not given	13	Ambient	99	43	75
Tomato	6.6	20	Ambient	93	1	9
Lycopersicon esculentum Mill.	6.6	16	−4	93	93	9
	5.0	20	Ambient	93	23	9
	5.0	20	−4	93	88	9
Wallflower *Cheiranthus cheiri* L.	Not given	6	Ambient	72	5	75

Table 5
LONGEVITY OF SEEDS STORED UNDER VARIOUS CONDITIONS IN MOISTURE BARRIER CONTAINERS MADE OF A VARIETY OF HEAT-SEALABLE MATERIALS

Species	Moisture (% when stored)	Storage			Germination (%)		Ref.
		RH (%)	Time (years)	Temperature (°C)	When stored	After storage	
		Barrier Material: Cellophane					
Bluegrass, Kentucky	3.8	50—60	2	1	75	65	54
Poa pratensis L.	3.8	80—95	2	10	75	5	54
	3.8	25—30	2	27	75	61	54
	3.8	20—80	2	16—35	75	59	54
	3.8	15	2	43	75	59	54
Cabbage	2.9	50—60	2	1	92	94	54
Brassica oleracea var. *capitata* L.	2.9	80—95	2	10	92	66	54
	2.9	25—30	2	27	92	92	54
	2.9	20—80	2	16—35	92	92	54
	2.9	15	2	43	92	74	54
Fescue, creeping red	3.6	50—60	2	1	95	85	54
Festuca rubra L.	3.6	80—95	2	10	95	2	54
	3.6	25—30	2	27	95	78	54
	3.6	20—80	2	16—35	95	75	54
	3.6	15	2	43	95	45	54
Onion	4.1	50—60	2	1	95	94	54
Allium cepa L.							
	4.1	80—95	1.5	10	95	22	54
	4.1	25—30	2	27	95	90	54
	4.1	20—80	2	16—35	95	72	54
	4.1	15	2	43	95	75	54

Table 5 (continued)
LONGEVITY OF SEEDS STORED UNDER VARIOUS CONDITIONS IN MOISTURE BARRIER CONTAINERS MADE OF A VARIETY OF HEAT-SEALABLE MATERIALS

Species	Moisture (% when stored)	Storage: RH (%)	Storage: Time (years)	Storage: Temperature (°C)	Germination (%): When stored	Germination (%): After storage	Ref.
Barrier Material: 1 mil Acetate/1 mil Foil/Vinyl Heat-Seal Coating (#114 Metalam)[a]							
Clover, crimson	5	90	8	32	72	37	39
Trifolium incarnatum L.	5	70	8	32	72	9	39
	5	90	11	21	72	76	39
	5	70	11	21	72	47	39
	5	90	11	10	72	69	39
	5	70	11	10	72	76	39
	5	95—100	11	20—30	72	43	39
	5	40—100	11	Rotating[b]	72	41	39
	7	90	6	32	77	26	39
	7	70	6	32	77	67	39
	7	90	6	21	77	63	39
	7	70	6	21	77	27	39
	7	90	11	10	77	64	39
	7	70	8	10	77	15	39
	7	95—100	2	20—30	77	10	39
	7	40—100	4	Rotating[b]	77	64	39
Papaya	—	35—40	6	5	68	56	18
Carica papaya L.							
Safflower	4	90	4	32	96	61	22
Carthamus tinctorius L.	4	70	4	32	96	69	22
	4	90	11	21	96	59	22
	4	70	8	21	96	66	22
	4	90	11	10	96	66	22
	4	70	11	10	96	61	22
	4	95—100	3	20—30	96	60	22
	4	40—100	6	Rotating[b]	96	65	22

	7	90	1	32	92	19	22
	7	70	2	32	92	1	22
	7	90	2	21	92	64	22
	7	70	4	21	92	2	22
	7	90	8	10	92	16	22
	7	70	8	10	92	83	22
	7	95—100	2	20—30	92	79	22
	7	40—100	2	Rotating[b]	92	42	22
Barrier Material: Metalized Polyester							
Clover, crimson							
Trifolium incarnatum L.	5	90	4	32	72	21	39
	5	70	4	32	72	51	32
	5	90	11	21	72	30	39
	5	70	11	21	72	30	39
	5	90	11	10	72	79	39
	5	70	11	10	72	71	39
	5	95—100	4	20—30	72	23	39
	5	40—100	8	Rotating[b]	72	31	39
	7	90	2	32	77	24	39
	7	70	6	32	77	22	39
	7	90	6	21	77	59	39
	7	70	6	21	77	62	39
	7	90	11	10	77	76	39
	7	70	11	10	77	75	39
	7	95—100	4	20—30	77	1	39
	7	40—100	6	Rotating[b]	77	48	39
Safflower	4	90	4	32	96	47	22
Carthamus tinctorius L.							
	4	70	6	32	96	28	22
	4	90	8	21	96	59	22
	4	70	8	21	96	8	22

Table 5 (continued)
LONGEVITY OF SEEDS STORED UNDER VARIOUS CONDITIONS IN MOISTURE BARRIER CONTAINERS MADE OF A VARIETY OF HEAT-SEALABLE MATERIALS

Species	Moisture (% when stored)	Storage: RH (%)	Storage: Time (years)	Storage: Temperature (°C)	Germination (%): When stored	Germination (%): After storage	Ref.
	4	90	11	10	96	95	22
	4	70	11	10	96	89	22
	4	95—100		20—30	96	81	22
	4	40—100	8	Rotating[b]	96	36	22
	7	90	1	32	92	17	22
	7	70	1	32	92	34	22
	7	90	2	21	92	69	22
	7	70	2	21	92	70	22
	7	90	11	10	92	47	22
	7	70	11	10	92	75	22
	7	95—100	2	20—30	92	81	22
	7	40—100	2	Rotating[b]	92	67	22
Barrier Material: Metalized Polyester							
Clover, crimson	5	90	4	32	72	21	39
Trifolium incarnatum L.	5	70	4	32	72	51	32
	5	90	11	21	72	30	39
	5	70	11	21	72	30	39
	5	90	11	10	72	79	39
	5	70	11	10	72	71	39
	5	95—100	4	20—30	72	23	39
	5	40—100	8	Rotating[b]	72	31	39
	7	90	2	32	77	24	39
	7	70	6	32	77	22	39

	7	90	6	21	77	59	39
	7	70	6	21	77	62	39
	7	90	11	10	77	76	39
	7	70	11	10	77	75	39
	7	95—100	4	20—30	77	1	39
	7	40—100	6	Rotating[b]	77	48	39
Safflower	4	90	4	32	96	47	22
Carthamus tinctorius L.	4	70	6	32	96	28	22
	4	90	8	21	96	59	22
	4	70	8	21	96	85	22
	4	90	11	10	96	95	22
	4	70	11	10	96	89	22
	4	95—100	3	20—30	96	81	22
	4	40—100	8	Rotating[b]	96	36	22
	7	90	1	32	92	17	22
	7	70	1	32	92	34	22
	7	90	2	21	92	69	22
	7	70	2	21	92	70	22
	7	90	11	10	92	47	22
	7	70	11	10	92	75	22
	7	95—100	2	20—30	92	81	22
	7	40—100	2	Rotating[b]	92	67	22
Barrier Material: Polyester							
Bluegrass, Kentucky	3.8	50—60	2	1	75	62	54
Poa pratensis L.	3.8	80—95	2	10	75	45	54
	3.8	25—30	2	27	75	59	54
	3.8	20—80	2	16—35	75	60	54
	3.8	15	2	43	75	59	54
Cabbage	3.1	50—60	2	1	96	95	54
Brassica oleracea var. *capitata* L.	3.1	80—95	2	10	96	94	54
	3.1	25—30	2	27	96	95	54
	3.1	20—80	2	16—35	96	93	54
	3.1	15	2	43	96	79	54
Fescue, creeping red	3.6	50—60	2	1	95	90	54
Festuca rubra L.	3.6	80—95	2	10	95	24	54

Table 5 (continued)
LONGEVITY OF SEEDS STORED UNDER VARIOUS CONDITIONS IN MOISTURE BARRIER CONTAINERS MADE OF A VARIETY OF HEAT-SEALABLE MATERIALS

Species	Moisture (% when stored)	Storage			Germination (%)		Ref.
		RH (%)	Time (years)	Temperature (°C)	When stored	After storage	
	3.6	25—30	2	27	95	82	54
	3.6	20—80	2	16—35	95	61	54
	3.6	15	2	43	95	39	54
Onion	4.1	50—60	2	1	95	94	54
Allium cepa L.	4.1	80—95	2	10	95	58	54
	4.1	25—30	2	27	95	90	54
	4.1	20—80	2	16—35	95	79	54
	4.1	15	2	43	95	76	54
Barrier Material: Thin Polyethylene Bags							
Bluegrass, Kentucky	3.8	50—60	2	1	75	58	54
Poa pratensis L.	3.8	80—95	2	10	75	39	54
	3.8	25—30	2	27	75	56	54
	3.8	20—80	2	16—35	75	59	54
Brussels sprouts *Brassica oleracea* var. *gemmifera* DC	6	58—83	6.50	11—19	71	10	71
Cabbage, red	3.3	50—60	2	1	94	94	54
B. oleracea var. *capitata* L.	3.3	80—95	2	10	94	91	54
	3.3	25—30	2	27	94	93	54
	3.3	20—80	2	10—35	94	89	54
	3.3	15	2	43	94	94	54
	6	58—83	4.75	11—19	95	95	71
Clover, red *Trifolium pratense* L.	6	58—83	5.50	11—19	92	85	70

Cowpea (southern pea) *Vigna unguiculata* (L.) Walp. subsp. *unguiculata*	Not given	Humid	2.25	Warm	84	69	46
Fescue, creeping red *Festuca rubra* L.	3.6	50—60	2	1	95	86	54
	3.6	80—95	2	10	95	35	54
	3.6	25—30	2	27	95	80	54
	3.6	20—80	2	16—35	95	59	54
	3.6	20—80	2	43	95	57	54
Onion *Allium cepa* L.	4.1	50—60	2	1	95	93	54
	4.1	80—95	2	10	95	84	54
	4.1	25—30	2	27	95	89	54
	4.1	20—80	2	16—35	95	75	54
	4.1	15	2	43	95	81	54
Rye *Secale cereale* L.	9	5—83	3 +	9—20	90	90	69
Ryegrass, annual *Lolium perenne* L.	6	58—83	6.66	11—19	94	92	70
Timothy *Phleum pratense* L.	7	58—83	6.66	11—19	98	94	70
Turnip *Brassica rapa* L.	8	58—83	6.5	11—19	98	94	71
Wheat *Triticum aestivum* L.	12	58—83	6 +	11—19	97	96	69
Barrier Material: 10 mil Polyethylene Bags							
Bluegrass, Kentucky *Poa pratensis* L.	8.10	80	2	1	86	84	15
	8.10	80	2	10	86	83	15
	8.10	20—40	2	21	86	81	15
	8.10	10—15	2	32	86	78	15
	8.10	4—6	2	43	86	77	15
	8.10	Ambient	2	Ambient	86	82	15
	6.70	80	2	1	86	84	15
	6.70	80	2	10	86	85	15
	6.70	20—40	2	21	86	84	15
	6.70	10—15	2	32	86	82	15
	6.70	4—6	2	43	86	84	15
	6.70	Ambient	2	Ambient	86	85	15

Table 5 (continued)
LONGEVITY OF SEEDS STORED UNDER VARIOUS CONDITIONS IN MOISTURE BARRIER CONTAINERS MADE OF A VARIETY OF HEAT-SEALABLE MATERIALS

Species	Moisture (% when stored)	Storage: RH (%)	Storage: Time (years)	Storage: Temperature (°C)	Germination (%): When stored	Germination (%): After storage	Ref.
	5.40	80	2	1	87	86	15
	5.40	80	2	10	87	85	15
	5.40	20—40	2	21	87	81	15
	5.40	10—15	2	32	87	81	15
	5.40	4—6	2	43	87	77	15
	5.40	Variable	2	Ambient	87	83	15
Fescue, creeping red	11.40	80	2	1	95	87	15
Festuca rubra L.	11.40	80	2	10	95	73	15
	11.40	20—40	2	21	95	13	15
	11.40	10—15	2	32	95	18	15
	11.40	4—6	0.25	43	95	4	15
	11.40	Ambient	2	Ambient	95	34	15
	6.40	80	2	1	72	53	15
	6.40	80	2	10	72	33	15
	6.40	20—40	2	21	72	15	15
	6.40	10—15	2	32	72	6	15
	6.40	4—6	1.5	43	72	2	15
	6.40	Ambient	2	Ambient	72	8	15
	3.20	80	2	1	81	65	15
	3.20	80	2	10	81	65	15
	3.20	20—40	2	21	81	41	15
	3.20	10—15	2	32	81	27	15
	3.20	4—6	2	43	81	17	15
	3.20	Ambient	2	Ambient	81	35	15
	Barrier Material: Polyethylene Laminated to Paper Bags						
Bluegrass, Kentucky	3.50	50—60	2	1	77	51	54
Poa pratensis L.	3.50	80—95	2	10	77	4	54

	3.50	25—30	2	27	77	48	54
	3.50	20—80	2	16—35	77	44	54
	3.50	15	2	43	77	44	54
Cabbage *Brassica oleracea* var. *capitata* L.	2.90	50—60	2	1	93	94	54
	2.90	80—95	2	10	93	47	54
	2.90	25—30	2	27	93	89	54
	2.90	20—80	2	16—35	93	87	54
	2.90	15	2	43	93	72	54
Fescue, creeping red	3.60	50—60	2	1	95	91	54
Festuca rubra L.	3.60	80—95	2	10	95	6	54
	3.60	25—30	2	27	95	80	54
	3.60	20—80	2	16—35	95	28	54
	3.60	15	2	43	95	30	54
Onion	3.8	50—60	2	1	93	94	54
Allium cepa L.	3.8	80—95	2	10	93	0	54
	3.8	25—30	2	27	93	89	54
	3.8	20—80	2	16—35	93	60	54
	3.8	15	2	43	93	74	54
Barrier Material: Saran®							
Cowpea (southern pea) *Vigna unguiculata* L. Walp. subsp. *unguiculata*	Not given	Humid	4.33	Warm	84	21	46
Tomato *Lycopersicon esculentum* Mill.	Not given	Humid	4	Warm	81	16	46

[a] Mention of trademark or proprietary product does not constitute a guarantee or warranty of the product by the U.S. Department of Agriculture, and does not imply its approval to the exclusion of other products that may also be suitable.

[b] Rotating between 20°C night to 30°C day and 5°C constant at 2-week intervals.

Table 6
EFFECT OF STORAGE CONDITIONS ON VIABILITY OF PELLETED CARROT, LETTUCE, AND ONION SEEDS

Species	Moisture content (%)	Storage			Germination (%)		Ref.
		RH (%)	Years	Temperature (°C)	Initial	After storage	
Carrot							
Daucus carota L.	Variable	40	3	5	83	75	86
		50	3	10	83	79	86
		70	3	10	83	62	86
Lettuce							
Lactuca sativa L.	Variable	40	3	5	96	96	85
		50	3	10	96	96	85
		70	3	10	96	32	85
Onion							
Allium cepa L.	Variable	40	3	5	94	96	86
		50	3	10	94	95	86
		70	3	10	94	83	86

REFERENCES

1. **Akamine, E. K.**, The effect of temperature and humidity on the viability of stored seeds in Hawaii, *Hawaii Agric. Exp. Stn. Bull.*, 90, 23, 1943.
2. **Allen, G. S.**, Storage behavior of conifer seeds in sealed containers held at 0°F, 32°F, and room temperature, *J. For.*, 55, 278, 1957.
3. Associated Seed Growers, Inc. (Asgrow Seed Co.), The preservation of viability and vigor in vegetable seeds, *Asgrow Monog.*, 2, 32, 1954.
4. **Barton, L. V.**, Storage of elm seeds, *Boyce Thompson Inst. Plant Res. Prof. Pap.*, 10, 221, 1939.
5. **Barton, L. V.**, The storage of citrus seeds, *Boyce Thompson Inst. Plant Res. Prof. Pap.*, 13, 47, 1943.
6. **Barton, L. V.**, Viability of seeds of *Fraxinus* after storage, *Boyce Thompson Inst. Plant Res. Prof. Pap.*, 13, 427, 1945.
7. **Barton, L. V.**, Effect of different storage conditions on the germination of seeds of *Cinchona ledgeriana* Moens, *Boyce Thompson Inst. Plant Res. Prof. Pap.*, 15, 1, 1947.
8. **Barton, L. V.**, Storage of seeds of the regal lily, *North Am. Lily Soc. Yearbook*, 47/48, 51, 1947; *Boyce Thompson Inst. Plant Res. Prof. Pap.*, 2, 45-51, 1948.
9. **Barton, L. V.**, Seed storage and viability, *Boyce Thompson Inst. Plant Res. Prof. Pap.*, 17, 84, 1953.
10. **Barton, L. V.**, Effect of subfreezing temperatures on viability of conifer seeds in storage, *Boyce Thompson Inst. Plant Res. Prof. Pap.*, 18, 21, 1954.
11. **Barton, L. V.**, Storage of seeds of *Lobelia cardinalis* L., *Boyce Thompson Inst. Plant Res. Prof. Pap.*, 20, 395, 1960
12. **Barton, L. V.**, Viability of *Pyrethrum* seeds, *Boyce Thompson Inst. Plant Res. Prof. Pap.*, 23, 267, 1966.
13. **Barton, L. V.**, The effect of storage conditions on the viability of bean seeds, *Boyce Thompson Inst. Plant Res. Prof. Pap.*, 23, 281, 1966.
14. **Barton, L. V.**, Effects of temperature and moisture on viability of stored lettuce, onion, and tomato seeds, *Boyce Thompson Inst. Plant Res. Prof. Pap.*, 23, 285, 1966.
15. **Bass, L. N.**, Packaging and storage of Kentucky bluegrass and creeping red fescue seed, *Proc. Assoc. Off. Seed Anal.*, 49, 173, 1960.
16. **Bass, L. N.**, Response of reed canarygrass *(Phalaris arundinacea* L.) seed to storage conditions, *Proc. Assoc. Off. Seed Anal.*, 57, 124, 1967.
17. **Bass, L. N.**, Response of seeds of 27 *Cucumis melo* cultivars to three storage conditions, *Proc. Assoc. Off. Seed Anal.*, 63, 83, 1973.
18. **Bass, L. N.**, Seed storage of *Carica papaya* L., *HortScience*, 10, 232, 1975.
19. **Bass, L. N.**, Longevity of seeds of two *Dimorphotheca* species, *J. Seed Technol.*, 2, 86, 1977.
20. **Bass, L. N.**, Sealed storage of crimson clover seed, *Seed Sci. Technol.*, 6, 1017, 1978.
21. **Bass, L. N.**, unpublished data, 1979.
22. **Bass, L. N. and Clark, D. C.**, Effects of storage conditions, packaging materials, and seed moisture content on longevity of safflower seeds, *Proc. Assoc. Off. Seed Anal.*, 64, 120, 1974.
23. **Bass, L. N., James, E., and Clark, D. C.**, Storage response of green and bleached lima beans (*Phaseolus lunatus* L.), *HortScience*, 5, 170, 1970.
24. **Bass, L. N. and Stanwood, P. C.**, Long-term preservation of sorghum seed as affected by seed moisture, temperature, and atmospheric environment, *Crop Sci.*, 18, 575, 1978.
25. **Bass, L. N., Toy, S. J., Sayers, R. L., and Clark, D. C.**, Storage of *Dimorphotheca sinuata* and *Osteospermum ecklonis* seed, *Proc. Assoc. Off. Seed Anal.*, 57, 67, 1967.
26. **Battle, W. R.**, Effect of scarification on longevity of alfalfa seed, *J. Am. Soc. Agron.*, 40, 758, 1948.
27. **Beattie, J. H. and Boswell, V. R.**, Longevity of onion seed in relation to storage conditions, *U.S. Dep. Agric. Circ.*, 512, 23, 1939.
28. **Beattie, J. H., Jackson, A. M., and Currin, R. E.**, Effect of cold storage and age of seed on germination and yield of peanuts, *U.S. Dep. Agric. Circ.*, 233, 12, 1932.
29. **Beattie, J. H. and Tatman, E. C.**, Long-term storage of parsnip seed, *Proc. Am. Soc. Hortic. Sci.*, 55, 435, 1950.
30. **Bhore, D. P., Bhapkar, D. G., and Chavan, V. M.**, Longevity of seed in Brinjal, *Indian J. Agron.*, 12, 51, 1967.
31. **Brison, F. R.**, Influence of storage conditions on the germination of onion seed. *Tex. Agric. Exp. Stn. Prog. Rep.*, 1492, 4, 1952.
32. **Brown, E.**, Preserving the viability of Bermuda onion seed, *Science*, 89, 292, 1939.
33. **Brown, E. O. and Porter, R. H.**, The viability and germination of seeds of *Convolvulus arvensis* L. and other perennial weeds, *Iowa Agric. Exp. Stn. Res. Bull.*, 294, 473, 1942.
34. **Buszewicz, G.**, The longevity of beechnuts in relation to storage conditions, *Proc. Int. Seed Test. Assoc.*, 26, 504, 1961.

35. **Canode, C. L.**, Germination of grass seed as influenced by storage condition, *Crop Sci.*, 12, 79, 1972.
36. **Chopinet, R.**, La conservation des semences en atmosphere seche; Son interet en selection, *13th Int. Hortic. Congr. (London)*, 1022-1043, 1952.
37. **Christidis, B. G.**, Seed vitality and other cotton characters as affected by the age of seed, *Plant Physiol.*, 29, 124, 1954.
38. **Clark, D. C. and Bass, L. N.**, Effects of time, temperature and relative humidity on the longevity of crambe seeds, *Proc. Assoc. Off. Seed Anal.*, 63, 148, 1973.
39. **Clark, D. C. and Bass, L. N.**, Effects of storage conditions, packaging materials, and moisture content on longevity of crimson clover seeds, *Crop. Sci.*, 15, 577, 1975.
40. **Clark, D. C., Bass, L. N., and Sayers, R. L.**, Storage of hemp and kenaf seed, *Proc. Assoc. Off. Seed Anal.*, 53, 210, 1963.
41. **Davies, W. E.**, Storage of clover seed. I. First interim report 1947—1954, *J. Br. Grassl. Soc.*, 11, 224, 1956.
42. **Davies, W. E.**, Storage of clover seed. II. Second interim report, 1955—1964, *J. Br. Grassl. Soc.*, 20, 180, 1965.
43. **Dillman, A. C. and Toole, E. H.**, Effect of age, condition, and temperature on the germination of flaxseed, *J. Am. Soc. Agron.*, 29, 23, 1937.
44. **Dorph-Petersen, K.**, How long do the various seed species retain their germination power?, *Int. Inst. Agric. Int. Rev. Sci. Pract. Agric. (N.S.)*, 2, 283, 1924.
45. **Dumlao, R. C. and Parker, M. B.**, The influence of cold storage conditions on the viability of rice and corn seeds, *Philipp. Agric.*, 44, 327, 1960.
46. **Edmond, J. B.**, Studies of storage of vegetable seed under warm-humid conditions, *Miss. Agric. Exp. Stn. Tech. Bull.*, 50, 24, 1962.
47. **Evans, G.**, The viability over a period of fifteen years of severely dried ryegrass seed, *J. Br. Grassl. Soc.*, 12, 286, 1957.
48. **Evans, G., Roberts, H. M., and Lewis, J.** Maintenance of life in seeds, *Welsh Plant Breed. Stn. Aberystwyth Rep., 1950—1956*, 152, 1958.
49. **Evans, H.**, Results of some experiments on the preservation of cacao seeds in viable condition, *Trop. Agric.*, 27, 48, 1950.
50. **Fifield, C. C. and Robertson, D. W.**, Milling, baking, and chemical properties of Marquis and Kanred wheat grown in Colorado and stored 25 to 33 years, *Cereal Sci. Today*, 4, 179, 1959.
51. **Goff, E. S.**, Comparative vitality of hulled and unhulled timothy seeds, *Wis. Agric. Exp. Stn. Annu. Rep.*, 7, 202, 1890.
52. **Gorman, L. W. and Greenwood, R. M.**, Effect of moisture content on the viability of Italian ryegrass seed in storage, *N.Z. J. Sci. Technol. Sect. A*, 33, 58, 1951.
53. **Grabe, D. F.**, Artificial drying and storage of smooth bromegrass seed, *Agron. J.*, 49, 161, 1957.
54. **Grabe, D. F. and Isely, D.**, Final report on improved packaging for expediting storage and marketing of seed, *Iowa Agric. Exp. Stn. Project* 1386, USDA Contract No. 12-25-010-554, 80, 1959.
55. **Haferkamp, M. E., Smith, L., and Nilan, R. A.**, Studies on aged seeds. I. Relation of age of seed to germination and longevity, *Agron. J.*, 45, 434, 1953.
56. **Hamilton, R. A.**, A study of germination and storage life of macadamia seed, *Proc. Am. Soc. Hortic. Sci.*, 70, 209, 1957.
57. **Hanson, C. H. and Moore, R. P.**, Viability of seeds of eight forage crop plants stored under subfreezing conditions, *Agron. J.*, 51, 627, 1959.
58. **Heidmann, L. J.**, Longevity of ponderosa pine seed, *Ecology*, 43, 344, 1962.
59. **Heit, C. E.**, Recent findings on laboratory germination and longevity of New Zealand spinach seed, *Proc. Assoc. Off. Seed Anal.*, 35, 115, 1945.
60. **James, E., Bass, L. N., and Clark, D. C.**, Longevity of vegetable seeds stored 15 to 30 years at Cheyenne, Wyoming, *Proc. Am. Soc. Hortic. Sci.*, 84, 527, 1964.
61. **James, E., Bass, L. N., and Clark, D. C.**, Varietal differences in longevity of vegetable seed and their response to various storage conditions, *Proc. Am. Soc. Hortic. Sci.*, 91, 521, 1967.
62. **Johnsen, T. N., Jr.**, Longevity of stored juniper seeds, *Ecology*, 40, 487, 1959.
63. **Joseph, H. C.**, Germination and keeping quality of parsnip seeds under various conditions, *Bot. Gaz.*, 87, 195, 1929; *Boyce Thompson Inst. Plant Res. Prof. Pap.*, 2, 115, 1929.
64. **Kearns, V. and Toole, E. H.**, Relation of temperature and moisture content to longevity of chewings fescue seed, *U.S. Dep. Agric. Tech. Bull.*, 670, 27, 1939.
65. **Kincaid, R. R.**, The effects of storage conditions on the viability of tobacco seed, *J. Agric. Res.*, 67, 407, 1943.
66. **Kincaid, R. R.**, Tobacco seed storage for 25 years, *Fla. Agric. Exp. Stn. Tech. Bull.*, 593, 10, 1958.
67. **Lewis, J.**, Longevity of crop and weed seeds. I. First interim report, *Proc. Int. Seed Test. Assoc.*, 23, 340, 1958.

68. **McAllister, D. F.**, The effect of maturity on the viability and longevity of the seeds of western range and pasture grasses, *J. Am. Soc. Agron.*, 35, 442, 1943.
69. **MacKay, D. B. and Flood, R. J.**, Investigations in crop seed longevity. II. The viability of cereal seed stored in permeable and impermeable containers, *Natl. Inst. Agric. Bot. J.*, 11, 378, 1968.
70. **MacKay, D. B. and Flood, R. J.**, Investigations in crop seed longevity. III. The viability of grass and clover seed stored in permeable and impermeable containers, *J. Natl. Inst. Agric. Bot.*, 11, 521, 1969.
71. **MacKay, D. B. and Flood, R. J.**, Investigations in crop seed longevity. IV. The viability of *Brassica* seed stored in permeable and impermeable containers, *Natl. Inst. Agric. Bot. J.*, 12, 84, 1970.
72. **MacKay, D. B. and Tonkin, J. H. B.**, Investigations in crop seed longevity. I. Analysis of long-term experiments with special reference to the influence of species, cultivar, provenance, and season, *Natl. Inst. Agric. Bot. J.*, 11, 209, 1967.
73. **Martin, J. A., Senn, T. L., Skelton, B. J., and Crawford, J. H.**, Response of okra seed to moisture content and storage temperature, *Proc. Am. Soc. Hortic. Sci.*, 75, 490, 1960.
74. **Moore, R. P.**, unpublished data, personal communication, 1973.
75. **Nakamura, S.**, The most appropriate moisture content of seeds for their long life span, *Seed Sci. Technol.*, 3, 747, 1975.
76. **Neal, N. P. and Davis, J. R.**, Seed viability of corn inbred lines as influenced by age and conditions of storage, *Agron. J.*, 48, 383, 1956.
77. **Nutile, G. E.**, Effect of relative humidity on the viability of lettuce and onion seeds during nine years of storage, *Newsletter Assoc. Off. Seed Anal.*, 31, 11, 1957.
78. **Nutile, G. E.**, Effect of desiccation on viability of seeds, *Crop. Sci.*, 4, 325, 1964.
79. **Pack, D. A. and Owen, F. V.**, Viability of sugarbeet seed held in cold storage for 22 years, *Proc. Am. Soc. Sugar Beet Technol.*, 6, 127, 1950.
80. **Pate, J. V. and Duncan, E. N.**, Viability of cottonseed after long periods of storage, *Crop Sci.*, 4, 342, 1964.
81. **Peel, A. C.**, The effect of storage on the viability of crop seeds, *Proc. Int. Seed Test. Assoc.*, 27, 893, 1962.
82. **Rao, V. N. M., Rao, I. K. S., and Hassan, M. V.**, Studies on viability in cashew, *Indian J. Agric. Sci.*, 47, 289, 1957.
83. **Rincker, C. M.**, Effect of frequent thawing on viability of red clover seed in cold storage, *Crop Sci.*, 14, 749, 1974.
84. **Robertson, D. W. and Thornton, M. L.**, Longevity of sugar beet seed, *J. Am. Soc. Sugar Beet Technol.*, 13, 177, 1964.
85. **Roos, E. E.**, Storage behavior of pelleted, tableted, and taped lettuce seed, *J. Am. Soc. Hortic. Sci.*, 104, 283, 1979a.
86. **Roos, E. E.**, 1979, unpublished data.
87. **Rudolf, P. O. and Clausen, K. E.**, Red pine seed germination after 30 years of storage, *J. For.*, 60, 128, 1962.
88. **Sifton, H. B.**, Longevity of the seeds of cereals, clovers, and timothy, *Am. J. Bot.*, 7, 243, 1920.
89. **Simpson, D. M.**, Long-time storage of cottonseed, *Agron. J.*, 49, 608, 1957.
90. **Soloveva, M. A.**, On conditions for prolonged storage of fruit crop seeds, *Agrobiologiyz*, 1, 81, 1953.
91. **Stevens, O. A.**, Germination studies on aged and injured seeds, *J. Agric. Res.*, 51, 1093, 1935.
92. **Tiedemann, A. R. and Pond, F. W.**, Viability of grass seed after long periods of uncontrolled storage, *J. Range Manage.*, 20, 261, 1967.
93. **Toole, E. H. and Toole, V. K.**, Relation of temperature and seed moisture to the viability of stored soybean seed, *U.S. Dep. Agric. Circ.*, 753, 9, 1946.
94. **Toole, E. H., Toole, V. K., and Nelson, E. G.**, Preservation of hemp and kenaf seed, *U.S. Dep. Agric. Tech. Bull.*, 1215, 16, 1960.
95. **Visser, T. and Tillerkeratine, L. M.**, Observations on the germination and storage of tea pollen and seed, *Tea Q.*, 29, 30, 1958.
96. **Weibull, G.**, The cold storage of vegetable seed — further studies, *Agric. Hort. Genet.*, 13, 121, 1955.
97. **White, D. G.**, Longevity of bamboo seed under different storage conditions, *Trop. Agric.*, 24, 51, 1947.

Tobacco

HANDLING, STORAGE, AND MARKETING PRACTICES FOR TOBACCO

Albert H. Graves

INTRODUCTION

Tobacco was on the continent when America was discovered. The early settlers used valuable consignments to bring people to America. The London Company in a letter dated September 11, 1621.

By this Shipp and Pinance called the Tyger, we also send as many maids and young women as will make up the number of fifty, with those twelve formerly sent in the Marmaduke, which we hope shalbe received with the same Christian pietie and chantie as they were sent from hence; the providing for them at their first landing and disposing of them in marriage (which is our chief intent), we leave to your care and wisdom, to take that order as may most conduce to their good, and satisfaction of the Adventures, for the charges disbursed in setting them forth, which coming to twelve pounds and upwards, they require one hundred and fiftie of the best leafe tobacco for each of them; and if any of them dye there must be a proportionable addition upon the rest; this increase of thirty pounds is weight since those sent in the Marmaduke, they have resolved to make, finding the great shrinkage and other losses upon the tobacco from Virginia will not leave lesse, which tobacco as it shalbe received, we desire maybe delivered to Mr. Ed Blany, who is to keep thereof a particular account. We have used extraordinary care and dilligence in the choice of them, and have received none of them we have not had good testimony of their honest life and cariadge, which together with their names, we send them enclosed for the satisfaction of such as shall marry them; for whose further encouragement we desire you to give public notice that the next spring we purpose to send over as many youths for apprentices to those that shall now marry any of them and make us due satisfaction.

As tobacco moved further West it had to be returned to the great center Jamestown; accordingly, various methods were adopted to get the tobacco to market, some of which was sent by boats or canoes down the rivers, while some was conveyed in carts and wagons, while another method was rolling in hoops. Time and experience led to the introduction of horses. One method of transportation to market was recorded:

The hogsheads, which are designed to be rolled in common hoops, are made closer in the joints than if they were intended for the wagon; and are plentifully hooped with strong hickory hoops (which is the toughest kind of wood), with bark upon them, which remains for some distance a protection against the stones. Two hickory saplings are affixed to the hogshead, for shafts by boring on auger-hole through them to receive the gudgeons or pivots, in the manner of a field rolling-stone; and these receive pins of wood, with square tapered points, which are admitted through square mortises made central in the heading, and driven a considerable depth into the solid tobacco. Upon the hind part of these hafts, between the horses and the hogshead, a few light planes are nailed, and a kind of little cart body is constructed of a sufficient size to contain a bag or two of provender and provision, together with an axe, and such other tools as may be needed upon the road, in case of accident. In this manner they set out to the inspection (market) in companies very often joining society with the wagons, and always pursuing the same methods of encamping.

The quantity has grown until currently about 2 billion lb are grown annually in the U.S. and Puerto Rico. With an average yield around 2000 lb/acre, we plant around one million acres annually bringing the grower over $2.25 billion annually for the sale of tobacco.[3]

J. H. Cyrus and R. L. Mozingo state, "There are five different kinds of tobacco grown in the U.S."[4] They are

1. Flue-cured, used principally in cigarettes
2. Burley, used in cigarettes, chewing, and smoking
3. Fire-cured, used in snuff and chewing
4. Cigar filler and binders, used in cigars
5. Cigar wrappers grown under shade for cigars

If you will notice on any package of cigarettes, you will see the word "blended" which means that a certain amount of several types of tobacco are mixed together to make the filler for the cigarette. Flue-cured tobacco is the meat or body and mildness of the blend; burley tobacco is added for filler and burning qualities and Turkish for spice or aroma. Most cigarettes contain these tobaccos in different proportions, but in standard brands, the real secret of one blend differing from anothers lies in the flavoring or casing that is put on the tobacco in the blending process.[3]

The principal flavorings used in the blending process are: rum, glycerin, saccharine, maple sugar, brown sugar, apple honey, white honey, and alcohol. So, if you prefer one brand of cigarettes to another, it could be the flavoring that makes the difference. The average proportions of tobacco in cigarette blends are

1. Flue-cured — 50%
2. Burley — 30 to 40%
3. Maryland and Burley — 6 to 10%
4. Turkish — 6 to 10%[4]

Flue-cured tobacco accounts for one half the tobacco purchased. In 1976 about 660,640 acres were planted yielding 1910 lb/acre for a total production of 1,124,185 lb for a total crop value of $1,452,884.[5]

WAREHOUSE DESIGNATION

Tobacco is a perishable product and must be handled carefully after harvest and curing. To eliminate long lines of farm trucks waiting at tobacco auction warehouses in the flue-cured area, growers designate during March which warehouse they want to sell their tobacco when the tobacco marketing season begins.

The percent of the marketing quota a grower can sell per sale can be calculated as follows:

1. Designations to marketing area ÷ total designations = % of total designations to that marketing area.
2. % of total designations to that × total set = % of total sets to be used in that marketing area.
3. Total weekly capacity ÷ 4-day sales = daily sales rate per week.
4. 110% national quota ÷ daily sales rate = season days.
5. Designations to marketing area ÷ % of total sets to be used in that marketing area = % of designations that each set will sell in that marketing area.
6. Designations to market ÷ % of designations that each set will sell in that marketing area = (computer factor) % of set that can sell pounds designated to that market.
7. (Computer factor) % of set that can sell pounds designated to that market × season days = number of sales that market can have that season.
8. Pounds designated to ÷ number of sales that market can have that season = pounds to be sold each sale.[5]

The growers receive their marketing card that lists their allocation of the marketing quota. They then designate[6] which market and which warehouse in that market that they want to sell their tobacco.

They can then be scheduled to the warehouse as follows.

SALE DATE ____________ ____________ SALE UNLOADING DATE ____________

GROWER'S NAME	NUMBER OF SHEETS OR POUNDS SCHEDULED	NUMBER OF SHEETS OR POUNDS ACCUMULATED	SCHEDULED UNLOADING TIME	REMARKS

FIGURE 1. Warehouse scheduling sheet.

SCHEDULING

Since several factors affect scheduling, final details are worked out only 1 week in advance of sale. The Industry-Wide Flue-Cured Marketing Committee may set the selling time for a belt for 1 week or several weeks. Because of the amount of tobacco already in the marketing system, this procedure helps to balance the quantity of tobacco to be sold with the quantity that processing plants can handle. The scheduler should plan to receive in a week only the amount of tobacco that will be purchased and loaded out during the week. Most warehousemen will receive only enough tobacco for two sales.

Before beginning his weekly scheduling, the scheduler must determine the rate at which tobacco can be received at the warehouse. An efficient warehouse operator can easily attain a receiving rate of 150 sheets per hour.

We recommend that the scheduler use scheduling sheets similar to the one shown in Figure 1. These sheets can be kept in a standard looseleaf notebook. If fieldmen are used, they may list the information on these sheets; however, the scheduler is responsible for the accuracy of this information.

Our study showed that the scheduler should confirm the schedule each Monday for the following week by contacting growers either by telephone or in person, making any necessary adjustments. Upon contacting the first growers, the scheduler fills out the appropriate spaces in the scheduling sheet to show the grower's name, the date and time of delivery, the number of sheets to be delivered, and the time of the sale during which this tobaco will be sold. The scheduler then fills out a scheduling card (Figure 2) and gives it to the grower as notice of when the grower is scheduled to deliver tobacco for a specific sale, and the amount of tobacco he is to deliver.

The scheduler continues to confirm schedules with the growers until approximately 90% of the sales allocation has been scheduled. The time increments in the schedule are broken on approximately 150 sheet intervals. The first 150 sheets scheduled will be unloaded during the first working hour; the second 150 sheets during the second working hour, and so on until the daily allotment is reached.

CAROLINA WAREHOUSE

Grower's Name ______________________

Number of Sheets ______________________

Unloading Date and Time ______________________

Sale Date ______________________

Sale ______________________

Scheduling Card

FIGURE 2. Scheduling cards to be filled out by scheduler and given to grower.

Table 1
CREW SIZE, TIME REQUIREMENTS, AND EQUIPMENT AND LABOR COSTS FOR RECEIVING 1000 SHEETS OF TOBACCO BY THE CONVENTIONAL METHODS

			Labor and equipment required		Labor and equipment and costs ($)		
Method	No. in crew	Elapsed time (hr)	Labor (man-hr)	Equipment (machine-hr)	Labor	Equipment	Total
Conventional method	18	7.17	129.06	—	226.27	43.00	269.27

As the scheduler contacts growers, he will complete scheduling sheets covering all of the days for 1 week in advance.

One week later, the scheduler should refer to his table of weekly group rotations (Table 1) and rotate the growers by groups. The scheduler also should rotate before scheduling for the second week the arrival times for growers for each day, so that each grower will have an equal opportunity to benefit from any advantage gained by a particular location of his tobacco on the sales floor or by a particular selling time.

SCHEDULING THROUGH FIELDMEN

Some tobacco markets receive growers' tobacco from a large geographic area, and the warehouseman (or scheduler) often is unable to contact each grower personally. Fieldmen, then, may be hired to help in contacting growers in the outlying areas.

A warehouseman who uses fieldmen should:

1. Determine the number of fieldmen necessary for his warehouse operation.
2. Assign each fieldman a specific geographic area in which to work.
3. Prepare a rotation (Figure 3) for each fieldman to provide for rotating the week and the day of the week that each grower group will deliver tobacco.
4. Assign each fieldman part of the daily sale allotment to be scheduled for delivery to the warehouse.
5. Assign each fieldman part of the daily receiving time.
6. Assign each fieldman a number (1 through 4) and instruct him to divide his customers into four groups (A-D, E-H, I-L, or M-P).

Week	Monday	Tuesday	Wednesday	Thursday
	Group	*Group*	*Group*	*Group*
1	A	B	C	D
2	B	A	D	C
3	D	C	B	A
4	C	D	A	B
5	A	B	C	D
6	B	A	D	C
7	D	C	B	A
8	C	D	A	B

FIGURE 3. Weekly rotation of groups through 8 weeks with all growers selling on the same number of Mondays.

Unloading week and fieldman (by No.)	Unloading time [1]	Unloading day			
		Monday	Tuesday	Wednesday	Thursday
		Group	*Group*	*Group*	*Group*
1st and 5th week:					
1..........	7:00 a.m. to 9:30 a.m.	A	B	C	D
2..........	9:30 a.m. to 12:00 p.m.	E	F	G	H
3..........	1:00 p.m. to 3:30 p.m.	I	J	K	L
4..........	3:30 p.m. to 6:00 p.m.	M	N	O	P
2nd and 6th week:					
4..........	7:00 a.m. to 9:30 a.m.	N	M	P	O
1..........	9:30 a.m. to 12:00 p.m.	B	A	D	O
2..........	1:00 p.m. to 3:30 p.m.	F	E	H	G
3..........	3:30 p.m. to 6:00 p.m.	J	I	L	K
3rd and 7th week:					
3..........	7:00 a.m. to 9:30 a.m.	L	K	J	I
4..........	9:30 a.m. to 12:00 p.m.	P	O	N	M
1..........	1:00 p.m. to 3:30 p.m.	D	C	B	A
2..........	3:30 p.m. to 6:00 p.m.	H	G	F	E
4th and 8th week:					
2..........	7:00 a.m. to 9:30 a.m.	G	H	E	F
3..........	9:30 a.m. to 12:00 p.m.	K	L	I	J
4..........	1:00 p.m. to 3:30 p.m.	O	P	M	N
1..........	3:30 p.m. to 6:00 p.m.	C	D	A	B

[1] Based on receiving of 150 sheets of tobacco per hour in a 10-hour work day.

FIGURE 4. Recommended schedule for the rotation of unloading times for 16 grower groups contacted by four fieldmen.

7. Instruct each fieldman to refer to the rotation provided to him (Figure 4) to determine the amount of tobacco that each grower in a group will sell each week, and then to proceed with his scheduling, accordingly.

The warehouseman provides each fieldman with two sets of forms for each sale day: (1) a scheduling sheet to list the growers in one group, the number of sheets the group will deliver, and the date and time of delivery and (2) a scheduling card to fill out for the grower.

Example

Assuming a warehouseman operates under an allotment to sell 275,550 lb of tobacco at each sale. He is allowed four sale days (Monday through Thursday) each week for an 8-week marketing season. The warehouseman converts the pounds of tobacco for sale to 1650 sheets of tobacco (275,550 pounds ÷ 167 lb/sheet). He decides to "book"

1500 sheets, which is approximately 90% of the sale, through four fieldmen. Each fieldman is responsible for scheduling the arrival of 375 sheets (1500 sheets ÷ four fieldmen). The warehouseman has handling equipment and labor available to receive and floor 150 sheets of tobacco each working hour; therefore, he can plan to receive the 1500 sheets of tobacco in the sale within ten working hours. On this basis, the 375 sheets of tobacco scheduled by each of the four fieldmen will represent 2½ hr of the total unloading time for the sale (10 hr ÷ four fieldmen).

The warehouseman prepares and distributes a table of recommended weekly group rotation to each of the four fieldmen. He then assigns each fieldman a number (1 through 4) and four groups of growers (A-D, E-H, I-L, or M-P). Using the table, fieldman No. 1 determines his part in scheduling for the entire marketing season in his assigned area. He schedules growers in group A to deliver a total of 375 sheets of tobacco between 7:00 a.m. and 9:30 a.m. on Monday of the first unloading week, to deliver the same amount between 9:30 a.m. and noon on Tuesday of the second week, and so on, according to the group rotation table provided him. Likewise, the other fieldmen determine their responsibility and proceed with their scheduling. They then contact the growers in their group and fill out their scheduling sheets and cards.

RECEIVING

For the scheduling system to operate successfully, the grower must not be allowed to enter the warehouse before his scheduled time. When he arrives, he should park his truckload of tobacco outside the warehouse taking special care not to block the entrance door. He then takes his scheduling card to the scheduling desk to determine his sequence of entry.

After the grower checks in with the scheduler, he drives his truck into the warehouse and stops at the desk near the entrance. He shows his scheduling card to the ticketmaker who ascertains from the card the number of sheets of tobacco the grower has scheduled. The grower then gives his ASCS marketing card to the ticketmaker, who imprints the marketing card information and the name and number of the warehouse on the tobacco sale bill (Figure 5). The ticketmaker completes the tobacco sale bill, except for weights of the sheets of tobacco. He then fills out a scale ticket (Figure 6) for each sheet to be sold, as indicated on the tobacco sale bill. This procedure minimizes delays at the scale when the grower's tobacco is weighed.

After leaving the ticketmaker's desk, and while waiting to move to the unloading dock, the grower should remove the tarp or plastic sheets and any unused tobacco replacement sheets from his truck so that the tobacco will be ready to unload. When the grower's truck is one or two trucks back from the unloading dock, he should arrange his tobacco sale bill and scale tickets to match the order in which the sheets are to be unloaded. He then gives them to the weighmaster.

The grower should be alert to position his truck for unloading as soon as the truck ahead is clear of the unloading dock. Failure to move forward promptly delays tobacco receiving more than any other factor and creates an uneven flow of tobacco through the receiving system.

As soon as the grower positions his truck at the unloading dock, he should leave his truck immediately and tell the unloading crew which sheets of tobacco to take off first and any other appropriate instructions.

After the last sheet is unloaded from the truck, the grower should drive his empty truck forward, away from the unloading dock, so that the next grower can drive his loaded truck into position without delay. After the grower parks his truck in the area provided, he can then return to the weighmaster and obtain either a customer's copy of the tobacco sale bill or a corresponding receipt form. Most warehouses require

BILL NO.

853

BALANCE BEFORE SALE
(Based on 110% of Quota)
Lbs.

TOBACCO SALE BILL

UNTIED ☐

TIED ☐

SALE DATE

STATE | COUNTY | FARM NO.

NAME AND ADDRESS OF WAREHOUSE

OPERATOR SHARE %

TENANT — FC NO. — SHARE %

NAME AND ADDRESS *(Exactly as shown on identification card)*

LINE	BASKET NO.	BUYER AND GRADE		LINE	POUNDS	PRICE		AMOUNT	
1				1				$	
2				2					
3				3					
4				4					
5				5					
6				6					
7				7					
8				8					
9				9					
10				10					
11				11					
12				12					

BALANCE AFTER SALE
(Based on 110% of Quota) →
Lbs.

← TOTALS → $

REMARKS:

G.A.C.C.

Whse. Charges

Auction Fees

Commissions (2½%) $

Advances

Tobacco Associates

$

DELIVERED TO STABILIZATION

POUNDS

AMOUNT
$

BILL NO.

853

WAREHOUSE COPY

FIGURE 5. Sample tobacco sale bill.

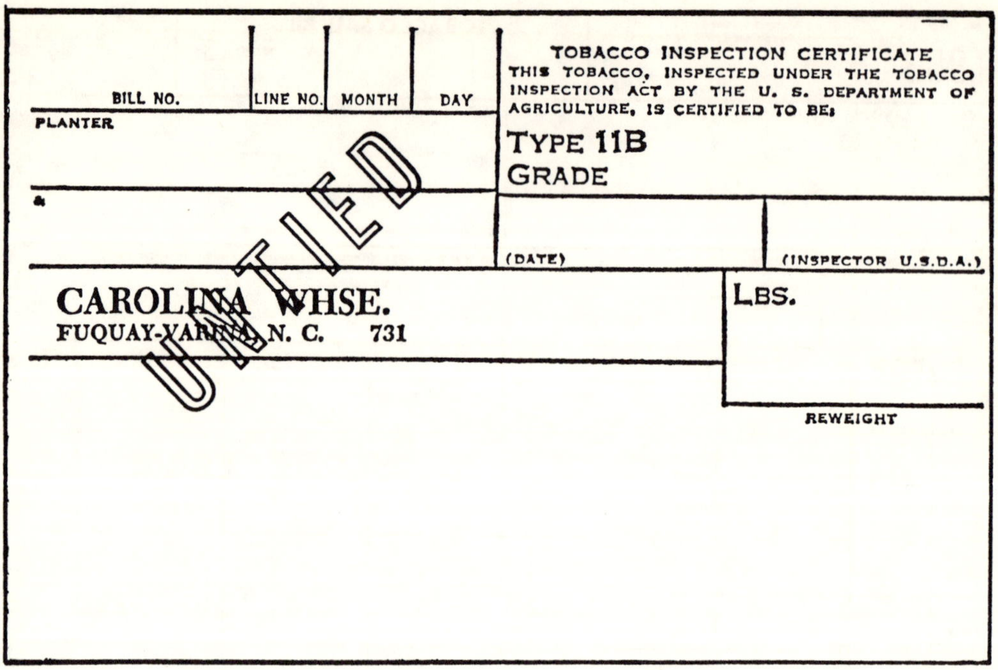
BILL NO. | LINE NO. | MONTH | DAY
PLANTER
&
TOBACCO INSPECTION CERTIFICATE
THIS TOBACCO, INSPECTED UNDER THE TOBACCO INSPECTION ACT BY THE U. S. DEPARTMENT OF AGRICULTURE, IS CERTIFIED TO BE:
TYPE 11B
GRADE
(DATE)
(INSPECTOR U.S.D.A.)
CAROLINA WHSE.
FUQUAY-VARINA, N. C. 731
LBS.
REWEIGHT
UNTIED

FIGURE 6. Scale ticket for one sheet of growers' tobacco.

verification before replacement sheets can be obtained from the sheet room. When the grower obtains the replacement sheets, he has completed the receiving operation.

Tobacco can be unloaded onto the sales floor using hand trucks as shown in Figure 7.

The time breakdown for an 18-man crew to unload three sheets of tobacco using handtrucks and baskets is shown in the following tabulation:

Productive Labor[a]

	Man-min
Unload sheets of tobacco (four-man crew)	5.16
Fill out headings on tobacco sale bill and scale tickets (weighmaster)	0.39
Weight three baskets of tobacco and record weights (weighmaster)	0.90
Push hand truck loads of tobacco to scales to sales floor and unload, and return to truck (12-man crew)	8.01
Position three baskets on sales floor and untie sheets as time permits (hook man)	1.17
Total productive labor	15.75

Unproductive Labor

Employees with hand trucks wait (12-man crew)	7.47
Total labor	23.22[a]

[a] Allowances for fatigue and personal time were 5% each, except for unloading, where 25% was allowed, and pushing hand trucks, where 10% was allowed. Elapsed minutes were 1.29.

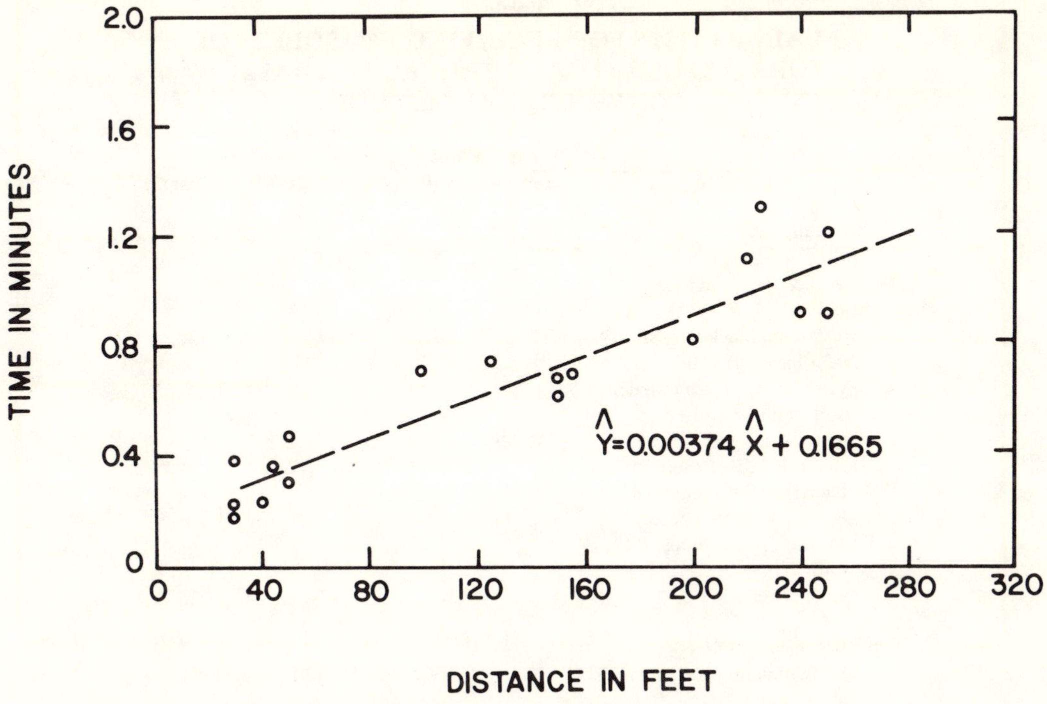

FIGURE 7. Time required to manually push a handtruck a specific distance.

The labor and equipment cost for receiving tobacco using hand trucks is shown in Tables 1 and 2.[8]

For a warehouse that wants more automation and lets conveyors do the work a powered belt conveyor receiving system was designed. It consisted of an unpowered pivot section to allow sheets to gravitate from the farm truck to the belt conveyor, two 20-ft powered belt conveyors, a 5-ft section of gravity roller equipment for the scale, and a special conveyor section that was attached to the forklift for carrying sheets of tobacco to the floor. This forklift conveyor section was long enough to carry three sheets at one time.

The pivot section was made by connecting to each end of a 10-ft section of gravity roller conveyor a 45° curved section with an inside radius of 2½ ft. The section was then mounted on heavy duty swivel casters with a pivot about which the entire section could rotate. The pivot was mounted solidly in the floor to prevent the section from bumping the conveyor belt.

Each steel conveyor roll was 2½ in. in diameter, 14-gauge, 36 in. long, with a 7/16-in. hex shaft. The rolls, each with a capacity of 200 lb, were spaced on 4-in. centers, and set high in a 3-in.,4.1-lb/ft structural channel frame. The forklift section was the same. The 20-ft belt conveyor consisted of the same basic frame with 1.9-in.-diameter rolls set high on 6-in. centers. A 190-rpm output drive unit with a 1-hp 220-V, three-phase AC 60-cycle reversing motor was used to power the 30-in., 3-ply, 28-oz rubber claw-top belting; with 1.9-in.-diameter return rolls mounted on 5-in. centers. Belt conveyors were controlled by magnetic reversing motor starters with foot switches to start and stop the belts.

All the warehouses using the belt conveyor receiving system used a chain hoist to unload the sheets from a truck and place them on the conveyor line. A 64-ft/min electric chain hoist equipped with a class-H motor (220 V, three phase) was mounted on a 20-ft 6-in. standard I beam. Since the hoist was operated almost continuously

Table 2
LABOR COSTS FOR RECEIVING 1000 SHEETS OF TOBACCO USING HAND TRUCKS AND BASKETS (CONVENTIONAL METHOD)

Operation	No. in crew	Elapsed time (hr)	Labor (man-hr)	Cost ($)
Unloading				
Place sheets of tobacco in baskets on hand trucks	4	7.17	28.68	45.8[a]
Trucking				
Push hand truck with sheet of tobacco to scale, to sales floor, return to truck and wait as required	12	7.17	86.04	137.66[a]
Weighing				
Fill out headings on tobacco sale bill, scale tickets and weigh tobacco	1	7.17	7.17	31.25[b]
Flooring; positioning and untieing	1	7.17	7.17	11.47[a]
Total			129. 06	226.27

[a] Labor rate at $1.60/hr.
[b] Weighmasters pay rate = $125/week ÷ 4 days/week × 7.17 hr/day = $4.36/hr.

and the control switch was operated nearly 10,000 times per day, a heavy duty motor was required.

The belt conveyor receiving system works like an assembly line. An operation occurs at each station simultaneously and sheets move in sequence past all stations as each preassigned function is performed. Descriptions of activities at each of these stations follows.

Positioning truck— When a grower enters the warehouse with a truckload of tobacco, a warehouse employee fills out headings on the sales bill and scale tickets and moves him forward in line. The grower then waits his turn for unloading. Immediately after the last sheet of tobacco is removed from the truck just ahead of the grower, he pulls his truck into the unloading position. He gives the sales bill and scale tickets to the weighmaster. After the grower's last sheet is unloaded his truck is immediately removed from the unloading dock to make room for the next truck in line. The grower normally moves his own truck; however, if he is occupied with the weighmaster one of the unloading crew moves the truck. Timely positioning of trucks in and out of the unloading dock is of paramount importance since any delay at this station disrupts the even flow of tobacco through the system.

Unloading sheets— A two-man crew using a chain hoist can efficiently unload sheets of tobacco from the growers truck. The work of these two men must be closely coordinated. One man standing in the truck hooking the sheets and one man standing on the warehouse floor operating the hoist controls can unload about four sheets per minute. The man in the truck hooks a sheet, then the crew member operating the control raises the sheet and simultaneously pulls the chain hoist, with sheet attached, toward the rear of the truck. The sheet acts as a pendulum and as the combined forces of gravity and momentum swing the sheet out beyond the rear of the truck and over the conveyor the hoist operator lowers the sheet onto the conveyor and the crew member in the truck quickly pulls the hook toward the front of the truck releasing the sheet. The sheet moves by gravity down the gravity roller conveyor section toward the belt conveyor. The crew member operating the controls then pushes the "down" button, letting out the chain as the crew member in the truck reaches to hook another sheet.

When the last sheet has been removed from the truck each crew member has an assigned responsibility that assists in speeding the positioning of the next truck. The man operating the controls pushes the pivot section (unloading dock) out of the traffic pattern in the driveway allowing the next truck to pull forward, while simultaneously holding the controls and hook with his right hand to keep them from becoming entangled with the truck being positioned for unloading. If the truck driver is in the truck that crew member helps roll the pivot section of the conveyor out of the driveway. With the unloading operation functioning properly there should be sufficient sheets stored on the belt conveyor to supply the scales during the time required to change trucks.

The crew should then be alert to direct the driver of the next truck into the unloading position. The crew member preparing to push the pivot section in position behind the truck is logically the person to tell the truck driver when to stop. Once the loaded truck is in position, the crew should immediately begin unloading the sheets to provide the weighman with a continuous supply of sheets.

Weighing— Using the belt conveyor system for receiving tobacco, the weighmaster controls the movement of tobacco sheets onto the scale with two footswitches. One switch moves the belt forward and positions a loaded sheet on the scale. The other switch reverses the belt and can move a sheet that may be resting on the scale, backward and clear of the scale. With these switches, the weighmaster has complete control of the weighing operation.

Loading the forklift— The weighmaster's helper stands beside the scale and operates a second belt conveyor that loads the forklift, which is used for carrying three sheets of tobacco, or the forklift may be loaded by the forklift operator using the pendant switch. This second belt conveyor removes the sheet from the scale and loads the forklift without additional help from the crew. This is not always possible with a gravity system.

Positioning sheets on warehouse floor— The "floor man" positions the sheets on the sales floor as they are removed from the forklift. At some warehouses the floor man staples the ticket on the sheets of tobacco as they are placed on the sales floor. At others, the weighmaster's helper staples the ticket to the sheet immedately after the tobacco is weighed.

The labor required for this operation is shown in Table 3. The equipment ownership and operating costs for this equipment are shown in Table 4. The combined labor and equipment cost for this operation is shown in Table 5.[10]

After the sale the tobacco must be picked up for the tobacco company that purchased the tobacco. The following equipment and procedure may be used to jack the tobacco.

A standard 3000-lb capacity forklift was used in conjunction with two powered belt conveyors supported by a frame attached to the forks of the truck. This unit, called a "sale breaker", was used to remove the sheets of untied flue-cured tobacco from the tobacco auction warehouse sales floor and to transport them to a jacking station. At the jacking station, the sheets of tobacco were placed on a pair of inclined, gravity, roller conveyors, pulled to the end of the conveyor, and stacked two high on jacks.

Each powered belt conveyor consisted of a 12-ft section of conveyor with a 4-in. structural steel channel frame, 36 in. long × 1.9-in.-diameter rollers with 7/16-in. hex shafts mounted on 6-in. centers. Belting consisted of 30-in.-wide, 3-ply, 28-oz, rubber-covered claw top belt.

Power for the belts was supplied by the hydraulic system on the forklift. Maximum belt speed with one belt running was 300 ft/min and 150 ft/min with both belts run-

Table 3
LABOR REQUIRED FOR EIGHT MEN TO RECEIVE THREE SHEETS OF UNTIED TOBACCO

Operation element	Labor required (man-min.)
Productive labor	
1 man writes identification and bookkeeping data on scale tickets and sales bill	0.45
2 man unloading crew places presheeted tobacco from farm truck onto conveyor line with chain hoist	1.56
Weighmaster's helper starts belt and bring sheet of tobacco onto scale	0.33
Weighmaster weighs tobacco listing weight on scale ticket and sales bill	0.36
Forklift loads and travels 150 ft, unloads three sheets of tobacco, and returns	1.24
Hook man positions three sheets on floor	0.84
Total productive labor	4.78
Unproductive labor	
Man filling out scale tickets and floor sheets waits	0.57
2-man unloading crew waits on positioning truck	0.48
Weighmaster's helper waits on sheets	0.69
Weighmaster waits on helper to bring sheets onto the scale	0.66
#2 forklift driver waits on weighman	0.80
Hook man waits on forklift	0.18
Total unproductive labor	3.38
Total labor, man-min	8.16
Elapsed minutes	1.02

ning. Maximum load conditions consisted of moving 600 lb of tobacco up an incline of 10° at a speed of 300 ft/min requiring 2.3 hp. The hydraulic motors selected to run the belts had a maximum rated capacity of 10.5 hp. The pump of the forklift hydraulic system that supplied fluid to the hydraulic motor had a rated output of 12 to 15 gal/min. With the hydraulic motors in a parallel circuit, each motor received 6 to 7 gal/min when running simultaneously. The motors selected produced approximately 895 in.-lb of torque and 125 rpm given a flow rate of 6 gal/min and a pressure of 750 lb/in.^{2} The torque required for the belt conveyors with a 6-in. drive sprocket was 870 in.-lb.

SALES BREAKING

The salebreaker carries sheets of tobacco to a jacking station consisting of two gravity roller conveyors which were aligned with the powered belt conveyors on the forklift. Each conveyor section was 12 ft long and 3 ft wide. Each roller was 2.5 in. in diameter, 36 in. long, with a 7/16-in. hex shaft. The rollers, each with a capacity of 200 lb were spaced on 4-in. centers and set high in a 3-in., 4.1-lb/ft structural frame. The supports for the conveyor sections were fabricated of 3 × 2 × 3/16 in. structural steel angles equipped with regular feet. One end of each conveyor was positioned 51 in. above the floor and the other end was 39 in. above the floor. The resulting downward slope of 4.75° was sufficient to permit moving each sheet with little effort, but was not sufficient to allow them to roll off the conveyor when left unattended.

Table 4
ESTIMATED ANNUAL COSTS OF OWNERSHIP AND OPERATION OF SELECTED ITEMS OF EQUIPMENT FOR UNLOADING SHEETS OF TOBACCO AT TOBACCO AUCTION WAREHOUSES BY ANNUAL VOLUME

Annual volume of tobacco and type of equipment	Number required	Ownership	Operating cost				Total annual cost
			Power	Maintenance	Total	
		($)	($)	($)	($)	($)
3 Million lb						
Belt conveyor	1	1055	6	17	23	1078
Chain hoist	1	261	1	50	51	312
Forklift	2	—	157	—	152	2157
					Total	3547
5 Million lb						
Belt conveyor[a]	1	1055	10	20	30	1085
Chain hoist[b]	1	261	2	75	77	338
Forklift[c]	2	—	262[b]	—	262	2262
					Total	3685
7 Million lb						
Bell conveyor[a]	1	1055	14	23	37	1092
Chain hoist[b]	1	261	2	100	103	364
Forklift[c]	2	—	367[b]	—	367	2367
					Total	3823

[a] Ten-year expected life, $6000 replacement cost, 10% depreciation.
[b] Five-year expected live, $900 replacement cost, 20% depreciation, 5% interest, 3% insurance, and 10% taxes.
[c] We are examining renting this equipment for 2.5 months, $400.00 per month = $400 × 2 forklifts × 2.5 months = $2000.

After the sale, the sheets are retied. The sale breaker starts in the center of rows 1 and 2, picks up, and transports six sheets of tobacco—three from each row—to a jacking station positioned about halfway down rows 1 and 2. The forklift moves into the two rows of sheets with the conveyor belts tilted forward permitting the moving belts to lift and roll under the sheets of tobacco as the forklift advances; while the forklift is transporting the tobacco back down the sales row, the forks are raised to the elevation of the gravity conveyors. The sheets of tobacco are moved onto the gravity conveyor sections by powered belt conveyors on the forklift. The cycle is then repeated until the sheets from rows 1 and 2 have been removed. The entire operation is continued until all the sheets of tobacco have been removed from the sale floor.

While the sale breaker reloads, four crew members at the distribution area move the sheets of tobacco from the gravity conveyors onto jacks. Very little labor is required, for the two men at the end of each gravity conveyor ramp, to handle the 200-lb sheets of tobacco on the down-sloping conveyors. The crew members place a jack on the end of the gravity conveyor ramp and guide the sheet onto the jack; the second sheet, if it belongs to the same company, is placed on top of the first sheet. In instances where consecutive sheets do not belong to the same buying company, the "single sheet" jack is moved aside. These singles remain out of the way until they can be matched with other sheets for the same company, jacked two high, and moved to the load-out area.

A jack is ready to move to a company loadout area after the second sheet is placed on the jack. Two men are assigned the responsibility of moving the loaded jacks to the loadout area. The distance required to push the jacks of tobacco are seldom over 100 ft. This is made possible by starting close to the loadout area (row 1 and 2) and moving the jacking station across the center of the warehouse. This short distance is

Table 5
COMBINED LABOR AND EQUIPMENT COSTS FOR UNLOADING SHEETS OF TOBACCO AT TOBACCO AUCTION WAREHOUSES BY VOLUME

Equipment Costs

Tobacco handled annually (lb)	Total annual cost($)	Cost per sheet (¢)	Cost per 1000 lb[a] (¢)
3,000,000	3,547	21	1.18
5,000,000	3,685	13	74
7,000,000	3,823	10	55

Labor Costs

Tobacco handled annually (lb)	Total annual cost ($)	Cost per sheet (¢)	Cost per 1000 lb[a] (¢)
3,000,000	2,280.00	14	76
5,000,000	3,800.00	14	76
7,000,000	5,320.00	14	76

Combined Labor and Equipment Costs

Tobacco handled annually (lb)	Total annual equipment cost ($)	Total annual labor cost ($)	Combined annual labor and equipment cost ($)	Cost per sheet (¢)	Cost per 1000 lbs ($)
3,000,000	3,547	2,280.00	5,827	35	$1.94
5,000,000	3,685	3,800.00	7,485	27	1.50
7,000,000	3,823	5,320.00	9,143	24	1.31

[a] The average weight of one sheet is assumed to be 180 lb.

significant; as the cluster of jacks around the loadout area increases in number, the sale breaker clears away rows of tobacco providing additional space for the increasing number of dollies waiting to be loaded out.

LOADOUT

Where the layout necessitates movement across long distances in the warehouse, a forklift may be used to push the tobacco from the jacking station to the loadout area.[11] Tobacco stored temporarily in the loadout area will have to be loaded onto semitrailers that may be performed as follows. During the past three marketing seasons a system for loading tobacco has been designed, built, modified, and used commercially. A standard 3000-lb capacity forklift with auxiliary attachment was used as the prime mover with a powered belt conveyor supported by a frame attached to the forks of the truck. This unit, called a "tobacco mover" was used to lift the sheets of tobacco from the auction warehouse floor, transport them to the semitrailer, and run each sheet onto the truck bed as needed by the crew loading the truck.

The powered conveyor consisted of a 12-ft section of conveyor with a 4-in. structural steel channel frame and 36-in. long × 1.9-in. diameter rollers with 7/16-in. hex shafts mounted on 6-in. centers. Belting consisted of a 30-in. wide, 3-ply, 28-oz, rubber-covered claw belt.

Power for the belt was supplied by the hydraulic system on the forklift. Maximum belt speed was 300 ft/min. Maximum load conditions consisted of moving 600 lb of tobacco up an incline of 10° at a speed of 300 ft/min and required 2.3 hp. The hydraulic motor selected to run the belt had a maximum rated capacity of 10.5 hp. The motor selected produced approximately 895 in.-lb of torque and 125 rpm given a flow rate of 6 gal/min and a pressure of 750 lb/in.^{2} The pump of the forklift hydraulic system that supplied fluid to the hydraulic motor had a rated output of 12 to 15 gal/min. The torque required for the belt conveyors with a 6-in. driven sprocket was 870 in.-lb.

The tobacco mover was used by the tobacco moving contractor in the Fuquay Tobacco Market during the 1973, 1974, and 1975 seasons. Marketwide tobacco moving activities were controlled from the control warehouse.

A semitrailer that arrived in the market usually went directly to the control warehouse for directions about where to load and when the crew would be available. The tobacco moving contractor was contacted and in turn he directed the driver of the semitrailer to the warehouse with the most tobacco of the grade being shipped (for example, the truck driver might be looking for a Phillip Morris grade purchased by J. P. Taylor).

During the 1973 tobacco marketing season sheets of tobacco were moved from the sale floor to the warehouse loadout door and piled near the door in random fashion. This method of operation necessitated the use of a two-man crew to position the sheets of tobacco in a straight line before the tobacco mover could lift them off the floor.

In 1974 the tobacco moving contractor removed sheets of tobacco from the mechanized sales system, and lined the sheets of tobacco by grade near the loadout door. That prepositioning eliminated the need for two men to position the sheets of tobacco in a straight line for the tobacco mover.

In 1975 sheets of tobacco were lined up by company and by grade as the tobacco was moved from the warehouse sales floor to the loadout dock. Upon arrival at the designated warehouse the semitrailer is positioned at the loadout door. Several operations are then necessary before the sheets are loaded onto the semitrailer: the scale ticket must be punched, marked, and the coupon pulled. For loading the semitrailer, the tobacco mover is moved forward to pick up three sheets. With the conveyor belt tilted forward the moving belt lifts and rolls under the sheets of tobacco as the forklift advances. The forks are raised while the forklift is transporting the sheets of tobacco to the semitrailer. The sheets of tobacco are lifted over the side of the semitrailer. The two men on the the semitrailer then position the sheets on the truck as the belt moves each sheet forward.[1]

MECHANIZED SALES

Warehouses that want to have an assembly line type operation permitting them to perform the operations of unloading, weighing, grading, selling and loading out in sequence may install a mechanized sales system. The following is a discussion of that operation that occurred in 1974.

Three receiving lines were utilized in the mechanized sales system. Each line consisted of a belt-conveyor receiving system, scales, a grading station, and an accumulation belt. Components of each belt-conveyor receiving station included a chain hoist, a gravity conveyor, two belt conveyors, and scales. The hoist was used to remove the sheets of tobacco from the grower's truck and to place them on the gravity conveyor. Sheets moved on the gravity conveyor to a belt conveyor that moved to the scales for weighing. The belt conveyor then moved the sheets to a section of gravity conveyor for grading. After each sheet of tobacco was graded, it was moved forward on the accumulator belt until all of a grower's crop had been graded. The three lines were identical except that the accumulator belts were of different lengths to accommodate different quantities of tobacco. A swing belt moved the sheets of tobacco from the three accumulator belts to a single sale belt. The accumulator belts and sale belt were controlled by the auctioneer at a rate between 0 and 825 sheets per hour (50 FPM). The last 10 ft of the sale belt sloped from a height of 10 in. to 33 in., so that the sheets were placed onto a belt 33 in. high. From the sale belt the tobacco moved onto a faster moving (100 FPM) belt, 75 ft long, which separated the sheets for ease of removal by the tobacco-moving contractor. A gravity conveyor at the end of the belt conveyor accumulated the sheets of tobacco that had not been removed from the line.

A detailed list of equipment used in the mechanized sales system is given in Table 6.

The Mechanized Sales System required a high degree of coordination between the personnel operating within various sections of the system. Not only was the coordination among personnel of several warehouses in the Fuquay-Varina market required, but also that between the three major segments of the industry: growers, warehousemen, and tobacco companies. The growers were told when to arrive at the warehouse with their tobacco in accordance with the sale schedules issued by the Industry Wide Marketing Committee and subsequently by the Fuquay-Varina Board of Trade. There were 83 growers who sold their tobacco at the Faquay Auction Center, and scheduling of these growers throughout the season was made as equitable as possible. Some growers sold all their tobacco at the Fuquay Auction Center; others sold only a small amount to "try out" the new system. Therefore, during the season, some growers sold only once, while others sold as many as 22 times. The average number of trips for all growers was eight. The quantity of tobacco they brought on each trip was also a function of the schedule established by the Industry Wide Marketing Committee. Each grower was scheduled to arrive at the warehouse approximately 20 min before the sale. Warehouse labor required to operator the mechanized sales system was approximately 24 man-hr/100,000 lb of tobacco. This is approximately one half the warehouse labor required in a conventional warehouse.

Men at the entrance door filled out the headings on the sale bills and on the scale tickets. The truck driver gave these bills and tickets to the weighmaster. Two men

Table 6
EQUIPMENT USED IN THE MECHANIZED SALES SYSTEM

Receiving systems — required
Chain hoist
Duff Norton — Tobacco Special-220 VAC/3P/60 Hz. 1/8 ton 64 ft/min
15-ft chain with to tobacco hook
11-ft push button cord (standard) or
14-ft push button cord (16 ft entrance door)
I Beam
6-in. standard, 20 ft length
Gravity conveyor
5-ft section (behind truck)
Conveyor rolls
2.5 in. in diameter, 36 in. long with 7/16 in hex shaft set high
Frame
3 in., 4.1 lb/ft structural steel
Supports — 4 required
3 in. × 2 in. × 3/16 in. structural steel

Belt conveyors — 2 required per receiving system
36¼-in. wide bed 20 ft length overall with #24 rolls — 1.9 in. diameter mounted on 6 in. centers — set high— 170 FPM — reversing
Return rolls #24 1.9-in diameter spaced on 5-ft centers
Logan Adapto, Drive No. 500 — two required
Drive pulley — 8-in. diameter — lagged crowned
End pulley — 4-in. diameter — machine crowned
Belting — 30-in. wide, 3-ply 28-oz rubber clawtop 45 ft long — each
Motors — 1½ hp, 175 rpm right angle gear motor totally enclosed for 220 VAC/3P/60 Hz.
Controls — magnetic reversing motor starter. NEMA type/enclosure NEMA size 0
Electric eyes — two for semi-automation
Scale section — 4 ft length of gravity as listed above under gravity conveyor
Switches — 4 start-stop foot switches (two for each conveyor)

Grading station — 3 required
10-ft section gravity roller conveyor as listed above under Gravity conveyor

Storage conveyors — 3 required
44¼-in. wide bed with 44-in. rolls — 1.9-in. diameter mounted on 6-in. centers set high — 50 FPM — nonreversing
Logan Adapto Drives
No. 650 drive pulley 10-in. dia meter, lagged crowned. End pulley 4 ¼-in. diameter machine crowned
No. 500 — 2 required
As listed above under belt conveyor
Belting
42-in. wide 3-ply 28-oz transmission 207 ft long (for 100 ft long overall with No. 650 drive)
42 in. wide 3 ply 28 oz transmission 152 ft long (for 73 ft 6 in. long overall with No. 500 drive)
42-in. wide, 3-ply 28-oz transmission 105 ft long (for 50 ft long with No. 500 drive)
Motors
1½ hp DC gear motors — 3 required
Electric eyes
Four for semiautomation

Swing belt
44¼-in. wide bed with 44-in. rolls — 1.9-in. diameter 16 ga., mounted on 6-in. centers set high in 4 in. — 5.4 # structural frame. 60 FPM nonreversing
Logan Adapto Drive
No. 350-SF drive pulley 6-in. diameter lagged crowned
End pulleys 3½-in. diameter machine crowned
Return rolls 1.9-in. diameter, 16 ga. on 5-ft centers

Table 6 (continued)
EQUIPMENT USED IN THE MECHANIZED SALES SYSTEM

Conveyor mounted on casters with pivot for directing sheets from storage conveyors with 24-in. elevation to 18-in. elevation for discharge onto sale belt
Motor ½ hp 70-rpm right angle gearmotor totally enclosed for 220 V/3P/60 Hz
Belting
42-in. wide, 3-ply, 28-oz transmission 29 ft long

Sale belt
44¼ in. wide bed with 44-in. rolls — 1.9-in. diameter, 16 ga., mounted on 6-in. centers, set high return rolls, 1.9 in. diameter, spaced on 5-ft centers
Logan Adapto Drive No. 650
Drive pulley — 10-in. diameter, lagged crowned
End pulley — 4 1/4-in. diameter, machine crowned
Elevation
12 in. top of belt for 100 ft 44-in. to top of belt discharge (17 in. elevation last 12 ft)
Belting
42 in. wide, 3-ply, 28-oz transmission 231 ft long

Jacking belt
44¼-in. wide bed with 44-in. rolls — 1.9-in. diameter, 16 ga. mounted on 6-in. centers set high return rolls, 1.9-in. diameter, spaced on 5-ft centers
Logan Adapto Drive No. 500
Drive pulley — 8-in. diameter — lagged crowned
End pulley — 4-in. diameter — machine crowned
Belting
42 in. wide, 3 ply, 28-oz transmission 89 ft long
Motor
1½ in., 113 rpm right angle gearmotor totally enclosed for 220 VAC/3P/60 Hz

removed sheets of tobacco from the truck and placed them on the receiving line; one man stayed on the truck and hooked the sheets while the other operated the chain hoist.

One weighmaster and weighmaster's helper were used on each receiving line. The weighmaster had the additional responsibility of stopping the receiving line if it became filled with tobacco. The helper had his usual responsibilities of removing tobacco from sheets that weighed over 250 lb (the maximum allowed) and of stapling the tickets to the weighed sheets. Also, he occasionally positioned a sheet on the scale if the electric eye failed to stop the sheet in the correct position. The warehouse had one employee on each line who untied the weighed sheets of tobacco. The belt conveyor moved the sheets from the scale to a gravity conveyor where the tobacco inspectors of the U.S. Department of Agriculture, Agricultural Marketing Service (USDA-AMS), placed an official grade on the tobacco for price support.

After the gravity conveyor was filled, the sheets were automatically advanced onto the accumulation belt. Each of the three accumulation belts was designed as temporary storage to hold a farmer's complete lot of tobacco, which could then be moved by the buyers before tobacco from one of the other accumulation belts was advanced. A swing belt transferred the sheets of tobacco from the accumulation belt to the sale belt. The swing belt was positioned by an operator at the control panel, who also controlled all functions of the system so that they flowed smoothly. Also, the auctioneer was able to control the sale belt and accumulator belt to assist in the viewing and selling of the tobacco. The ticket marker, bookman, "pull boys", "tag boys", and retiers downstream of the buyers then completed the sale. (The sale bills which had been hand carried to the bookman were hand carried to the office). Sheets were removed from the line manually if the grower wanted to review them or reject the sale bid.

Two men were required for jacking tobacco.* Loaded jacks were pushed to the loadout door by forklift. For buying companies that did not use jacks, personnel of the tobacco-moving contractor aligned the sheets directly on the floor beside the conveyor according to grade in groups of three sheets per grade. The tobacco mover (12-ft-long belt conveyor attached to the forks of a forklift) then transported these sheets, three at a time, to the loadout door where they were ready to be placed on a semitrailer.[12]

CIGAR TOBACCO

This is the land — a sandy strip of the Connecticut River Valley 75 mi long by 20 mi wide — where, long before the white man arrived on this continent, the Indians raised tobacco. Of course, the wild tobacco seed indigenous to this continent produced a heavy, black leaf, the smoking of which could hardly have come under the heading of pleasure. In fact,the Indians usually smoked tobacco as part of a ceremonial rite, generally associated with their peace conferences. Judging by what we can learn from the early historians, the overpowering virility of this crude leaf usually helped the tribesmen win at least a temporary peace — it left them far too weak for further battle![13]

After the tobacco has all been stripped and baled the farmer's work is finished and the tobacco is sold and delivered to a tobacco firm. Presently there is only a handful of large firms but previously, before the invention and widespread use of cigar making machines, there were many small, independent dealers and packers. A "packer" would buy a number of crops from local farmers and have them deliver their tobacco to his warehouse, where he would case it. This meant his workers would take the hands of tobacco out of the bales the farmer made, pack it in large wooden cases, stack and store the cases for the prescribed time until the tobacco had "sweated", a form of changing or aging. Then the cased tobacco was ready to be used by independent local cigar makers or a cigar factory, which may have employed up to 50 or more workers.[14]

REFERENCES

1. **Billings, E. R.,** *Tobacco: Its History, Varieties, Culture, Manufacture and Commerce,* American Publishing, Hartford, Conn., 1875, 57.
2. **Billings, E. R.,** *Tobacco: Its History, Varieties, Culture, Manufacture and Commerce,* American Publishing, Hartford, Conn., 1875, 73.
3. **Anon.,** Statistical Bulletin No. 605, Agricultural Marketing Service, U.S. Department of Agriculture, Washington, D.C., 1978, 6.
4. **Cyrus, J. H. and Mozingo, R. L.,** Tobacco Story, North Carolina Department of Agriculture, Raleigh, 1966, 3.
5. **Edwards, L.,** unpublished data, 1979.
6. **Davis, R. B., Jr.,** U.S. Auction Problems from a Farmers Viewpoint, Industry-Wide Flue-Cured Marketing Committee, Virginia Department of Agriculture and Commerce, Richmond, 1973.
7. **Graves, A. H. and Forrest, K. R.,** Scheduling the Arrival of Tobacco at Tobacco Warehouses, USDA-ARS-S-12, U.S. Department of Agriculture, Washington, D.C., 1973, 4.
8. **Graves, A. H.,** Improving the Receiving of Tobacco at Auction Warehouses, USDA- ARS-S-28, U.S. Department of Agriculture, Washington, D.C., 1973, 3.
9. **Graves, A. H., Sowell, R. S., and Cyrus, J. H.,** Receiving sheets of untied tobacco at auction warehouses with use of a powered belt conveyor, *Tobacco Sci.,* 22, 12, 1978.

* Jacking is the stacking of no more than two sheets of tobacco belonging to the same company or jack (square frame with casters under each corner).

10. **Graves, A. H.,** Mechanized system for removing sheets of untied tobacco from warehouse sales floor after auction, *Tobacco Sci.,* 17, 28, 1973.
11. **Graves, A. H.,** Mechanized system for loading sheets of tobacco onto semi-trailers, *Tobacco Sci.,* 20, 94, 1976.
12. **Graves, A. H. and Sowell, R. S.,** Mechanized sales of untied tobacco delivered in sheets to auction warehouses, *Tobacco Sci.,* 20, 67, 1976.
13. **Anon.,** Tobacco-Land in Old New England, Imperial Agricultural Corporation, Hartford, Conn.
14. **Kline, John B.,** Tobacco Farming and Cigar Making Tools, published by John B. Kline, 1975.

CONTROL OF INSECTS IN STORED TOBACCO

D. P. Childs

Tobacco is the most widely grown commercial non-food plant in the world.[1] Curing, storing, processing, and manufacturing tobacco is one of the oldest and largest industries in the U.S. Most tobaccos are held in storage for 1.5 years or more for natural aging. During this storage period and throughout manufacture and marketing of tobacco products, the commodity is subject to attack by stored-product insects. Listed in the U.S.D.A. Handbook No. 233 are 27 insect species, in addition to the cigarette beetle, *Lasioderma serricorne* (F.) and the tobacco moth, *Ephestia elutella* (Hübner), that have caused some injury to post harvest tobacco.[2] Of these pests, only the cigarette beetle and tobacco moth are of major economic importance. In 1973 these insects caused a loss of $1748.8 million to postharvest tobacco in the U.S.[3]

The cigarette beetle is the leading insect pest of stored tobacco. It is distributed worldwide primarily because of its adaptation to a very wide range of host foods. Howe believes this species has probably the most varied taste in food of all storage insects.[4] Fletcher lists 57 different items that have been reported damaged by the cigarette beetle.[5] Most of these are of plant materials, but also included are some types of animal matter and miscellaneous items such as books, edible fungus, gun wads, pyrethrum flowers and powder, rugs, and upholstery.

The tobacco moth is also an international pest of postharvest products. It does not feed on as many different types of products as the cigarette beetle. The webbing secreted by the larva with attached excrement depreciate in value the postharvest product beyond the amount of product actually destroyed. The tobacco moth does not thrive on low-sugar, high-nicotine tobaccos and is seldom known to damage manufactured tobacco products. This insect is frequently called the chocolate or warehouse moth, and is destructive to chocolates as well as cacao beans. Munroe reported that the tobacco moth (warehouse moth) is one of six stored product insects established and resident in warehouses located in the U.K.[6]

Sanitation

Accumulation of tobacco scrap or other vegetable matter in crevices, corners, or silts and ledges, in machinery, electrical panel boxes, packaging materials, etc. provide food and refuge for insect pests of stored tobacco. Insects will migrate from the scrap and debris to tobacco whether it is stored on the farm, in warehouses, factories, or as a manufactured product at the retailer. Cleaning the sites by brushing is helpful, but often is inadequate. Moving the scrap tobacco from machinery with an air stream does little more than transfer the scrap to a less visible area. Vacuum cleaners are needed to remove the debris from inaccessible spots located in and around machinery. Rooms with stored tobacco must be kept tidy. Tobacco sheets, casing material, and machinery should not be stored in factories where they can accumulate dust and provide sites for an insect infestation.

Insect traps

Light traps are of little value as a control measure in tobacco warehouses and manufacturing plants. They are indispensable in determining where insects are located and also give an indication of the insect population. The trap used most frequently by the tobacco industry is the suction/light type described in U.S.D.A. Handbook No. 233.[2] The electric-grid trap of design similar to that described by Fletcher and Long is more efficient than the suction/light type.[7] It is recommended for use in factories where insect population density is consistently low. The light source in either type of trap should be from fluorescent blacklight lamps. In heated buildings the traps should be operated continuously, and operated in unheated storage warehouses from early spring until late fall.

Temperature

Recent technology for accelerating redrying leaf tobacco and strips for storage has reduced the thermal efficiency of this process for destruction of insect life. Fletcher et al.[8] found that cigarette beetles (all stages) infesting tobacco strips may survive the redrying process currently practiced by most tobacco companies. They report that the effectiveness of this process could be nearly 100% if tobacco was exposed to high temperature in the heat chambers for long periods.

Many tobacco companies have changed to a vacuum-steam flow process for moisture conditioning tobacco leaf and lamina prior to manufacture. This process conditions the leaf more uniformily and at a lower temperature than the thermal-vacuum process. It destroys insect life at moderate temperature (49 to 60°C) in exposures of 1 to 25 min.[9]

Cold or cool storage of tobacco prevents or checks cigarette beetle and tobacco moth infestations. The moth is more tolerant of low temperature than the beetle, but both insects become relatively inactive at 15°C. If the tobacco is held long enough at low temperature, the insects will eventually die. Cigarette beetle infestations in tobacco hogsheads are killed in 12 weeks at 4.4°C, 20 weeks at 7.2°C, and 32 weeks at 8.9°C.[10] Fourth instar larvae and adults are most resistant to low temperatures. At lower temperatures, mortality occurs in less time. However, heat transfer within the commodity is the criterion that determines the length of exposure needed to kill the insects. In bales or cases of leaf tobacco, the cigarette beetle is killed in 3 to 5 days at −23°C.[11]

Biological Control

The cigarette beetle and tobacco moth are vulnerable to most predators and parasites found in tobacco warehouses. These natural enemies undoubtably destroy many insect pests but are of little value in practical control. Howe has cited a large number of references that identify predators and parasites of the cigarette beetle.[4] Munro reported that Richards and Waloff found 90% destruction of tobacco moth larvae infected with bacterium *Bacillus thuringiensis* (BT).[6] However, these insects were found in heavily infested tobacco stocks and the moth population was still able to maintain itself in the warehouse at about 300,000 individuals. Mistric reported that tobacco leaf treated with BT (8 ppm active Dipel®) is protected from damage by the tobacco moth for not less than 1 year.[12] His experiments were oriented towards protection of tobacco stored on the farm, and after 1 year value of the treated tobacco was $109/cwt while nontreated tobacco brought $40.

Chemical Control

Aerosols and fumigants are used extensively for control of the cigarette beetle and tobacco moth in storage warehouses and manufacturing plants. Aerosols will kill only the insects located in the free space of the building, so treatment should coincide with the insects maximum activity period. For the beetle and moth, this period occurs in the evening generally between 5 and 10 p.m. The safest aerosol formulation contains pyrethrum as its active ingredient, with or without a synergist. The other major insecticide applied as an aerosol is dichlorvos, an organophosphate compound. The vapor-phase action of dichlorvos improves its potential for uniform distribution in the free space of a storage or factory. It controls both pests well and against the moth its continued use has eliminated the insect as a serious pest.

Fumigants, at atmospheric pressure or in vacuum, have been used for several years to kill insects in stored and manufactured tobacco products. As the cost of tobacco increases, there has been greater use of fumigants, particularly for control of insects in tobacco warehouses. Fumigants used most frequently by the industry are methyl bromide and phosphine. Generally, methyl bromide is used when it is expedient for the exposure period not to exceed 48 hr (4 hr under vacuum) or temperature of the tobacco is below 15°C. Phosphine may be used only at atmospheric pressure and with exposure periods of 96 hr. Use of shorter exposure periods reduces confidence in a phosphine fumigation. Phosphine readily penetrates units as large as a tobacco hogshead and the residue remaining is normally less than 8 ppb. Data presented in Table 1 are intended as a guide for use of aerosols and fumigants for control of the cigarette beetle and the tobacco moth.

Chemical pesticides used improperly can be injurious or fatal to man. Follow the directions and heed all precautions given on the table.

Table 1
CHEMICAL CONTROL OF THE CIGARETTE BEETLE AND TOBACCO MOTH IN FARM STORAGES, REDRYING PLANTS, WAREHOUES, FACTORIES, AND MANUFACTURED PRODUCTS

Treatment and storage	Insect	Insecticide	Dosage (active ingredient/ 28.3 m^3 unless otherwise stated)
Aerosol			
On farm	Beetle and moth	Dichlorvos	0.5 to 1.0 g weekly or as needed
	Moth	Pyrethrins	1% formulation by wt.; 3.8 ℓ/ 5664 m^3, weekly or as needed
		Pyrethrins + piperonyl butoxide	0.75% pyrethrins + 3.75% piperonyl butoxide by wt; 3.8 ℓ/5664 m^3, weekly or as needed
	Beetle	Pyrethrins	1% formulation by wt.; 3.8 ℓ/ 1133 m^3, weekly or as needed
		Pyrethrins + piperonyl butoxide	0.75% pyrethrins + 3.75% piperonyl butoxide by wt; 3.8 ℓ/1133 m^3, weekly or as needed
Factory, redrying plant and warehouse	Moth	Dichlorvos	1.0 g, application once per week
		Pyrethrins	Same as on farm storage
		Pyrethrins + piperonyl butoxide	Same as on farm storage
	Beetle	Dichlorvos	0.5 g daily after 5 p.m. or end of workday
Fumigation			
Factory	Beetle and moth	Methyl bromide	0.68—0.90 kg, 36—48 hr exposure
	Beetle and moth	Methyl bromide in atmospheric chamber (bales, cases and hogsheads)	0.68 kg, 48 hr exposure at 21°C or above; 0.90 kg, 48 hr exposure at 7—20°C
	Beetle and moth	Methyl bromide under tarpaulin (bales, cases and hogsheads)	0.68—0.90 kg, 48 hr exposure at 21°C or above
	Beetle and moth	Methyl bromide in vacuum chamber (bales, cases and hogsheads)	1.36 kg, 4 hr at 21°C or above; 1.81 kg, 4 hr at 7—20°C
		Phosphine under polyethylene tarpaulin or in atmospheric chamber (bales, cases and hogsheads)	Bags, pellets, or tablets; follow dosage given on manufacturer's label; 7 days at 10—15°C, 6 days at 16—20°C, 4 days above 20°C
Warehouse	Beetle and moth	Phosphine	Bags, pellets, or tablets; follow dosage given on manufacturer's label; 7 days at 10—15°C, 6 days at 16—20°C, 4 days above 20°C
Cigars	Beetle	Methyl bromide in vacuum chamber (in boxes, or cartons)	1.36 kg, 4 hr at 21°C or above
	Beetle	Phosphine under polyethylene tarpaulin or in atmospheric chamber	Bags, pellets, or tablets; follow dosage given on manufacturer's label; 6 days at 16—20°C, 4 days above 20°C

REFERENCES

1. **Akehurst, B. C.,** *Tobacco,* Humanities Press, New York, 1968, 3.
2. **Anon.,** Stored tobacco insects-biology and control. *U.S.D.A. Dep. Agric. Agric. Handb.,* 233, 1972.
3. **Anon.,** Insect control in marketing, Rep. No. 20620, Agricultural Research Service, U.S. Department of Agriculture, Washington, D.C., 1976, 216.
4. **Howe, R. W.,** A laboratory study of the cigarette beetle, *Lasioderma serricorne* (F.) (Col. Anobiidae) with a critical review of the literature on its biology, *Bull. Entomol. Res.,* 48, 9, 1957.
5. **Fletcher, L. W.,** unpublished data.
6. **Munroe, J. W.,** *Pests of Stored Products,* Hutchinson, London, 1966, 53, 62.
7. **Fletcher, L. W. and Long, J. S.,** Evaluation of an electric-grid light trap as a means of sampling populations of the cigarette beetle, *Tobacco Sci.,* 17, 37, 1973.
8. **Fletcher, L. W., Overby, J. E., Childs, D. P., and Dean, T. A.,** Redrying of tobacco strips and its effect on mortality of cigarette beetles, *Tobacco Sci.,* 19, 35, 1974.
9. **Childs, D. P.,** Effect on the cigarette beetle of flowing steam vapor under vacuum, *Tobacco Sci.,* 9, 56, 1965.
10. **Childs, D. P., Overby, J. E., and Watkins, B. J.,** Low temperature effect on cigarette beetle infestation in tobacco hogsheads. *J. Econ. Ent.,* 61(4), 992, 1968.
11. **Tenhet, J. N., Bare, C. O., and Childs, D. P.,** Cold storage and cool storage of tobacco to control the cigarette beetle, *Tobacco Sci.,* 1, 169, 1957.
12. **Mistric, W. J., Jr.,** Tobacco moth control on farm-stored tobacco, in Proc. Entomol. Sect. 27th Tob. Work. Conf., Atlanta, 1977, 16.

Cotton

GINNING, CLEANING, HANDLING, AND STORAGE OF COTTON

Beverly G. Reeves and Warren E. Garner

THE GINNING PROCESS

Eli Whitney started work on development of the saw gin in 1792 at Savannah, Ga., and the gin he patented in 1794 made it possible to separate upland lint from seed about 100 times as fast as it could be done by hand. His gin removed the fibers from the seed by means of spikes on a cylinder. The spikes pulled lint through a slotted plate. The slots were too narrow for the seeds to pass and a revolving brush removed the lint from the spikes. Many refinements have been made, but the basic principle remains unchanged today.

The principle of Eli Whitney's gin, when applied to absolutely clean seed cotton, is as satisfactory today as in the 1790s, but the ginning problems of today are much more complicated. So, the modern cotton gin is equipped with machines employing several different physical principles that remove green bolls, dry the fibers, scrub out dirt and leaves, separate the sticks and burs, replace moisture if necessary, remove lint from seed, straighten, parallelize, and smooth the fibers, and bale the lint.

The machines to perform these functions have been developed and adapted through carefully planned research by both private concerns and public agencies for the purpose of providing the highest quality lint in less ginning time and at less ginning expense. Today's more expensive and more complicated gin is the industry's answer to much more complicated processing problems. The principal causes of these problems are the twin jinxes of excess trash and excess moisture, due primarily to labor-saving practices employed during harvest. As mechanical harvesting has increased, ginning problems have multiplied.

At the same time, careful research has shown that the quantity of trash is not the whole problem. For instance, 600 lb of dry burs (hulls) could be removed from a bale of hand-snapped seed cotton very efficiently in 1940, but we could not properly remove 50 lb of green leaf and stems with the ginning machinery available at that time. Therefore, the rapid advance of cotton mechanization in the late 1940s made it necessary to reexamine the cotton ginning systems, with special attention to the necessity of identifying the characteristics of all types of foreign matter found in seed cotton.

Machinery had to be designed not just to remove pounds of trash, but to deal effectively with the specific types of trash that are carried in mechanically harvested seed cotton. These trash types include green bolls, burs, sticks, stems, grass, bark, large leaf, small leaf, sand, dirt, pin trash, and motes. The whole problem of adequately removing the trash also involves properly handling that which is broken up or "manufactured" in the ginning process as well. In other words, machines had to be designed to remove all of the various types of trash and the excess moisture while still protecting those fiber properties so important to cotton quality and end use. Such machines have been designed, are in general use, and will be discussed in detail.

Green Boll Traps

Early season machine-stripped cotton generally contains some green, unopened bolls. As the first step in ginning, these should be separated from the seed cotton. Otherwise they will be broken open, exposing the juicy fiber and soft seed which will cause clogging of saw teeth and a buildup of sticky residue on interior machinery surfaces. The ginner should encourage his farmer customers to follow recommended production guidelines and delay the application of a desiccant until all bolls are mature and 90% are open. Then the green boll separator on the stripper should be adjusted for maximum separation in the field. If the separated green bolls are collected and left in the sun to dry, they will open and can be ginned later.

The green boll traps used in gins and the green boll separators used on stripper harvesters are designed to capitalize on the difference in the density and the aerodynamic characteristics of opened and unopened bolls.

Research at the U.S. Department of Agriculture Cotton Ginning Research Laboratory at Stoneville, Miss. has shown that the floating velocity of material in an air pipe is dependent upon the cross-sectional area of the material and its density. Green bolls are heavy and more streamlined, so that a much greater air velocity is required to convey them than is required for fluffy locks of seed cotton or whole open bolls of mature cotton. By using the formula

$$V = 3250\sqrt{S \times D}$$

where V = vertical floating velocity in feet per minute; S = the specific gravity of the material; and D = the diameter in inches of the equivalent cross section of the particle. It is calculated that the floating velocity of bolls is 3500 ft/min, while the floating velocity for seed cotton is 1300 ft/min. Thus to drop green bolls out of an air stream, the air velocity at the trap outlet must be less than 3500 ft/min but more than 1300 ft/min so that the seed cotton can be carried through the trap. As a rule of thumb, the air velocity in a green boll trap should not be less than 1500 ft/min. Under this condition the only green bolls that will be carried through are those entangled in a large mass of seed cotton, so the cotton should be spread out as much as possible to minimize the size of the wads of cotton.

The original green boll traps were called hopper-type traps. They were relatively simple in design and are still in common use. The hopper-type green boll trap is normally installed in the wagon telescope air line, but it is not altogether satisfactory because of the high air velocity, minimum spreading of the cotton, and unloading capacity reduction due to the air bleed. So, public and private work has continued in an effort to develop a more acceptable unit.

The Oklahoma Cotton Ginning Research Laboratory developed a green boll trap that is designed to meter the flow cotton onto a conveyor belt in a thin, preferably single-layer bat. A suction duct is provided to elevate mature, open cotton and leave the green bolls on the belt to be delivered to a suitable holding bin.

Another type of trap under development at the South Plains Cotton Ginning Research Laboratory, Lubbock, Tex. employs air jets to convey seed cotton across a stilling chamber with a grated bottom which allows green bolls to fall out of the stream of cotton.

The Oklahoma Cotton Ginning Research Laboratory type of boll trap is not normally used in conjunction with a suction wagon unloading system because a second suction fan is required. However, this type of trap is gaining acceptance in installations where module break-up units are used. The cotton is delivered mechanically by the break-up to a belt conveyor and a suction duct is used to pick up and convey the cotton into the gin plant. Green boll separation then becomes a logical part of this materials handling scheme.

Cotton Moisture Control

The two principal reasons for controlling cotton fiber moisture content during ginning are to assure smooth operation of the processing machinery and to obtain the optimum fiber quality from the seed cotton.

Cotton with too high a moisture content will not readily separate into single locks and will form wads that may choke, warp, or break ginning machinery. Cotton too low in moisture may also cause machinery stoppages by clinging to metal surfaces as a result of static electricity generated on the fibers.

Damp cotton does not clean as well as dry cotton. Before the widespread use of lint cleaners, some ginners sought to obtain the highest grades possible by drying the cotton to moisture levels that are now considered unsafe for fiber length preservation. Research has established that the apparent strength of cotton fibers is directly proportional to fiber moisture content and is therefore greater at higher moisture levels. Consequently, as fiber moisture content is lowered, as by drying, the apparent strength is reduced and the frequency of fiber breakage during ginning is increased.

Fiber moisture content from 6.5 to 8% gives smooth saw gin operation and profitable lint grades while holding the fiber breakage rate to a reasonably low level. Although cotton can be processed at moisture levels higher than 8% and lower than 6.5%, the risk of operational or quality control problems increases as the fiber moisture content departs from the range. In roller ginning, a fiber moisture content of 5 to 6% is desirable.

Cotton is a hygroscopic material. Consequently, its natural moisture content varies in relation to the relative humidity of the surrounding air. A typical relationship between raw cotton fiber moisture content and ambient relative humidity is shown in Figure 1.

Cotton harvested during periods of high humidity may arrive at the gin with a moisture content as high as 12% or more, and cotton harvested during periods of low humidity may contain fiber moisture of 4% or less. For these reasons, gins should be equipped to add as well as remove moisture from the cotton being processed.

Driers

The first heated-air driers for use in cotton ginneries were developed by the U.S. Department of Agriculture in the period 1926 to 1932. These included moving trays and drag belts, but the most successful and widely used drier has been the shelf type in which cotton is conveyed through the drier by the heated air. Other types of driers, such as counterflow, mechanical reel, screw conveyor, and combination cleaner-driers, have also been used.

The physical dimensions of tower driers vary among manufacturers; most conventional tower driers are 4 to 6 ft wide, about 6 ft long, and 17 to 20 ft high. The number of shelves varies from about 16 to 24, and the spacing between shelves varies from 8½ to 10 in. Larger driers are available, but most gin driers in use today fall within the limits given.

The cotton-handling capacity of tower driers may range from 8 to 30 bales/hr depending upon drier width, condition of the cotton, and quantity of air moving through the system.

Spindle-harvested cotton processes faster through tower driers than stripped or snapped cotton because it contains less waste. Rain-wet and other relatively dense cottons must be handled with care to prevent choking because of temporarily overloading the conveying system. The most critical locations in the drying system are in the pipe immediately below the bulk feed dropper and in the first few shelves at the top of the tower, where the air velocity is suddenly lowered as it moves from the conveying pipe into the drier. In commercial practice, air velocity in the cotton-conveying pipe ranges from 4000 to 5000 ft/min and in the driers, the velocity may be reduced to only one half of the conveying velocity.

There is some slippage between cotton and air, this being greatest for damper, heavier, denser cottons. As the cotton is opened and fluffed by moisture removal and me-

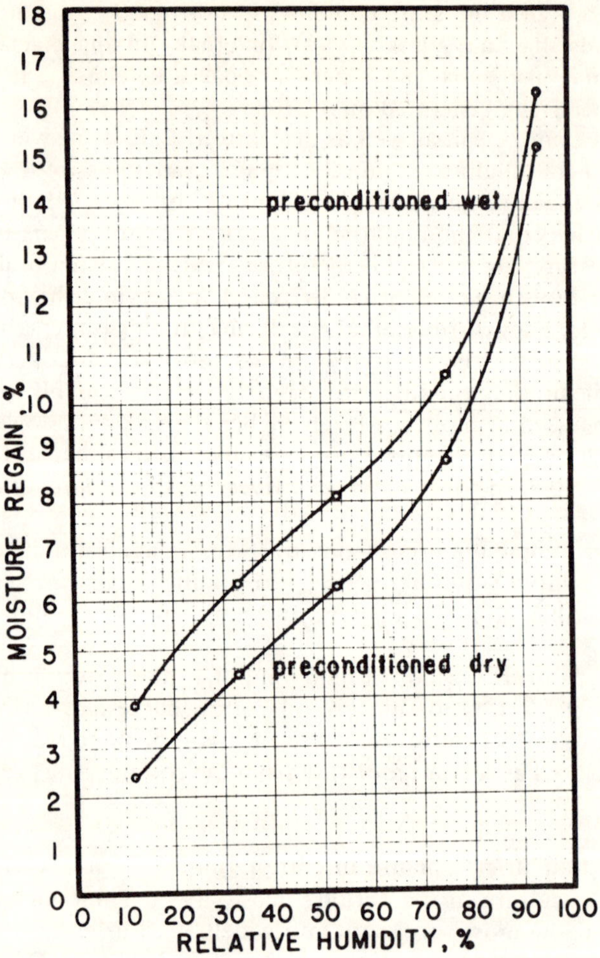

FIGURE 1. Relationship of cotton fiber moisture content (dry basis) to ambient relative humidity at 70°F.

chanical action, it is more easily airborne, and its final velocity more nearly approaches that of the conveying air. Cotton is seldom in a gin drying system for as long as 15 sec; 7 to 10 sec is common.

The first driers were heated by steam coils with drying air temperature initially at about 160°F. But cotton roughly harvested and harvested during periods of high humidity called for removal of more moisture, requiring the use of higher drying temperatures, more than one stage of drying, or both. Modern drying systems are capable of serious overdrying and must be regarded as quality hazards if they are not used properly. Most of the moisture removal takes place in the first 3 sec after the cotton is exposed to the heated air, and drying-air temperatures in excess of 350°F at the air-cotton mixpoint may have serious adverse effects on fiber quality. Therefore, the use of two or more stages of drying at relatively low temperatures is preferable to a single stage at higher temperature.

The typical gin drying system consists of a heater, one or more fans, piping, shelf drier, and a seed-cotton cleaner that serves the dual purpose of separating the warm moist air from the cotton and cleaning the cotton.

There are wide variations in the layout of gin plants, so that air requirements and drying exposure periods may be expected to differ from gin to gin. Because of these wide variations, specific operating temperatures are not recommended (except that in no case should the temperature in any portion of the drying system exceed 350°F). The moisture content of the fiber after drying should be used as the guide for regulation of temperature.

The location of temperature control sensors is important. Many bales have been overdried and some have been set on fire through overheating, even though the sensing device indicated a safe drying-air temperature. When the temperature sensor is located well into the drying system—near the bottom of the drier, for example—the temperature at the air-cotton mixpoint may be considerably higher than that indicated by the sensor.

Mixpoint temperatures as high as 450°F have been observed with a control point temperature of about 180°F. It is recommended that the temperature control sensor be located in the hot-air line just ahead of the mixpoint. If this is not practicable, then a temperature high-limit switch at this location, set for 350°F, is recommended to prevent scorching and ignition of cotton in the drying system.

Gin drying systems originally consisted of a single processing line, with machinery sized to meet production rate requirements, but modern hourly processing rates may often require dividing the seed-cotton stream for adequate drying. The ginning machinery manufacturers' recommendations should be followed with respect to equipment size and processing rate.

The pace of modern ginning is too rapid for ginners to examine each load of cotton and adjust the drying system accordingly; all U.S. ginning machinery manufacturers offer some type of automatic drier control. These controls permit the plant to reduce the variability of moisture content of the finished bales and may offer considerable energy savings by adjusting the heater fuel to the amount actually needed to remove excess moisture from the cotton and by cutting the burners back to low flame when no cotton is in the drying system.

A study of heat use by a tower drier at the U.S. Cotton Ginning Research Laboratory, Stoneville, Miss. showed that a drying system having uninsulated pipe and an uninsulated steel tower drier used only about 16% of the heat supplied by the air heater to remove moisture from the cotton. The remainder of the added heat was lost by radiation or passed from the system as warm air.

The amount of fuel used by gin drying systems may be estimated in several ways. The simplest method is to relate fuel purchases to bales ginned, making allowance for fuel not used by the cotton driers. When the volume of air moving through the drying system is known (see "Fans and Piping"), a thermometer in the air line near the heater will provide a basis for estimating fuel consumption. From Figure 2, the heat added to the air may be read on the vertical axis for a known air volume at a given temperature. For example, if the drying system is pulling 5000 ft^3/min and its temperature near the heater is 300°F, then the heat being supplied is estimated at about 1.25 million Btu/hr. Assuming a combustion efficiency of 95% and using the heat values of Table 1, the quantity of fuel required to yield 1.25 million Btu is about 1300 ft^3 of natural gas or about 14.4 gal of commercial propane.

A sampling of 48 cotton ginneries in Arkansas and Missouri in 1960—1961 showed that an average of about 430,000 Btu was required to dry a bale of cotton. In terms of quantity of fuel consumed, this was about 381 ft^3 of natural gas or about 4.4 gal of the conventional butane-propane mixture per bale.

Moisture Restoration

Since the fiber breakage rate during ginning is inversely proportional to the fiber moisture content at the gin stand and lint cleaners, the number of fibers that break

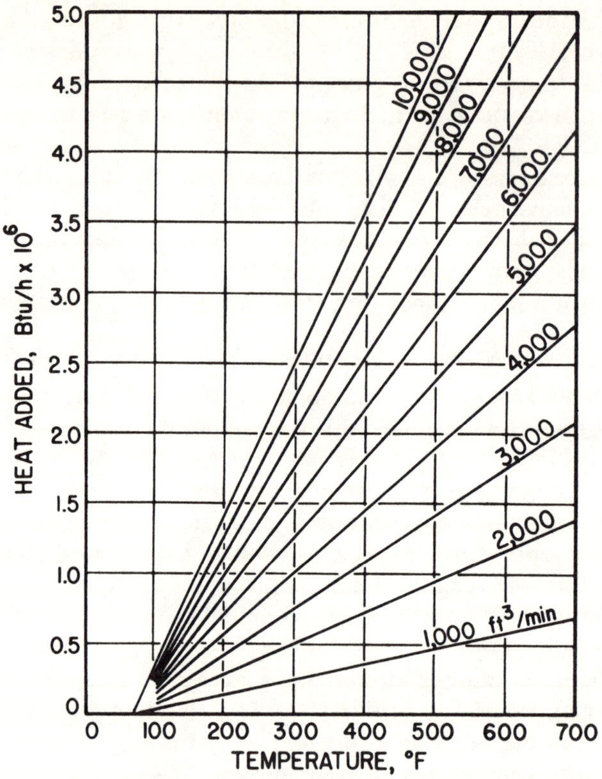

FIGURE 2. Heat-to-temperature relationships for air moving through a gin drying system from an initial temperature of 70°F.

Table 1
NOMINAL HEAT CONTENT FOR SOME FUELS USED IN DRYING SYSTEMS AT GINNERIES

Fuel	Heat content
Natural gas	1,000 Btu/ft³
Commercial butane	102,600 Btu/gal
Commercial propane, HD 5	91,500 Btu/gal
Propane/butane (40/60)	97,800 Btu/gal
Fuel oil, CS No. 2	140,000 Btu/gal

may be minimized by conducting the fiber-seed separation and lint-cleaning processes at as high a moisture level as practicable.

Much of the American cotton crop is harvested during low humidity periods and often arrives at the gin with fiber moisture from 4 to 5%. The average fiber length of such cotton may be improved by adding moisture before fiber-seed separation and lint cleaning, by reducing the number of fibers that break in the gin stand and lint cleaners. However, restoration of moisture to ginned lint will not improve fiber length. Other benefits resulting from moisture restoration include reducing the static electricity level of the cotton and reducing the amount of power required to pack and press a bale of cotton (see "Packaging"). Methods of moisture restoration have included introducing humid air from vapor generators into extractor-feeders, feeder chutes, and lint flues; creating a fog or mist through which seed cotton must fall or be conveyed; and spraying moisture on or exposing ginned lint to humid air as it passed from the final lint condenser to the press box.

Exposure of seed cotton with less than 5% fiber moisture can help preserve fiber length during separation of lint and seed in the gin stands and during lint cleaning. The addition of moisture at the lint slide reduces the resiliency of the lint. This makes it easier to press the cotton into a bale and reduces the likelihood of the bale ties breaking when the bale is removed from the press.

There is a practical physical limit to the quantity of moisture that may be added to seed cotton. Wetting of the cotton by uncontrolled emission or by unexpected condensation within machinery and pipes must be prevented or chokages will result. Gin stand operation becomes irregular and may cease altogether if liquid water pervades the seed-cotton mass. Cotton with fiber moisture of 9% or more may not gin smoothly and will not process properly through the lint cleaners. Thus the recommended fiber moisture level of 6.5 to 8% has a gin plant production aspect as well as a fiber quality aspect.

A development of the Southwestern Cotton Ginning Research Laboratory at Mesilla Park, N.M. called the monoflow cotton conditioning system, is based on the hygroscopic nature of cotton and its tendency to approach moisture equilibrium with the surrounding air. In the monoflow system all the conveying air is maintained at a specified relative humidity. This causes overwet cotton to give up moisture and overdry cotton to absorb moisture toward a preselected target moisture level of about 7%.

An automatic moisture control system that dries with heated air and adds moisture on demand from a single measurement at the seed cotton input hopper has been demonstrated at the U.S. Cotton Ginning Research Laboratory, Stoneville, Miss. The system has been modified to include fog nozzles in the distributor for moisture addition at high ginning rates.

Provision for moisture restoration should be a part of plans for new gin plants or for remodeling older plants. Moisture restoration facilities will ensure that ginneries are able to maintain the inherent cotton fiber length properties in spite of harvesting and ginning weather conditions.

Cleaning and Extracting

The need for seed-cotton cleaning and extracting equipment in the gin has developed gradually through the years to cope with rougher harvesting methods. The development of the seed-cotton drier has greatly expanded the use of cleaning machinery in the gin because dry cotton can be cleaned much more readily than damp cotton.

The terms "cleaning" and "extracting", as used in the ginning industry, refer to processes normally performed by separate and distinct machines, although some machines may do both cleaning and extracting. Cleaners usually employ a series of beater cylinders and concave screens or grid bars. These cylinders, with a top speed of about 2000 ft/min, convey the cotton through the machine, scrubbing the locks of cotton over the concave surfaces and permitting the trash to sift out. The size of the openings limits the kind and size of trash that can be scrubbed through. Therefore, only the smaller particles of broken leaf, dirt, and small sticks and stems can be removed with a minimum loss of cotton locks. Extractors accomplish trash removal by a toothed cylinder that seizes the cotton locks and slings the trash off by centrifugal force or removes it by striking against stripper rollers or grid bars. Clearances are large enough to convey the locks of cotton safely through, but the burs and larger pieces of trash are stripped off. The trash is directed to augers for discharge, while reclaimers pick up waste cotton locks and return them to the main cotton stream.

Cleaners and extractors can be either bulk machines or unit machines. As bulk machines, they handle all cotton being processed through the gin and are normally located

Table 2
FOREIGN MATTER IN BALES OF COTTON HARVESTED BY THREE METHODS

	Pounds per bale		
Foreign matter	Machine-picked	Machine-stripped	Machine-scrapped
Hulls	29	450	329
Sticks and stems	9	115	143
Leaves and dirt	43	135	398
Total	81	700	870

From Pendleton, A. M. and Moore, V. P., Ginning Cotton to Preserve Fiber Quality, Ext. Serv. Rep. ESC-560, U.S. Department of Agriculture, 1967.

in the overhead preceding the distributor. Unit machines serve only one gin stand and usually combine the process of cleaning and extracting, plus regulating the flow of cotton to the gin stand.

Cleaning is necessary in a gin to improve both gin stand performance and lint grade, but there are practical limits to both. Removal of trash reduces wear and tear on the gin stand, and cleaning improves gin stand efficiency by presenting cotton to it in small, uniform units. Removal of trash is also associated with grade improvement, but there exists a point of diminishing return where the benefits of further trash removal are offset by fiber and cotton-seed damage and excessive loss of weight. Most modern gins contain cleaning equipment to handle the most severe trash condition that is expected in their areas. However, actual use of cleaning equipment should be based on the incoming trash content of the cotton, and cleaner cottons should not be processed through every cleaning machine in the gin just because they are available. Bypass valves should be provided to omit certain machines for the cleaner cottons; trash removal should be restricted to that which is necessary to produce the grade determined by the color of the cotton. Further cleaning reduces the weight without increasing the value of the bale.

The range and type of trash in seed cotton varies throughout the Cotton Belt. Trash can be characterized as bolls, burs or hulls, sticks and stems, leaves, motes, grass, weeds, dirt, pin trash, and even tramp metal. Table 2 gives typical trash contents by types for cottons harvested by three machine methods. These methods account for over 99% of the cotton harvested in the U.S.

According to the 1978 Cotton Gin Equipment Survey by the Agricultural Marketing Service, U.S. Department of Agriculture, there were 2726 cotton gin batteries in the U.S. Of these, 55% were equipped with bur machines and 67% were equipped with stick and green-leaf machines. Earlier reports had shown that practically all U.S. gins are equipped with cylinder cleaners and that their use is stable; however, the use of bur machines is diminishing and the use of stick and green-leaf machines is increasing. Another type of machine whose use is expanding is the modified bur machine-stick machine combination unit. This is a recently developed machine that is highly effective in removing trash from machine-stripped and difficult-to-clean machine-picked cotton.

Cylinder Cleaners

Fine trash, leaves, and dirt are removed from seed cotton by machines generally classified as cleaners, or specifically as cylinder cleaners. The cylinders are usually spiked drums that open and clean the cotton by conveying and scrubbing it across surfaces with openings that allow the trash to sift through. The cleaning surface configuration leads to further classification of cylinder cleaners as either screen, grid bar, or revolving screen (impact). The screens may be either woven wire or perforated metal. A typical screen is made of 2 × 2 mesh, 0.105 woven galvanized wire cloth. Screen cleaners have largely been replaced by more efficient grid-bar cleaners that have 3/8-in. diameter rods spaced 5/16 to 5/8 in. apart. In the revolving screen, or impact, cleaner the cotton is conveyed across a series of revolving serrated disks.

Cylinder cleaners may also be classified as to the way they are normally used in a cotton gin. In this respect they are either air-line, hot-air, or gravity-fed cleaners.

Air-line cleaners may have cylinders of the spiked-drum or crusher-plate type that permit only a horizontal flow of the seed cotton from the inlet to the outlet of the machine. They allow both air and seed cotton to pass entirely through the cleaner. Air-line cleaners are popular in the high plains of Texas and in western Oklahoma as an effective means for removing sand from cotton and for breaking the bolls for further cleaning.

Hot-air and gravity-fed cylinder cleaners may have the cylinders arranged in an inclined or horizontal position. Hot air may enter the cleaner through a blow-in or push-pull fan arrangement. In many cases the first cleaner serves as a cleaner separator following the first drier. It may discharge into an extractor or a gravity cleaner for further cleaning.

For gins having a capacity of about 12 to 14 bales/hr, two 6-ft-wide cleaners are usually installed in parallel, with each cleaning half the seed cotton under a split-stream arrangement.

The efficiency of cylinder cleaners is generally low when considered as individual machines; the efficiency is calculated from seed-cotton fractionation tests before and after cleaning. However, they are not normally used alone, but in combination with other machines, and serve a most useful function in opening the cotton or breaking up large wads and removing fine trash such as leaves, sand, and dirt. Since they do remove trash, they can improve grade as more cylinders of cleaning are added. Naturally, higher cleaning efficiencies are associated with cotton of higher trash content.

In early U.S. Department of Agriculture research tests the cleaning efficiency of a seed-cotton cleaning setup consisting of a tower drier at 160°F, a bur machine, a seven-cylinder cleaner, and an extractor-feeder was reported to be 57.7%. When the cylinder cleaner was placed ahead of the bur machine in the system, cleaning efficiency increased to 71.8%. In later tests, the total cleaning efficiency of overhead seed cotton cleaning and extracting systems for machine-picked cotton has been reported as 54 to 86%. The trash removed by a seven-cylinder cleaner has been reported to be 11.7 to 14.7/lb/1000 lb of seed cotton, from machine-picked cotton with an input trash content of 4.6%. In another test the individual cleaning efficiency of a six cylinder cleaner was reported as 18.0 to 23.4%, on machine-picked cotton with input trash content of 7.0 to 17.3%.

Specifications of commercially available cylinder cleaners, as furnished by the manufacturers, are given in Table 3.

Extractors

Burs and sticks are extracted from seed cotton by specially designed equipment generally classified as extractors. The trade usually refers to extractors by type, i.e., bur machine, stick and green-leaf machine, and extractor-feeder.

Bur Machine

The bur machine was developed in the 1920s in response to hand snapping and mechanical stripping of cotton in Texas. The bur machine has been widely adopted by the ginning industry, especially in stripper-harvesting areas. In 1978, 55% of all U.S. gins and 88% of all Texas gins had bur machines.

Table 3
SIZES AND CHARACTERISTICS OF COMMERCIALLY AVAILABLE SEED-COTTON CYLINDER CLEANERS

Manufacturer	Type	Width (ft)	Cleaning cylinders (no.)	Speed of cylinders (r/min)	Manufacturer's recommended capacity (bales/hr)	Power requirement range (hp)	Manufacturer's recommended capacity (bales/hr/ft of width)	Power requirement (hp/ft of width)
Bush Hog/Continental	Inclined, gravity	6	6	400—500	10—15	10—15	1.7—2.5	1.7—2.5
		8	6	400—500	15—20	15—20	2.0—2.5	2.0—2.5
		10	6	400—500	20—30	20—25	2.0—3.0	2.0—2.5
	Impact, gravity	6	7	600 upper, 400 lower	10—15	10—15	1.7—2.5	1.7—2.5
		8	5	600 upper, 400 lower	15—20	15—20	2.0—2.5	2.0—2.5
		10	5	600 upper, 400 lower	20—30	20—25	2.0—3.0	2.0—2.5
Hardwicke-Etter	Air line	4—5	4 or 6	400—650	10—15	7.5	1.3—2.0	—
	Inclined, hot air and gravity	6	7	407—450	12	15	2.0	2.5
		8	7	407—450	16	20	2.0	2.5
		12	7	407—450	25	30	2.1	2.5
Horn and Gladden	Inclined, hot air and gravity	6	7	455	12	20	1.5—2.0	3.3
		8	7	455	16	25	1.5—2.0	3.3
	Air line	6	4	455	12	15	2.0	2.5
Lummus			6	450—500	6—8		2.0	2.0[a]
	Inclined, hot air and gravity	6	6	450—500	9—12	15	2.0	2.0[a]
		8	6	450—500	12—16	20	2.0	2.0[a]
		12	6	450—500	18—24	30	2.0	2.0[a]

Murray	Inclined, reclaimer	6	6	450	9—12	15	2.0	2.0[a]
		8	6	450	12—16	20	2.0	2.0[a]
		12	6	450	18—24	30	2.0	2.0[a]
	Air line	4—5	4	400-500	10—15	7.5	1.3—2.0	
	Inclined, hot air and gravity	6	7	455	12	10—15	2.0	1.7—2.5
		8	7	455	16	15—20	2.0	2.0—2.5
		10	7	455	24	20—25	2.4	2.0—2.5

[a] Includes vacuum for hot-air cleaner. Applies to 6-cylinder cleaner only.

The bur machine utilizes a dislodging or stripping principle for extracting burs and sticks from seed cotton. Seed cotton is fed to a large-diameter saw cylinder by a kicker conveyor, which is a conventional auger with angle-iron flippers attached to the conveyor flighting. Seed cotton adheres to the saw teeth and is carried past a stripping point near the top of the cylinder. At this point a flighted stripper roller, which clears the tips of the saw teeth by ¼ to ½ in., dislodges burs and sticks from the surface of the cylinder and discharges this material back into the kicker conveyor. The kicker conveyor moves the dislodged material (including burs, sticks, fine trash, and some seed cotton) to one end of the machine, where it falls onto a spiked conveyor. The spiked conveyor moves the dislodged material back along the entire length of the saw cylinder so that seed cotton can be reclaimed from the dislodged material. Fine trash and dirt sift through a screen bottom to another auger located underneath the spiked conveyor.

Bur machines are usually employed about midway through the seed-cotton cleaning and drying processes. Normally, a bur machine is preceded by one stage of drying and a cylinder cleaner. They can be employed in single units or, for high-capacity gins, as multiple units in parallel, but are seldom operated in tandem.

Bur machines manufactured by the various gin machinery companies are similar in design and size. The saw cylinder is approximately 30 in. in diameter and operates at a speed of 110 to 140 r/min. The stripper roller is approximately 12 in. in diameter and operates at a 4:1 speed ratio with the saw cylinder. The operating speeds of the kicker conveyor and doffing brush are approximately equal to that of the stripper roller. Bur machines are supplied in lengths of 10, 14, and 18 ft, and have capacities of about ½ bale/hr/ft of length. Approximately 3/4 hp/ft of length is required to operate these machines.

The cleaning efficiency of a bur machine is dependent to a large degree upon the amount of trash in the seed cotton. Highest efficiencies are obtained when cleaning high-trash-content cotton. When extracting machine-stripped cotton, the bur machine can be expected to be about 65% efficient in removing burs and 35% efficient in removing sticks. Fine-trash removal by a bur machine is very low. In many cases, seed cotton will contain more fine trash after extracting with the bur machine than before extracting, because the bur machine tends to pulverize the large trash particles and produces more fine trash than it removes. This situation is particularly noticeable when processing seed cotton containing very dry and brittle burs and sticks.

The cleaning efficiency of the bur machine on machine-picked cotton is lower than it is on stripped cotton. The bur machine can be expected to remove 7 to 12% of the total trash from machine-picked cotton. Fine-trash removal from machine-picked cotton is also low.

Stick and Green-Leaf Machine

The stick and green leaf machine was developed by the U.S. Department of Agriculture in the early 1950s to meet the more stringent cleaning requirements imposed on gins of that era by the rapidly expanding practice of harvesting cotton mechanically. The basic principle introduced by the U.S. Department of Agriculture was rapidly adopted and further developed by gin machinery manufacturers. Early stick and green-leaf machines were sold as attachments for bur machines, but these attachments have been gradually replaced with individual machines commonly referred to as stick machines, stick removers, sling-off machines, or stick and green-leaf machines. The principle has also been incorporated into modern high-capacity extractor-feeders. In 1978, 67% of all U.S. cotton gins were equipped with at least one stick machine.

Stick machines utilize the sling-off action of high-speed saw cylinders to extract foreign matter from seed cotton by centrifugal force. Seed cotton is fed onto the saw cylinder, either by air or by gravity, and wiped onto the saw teeth by one or more stationary wire brushes. Foreign matter is slung off the saw cylinder by centrifugal forces 25 to 35 times that of gravity. During this process, some seed cotton is extracted with the trash. Grid bars or additional wire brushes are strategically located about the periphery of the saw cylinder to control this loss of seed cotton. However, it has been found that some loss of seed cotton is unavoidable if satisfactory cleaning efficiencies are attained. Additional saw cylinders are used to reclaim the seed cotton extracted with the trash. Reclaimer saw cylinders resemble the sling-off saws, but usually operate at slower speeds and are equipped with more grid bars.

Stick machines are usually preceded by one or two stages of drying and at least one stage of cleaning. They may be fed by air or gravity; the trend in recent years has been to gravity feeding by a separator or cylinder cleaner. Gravity feeding is more uniform and less troublesome than air feeding, and a cylinder opens stripped cotton for more efficient cleaning by the stick machine and reduces seed cotton losses. Depending upon capacity requirements, stick machines may be employed as single or multiple parallel units.

Table 4 summarizes the sizes and characteristics of commercially available stick machines. Machines are available in widths of 6, 8, 10, and 12ft for approximate single-unit capacities of 9, 12, 15, and 18 bales/hr, respectively. Some models of stick machines have two sling-off saws, while others use only one. All models have at least one reclaimer saw. The trend in recent years has been to use two large-diameter (16 to 18 in.) sling-off saws and one reclaimer saw.

Most stick machines use a stationary wire brush to implant the cotton fibers firmly into the saw teeth. Recently the manufacturers have begun using a feeding method similar to a bur machine. This type of machine model, which was designed for stripper and ground-salvaged cotton, also contains a stationary stripper in addition to brushes and grid bars This machine is a combination of a modified bur machine section and a stick machine section.

Cleaning efficiencies of stick machines vary widely, depending upon the condition and trash content of seed cotton. For machine-stripped cotton, commercial stick machines can be expected to remove approximately 65% of the burs, 50% of the sticks, and 10% of the fine trash.

The efficiency for machine-picked cotton is about 13 to 20% for total trash in the seed cotton.

Extractor-Feeder

Gin stand feeders with extracting capabilities have been used since the early 1900s. Early extractor-feeders were based on the stripping or dislodging principle utilized by bur machines. However, after development of the stick machine, most manufacturers abandoned the stripping principle in favor of the stick machine's more efficient sling-off principle. Also, the introduction of high-capacity gin stands in the late 1950s forced manufacturers to simplify and streamline feeders in order to achieve desired capacities. This was accomplished by adopting the sling-off principle of a stick machine.

The primary function of a modern high-capacity extractor-feeder is to feed seed cotton to the gin stand uniformly and at controllable rates, with extracting and cleaning as a secondary function. The feed rate of seed cotton is controlled by the speed of two star-shaped feed rollers located at the top of the feeder directly under the distributor hopper. These feed rollers are powered by variable-speed hydraulic or electric motors, controlled manually or automatically by various interlocking systems with the gin stand. The drive may be designed to start and stop automatically as the gin breast is engaged or disengaged; the system may also be designed to stop feeding seed cotton in cases of gin stand overloads or underloads. Many of the systems are designed to

Table 4
SIZES AND CHARACTERISTICS OF COMMERCIALLY AVAILABLE STICK AND GREEN-LEAF MACHINES

Manufacturer and make	Widths available (ft)	Primary sling-off saw cylinder		Approxiate capacity/ft of width (bales/hr)	Hp requirement per ft of width	Feeding method
		Diameter (in.)	Speed (r/min)			
Bush Hog/Continental®						
Little David	6, 8	13¾	363	1—2	1.7	Air or gravity
Super II	6, 8, 10	16	320	1.5—2	1.9—2.5	Air gravity
Super III	8, 10	16	340	1.5—2	1.9—2.5	Gravity
Stripper and Super III	8, 10	30	110	1.5—2	3.3—3.8	Gravity
Hardwicke-Etter						
Hustler®, AG	6, 8	18	300	1	1.7—1.9	Air or gravity
Hustler®, G	6, 8	18	335	1—2	1.7—1.9	Gravity
Hustler®, G-2	6, 8, 12	18	378	1.3—1.5	2—2.5	Gravity
Horn and Gladden						
Rescuer 220 Series	6, 8	20	402	1.5—2	1.9	Gravity
Rescuer 320 Series	6, 8	20	402	1.5—2	2.5	Gravity
Rescuer 3000	8, 10	30	150	1.5—2	2.5	Gravity
Rescuer 4000	8, 10	30	150	1.5—2	3.1	Gravity
Lummus						
Little Giant®	6, 8, 12	17½	260—350	1.5—2	2—2.5	Air or gravity
S. & G.H. Machine	6, 8, 12	30	140—150	1.5—2	3.3—3.8	Air or gravity
Murray®						
H-L-S-T	6, 8, 10	20	330	1.6—1.8	1.9—2.5	Air or gravity
Compac® Machine	6, 8	20	330	1.6—1.8	1.9—2.5	Gravity
Bur Extractor and H-L-S-T limited quantity	6, 8, 10	30	153	1.5—2.0	3.9—4.25	Gravity

[a] Air or gravity for 6- and 8-ft widths, gravity only for 10-ft width.

maintain constant seed-roll densities. This is usually accomplished by modulating the speed of the feed rolls in response to feedback control signals from the gin stand, generated by (1) monitoring the power consumption of the electric motor driving the gin stand, (2) measuring displacements of the cove board in a seed-roll box, or (3) monitoring the pressure required to drive a hydraulically powered seed-roll agitator.

Saw Gin Stands

The gin stand is the heart of the gin plant. The capacity of the plant and the quality and potential spinning performance of the ginned lint depend on the stand's being in good condition and in proper adjustment. The gin stand can affect every commonly measured fiber property except fiber strength and micronaire.

All stands are adjusted before leaving the factory, but during handling and shipping the settings may change. It is therefore important even in new gins, to check all settings and to make necessary adjustments in accordance with the manufacturer's instructions. Machinery manufacturers will furnish up-to-date manuals that give detailed instructions on installing, adjusting, and operating their gin stands.

When installing new equipment, the gin stand should first be placed in correct alignment with the other machinery, leveled, and bolted to the floor before any adjustments are made. After checking the settings and making the necessary adjustments, the gin breast is placed in the ginning position and the saw cylinder rotated by hand to make sure that all parts are clear and that no foreign objects are within the gin stand. Before running the gin stand at full speed, "jog" the drive motor to see if all cylinders of the gin stand are free to turn. Then turn the motor on and check the saw speed. It is very important that the saw speed be very close to the recommended speed, since as little variation as 20 to 25 revolutions per minute above or below the manufacturer's recommendations can make an appreciable difference in performance.

Gin manufacturers have made their gin stands as safe to operate as they know how; however, gin stand operation still can be hazardous unless proper precautions are taken. Other than the manipulation of the normal operating controls, no adjustments or maintenance should be attempted while the gin stand is in operation. The gin stand should always be stopped and the power turned off and locked out before making adjustments to the settings or attempting maintenance work.

Gin Breast

After installing new stands or repairing old ones, be sure that the lateral adjustment of the breast is correct. The saws should be positioned in the center of the rib slots. After visual inspection, the saw cylinder should be rotated slowly by hand to make certain that none of the saws rub the ribs.

When the gin breast is removed from a gin stand, it should be identified with that gin stand, and special care should be taken to be sure the breast is reinstalled on the same gin stand from which it was removed.

When replacing a broken rib, care should be taken to install the replacement so that the saw opening is the same on each side. It is generally more satisfactory to have worn ribs replaced by the manufacturer or a competent gin service company. A matched set of ribs installed by well-equipped and experienced personnel will usually give better service than unmatched sets installed without proper jigs.

Never move the breast into ginning position when there is seed cotton in the roll box, unless the saws are running. To do so may damage the ribs or buckle the saws.

On many gins the picker roller is adjustable. In some cases it can be adjusted by means of a ratchet lever while the gin is in operation. Generally speaking it should be set as far away from the saw or huller ribs as possible without dropping cotton. The closer the picker roller is set to the huller ribs, the less space there is for hulls to fall out and the more the saw will break the hulls up and pull small pieces into the roll box with the cotton.

The seed fingers, or lambrequin, should be set as widely open as possible but close enough so that the seed will be clean. When building up a new roll, close the seed fingers and then slowly open them as widely as possible, consistent with clean seed, to allow the cleaned seed to fall out of the roll box. Holding seed in the roll box longer than necessary will reduce the ginning rate and cause tight seed-roll operation, which reduces the value of the cotton.

Gin Saws

Saws should be checked to make sure they are properly trained and are running in the middle of the rib slots at the recommended speed. If the distance from the point where the saws project through the ginning rib to the top of the rib is not correct, capacity will be reduced and fibers will be damaged. The distance the saws project through the huller ribs should also be checked. If the saws project too far, an excess of hulls and sticks will be pulled into the roll box. If the saws do not project far enough, ginning capacity will be reduced, seed cotton will fall out of the front, and the huller ribs may choke.

The pitch and shape of the saw teeth are also important in maintaining capacity and cotton quality. To ensure good ginning, the teeth must pass through the ribs at the proper angle. The point of the tooth should enter the ginning rib slightly ahead of the throat. If the saws are improperly filed or the saw-rib relationship is improperly adjusted so that the throat of the tooth enters the rib ahead of the point, the resulting cutting action will reduce capacity and break fibers and may cause chokages at the top of the ginning ribs. In many instances, the moting action of gin stands is affected by the saw teeth being tilted too far forward or backward.

Saws should be examined frequently, and bent teeth should be straightened or even broken off so that lint will not remain hung in them. Lint that cannot be doffed will tend to collect in the low part of the gin ribs. If this lint is allowed to accumulate the friction of the saw will cause it to catch fire and can damage the saws. A heat-damaged saw that wobbles cannot be retrained and should be replaced or cut out immediately. Lint collected in the bottom rib slot can also be grabbed by the saw when the stand is disengaged, possibly breaking the ribs and damaging the saws.

The number of bales that can be ginned between changes or sharpening of the saws depends on the type of cotton being handled and on the material the saws were made from. Rough cotton causes more wear on the gin saw than clean cotton, regardless of the type of steel used. Modern gin saws are made from stronger, more durable steel than older saws, which has resulted in greatly increasing the useful life of the gin saw. The high-capacity gin stands of current and older design are equipped with such durable saws that they can gin 2000 to 3000 bales per stand before changing. Processing trashy cotton with minimum overhead cleaning will make more saw changes necessary. The general practice in high-capacity ginning is to replace rather than attempt to sharpen worn saws. Sharpening of saws is still essential for the older outfits, especially those equipped with 80-, 90-, and 120 12-in.-diameter saw stands. Usually, saws of these older model stands should be sharpened after ginning 400 to 900 bales per stand, depending on the number of saws per stand and the type of cotton being ginned. Sharpening reduces the diameter of the saw. After several sharpenings, gin capacity will have been so reduced that new saws are needed. Many ginners find it uneconomical to continue to use saws that have been sharpened more than three or four times. Saws should be sharpened by the manufacturer or a competent gin service company. However, if a ginner elects to sharpen his own saws, he should take extreme care to maintain the original tooth shape and pitch. The edges of the teeth should be deburred after the sharpening.

When changing saws, it is best to keep saws of the same diameter on a mandrel. If new saws are mixed with saws that have been reduced 1/32 or 1/16 in. in diameter, it will be impossible to adjust the gin breast properly to the saws. At least one spare saw cylinder should be on hand during the ginning season.

Doffing System

Brushes

For proper doffing, the brush should mesh to the depth of the saw tooth. Belts driving the brush must be kept tight to maintain proper speed for doffing and to provide sufficient air velocity in the lint flue to prevent backlash.

The brushes should be examined periodically and replaced if the bristles are badly worn. They should be returned to the manufacturer for repair if facilities are not available in the field for rebalancing them. Brushes that are out of balance will cause excessive vibration and bearing wear.

When replacing brushes in the stand, take care to adjust the shaft and bearings to eliminate lateral motion or "end play" and to check the setting of the brush to the saw, mote board, and cutoff plate according to the manufacturer's recommendations.

Airblast Nozzles

The manufacturer gives specific instructions for setting the airblast nozzle for each model of gin stand; follow these instructions to ensure proper doffing and moting. Keep the nozzle clean and free of tags, and avoid damaging the nozzle or the gin saws when removing a saw cylinder. Many tags may be prevented by removing the cap from the end of the airblast trunk periodically and starting the fan to blow out the accumulation of dust and fly.

Check the airblast pressure frequently to make certain that correct doffing pressure is being maintained. For most gins, the correct pressure will be in the range of 10 to 19 in. of water. Sufficient pressure should be maintained to give proper doffing without wasting power. Manufacturers can supply an adjustable airblast fan valve by which the air pressure can be readily adjusted. It is important that the airblast fan inlet be equipped with a screen to keep foreign matter out of the fan. It is equally important that the screen be kept clean.

Moting System

Modern saw gin stands are equipped with both overhead and gravity moting systems. The seals on the overhead moting system, whether dropper wheel or roller type, should be kept in good condition. In some gins, pressure is maintained in the overhead moting chamber, and the system will not operate properly if the seals leak excessively. Honeydew and green wet lint sometimes cause motes to be sticky and build up on the moting bars, wiper flights, and roller seals, drastically reducing moting system effectiveness. Under extreme conditions the buildup will cause chokage in the gin ribs. When sticky motes are encountered, clean the moting bars and seals frequently with fine steel wool, and coat the surfaces with a light textile oil, silicone, or Teflon® spray.

Proper adjustments are also important on gins having a mote board in the gravity moting system. To adjust the mote board, open it slowly until it starts dropping lint, then close it slightly. The wider the mote board can be opened without losing lint, the more effectively the system will operate.

Ginning Capacity

The high-capacity saw gin stands now on the market are the result of years of research. They will give good service as long as they are kept in good condition and in proper adjustment. Manufacturer's capacity recommendations for 1962 and to 1979 model gin stands are given in Table 5. The gins can be made to gin faster, but the quality of the cotton will be reduced.

Roller Gin Stands

Although American-Pima cotton production is limited to less than 100,000 bales/year — or about 0.7% of the American upland production — the price averaged 23.31 ¢/lb more than the American upland price during the 5-year period, 1973 to 1978. Therefore there are certain markets that demand this silky-fine-fibered, extra-long staple type of cotton, and ginning it is an important part of the total cotton industry. Of the 2700 operating gin batteries in the U.S. in 1978, 43 were equipped with roller gin stands.

Although for many years American-Pima cotton was given special care in harvesting and ginning, research has shown that it can be mechanically harvested, dried, cleaned, and handled much the same as American upland cotton. However, it is roller ginned. Tests have shown that saw ginning increases the neps approximately threefold and reduces the staple length approximately 3/32 in. Prior to the development of Pima S1, it was practically impossible to gin American-Pima cotton on a conventional saw gin. The seed of Pima was so smooth and the fiber was so easily removed from the seed that a seed roll could not be maintained. Pima S1 had a semifuzzy seed as do the

Table 5
SAWS AND CAPACITIES OF VARIOUS MAKES AND MODELS OF SAW GIN STANDS RECOMMENDED BY MANUFACTURERS[a]

Manufacturer	Model year	Saw cylinders (no.)	Saw diameter (in.)	Saws (no.)	Capacity (bales/hr)
Bush-Hog/Continental®	1962	1	12	120	3—3.5
	1962	1	16	79	4
	1962	1	16	119	6
	1973	1	16	93	5
	1973	1	16	141	7.5
	1979[b]	1	16	141	12
Hardwicke-Etter Co.	1958	1	12	100	3
	1959	1	12	120	3½
	1962	2	[c]	178	5
	1963	2	[c]	224	6½—7
	1974	2	[c]	252	8
	1974	2	[c]	200	6
Lummus Industries, Inc.	1962	1	12	88	4—5
	1973	1	12	128	8
	1977	1	12	108	6.5
	1977	1	12	158	10
Murray®-Piratininga	1962	1	12	120	3
	1962	1	18	80	4—5
	1968	1	18	94	5—6
	1968	1	18	120	6
	1973	1	18	142	8

[a] Data are averages and will vary considerably for different types of cotton.
[b] Equipped with a seed tube in the seed roll box.
[c] Bottom saw, 12 in.; top saw, 11¼ in.

current varieties Pima S4 and Pima S5, so they can be saw-ginned using special adjustments for comparison tests. The industry has long wanted to saw gin American-Pima cotton because roller ginning is a slower process than saw ginning. However, the reductions in quality cannot be tolerated, so roller ginning remains a part of the industry.

The rotary knife roller gin stand was developed to give the industry a higher capacity machine capable of delivering desirable fiber quality. The U.S. manufacturers of roller gins have shifted their production to the rotary knife principle and most McCarthy-type roller gin stands in the U.S. have been replaced with the new high capacity units. Table 6 shows the specifications and recommended capacities of current model rotary knife commercial gin stands.

American-Pima cotton is characteristically a late maturing cotton. It is also unique in that chemical defoliants are not effective; however, since it matures late and is grown under irrigated conditions, a high degree of natural defoliation can be achieved under proper fertility and water management. With the changeover to mechanical picking, the crop arrives at the gin with about the same moisture and trash content as the American upland crop. Roller gins are equipped with the same basic overhead drying, cleaning, and extracting equipment as saw gins. In fact, some ginning companies have arranged their facilities so that the overhead equipment can be used for saw or roller ginning by directing the flow of seed cotton to the desired gin stands at the seed cotton distributor.

Feeding seed cotton to a rotary knife roller gin is a unique problem due to the principle of operation of the machinery. On the Whitney saw gin and the McCarthy roller

Table 6
ROLLER GIN STAND SPECIFICATIONS AND CAPACITIES RECOMMENDED BY MANUFACTURERS

	Model	Roller length (in.)	Roller diameter (in.)	Roller (r/min)	Capacity bales/hr
Continental®	Rotobar	40	15	110	1.0
Hardwicke-Etter	1962 Hi-Cap	40	15	80	5/8—7/8
Murray®	1977 RG-40	40	15	86	0.6—1.2
Lummus	1979	40	15	100	1.0—1.6

gin, seed cotton is held at the ginning point until all fibers are removed, then the clean seed is allowed to drop out of the process.

On the rotary knife gin, the fiber clings to the surface of the roller and is pulled past the stationary knife. The rotary knife strips away the seed sometimes with unginned fiber, and it is carried over into the seed conveyor. Partially ginned seed, carry-over cotton, is reclaimed and recirculated until it is thoroughly ginned. This carry-over characteristic requires that seed cotton be fed to the stand in a continuous one-layer sheet. Ideally, this sheet would be one unit thick or one seed with its attached fiber, but practically, if the sheet could be one lock thick and all the seed in each lock could hit the roller in one plane, then the fiber of each seed would be in contact with the roller and would be removed. An experimental feeder to accomplish this has been developed at the U.S. Department of Agriculture Southwestern Cotton Ginning Laboratory. It is designed to throw single locks of seed cotton onto the roller of the rotary knife gin rather than letting the cotton slide down a feeder apron. The experimental feeder, compared to a conventional feeder, greatly increases the capacity of the rotary knife roller gin stand.

Any portion of the roller at the ginning point not covered with cotton is nonproductive, so uniform single-layer feeding is basic to efficient rotary knife roller gin stand ginning. Efficiency can be further improved by increasing the speed of the roller and the stationary knife pressure on the roller. As these changes are made, the chances for chokage increase and closer supervision is required. From a practical standpoint, rotary knife roller gins are starved to avoid process interruption due to chokage that may result in roller damage. Today the manufacturers are offering automatic controls such as stationary knife pressure relief to avoid heat damage to the roller when cotton is not available to correct this situation.

Originally, roller gin rollers were covered with walrus leather, but today rubberized canvas water pump packing material is the accepted roller material. This material should have 16 to 18 layers of fabric per inch of thickness. Its hardness should be in the range of 50 to 63 — with 56 the norm — measured by a type DO durometer. It can be wound around a wooden core or it can be fabricated as discs stacked on a square steel core. Fiber bristles of the fabric should project 0.02 to 0.04 in. beyond the rubber surface. This type of surface allows the cotton fiber to become embedded in the roller surface and then be pulled from the seed by the action of the stationary knife and the roller.

Mote grooves are sometimes cut into the roller surface on a 30° spiral pattern to temporarily relieve the pressure on the roller and allow motes to pass through the ginning point with the fiber. Otherwise, motes would tend to build up on the stationary knife and cause chokage. Mote grooves can be cut or recut after wear with a portable electric saw, a hand-held vee type wood carving tool, or a special precision lathe attachment developed at the U.S. Department of Agriculture Cotton Ginning Laboratory at Mesilla Park, N.M. The portable tools are not satisfactory because the grooves cannot be cut with proper precision.

Static electricity is a problem in roller ginning due to friction among the stationary knife, the cotton, and the ginning roller. Sufficient static charge is generated in the lint during ginning to cause it to adhere to the roller. This can damage the knife and roller. Various methods for control have been tried. Spinning mill type misting nozzles have been found unsatisfactory due to accumulation of moisture on the machinery. High potential ionization bars have been used, but they are not satisfactory because ionization could not be accomplished at a location that will allow the roller to shed the cotton satisfactorily. An antistatic agent, Tween-20 ®, is also effective in controlling static. Some ginners apply this material directly to the roller twice a day, using about 1 gal of Tween-20 ® per 100 bales of cotton.

Since some of the seed cotton is carried over the ginning point by the rotary knife, a means for reclaiming this cotton must be provided. In a system developed at the Southwestern Ginning Research Laboratory, the seed and seed cotton are carried by a conveyor belt under the stands to a seed dropper at the end of each battery. During the travel down the belt, the seed tends to settle out, leaving the seed cotton riding free on top of the seed. This seed cotton is picked up by a reclaimer saw and scrubbed over 3/8-in. grid bars spaced approximately 1½ in. on centers with ½ in. clearance between the grid bars and the reclaimer saws. The seed dislodged by the first saw falls back onto the conveyor belt and that dislodged by the second saw drops through a sheet metal chute directly into the seed dropper. The reclaimed seed cotton is sucked off the upper saw by a direct suction to the suction line separator immediately before the separator. The clean seed is then delivered to a small-pipe pneumatic seed line. The seed and seed cotton from an auger system can be passed through a reclaimer unit between the auger outlet and the seed dropper that feeds the small-pipe pneumatic seed line. Since some of the seed cotton is hard lock or immature cotton, it is advisable to feed the reclaimed cotton back into the system during ginning rather than accumulate it to be fed back into the system from the overflow. Feeding this reclaimed cotton as overflow cotton could conceivably result in a two-sided bale.

Lint Cleaning in Saw Gins

Lint cleaners were developed specifically for removing leaf particles, motes, grass, and bark left in the cotton by seed-cotton cleaners and extractors. They have been developed and improved in conjunction with the transition from manual to mechanized harvesting of the cotton crop, which is virtually complete. Nearly all gin batteries in the U.S. have lint-cleaning facilities, and over 85% of all gin batteries with lint cleaners have two or more stages of cleaning. The lint cleaners now being marketed are of two general types, the flow-through air type and the controlled-batt saw type.

Flow-Through Air Lint Cleaner

The flow-through air lint cleaner, commercially known as the air jet, has no saws, brushes, or moving parts. It is usually installed immediately behind the gin stand. Loose lint from the gin stand is blown through a duct within the chamber of the cleaner. Air and cotton moving through the duct make an abrupt change in direction as they pass across a narrow trash-ejection slot. Foreign matter that is heavier than the cotton fibers and not too tightly held by fibers is ejected through the slot in inertial force. The amount of trash taken out by the air jet is controlled by opening and closing this cleaning slot (which can be completely closed). Vane-axial fans pull the air and lint to suction condensers, where air and cotton are separated. A critical factor in the operation of air cleaners is that a vacuum of 2 to 2½ in. of water must be maintained in the duct on the discharge side of the cleaner.

Flow-through air lint cleaners are less effective in improving the grade of cotton than saw lint cleaners, but they also remove less weight from the bale. Fiber length and strength are unaffected by the air lint cleaner.

Controlled-Batt Saw Lint Cleaner

The controlled-batt saw cleaner is now the most common in the ginning industry. Lint from the gin stand or another lint cleaner is formed into a batt on a condenser screen drum. That batt is then fed through one or more sets of compression rollers, passed between a very closely fitted feed roller and feed plate or bar, and fed onto a saw cylinder. Each set of compression rollers rotates slightly faster than the preceding series and produces some thinning of the batt. The feed roller and plate grip the batt so that a combing action takes place as the saw teeth seize the fibers; the feed plate clears the saw by about 1/16 in. The teeth of the saw cylinder convey the fibers to the discharge point. While the fibers are on the saw cylinder, which may be 10 to 17 in. in diameter, they are cleaned by a combination of centrifugal force, scrubbing action between saw cylinder and grid bars, and gravity assisted by an air current. The fibers may be doffed from the saw teeth by a revolving brush, airblast, or air suction. When brush doffing, the ratio of the tip speed of the brush to the tip speed of the saw varies from about 1.5:1 to 2:1. The principal features and settings of saw lint cleaners are given in Table 7, and their recommended installations and power requirements are given in Table 8.

An important factor in the operation of the cleaner is the condition of the batt as it is fed to the saws. If the batt is thicker on one side than on the other, or if it is too thin or broken, the lint cleaner will not operate properly. Poor batt conditions can also cause chokages and may damage the equipment. A good batt can be obtained only by having the proper air balance between the gin stand, the condenser, the duct fan, and the lint flue. Axial or vane-axial fans, often referred to as clean-air fans, are usually used to discharge air from lint cleaner condensers.

Combing Ratio and Saw Speed

In saw lint cleaners the uniformity and thickness of the batt and the manner in which it is delivered to the saw teeth are important factors in the effective operation of the cleaner. These factors are controlled by the manufacturer, but in some cases may be modified by the ginner by regulating the combing ratio and saw speed of the lint cleaner. Combing ratio is the ratio of the rim speed of the saws to the rim speed of the spline feed roller.

Increasing the combing ratio and saw speed increases the cleaning efficiency of the cleaner, but can affect the spinning qualities of the cotton. Most lint cleaner installations operate within a combing ratio range of about 16 to 28 at saw-cylinder speeds of 800 to 1200 r/min (2515 to 5025 ft/min saw-tip speed) for normal feed rates. The ginner should not attempt to change these settings without first consulting the manufacturer. Higher ratios and saw speeds will result in high fiber breakage and nep counts, while lower ratios and speeds considerably decrease cleaning efficiency.

Multiple Lint Cleaning

Lint cleaners are referred to as unit or bulk (battery), depending on whether they process lint from one or more gin stands. The unit cleaner is located behind a gin stand and receives lint only from that stand. This is always the case for the air lint cleaner in saw gins. A bulk cleaner receives lint from two or more gin stands. Two lint cleaners, either unit or bulk, placed in series so that the same lint passes through both of them results in what is variously called tandem, dual, or double lint cleaning.

Table 7
PRINCIPAL FEATURES AND SETTINGS OF SAW LINT CLEANERS

Manufacturer and make	Saw diameter (in.)	Saw speed (r/min)	Grid bars (no.)	Feed-roller diameter (in.)	Settings[a]			Doffing system
					Feed plate to saws (in.)	Feed roller to feed plate	Grid bar to saw (in.)	
Bush-Hog/Continental® Gin Sixteen D								
66-in.	16	800—850	8	4 7/16	1/16	0.020-in., spring-loaded	1/16	Brush, 17¾-in. diameter
94-in.	16	866	8	4 7/16	1/16	0.020-in., spring-loaded	1/16	Brush, 17¾-in. diameter
Model 580	12	700—800	5	2 5/8	1/16	0.0187-in.	1/16	Brush, 17¾-in. diameter
DFB	16	1000	8	4 7/16	1/16	Spring-loaded, 150-lb tension	1/16	Brush, 17¾-in. diameter
Rebel and Commander®	14	800—1000	5	4½	1/16	0.010-in., floating-spring-loaded	1/32 top, 3/32 heel	Brush, 18-in. diameter
Cleanmaster	14	800	5	4½	1/32	0.010-in., floating-spring-loaded	1/32 top, 3/32 heel	Brush, 18-in. diameter
Super Cleanmaster	14	800—1050	5	4½	1/16	0.010-in., floating-spring-loaded	1/32 top, 3/32 heel	Brush, 18-in. diameter
Constellation	16	800—1000	5	4½	1/16	0.010-in., floating-spring-loaded	1/32 top, 3/32 heel	Brush, 18-in. diameter
Super Constellation	16	800—1000	5	4½	1/32	0.010-in., floating-spring-loaded	1/32 top, 3/32 heel	Brush, 18-in. diameter
Revelation®	16	800—1000	6	4½	1/16			
Super Revelation®	16	800—1000	5	4½	1/16	0.010-in., floating-spring-loaded	1/32 top, 3/32 heel	Brush, 18-in. diameter

Lodestar	16	800—1000	5	4½	1/16	0.010-in., floating-spring-loaded	1/32 top, 3/32 heel	Brush, 18-in. diameter
Hardwicke-Etter Co.								
Challenger®								
Model 650	13¼	1100	5	4 7/16	1/16	Floating-spring-loaded	1/16	Brush, 18-in. diameter at 1450 r/min
Model 840	13¼	1100	5	4 7/16	1/16	Floating-spring-loaded	1/16	Brush, 18-in. diameter at 1450 r/min
Conqueror								
Model 650	13¼	1100	5	4 7/16	1/16	Floating-spring-loaded	1/16	Brush No. 1, 12-in. diameter; Brush No. 2, 18-in. diameter
Model 840	13¼	1100	5	4 7/16	1/16	Floating-spring-loaded	1/16	Brush No. 1, 12-in. diameter; Brush No. 2, 18-in. diameter
Model 940	16	850	7	4 7/16	3/32	Floating-spring-loaded	1/16	Brush, 18-in. diameter 1520 r/min
Model 941	16	1030	5	4 7/16	3/32	Roll fixed bar, 28 sections, spring-loaded	1/16	Brush, 18-in. diameter 1620 r/min
Lintmaster	16	1000—1500	5	3 5/8	1/16	Roll fixed bar, 23 sections, spring-loaded	1/16	Air blast, 14- to 16-in. (water gage)
Horn and Gladden								
Super 66	16	800—1050	5	4 7/16	1/32	0.010 in., floating-spring-loaded	1/32 top, 3/32 heel	Brush, 18-in. diameter
Super Twin 66	16	800—1050	5	4 7/16	1/32	0.010 in., floating-spring-loaded	1/32 top, 3/32 heel	Brush, 18-in. diameter
Super 86	16	800—1050	5	4 7/16	1/32	0.010 in., floating-spring-loaded	1/32 top, 3/32 heel	Brush, 18-in. diameter

Table 7 (continued)
PRINCIPAL FEATURES AND SETTINGS OF SAW LINT CLEANERS

Manufacturer and make	Saw diameter (in.)	Saw speed (r/min)	Grid bars (no.)	Feed-roller diameter (in.)	Settings[a] Feed plate to saws (in.)	Settings[a] Feed roller to feed plate	Settings[a] Grid bar to saw (in.)	Doffing system
Super Twin 86	16	800—1050	5	4 7/16	1/32	0.010 in., floating-spring-loaded	1/32 top, 3/32 heel	Brush, 18-in. diameter
Lummus® Industries, Inc.								
Model 86	16		6	5 3/8	1/16	0.005-in., floating-spring-loaded	1/32 top, 3/16 heel	Brush, 15-in. diameter; or air suction, 2-in. vacuum (water gage)
Model 108	16	800—1200	6	5 3/8	1/16	0.005-in., floating-spring-loaded	1/32 top, 3/16 heel	Brush, 15-in. diameter; or air suction, 2-in. vacuum (water gage)
Murray®-Carver								
Combing	12	700—980	8	3	0.050—0.060	1/16-in. carding 0.010, grooved	0.050—0.075	Brush
Big 60	14¾	1000—1400	4	4½	.050—.060	1/16-in., carding 0.010, grooved	0.050—0.075	Brush
Big 72	17	880—1200	4	4½	.050—.060	1/16-in., carding 0.010, grooved	0.050—0.075	Brush
Big 84	17	880—1200	4	4½	0.050	1/16-in., carding 0.010, grooved	0.050—0.075	Brush
Seventeen	17	800	6	4½	.060	0.010-in.	0.050—0.075	Brush

[a] All makes of machinery have a 1/16-in. feed-roller-to-saw setting.
[b] A speed of 1000 r/min is most popular.

Table 8
POWER REQUIREMENTS AND RECOMMENDED INSTALLATION OF SAW LINT CLEANERS

Make	Condenser diameter (in.)	Horsepower	Clean-air fan: Type and size[a] (in.)	Clean-air fan: Horsepower	Recommended installation	Capacity (bales per hour)
Bush-Hog/Continental Gin						
Sixteen D						
66-in.	24	5 per unit	V.A., 26	15, 20, or 25	Unit behind each gin stand, single and tandem	4—6
94-in.	24	10—20	V.A., 26, 1 or 2 per battery	25	Unit behind each gin stand, single and tandem	8
Model 580	20	3 per unit	Prop., 18	15	Unit	1—2
DFB	[b]	3		20	Battery units	6—8
Rebel	40	5	C.A., 36	20	Battery unit	3—4
Commander	40	5 or 7½	C.A., 36	20	Battery unit or tandem, split stream	6—8
Cleanmaster	40	5—10	C.A., 36 or 42	20	Battery unit or tandem	5
Super Cleanmaster	40	7½	C.A., 36 or 42	Two 20s	Battery unit or tandem, split stream	10—12
Constellation	50	7½—10	C.A., 42	25	Battery unit or tandem	7—8
Super Constellation	50	7½—10	C.A., 42, 2 per battery	Two 25s	Battery unit or tandem, split stream	18—20
Revelation	50	10	C.A., 42	25	Battery unit, including battery condenser	7—8
Super Revelation	50	10	C.A., 42, 2 per battery	Two 25s	Battery unit, split stream, including battery condenser	18—20
Lodestar	24	10	V.A., 26	15	Unit behind each gin stand, tandem	7
Hardwicke-Etter Co.						
Challenger						
Model 650	20	7½ per unit	V.A., 21 unit	15	Unit behind each gin stand, single and tandem	5—6
Model 840	20	10 per unit	V.A., 21 unit	20	Unit behind each gin stand, single and tandem	7—8

Table 8 (continued)
POWER REQUIREMENTS AND RECOMMENDED INSTALLATION OF SAW LINT CLEANERS

Make	Condenser diameter (in.)	Clean-air fan		Horsepower	Recommended installation	Capacity (bales per hour)
		Horsepower	Type and size[a] (in.)			
Conqueror						
Model 650	20[c]	3 H.P. bottom 18 7½ H.P. top 21	V.A., one 18 and one 21	25	Unit behind each gin stand for two-stage lint cleaner	5—6
Model 840	20[c]	5 HP. botton 18 10 H.P. top 21	V.A., one 18 an one 21	30	Unit behind each gin stand for two-stage lint cleaner	7—8
Model 940	30	15 per unit	V.A., 26 unit	25	Unit behind each gin stand, single or tandem	8—9
Model 941	30	15 per unit	V.A., 26 unit	25	Unit behind each gin stand, single or tandem	8—9
Lintmaster	30	5	C.A., 28	25	Battery unit single, tandem, or split stream	6—8, per unit
Horn and Gladden						
Super 66	50	10	V.A., 29 per unit	20	Unit behind each gin stand, single and tandem	6—8
Super Twin 66	50	30	V.A., 36	Two 20s	Battery	12—16
Super 86	50	15	V.A., 29 per unit	25	Unit behind each gin stand, single and tandem	10—12
Super Twin 86	50	50	V.A., 42	Two 25s	Battery	20—24
Lummus Industries, Inc.						
Model 86	30	5—20 as required	V.A., 24, per single or double machine	25, air; 30, brush; per machine	Behind each gin stand, single and tandem, or split systm	6—9 per machine, 3—4½ when split
Model 86	30	5—20 as required	V.A., 24, per single or double machine		Battery units, single and split stream	6—9 per machine

Model 108	30	5—25 as required	V.A., 24, per machine; V.A., 36 per two machines	40		Behind each gin stand, single and tandem, or split stream	8—12 per machine, 4—6 when split
Model 108	30	5—25 as required	V.A., 24, per machine; V.A., 36 per two machines	40		Battery units, single and split stream	8—12 per machine
Murray-Carver							
Type F combing lint cleaner	20					Unit behind each stand	2—4

[a] All fans are axial-flow type. Manufacturer's description is shown (V.A. = vane axial, Prop. = propeller, C.A. = clean-air).
[b] Battery.
[c] Two condensers.

For high-capacity saw gin stands batteries, the bulk cleaners are sometimes installed back-to-back with a common lint flue. In that way the lint from all the gin stands is divided ("split stream") and each saw receives only half of the cotton. The number of stages of saw cleaning refers to the number of saws over which the fibers pass, i.e., tandem-split stream lint cleaning.

Lint Cleaner Waste

The amount of material removed by lint cleaning depends on harvesting practices and grades of cotton being ginned. With multiple stages of saw lint cleaners, the first cleaner removes the most weight; the second, about half as much as the first; and the third, about half as much as the second. Typical quantities of waste removed by one, two, and three lint cleaners are 11, 18, and 22 lb for cotton graded as Strict Low Middling before cleaning; corresponding weights are 19, 25, and 36 lb for cotton classed as Strict Good Ordinary.

Lint cleaners remove a greater percentage of lint from the cleaner cottons. Nonlint content of the lint-cleaner waste for one, two, and three stages of lint cleaning averages about 75, 70, and 67%. This material is composed of motes, leaf particles, grass, bark, and green leaves in varying amounts, depending on the condition of the cotton.

The sale of reclaimed gin waste may provide income for cotton gin owners. The bedding, automotive, and furniture industries use large quantities of cotton batting composed of about 60% linters, and 40% cotton waste. If waste materials must be cleaned to make them marketable, and if the premium is less than 1¢, the weight loss in cleaning should not exceed 20% of the original weight.

Cotton Quality

Perhaps the best index to cotton quality is the performance of the fibers during spinning at the mill. Increasing the number of saw lint cleaners at the gin decreases the manufacturing waste during spinning, but often has the adverse effect of increasing neps in the card web and lowering the yarn strength and appearance. A decline in appearance is greater for the finer carded yarns. From a spinning standpoint, the use of more than two saw lint cleaners in series is discouraged.

Bale Value

Lint cleaning generally improves the grade classification of the lint. As the number of lint cleaners increases, grade tends to increase. However, each succeeding cleaner gives less grade improvement than the preceding one. In addition, lint cleaners blend Light-Spotted cottons so they pass into the White grades and decrease the number of bales which are reduced in grade because of the grass and bark content.

But when grades are improved, bale weights are reduced and staple length may decrease; and these opposing factors affect bale value. In some cases the offsetting losses cause the bale value to be reduced by lint cleaning.

When price spreads between grades are small, the grower can obtain maximum bale value most often by using one saw lint cleaner on early-season clean cottons and two stages of lint cleaning on late-season, more trashy, or Light-Spotted cottons. It is recommended that no lint cleaning be used on cottons graded as high as Middling White. Application of these recommendations would depend on the ability of the ginner to evaluate the type of cotton he is processing.

Operation and Maintenance

The ginner and crew must know their machines — how they work and the required operating settings and speeds. For best performance and service, the recommendations of the manufacturer as provided in service manuals and parts catalogs should be carefully followed. These provide maintenance procedures and trouble-shooting guides, close adherence to which will minimize the time required to keep machines in normal operation. When in doubt regarding appropriate action, contact the manufacturer's nearest office.

Maintenance of the condenser will require occasional replacement of the rubber flashing around the drum and across doffing rollers. The feed roller rubber flashing will wear and should be replaced to prevent air leakage at this point.

Maintain normal operating tension for V-belts and chain drives. Do not forget the hidden V-belt drive to the vane-axial fan in the condenser dust flue. When replacing badly worn chains, it is good policy also to replace sprockets on which they operate because their worn teeth will not mate well with the new chain. Test regularly for loosened set screws, bolts, and jam nuts. Before tightening any bolt responsible for holding a clearance, gage and set the clearance to comply with recommended dimension.

Most ball bearings are the sealed type and are packed with high-temperature grease; they should need no further lubrication for the life of the bearing.

To start the lint-cleaning system, first turn on the trash fan. Be sure the lint cleaner and the condenser mounted on the lint cleaner are in operation before energizing the vane-axial fan in the receiving condenser dust flue. Shutdown of the lint-cleaning system should be in reverse order, first stopping the flow of cotton through the machinery. Always keep safety guards and doors closed and fastened during operation.

The sequence of the emergency stopping procedure depends entirely on the position of the operator. If he is nearer the machine than the control console, the belt release should be pulled before the "Stop" button is pushed. Any obstruction that will tag cotton should be removed or smoothed. In flow-through air cleaner installations, inspect each day for tags in the lint flue between gin stand and lint cleaner. Periodic inspection of saw cylinders is urged. Should trash become embedded between teeth, remove it with a stiff brush (but only when the machine is completely stopped). Vane-axial fans should be kept free of lint tags, which reduce fan efficiency and can cause friction fires.

Following a regular service schedule based on the manufacturer's recommendations will enable the ginner to keep his lint cleaner operating efficiently with minimum overall maintenance.

Lint Cleaning in Roller Gins

Today, lint cleaning of American-Pima cotton is a standard practice in U.S. roller gins. The equipment used is a combination of either an Aldrich mill type beater opener and a Lummus air jet lint cleaner or a modified seven cylinder inclined cleaner and a Lummus air jet lint cleaner.

Either of these systems are very effective in improving the lint grade by removing motes, broken seed, fiber entanglements caused by the machine pickers, and pin trash that had not been removed by the seed cotton cleaning machinery.

The U.S. Department of Agriculture Southwestern Cotton Ginning Research Laboratory, Mesilla Park, N.M. is carrying on cooperative research with commercial gins to further evaluate the use of these systems on machine picked American-Pima cotton.

Packaging of Lint Cotton

Packaging is the final step in the processing of lint cotton at the gin. The packaging system consists of a lint condenser, lint slide, lint feeder, tramper, bale press, and a strapping subsystem. The bale press consists of a frame, one or more rams, and a hydraulic system. The strapping subsystem may be entirely manual, semiautomated, or fully automated.

Bale presses are described primarily by the density of the bale that they produce, such as low, standard, or universal. There are other descriptions, such as up-packing, down-packing, fixed-box, and doorless presses. Regardless of description, they all accomplish the same thing — packaging lint cotton so that it can be handled in the channels of trade.

Battery Condenser

Condensers employ a slow-turning, screened drum on which the ginned lint forms into a batt. The batt is discharged between doffing rollers to the lint slide. Conveying air passes through the screen on the drum and is usually discharged out one end of the drum through an air duct and into the atmosphere.

Most condensers employ a suction fan in the air discharge line to overcome the static pressure loss in the system. This also makes it possible to handle more lint with a smaller condenser and lint flue system.

The speed of the condenser drum should be adjusted to the capacity of the gin so that a smooth, solid batt is formed. If the drum runs too fast, breaks will occur in the batt; if too slow, the batt will be too thick, causing high static pressure loss and possible chokage. The recommended condenser drum speed is 6 to 15 rpm.

Uneven formation of the batt in the condenser is one of the primary causes of big-ended bales. Big-ended bales are formed when more lint is deposited at one end of the press box than at the other. During the pressing operation, the increased density or weight at one end of the bale forces the follow block into an inclined position, thus forming wedge-shaped bales. Press hands may leave slack ties on the light end of the bale, which prevent the uneven weight from being apparent when the bale is released from the press.

Some bales appear to be big-ended, although they are not. For instance, a hydraulic ram can slip after the bale is pressed and while the ties are being applied. Between the time the first and last ties are applied, the bale becomes larger at one end than at the other. The dangers of tramper breakage and press distortion are great with uneven bales. A ginner cannot afford to risk a broken ram or sprung center column, particularly when uneven packing can be eliminated.

The manner in which the lint is deposited on the condenser drum can cause a big-ended bale. Lint cotton is conveyed to the condenser from the gin stands or lint cleaners by a continuous air current. Any deviation in the airflow to the condenser will automatically disrupt the flow of the lint cotton. Lint flues that are bent and battered can sometimes affect the flow of air so that lint will not travel through them uniformly.

Sheet-metal deflectors are used in lint flues to provide uniform distribution of the lint on the condenser drum to prevent or correct big-ended bales. Foreign matter such as tags, rivet heads, rust, or protrusions can often cause deflection of the air current, resulting in uneven batts on the screen of the condenser.

Condensers in poor operating condition, flashings in bad condition, and nonuniform drum resistance may cause more cotton and air to be deflected to one end of the drum and thus produce big-ended bales. Daily inspection of the condenser and its vent pipes is a good preventive measure and will reduce the number of big-ended bales.

Battery condensers that discharge the conveying air from only one side are sensitive to the volume of airflow. Excess airflow will cause the batt to form on the drum nearest the air discharge and cause a big-ended bale. This problem can usually be corrected by reducing the air volume; however, installation of a larger condenser may be necessary. Use of a double manifold condenser, similar to a lint cleaner condenser, would be another way to correct the problem of big-ended bales and, at the same time, produce a more uniform batt of cotton for the press.

Lint Slides

The lint slide is a sheet metal trough (approximately 54 in. wide) which connects the battery condenser to the lint feeder on the tramper. It is installed on an angle of 40 to 45° to ensure movement of the lint. The length of the lint slide is based on the capacity of the ginning system and the time required to turn the press between bales. The following is a general guide for selecting lint slide lengths based on ginning capacity:

Maximum ginning capacity (bales/hr)	Lint slide length (ft)
6	10
12	12
18	14
24	16
30	18

Lint Feeder

The lint feeder is a device for moving lint from the lint slide into the charging box of the press. There are four basic types of feeders: (1) the revolving paddle kicker, (2) the belt feed used in conjunction with the kicker, (3) the lint pusher, and (4) the air-suction feed.

All of these devices are intended to deposit lint into the charging box with a fast but gentle action, without agitating, turning, rolling, tumbling, tearing, or otherwise breaking up the batt as it is received from the condenser. A smoother lint sample will result if this can be accomplished.

Rolling bales are caused primarily by faulty action of the kicker device. Proper adjustment of the kicker speed and synchronization with the tramper is necessary to achieve good results. The speed or rate of sweep is the principal factor affecting the front-to-rear distribution of lint in the press box. If the discharge is too strong, more lint will be deposited in the front of the box than in the rear. The reverse is true if the discharge is too weak. Variations in tramper speed seem to have little effect on uniformity of bale layering. However, if the kicker is driven from the tramper, a change in tramper speed will affect unevenly packed bales because of the change in kicker speed.

Sometimes rolling bales are produced when the lint slide is too short or has too much slope. The lint slide shape can be adjusted to prevent rolling bales. The ginner, however, is still responsible for making reasonable efforts to have cotton delivered in an even flow into the gin press box.

Tramper

The purpose of the tramper is to pack the lint into the press box, thus putting it under light compression. A good tramper combines ruggedness with simplicity. It must be regulated to permit clearing the lint slide between strokes. A kicker device deposits lint in the press box between strokes of the tramper. The speed of the kicker is related to the time the tramper is up, allowing cotton to be "kicked" into the top of the press box.

Trampers are available in different types, such as mechanical double-chain and hydraulic. Regardless of type, care should be taken to prevent contamination of lint beneath the tramper by lubricants or hydraulic fluid that might drip from the tramper mechanism. Motors of 10 to 15 hp, equipped with a fail-safe brake, are usually required for trampers.

Approximately 10 tramper strokes per minute is a rate recommended for high-ca-

Table 9
TYPES AND DIMENSIONS OF THE FOUR MOST COMMON BALES PRODUCED AT COTTON GINS

	Bale dimensions (in.)		
Bale type	Length	Width[a]	Thickness
Flat	55	28	36—48
Modified flat	55	25	36—48
Gin standard	55	20—21	30—36
Gin universal	55	20—21	26—30

[a] Measured at the bale tie.

pacity gins (12 to 15 bales/hr). For gins of smaller capacity, the tramper speed should be reduced proportionately. The tramper stroke rate is governed by the ginning rate for a particular gin. To determine the number, observe the cotton in the lint slide between strokes. If the lint feeder cleans the slide for each stroke, the number of strokes is correct. However, if the storage space is filled between the kicker and the front gate and the kicker is carrying cotton around between strokes after the space under the tramper block is filled, add another stroke to the tramper by changing the ratio of the tramper driving and driven pulleys.

Bale Press

There are four types of gin presses that produce bales of different sizes and densities — flat, modified flat, gin standard, and gin universal (Table 9). The theoretical densities in bales produced by these presses are shown in Tables 10 to 12.

Flat-bale presses are being phased out in favor of modified flat-bale presses so that a universal bale (55 × 25 × 21 in.) can be made at the compress without applying side pressure to the bale. A flat-bale press (27 × 54 in.) can be easily converted to a modified flat press (24 × 54 in.).

Steps in modifying a flat-bale press are as follows:

1. The 3-in. reduction in box width can be obtained by bolting finished 2- × 6- or 8-in. lumber to the inside walls. Use C-grade kiln-dried pine or clear-grade oak with a finished thickness of 1½ in. Check dimensions for dog spacing and press box construction in order to purchase lumber most economically.
2. Bolt each board to side walls with two cadmium-plated 3/8-in. carriage bolts near both ends of each board. Notch board to allow clearance for dogs. Extend dog slots downward if necessary to clear the extended dogs.
3. Lengthen dogs 1½ in. to maintain original penetration into the box by welding metal of the same thickness to end of dogs and grinding smooth.
4. Reduce both follow blocks 3 in. in width by cutting 1½ in. off each end of cross channels. Re-cover channel ends with metal to keep bale strap guides intact, and reduce bagging retainer loops on the platen by 1½ in. on each side.
5. Reduce tramper box by 3 in. Where possible, cut tramper platen width 1½ in. on each side. In some models, the full 3 in. may have to be removed from the front side of the platen and the whole tramper moved 1½ in. forward in order to stay on center with the press box. Make a thorough study of your particular press to see exactly what pieces will have to be cut or moved to maintain proper clearances.

Table 10
TYPICAL DENSITIES OF BALES FROM A FLAT-BALE PRESS[a]

Bale dimensions[b] (in.)	Volume (ft^3)	Density (lb/ft^3) at bale weight (lb)					
		460	480	500	520	540	560
55 × 28 × 36	32.1	14.3	15.0	15.6	16.2	16.8	17.5
55 × 28 × 37	33.0	14.0	14.6	15.2	15.8	16.4	17.0
55 × 28 × 38	33.9	13.6	14.2	14.8	15.4	15.9	16.5
55 × 28 × 39	34.8	13.2	13.8	14.4	15.0	15.5	16.1
55 × 28 × 40	35.6	12.9	13.5	14.0	14.6	15.1	15.7
55 × 28 × 41	36.5	12.6	13.1	13.7	14.2	14.8	15.3
55 × 28 × 42	37.4	12.3	12.8	13.4	13.9	14.4	15.0
55 × 28 × 43	38.3	12.0	12.5	13.0	13.6	14.1	14.6
55 × 28 × 44	39.2	11.7	12.2	12.8	13.3	13.8	14.3
55 × 28 × 45	40.1	11.5	12.0	12.5	13.0	13.5	14.0
55 × 28 × 46	41.0	11.2	11.7	12.2	12.7	13.2	13.7
55 × 28 × 47	41.9	11.0	11.5	11.9	12.4	12.9	13.4
55 × 28 × 48	42.8	10.8	11.2	11.7	12.2	12.6	13.1

[a] Press box area: 54 × 27 in. = 1458 $in.^2$.
[b] Width and thickness measured at the bale tie.

Table 11
TYPICAL DENSITIES OF BALES FROM A MODIFIED FLAT-BALE PRESS[a]

Bale dimensions[b] (in.)	Volume (ft^3)	Density (lb/ft^3) at bale weight (lb)					
		460	480	500	520	540	560
55 × 25 × 36	28.6	16.1	16.8	17.5	18.2	18.9	19.1
55 × 25 × 37	29.4	15.6	16.3	17.0	17.7	18.3	19.0
55 × 25 × 38	30.2	15.2	15.9	16.5	17.2	17.9	18.5
55 × 25 × 39	31.0	14.8	15.5	16.1	16.8	17.4	18.0
55 × 25 × 40	31.8	14.5	15.1	15.7	16.3	17.0	17.6
55 × 25 × 41	32.6	14.1	14.7	15.3	15.9	16.6	17.2
55 × 25 × 42	33.4	13.8	14.4	15.0	15.6	16.2	16.8
55 × 25 × 43	34.2	13.4	14.0	14.6	15.2	15.8	16.4
55 × 25 × 44	35.0	13.1	13.7	14.3	14.9	15.4	16.0
55 × 25 × 45	35.8	12.8	13.4	14.0	14.5	15.1	15.6
55 × 25 × 46	36.6	12.6	13.1	13.7	14.2	14.8	15.3
55 × 25 × 47	37.4	12.3	12.8	13.4	13.9	14.4	15.0
55 × 25 × 48	38.2	12.0	12.6	13.1	13.6	14.1	14.7

[a] Press box area: 54 × 24 in. = 1296 $in.^2$.
[b] Width and thickness measured at the bale tie.

Existing hydraulic and mechnical press equipment should have sufficient strength if the bales are tied out 3 to 5 in. thicker. If there is any doubt that any part of the tramper or press should be strengthened structurally or horsepower increased, it is suggested that the original manufacturer of the units be consulted. This is especially important on some models of down-packing presses with short boxes. The long-range goal is to have only two types of bales produced at the gin — gin universal and modified flat, which is recompressed at the compress to universal density.

Table 12
TYPICAL DENSITIES OF BALES FROM STANDARD- AND UNIVERSAL-DENSITY PRESSES[a]

Bale dimensions[b] (in.)	Volume (ft^3)	Density (lb/ft^3) at bale weight (lb) 460	480	500	520	540	560
55 × 20 × 25	15.9	28.9	30.2	31.4	32.7	33.9	35.2
55 × 20 × 26[c]	16.6	27.8	29.0	30.2	31.4	32.6	33.8
55 × 20 × 27[c]	17.2	26.8	27.9	29.1	30.3	31.4	32.5
55 × 20 × 28[c]	17.8	25.8	26.9	28.1	29.2	30.3	31.4
55 × 20 × 29	18.5	24.9	26.0	27.1	28.2	29.3	30.3
55 × 20 × 30[d]	19.1	24.1	25.1	26.2	27.2	28.3	29.3
55 × 20 × 31[d]	19.7	23.3	24.3	25.3	26.4	27.4	28.4
55 × 20 × 32[d]	20.4	22.6	23.6	24.5	25.5	26.5	27.5
55 × 20 × 33	21.0	21.9	22.8	23.8	24.8	25.7	26.7
55 × 20 × 34	21.6	21.3	22.2	23.1	24.0	24.9	25.9
55 × 20 × 35	22.3	20.6	21.5	22.4	23.3	24.2	25.1
55 × 20 × 36	22.9	20.1	20.9	21.8	22.7	23.6	24.4
55 × 21 × 25	16.7	27.5	28.7	29.9	31.1	32.3	33.5
55 × 21 × 26[c]	17.4	26.5	27.6	28.8	29.9	31.1	32.2
55 × 21 × 27[c]	18.0	25.5	26.6	27.7	28.8	29.9	31.0
55 × 21 × 28[c]	18.7	24.6	25.6	26.7	27.8	28.9	29.9
55 × 21 × 29	19.4	23.7	24.8	25.8	26.8	27.9	28.9
55 × 21 × 30[d]	20.1	22.9	23.9	24.9	25.9	26.9	27.9
55 × 21 × 31[d]	20.7	22.2	23.2	24.1	25.1	26.1	27.0
55 × 21 × 32[d]	21.4	21.5	22.4	23.4	24.3	25.2	26.2
55 × 21 × 33	22.1	20.9	21.8	22.7	23.6	24.5	25.4
55 × 21 × 34	22.7	20.2	21.1	22.0	22.9	23.8	24.6
55 × 21 × 35	23.4	19.7	20.5	21.4	22.2	23.1	23.9
55 × 21 × 36	24.1	19.1	19.9	20.8	21.6	22.4	23.3

[a] Press box area: 54 × 20 in. = 1080 in.^{2}.
[b] Width and thickness at the bale tie.
[c] Suggested universal-density bale size.
[d] Suggested standard-density bale size.

Hydraulics

The hydraulic requirement to press lint cotton depends on the moisture content of the lint and the density to which it is to be pressed. Typical densities encountered when pressing flat, modified flat and standard or universal bales are shown in Tables 13, 14, and 15, respectively.

The force required to compress lint cotton of given densities at various moisture contents may be predicted from the equation

$$\log F = 2.0929 - 0.0313M + 2.4469 \log_{10} p$$

where F = compressive force (lb), M = lint moisture content (wet basis) (%), and p = density (lb/ft^3).

The results of evaluating the prediction equation are shown in Figure 3 for a density range of 10 to 45 lb/ft^3 and lint moisture content of 3, 6, and 9%. Extrapolation beyond these values cannot be supported by research data at this time.

The theoretical force available from selected single-acting hydraulic cylinders is shown in Table 16. The hydraulic requirement can be determined from these tables for almost any bale press.

Table 13
DENSITIES PRODUCED BY PRESSING IN FLAT-BALE PRESSES[a]

Platen separation (in.)	Density in press (lb/ft³) at bale weight (lb)					
	460	480	500	520	540	560
26	21.0	21.9	22.8	23.7	24.6	25.5
27	20.2	21.1	21.9	22.8	23.7	24.6
28	19.5	20.3	21.2	22.0	22.9	23.7
29	18.8	19.6	20.4	21.3	22.1	22.9
30	18.2	19.0	19.8	20.5	21.3	22.1
31	17.6	18.4	19.1	19.9	20.6	21.4
32	17.0	17.8	18.5	19.3	20.6	20.7
33	16.5	17.2	18.0	18.7	19.4	20.1
34	16.0	16.7	17.4	18.1	18.8	19.5
35	15.6	16.3	16.9	17.6	18.3	19.0
36	15.1	15.8	16.5	17.1	17.8	18.4
37	14.7	15.4	16.0	16.7	17.3	17.9
38	14.3	15.0	15.6	16.2	16.8	17.5

[a] Press box area of 1458 in.².

Table 14
DENSITIES PRODUCED BY PRESSING IN A MODIFIED FLAT-BALE PRESS[a]

Platen separation (in.)	Density in press (lb/ft³) at bale weight (lb)					
	460	480	500	520	540	560
26	23.6	24.6	25.6	26.7	27.7	28.7
27	22.7	23.7	24.7	25.7	26.7	27.7
28	21.9	22.9	23.8	24.8	25.7	26.7
29	21.1	22.1	23.0	23.9	24.8	25.7
30	20.4	21.3	22.2	23.1	24.0	24.9
31	19.8	20.6	21.5	22.4	23.2	24.1
32	19.2	20.0	20.8	21.7	22.5	23.3
33	18.6	19.4	20.2	21.0	21.8	22.6
34	18.0	18.8	19.6	20.4	21.2	22.0
35	17.5	18.3	19.0	19.8	20.6	21.3
36	17.0	17.8	18.5	19.3	20.0	20.7
37	16.6	17.3	18.0	18.7	19.5	20.2
38	16.1	16.8	17.5	18.2	18.9	19.6

[a] Press box area of 1296 in.².

Example

Determine the diameter of ram and hydraulic requirement to compress 500 lb of lint at 3% moisture content in a modified flat press to a platen separation of 37 in. in 25 sec. Assume a ram travel of 7 ft and hydraulic pressure not to exceed 2800 lb/in.² gage.

Solution

Table 14 indicates that a density of 18 lb/ft³ would result in a modified flat press box at a platen separation of 37 in. for a 500-lb bale. Figure 3 indicates that lint cotton at 3% moisture would require aproximately 120,000 lb of force for compression to 18 lb/ft³ density. From Table 16 we find that an 8 3/8-in.-diameter ram will develop

Table 15
DENSITIES PRODUCED BY PRESSING IN STANDARD- AND UNIVERSAL-DENSITY PRESSES[a]

Platen separation (in.)	Density in press (lb/ft³) at bale weight (lb)					
	460	480	500	520	540	560
17	43.3	45.2	47.1	48.9	50.8	52.7
18	40.9	42.7	44.4	46.2	48.0	49.8
19	38.7	40.4	42.1	43.8	45.5	47.2
20	36.8	38.4	40.0	41.6	43.2	44.8
21	35.0	36.6	38.1	39.6	41.1	42.7
22	33.5	34.9	36.4	37.8	39.3	40.7
23	32.0	33.4	34.8	36.2	37.6	39.0
24	30.7	32.0	33.3	34.7	36.0	37.3
25	29.4	30.7	32.0	33.3	34.6	35.8
26	28.3	29.5	30.8	32.0	33.2	34.5
27	27.3	28.4	29.6	30.8	32.0	33.2
28	26.3	27.4	28.6	29.7	30.9	32.0

[a] Press box area of 1080 in.².

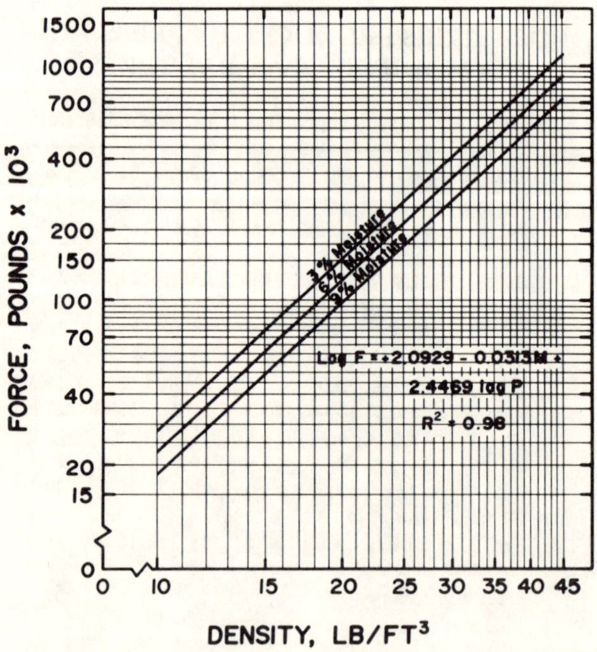

FIGURE 3. Predicted force to compress lint cotton to give densities at three moisture contents.

137,721 lb of theoretical force at 2500 lb/in.² gage. Also from Table 16 we find that an 8 3/8-in.-diameter ram would require 2.9 gal of oil/ft of stroke. The quantity of oil required per 7-ft stroke would be 2.9 × 7 = 20.3 gal. The hydraulic pump requirement can be determined by multiplying the oil required per stroke by the time in minutes — 20.3 × (60/25) = 48.7 gal/min; therefore a 49-gal/min pump would be required to raise the ram in 25 sec.

Another approach to determine the hydraulic capability for packaging lint cotton is

Table 16
THE THEORETICAL FORCE AVAILABLE FROM SELECTED SINGLE-ACTING HYDRAULIC CYLINDERS

Cylinder diameter (in.)	Oil requirement per foot of stroke (gal)	Theoretical force (lb) at hydraulic pressure (lb/in.^{2})							
		1,500	2,000	2,500	3,000	3,500	4,000	4,500	5,000
8 3/8	2.9	82,633	110,177	137,721	165,265	192,809	220,353	247,897	275,442
9	3.3	95,426	127,235	159,043	190,852	222,660	254,469	286,278	318,086
10	4.1	117,810	157,080	196,350	235,619	274,890	314,159	353,429	392,699
11	4.9	142,550	190,066	237,583	285,100	332,616	380,133	427,649	475,166
12	5.9	169,646	226,195	282,743	339,292	395,841	452,389	508,938	565,487
13	6.9	199,095	265,460	331,825	398,190	464,555	530,920	597,285	663,650
14	8.0	230,907	307,876	384,845	461,814	538,783	615,752	692,721	769,690
15	9.2	265,072	353,429	441,787	530,144	618,501	706,858	795,216	883,573
16	10.4	301,593	402,124	502,655	603,186	703,717	804,248	904,779	1,005,310
17	11.8	340,470	453,960	567,450	680,940	794,430	907,920	1,021,410	1,134,900
17 3/8	12.3	355,657	474,209	592,761	711,313	829,865	948,417	1,066,970	1,185,522

to pick a ram size and determine the bale size that can be produced in a specific type of press. For multiple-ram presses, multiply theoretical force and oil requirements in Table 16 by the number of rams.

Example

Determine the capability of a universal-bale press having two 14-in. diameter rams with a 7½-ft stroke required to obtain a 20-in. platen separation, and a 360-gal/min hydraulic system that is set to unload at 3500 lb/in.^{2} gage.

Solution

From Table 16 we find that two 14-in. diameter rams will deliver 1,077,566 lb of force at 3500 lb/in.^{2} gage. From Figure 55 we find that lint cotton at 3% lint moisture can be compressed to a density of 45 lb/ft^3 with that force. Referring now to Table 15 we can determine the capabilities of the press. This press is capable of pressing a 560-lb bale to a platen separation of 20 in. The time required for ram travel can be determined by multiplying the quantity of oil required per foot of stroke (from Table 16) by the length of stroke and dividing by the pump capacity.

$$\frac{(2)\ (8\ \text{gal/ft})\ (7.5\ \text{ft})}{360\ \text{gal/min}}$$

$$= 0.33\ \text{min or}\ (0.33)\ (60) = 20\ \text{sec}$$

The final bale size, after it is released from the press, will depend upon the bale-restraining material.

Hydraulic systems should be designed to meet the most adverse conditions anticipated, which would be the heaviest bale at the lowest moisture content and packaged at the smallest size desired.

Bale Ties

After the bale is compressed to a given density, bale ties are applied around the circumference of the bale to restrain the lint cotton within certain dimensions. Bale ties are normally flat metal bands, either hot- or cold-rolled steel, or wire, and are placed at intervals along the length of the bale. Usually six, eight, or nine ties per bale are used. The resilient force of the lint cotton produces a resulting force on the bale ties, which is a function of the density to which the bale is compressed before it is released from the press, the density at which it is restrained after it is released from the press, and the moisture content of the lint cotton.

The difficulties encountered in calculating the true densities involved in packaging lint cotton necessitate the use of more easily obtainable variables. The factor "density compressed to" can be replaced by using the platen separation to which the bale is compressed, the quantity of cotton in the bale, and the cross-sectional area of the press. For a given bale weight in a particular press, the platen separation alone may be used. The factor "density restrained at" may be replaced by the quantity of cotton in the bale, the bale cross-sectional area, and the bale circumference. For a given bale weight produced in a particular press, the bale circumference alone can be used.

Typical bale circumferences for gin-packaging systems are given in Table 17. Experimental work at the U.S. Cotton Ginning Research Laboratory at Stoneville, Miss. has related the bale tie force to the final platen separation pressed to, the bale circumference (measured at the tie), and the lint moisture content. The force on the ties has been found to be different for each of the six to nine ties. The bale tie with the highest force exerted on it has been found to represent 15 to 20% of the total

Table 17
APPROXIMATE BALE CIRCUMFERENCES FOR BALES PACKAGED IN GIN PRESSES[a]

Press box dimensions[b] (in.)	Bale width (in.)	Bale thickness (in.)	Tie length (in.)
27 × 54	28	36	110.0
	28	38	114.0
	28	40	118.0
	28	42	122.0
	28	44	126.0
	28	46	130.0
	28	48	134.0
24 × 54	25	36	107.0
	25	38	111.0
	25	40	115.0
	25	42	19.0
	25	44	123.0
	25	46	127.0
	25	48	131.0
20 × 54	21	26	84.0
	21	28	88.0
	21	30	92.0
	21	32	96.0
	21	34	100.0
	20	26	82.5
	20	28	86.5
	20	30	90.5
	20	32	94.5
	20	34	98.5

[a] Length and thickness measurements were made at the ties.

[b] Dimensions are approximate.

force exerted on all the ties of a bale. The bale tie force distribution for one gin universal bale press is shown in Figure 4.

The force per tie has been found to range from less than 100 to more than 4000 lb. Flat and modified flat bales are normally packaged with six steel bands having a tensile strength of 2500 to 2800 lb or wire having a strength of 2500 to 2800 lb or wire having a strength of 3400 lb. Gin standard bales are normally packaged with eight steel bands having a tensile strength of 2500 to 2800 lb and special buckles or wire having a tensile strength of 3400 lb. Gin universal bales are normally packaged with eight flat, cold-rolled steel bands having a tensile strength in excess of 4000 lb.

The relationship between the maximum force per tie, the platen separation to which a bale is pressed before applying the ties, and the bale circumference at the tie after the bale is released from the press is shown in Figure 5 for 500-lb bales at 5% lint moisture packaged in a gin universal press.

Example

Determine the maximum tie force exerted on one of eight equally spaced bale ties on a 500-lb bale at 5% lint moisture packaged in a gin universal press. The bale circumference at the tie is 85 in.

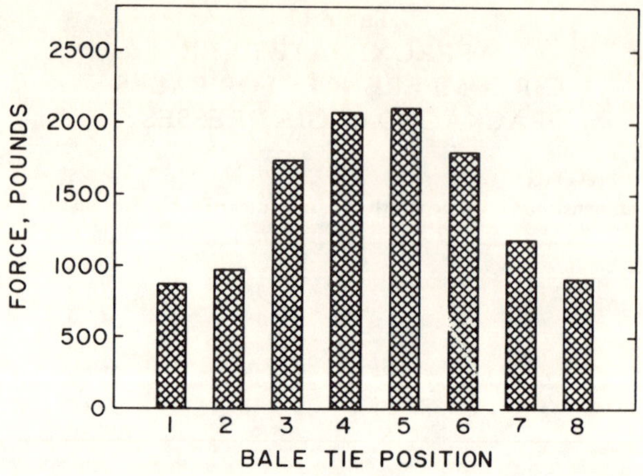

FIGURE 4. Bale tie force distribution for one gin universal bale press.

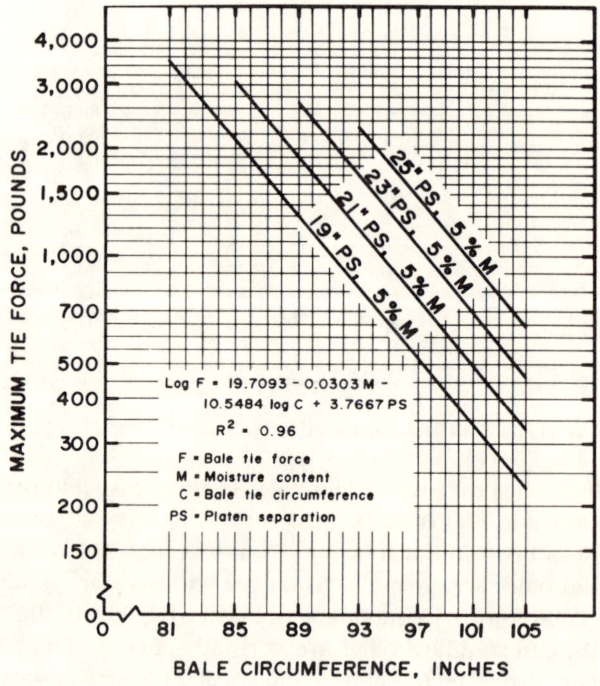

FIGURE 5. Maximum tie force per tie exerted by 500-lb bales of Upland cotton at 5% lint moisture content.

Solution

Refer to Figure 4 and note that 3000 lb of force is exerted when the bale is compressed to a platen separation of 21 in. and 2050 lb when the bale is compressed to 19 in.

A change in platen separation of only 2 in. resulted in a 32% reduction in bale tie force.

Variation in the bale circumference also has a pronounced effect on bale tie force. When a 500-lb bale at 5% lint moisture is compressed to a platen separation of 19 in. in a gin universal press, a maximum tie force of 2350 or 1650 lb results, depending on whether restrained bale circumference is 84 or 88 in., respectively (Figure 5). A change in bale circumference of only 4 in. resulted in a 30% reduction in bale tie force. Note that 4 in. of bale circumference produces a 2-in. change in bale thickness (Table 17).

The maximum tie forces exerted by bales produced in flat and modified flat presses vary in much the same manner as in gin universal presses, although the forces involved are lower. The maximum tie force exerted by bales can be reduced by uniform distribution of the lint cotton in the press box. When the lint cotton is very unevenly distributed in the press box, one tie may have 300% more force exerted on it than another tie on the same bale. The force used to compress lint cotton and the tensile strength of the bale ties used to restrain lint cotton can be varied to compensate for low moisture on large bales. Special care must be taken to ensure that the bale ties are not overstressed. Broken bale ties can be quite expensive to replace since the bale must be repressed before a new tie can be applied.

MATERIALS HANDLING

Seed Cotton Unloading Systems

Suction-Telescopes

Seed cotton is a nonflowing material, and since the early days of ginning system development, it has been unloaded from farmers' transport units (wagons and trailers) with a pneumatic suction system. The pick-up point of this type of system is equipped with a telescoping duct to allow removing the seed cotton from the top of a loaded trailer 6 to 8 ft deep. This telescoping duct is rigged as a free-swinging traveling device so the operator can reach all parts of a trailer, i.e., 8 ft wide × 40 ft long × 8 ft deep. Recent commercial units are hydraulically controlled from a remote operator's console.

Module Break-Ups

With the development of the seed cotton module system, Texas A&M University and Cotton Incorporated developed the module breakup unit. This apparatus was based on a basic research study done at the U.S. Department of Agriculture Cotton Ginning Research Laboratory, Stoneville, Miss. which showed that seed cotton can be metered with variable speed feed rolls. This principle was commercialized as a seed cotton input control unit and installed in most gins as the need for drying and cleaning escalated with mechanical harvesting. A companion research project at Stoneville showed that cotton can be fed to an input control from a live bottom hopper. Thus, the module break-up unit is a combination of a live bottom hopper and a variable speed seed cotton input control or spiked roll type of break-up unit. The seed cotton is delivered by the break-up unit onto a conveyor belt or into a pneumatic conveyor.

Trailer Dumps

Pneumatic suction unloading systems require a large amount of power and with the development of the module break-up units, interest has developed in dumping cotton trailers directly into the live bottom hopper of a break-up unit. These units consist of a tilting platform that the loaded trailer is pulled onto. After it is secured to the tilting platform, a mechanical or hydraulic system is used to lift and tilt the platform around a pivot point at the top of the receiving hopper. The cotton is thus dumped into the receiving hopper of the break-up which meters the cotton into the gin.

Separators and Droppers

The main function of the separator in the gin outfit is to separate the seed cotton from the conveying air of the suction unloading system. Due to its location in the machinery sequence, the separator is subjected to the most severe operating conditions of uneven material flow and negative (suction) pressure. High-capacity ginning has made the operating conditions more severe and has brought about heavier and tighter construction in separator design.

Separators of present manufacture are of four basic types: (1) those in which the main flow of seed cotton does not come in contact with a stationary screen and in which the screen is cleaned by rubber wiper flights attached to a revolving reel or cleaner-type cylinders; (2) those in which the seed cotton is directed onto a stationary screen and in which the screen is kept clean by rubber wiper flashing attached to a revolving reel; (3) those in which the seed cotton is directed over and around a revolving screen drum and in which the screen is kept clean by centrifugal force of the rotating drum; and (4) those in which the seed cotton is directed diagonally onto a stationary grid bar screen and in which the screen is cleaned by wiping action of the incoming seed cotton and air.

The first and second types of separators have rubber wipers on the rotating member or members. These wipers, which clean the screen, require little maintenance unless they are subjected to high temperature, as is the case when the separator is used as the receiving unit in the drying system. However, the wipers should be inspected periodically for wear and damage. The wipers are made with slotted fastening holes and can be adjusted to the screen as wear occurs. If the wipers have become set in a cupped position away from the screen, they can sometimes be removed and turned over to give further service. Excessively worn and broken wipers should be replaced.

The upper section of the third type of separator requires little maintenance other than periodic inspection of the seal flashing to see if it is sealing the ends of the screen drum properly. Worn or damaged flashing should be replaced.

The maintenance required for the upper section of the fourth type of separator is mainly periodic inspection to see if the grid-bar section is being properly cleaned by the wiping action of the incoming seed cotton and air. The spread of seed cotton over the entire width of the screen is important and is obtained by adjusting a dovetail spreader in the inlet transmission. The grid-bar screen should be adjusted so that the ends of the grid bars clear the back baffle by approximately 2 in.

The rubber flights in the lower, or vacuum, section of all separators require more attention and more frequent adjustment or replacement than do the rubber wipers in the upper screen section of the separator. To minimize air leakage into or out of the separator, these flights must be maintained in proper adjustment and good condition. When separation choking occurs, it is usually because the flights have become so worn and damaged that they are not properly sealing against the side scrolls. Under this condition, the air leakage passing the flights may cause an updraft of sufficient force to float the seed cotton in the upper section and prevent it from discharging through the vacuum section.

Vacuum wheels, or droppers, are used in the gin outfit to meter seed cotton, seed, and trash into a conveying airstream. They are also used as a sealer to prevent escape of air while the seed cotton or seed is metered from an airstream into an "out of line" type of gin machine.

Most droppers are identical in construction and maintenance requirements to the lower or vacuum section of the separator. When adjusting or installing flights, the manufacturer's instructions should be followed carefully. Generally, flights should be set so that the edges rub securely against the side scrolls. They should be set out sufficiently so that when they pass the side scroll they will bend approximately ½ in. and not more than ¾ in. from a straight line. They should not be required to bend or flex unduly. When the rubber flights are checked or replaced, the rubber flashings or wipers on the ends of the vacuum cylinder should also be inspected and replaced if they are damaged or worn. Some of the latest models of seed-cotton droppers have quick-change rubber flights that make repair and maintenance quick and easy. Most new drying systems incorporate a deflector dropper, commonly called a flight-saver, between the seed-cotton dropper and the hot-air stream. The flight-saver deflects most of the hot air away from the seed-cotton dropper, lengthening the life of the rubber flight. Seed-cotton droppers used in the drying system should be equipped with heat-resistant flights.

Trash and seed droppers are smaller in diameter and length than seed-cotton droppers. The main difference is that the seed dropper has more flights than the trash dropper. The additional flights are necessary to obtain sufficient sealing power for feeding into high-pressure seed lines. Trash and seed are very abrasive, and the flights in these droppers will require more frequent inspection and maintenance than those of the seed-cotton dropper.

A special type of seed dropper is being used in some modern high-capacity gin plants. This dropper has a close-fitting, all-steel, eight-blade rotor or wheel, which requires no flexible flights or flashing. It operates very efficiently in high-pressure seed systems using positive-pressure blowers.

Capacity and power requirements for various sizes of separators and droppers manufactured by one of the major gin machinery companies are given in Table 18. These data are averages and will vary considerably for different types of cotton. The manufacturer should always be consulted in making capacity and power estimates for separators and droppers.

Separators and droppers used in gin pneumatic systems may constitute a major source of power waste. Continuous upkeep, inspection, and repair are necessary to reduce air leakage. Even new and well-sealed separators have significant intake and leakage of air through the vacuum dropper section; a 35% loss at the separator is not uncommon. A periodic check of all separators and droppers will greatly reduce downtime from chokages caused by improper settings or worn parts in the separator or dropper.

Seed Cotton Input Control

Early observations by ginners and engineers showed that several aspects of ginning would be improved if controlled feeding of seed cotton to the gin by way of overhead cleaning machinery could be satisfactorily obtained. Damage to ginning machinery and lost ginning time (downtime) result from overfeeding — especially from feeding large, tight wads of seed cotton from the wagon directly into the overhead equipment. Such damage and loss have always contributed directly to increased ginning cost, and their prevention was one of the principal objectives of improved feed control.

Another desired operational improvement was regulation of the rate of feed to maintain an even flow of cotton to all gin stands in the battery without permitting a large quantity of overflow. Uncontrolled feeding, which occurs when the wagon telescope is connected directly to the overhead machinery, results in either too little or too much cotton arriving at the gin stands. Too little cotton results in reduced ginning capacity, whereas too much results in reduced overhead cleaning efficiency plus a large overflow accumulation.

In early methods, overflow cotton was often rerouted through the overhead machinery where it was again dried and cleaned, sometimes to the detriment of fiber quality and always at greater expense to the producer. In this situation, the removal of additional weight through drying and cleaning does not increase bale value, inasmuch as this value is determined by the lowest quality cotton in the bale.

Other benefits expected of controlled and even feeding were those relating to drying and cleaning efficiency. The most practicable efficiency for a specific gin is realized

Table 18
CAPACITIES AND POWER REQUIREMENTS FOR VARIOUS SIZES OF SEPARATORS AND DROPPERS[a]

Machine	Inside width (in.)	Dropper diameter (in.)	Recommended capacity[b]	Motor required (hp)	Dropper speed (r/min)
Separator[c]	72	22½	1—8 b/hr	10	65/70
	72	22½	8—16 b/hr	15	65/70
	72	34	16—24 b/hr	20	65/70
	96	34	24—35 b/hr	25	65/70
Seed-cotton dropper	72	22½	8—16 b/hr	10	65/70
	72	34	16—24 b/hr	15	65/70
	96	34	24—35 b/hr	20	65/70
Seed droppers	12	16	235 lb/min	1	50/55
	18	16	352 lb/min	2	50/55
	24	16	470 lb/min	3	50/55
Trash droppers	12	16	120 lb/min	1	50/55
	18	16	180 lb/min	2	50/55
	24	16	240 lb/min	3	50/55

[a] Courtesy of Lummus Industries, Inc., Columbus, Ga.
[b] b/hr = bales per hour.
[c] When used as live overflow separator, only 10-hp motor required.

when seed cotton is fed into the machinery at an even rate, just fast enough to keep all the gin stands in operation. It is desirable that the seed cotton fed be as nearly and uniformly in the single-lock state as possible, so that the maximum potential of the drying and cleaning system can be utilized.

Early Developments

From the time Eli Whitney and Hogden Holmes developed the saw gin until after the Civil War, cotton brought to the gin was unloaded by hand and then fed into the gin stand by hand at a rate determined by the capacity and characteristics of the specific gin. In the early 1870s, a feeder was invented for even feeding of cotton into the gin, but the feeder hopper was filled by hand.

In 1884 R.S. Munger was granted a patent for an outfit that included a pneumatic system for unloading cotton wagons. The gin feeders received cotton from a belt conveyor into which the penumatic unloader emptied. The rate of unloading was controlled manually by the ginner or left to the judgment of the suction-feeder operator. Later patents were granted on devices that improved the Munger system by regulation of the unloading by flexible check valves located about and in the gin feeders. These pneumatic elevators, first used in 1887, provided intermittent suction as required by the gin feeder; there was no overflow, because the unloading from the wagon was regulated by the gin feeder itself.

When overhead, or bulk, seed-cotton cleaning machinery came into general use, the pneumatic chute regulator was found to be no longer practicable. For several years, the conveying and distributing of cotton to gin feeders was accomplished by conveyor belts; in more recent years, these functions have been performed by the screw conveyor. However, it was found that both conveying arrangements, located between the overhead machinery and the gin feeders, delivered cotton faster than it could be ginned. As a result, the excess cotton flowed into an overflow pen, to be picked up and ginned after the wagon was unloaded.

To meet these needs a research project was begun at the U.S. Department of Agriculture Cotton Ginning Research Laboratory, Stoneville, Miss. in 1950. A seed cotton input-control unit was developed. The unit consisted of a surge bin or hopper, with variable speed feed rollers located in the bottom of the hopper to feed the cotton into the gin conditioning and cleaning system. The hopper was large enough to hold one third of a bale of seed cotton. To prevent overfilling the hopper, pressure switches were installed in the hopper to control the valve in the unloading fan line.

The unit could be adjusted to slightly overfeed the gins and a slight overflow would build up in the overflow pen. This cotton could be picked up with an overflow suction between trailer loads of seed cotton to reduce lost time while changing trailers. The overflow cotton was directed to the distributor to avoid overcleaning. The U.S. Department of Agriculture seed cotton input control unit has been commercialized and practically all gins today are so equipped.

Fans and Piping

Pneumatic systems consume from 40 to 60% of the total power required in a modern cotton gin. These systems are used to convey seed cotton from truck, trailer, or storage; operate cotton conditioners or driers; supply air to the doffing nozzles of airblast gins; convey cotton from point to point in the ginning system; and convey seed, trash, and dirt.

Air Facts Ginners Should Know

There are a few major definitions and facts that a ginner should keep in mind when dealing with pneumatic handling systems, in order to conserve power and keep them operating properly.

Table 19
RESISTANCE OF 90° ELBOWS (ROUND DUCT)[a]

Throat radius of elbow in pipe diameters	½	¾	1	1¼	1½	2	3
Diameters of straight pipe offering equivalent resistance	17	14	12	11	9.7	8.5	6.5

[a] For bends of less than 90° the equivalent resistance is in proportion to the arc. For example, a 60° bend will have 2/3 the equivalent resistance. Example: A 90° elbow in 24-in. round pipe with a 24-in. throat radius will have the equivalent of 12 diameters, or 24 ft, of straight duct.

1. "Static" or resistance pressure in piping systems corresponds to blood pressure in the human body. "Velocity" pressure in piping systems is like the heartbeat that causes blood to flow; it conveys the material in a pipe. Static pressure plus velocity pressure make up the total pressure against which a fan must operate.
2. Few open-wheel cotton gin fans have an efficiency greater than 50%. Rembert-type fans, with perforated disks that close the wheel, have approximately 35% efficiency.
3. Where pipe lengths are equal and where equal flow of air is handled, smaller pipes require more power and cause greater friction resistance or static pressure than do larger pipes. Table 19 shows the equivalent straight-pipe resistance for 90° elbows.
4. Piping velocities for handling dry seed cotton should be held below 4500 ft/min to avoid cracking or shattering seed where the material hits fan disks, sharp-turn elbows, separator reels, and other objects (Table 20).
5. Airflow calculations should include leakage as well as flow in pipes. Up to 30% leakage in the vacuum wheel of the separator must be allowed for at the fan. Tower driers also frequently leak, and stopping air leaks often saves money for gin owners.
6. Normal operating speeds vary, usually from 1400 to 3000 r/min, but for safety the tip speed of the wheel should not exceed 18,000 ft/min. Forward-pitch fan wheels are sensitive to speed changes and therefore usually waste more power than do straight-blade wheels. Dampers should be provided on the intakes of airblast fans. By use of slides or other good dampers on fan intakes, a substantial saving in power consumption is often possible. Cutting holes in pipes to relieve pressures or to increase airflow is usually unsatisfactory.

Large-diameter fan wheels operating at low speeds are generally more efficient and produce less noise than small-diameter fan wheels at high speeds. Since cotton gin fan casings have a large clearance, it is possible to adapt three sizes of wheels to the fan, namely, an oversize, a standard, or an undersize wheel. Thus a No. 40 fan can be converted to a No. 35-40 or No. 40-45 fan by changing the wheel. The true size of any fan is the size of its wheel.

Piping

A practical pneumatic system must convey material efficiently from point to point. In cotton gins, air is directed through pipes that vary in diameter from 6 to 36 in. made of galvanized iron sheet from 24 to 36 gage. Piping systems are an important part of the gin and should be carefully planned and maintained. Some good rules to follow are

Table 20
VOLUME OF AIRFLOW RATE IN VARIOUS SIZES OF PIPE BASED ON MEAN LINEAR VELOCITY OF 4,500 FT/MIN

Pipe dimensions		
Diameter (in.)	Area[a] (ft)	Volume flow rate[b] (ft^3/min)
5	0.136	614
6	0.196	884
7	0.267	1,203
8	0.349	1,571
9	0.442	1,988
10	0.545	2,454
11	0.660	2,970
12	0.785	3,534
13	0.922	4,148
14	1.069	4,811
15	1.227	5,522
16	1.396	6,283
17	1.576	7,093
18	1.767	7,952
19	1.969	8,860
20	2.182	9,817
21	2.405	10,824
22	2.640	11,879
24	3.142	14,137
26	3.687	16,591
28	4.276	19,242
30	4.909	22,089

[a] Area (square feet) = (πr^2) 144 = 0.7854d,[b] where r = radius in inches and d = diameter in inches.

[b] Volume flow rate (ft^3/min) = 4,500× area (ft^2).

1. Make the piping as simple and direct as possible, eliminating unnecessary elbows and valves.
2. Keep all joints airtight and rigid to prevent air leakage. This saves horsepower.
3. For good suction in seed-cotton pipe, maintain velocities of 4000 to 5000 ft/min. Telescope diameters are usually 12 to 24 in., depending on the seed-cotton requirements of the ginning system.
4. For conveying trash, keep the pipe diameter between 8 and 16 in., and for seed, between 10 and 13 in. The velocity in the pipe should be about 4500 ft/min for seed cotton or trash, and at least 5300 ft/min for seed. (This rule does not apply to small-pipe seed-blowing systems where the velocity is held to approximately 4500 ft/min, while the pressure is greatly increased and the volume of air is reduced.)
5. Do not allow piping for cottonseed or seed cotton to slope downward in the direction of flow; downward slopes may cause chokages.
6. A Rembert-type of fan is simple and economical for handling seed cotton in storagehouse operations. However, a standard separator and fan system with

Table 21
SUGGESTED SEED-COTTON PIPE SIZES FOR VARIOUS GINNING CAPACITIES

Ginning capacity (bales/hr)	Seed cotton pipe diameters (in.)	
	Unloading[a]	Drying systems
6—10	13	14
11—15	14	13 or two 12
16—20	16	18 or two 13
21—25	18	20 or two 14
26—30	20	24 or two 17

[a] Conveyor pipes from the telescope to the separator.

well-designed piping is more efficient and provides greater capacity than the Rembert-type fan system.

7. Approximately 20 ft^3 of air/lb of material should be sufficient for series fans on the unloading system and push-pull fans on the drying systems. This volume will vary depending on static pressure demands on the fan.

Air velocities required for efficient conveying of the materials handled at gins are

	ft/min
Seed cotton	4000—5000
Seed cotton in tower drier	1500—2000
Dried seed cotton (from machine to machine)	4000—5000
Seed	5000—5500
Hulls and trash	4000—5000
Lint cotton	1500—2000

Keep in mind that the flow of air in a pipe is similar to the flow of water in a pipe, although flow rate is measured in cubic feet per minute instead of gallons per minute. Pressures within the air pipes are expressed in inches of water measured on a U-tube gage or manometer, rather than in pounds per square inch. Most cotton gin centrifugal fans operate against static pressures or resistance below 1 lb/in.^{2} (27.8 in. of water).

The number, size, and operating rate of gin stands will determine the required rate of flow of seed cotton to the gin (usually from 8 to 30 bales per hour). This flow rate and the layout of the gin floor should determine the dimensions and arrangement of the piping (Table 21). The air pressures needed to overcome the resistance of equipment and pipes and the quantity of cotton needed to supply the gins will determine what fan capacity is required.

Air Measurements

Even when fan speeds, wheel diameter, and other factors are known, it is difficult to predict fan performance accurately. The only reliable way to determine how a fan is performing and to determine the characteristics of a system is by making air readings. Pneumatic systems in gins vary greatly, and no two installations are identical. However, the necessary determinations can be made with a relatively simple kit of tools and instruments containing the following items:

1. A 5/16-in.-diameter standard or a 1/8 in.-diameter pocket-size pitot tube. The 1/8-in.-diameter pitot tube is accurate enough for all practical purposes in round pipes up to 24 in. in diameter. It requires a hole in the pipe only 9/64 in. in diameter, which is not large enough to require patching.
2. A manometer or U-tube to measure velocity pressure (P_v) and static pressure (P_s). A 24-in. manometer should be sufficient to check fan system characteristics, with the exception of static pressure in a series fan unloading system.
3. Two 5-ft lengths of rubber or plastic hose to connect the pitot tube to the manometer.
4. A speed indicator for measuring fan speeds.
5. A measuring tape for determining pipe diameters.
6. Tables of airflow for various pipe diameters and velocity pressures.

To find the volume of air, velocity, pressures, and horsepower in a piping system, readings must be taken at the fan inlet and discharge. Since a fan may be used either for suction or for blowing through pipes, the static readings are taken at both the fan inlet and the fan outlet to find the overall static pressure. An outlet reading is all that is necessary on a fan used only for blowing. Readings may also be taken wherever information is needed, such as at troublesome points in suction, drying, and airblast piping systems, and at separator inlets and outlets to determine leakage. Test holes should be located approximately 8½-pipe diameters downstream and 1½-pipe diameters upstream from elbows and valves when possible. Gin fans usually operate at static pressures ranging from 5 to 26-in. of water.

The following are directions for using the test kit:

1. Fill the manometer half full of the liquid (water or oil) recommended by the manufacturer. An oil manometer is preferable to prevent damage from freezing in cold weather operations.
2. Connect the manometer to the pitot tube with the two lengths of flexible hose.
3. Insert the pitot tube into the test hole, until the end reaches the center of the pipe. The stem of the pitot tube should be kept at a right angle to the pipe, while the nose of the tube points upstream into the air current. Measure the difference in liquid levels between the two columns of the manometer. This reading is the velocity pressure (P_v). The liquid column levels usually fluctuate; therefore, read the differences between the average high and the average low.
4. Disconnect the impact pressure hose from the pitot tube and leave the static pressure outlet connected (the static pressure outlet on the pitot tube is perpendicular to the stem). Insert the pitot tube into the pipe in the same manner as described in Item 3 above, and measure the difference in liquid levels. This reading is the static pressure (P_s); it may be checked by withdrawing the pitot tube and holding the end of the flexible hose over the test hole in the pipe. The two P_s readings should be nearly the same.
5. Repeat procedure 4 on the opposite side of the fan and add the two to obtain the overall P_s.

After the P_v reading has been obtained, the velocity of the air and the volume of air the fan is delivering can be obtained from Table 22. The left-hand column of this table gives P_v, and the second column gives the velocity that corresponds to the P_v regardless of pipe diameter. The body of the table gives the volume for any given P_v reading for 0.1 to 6.0 in water and pipe sizes from 5 to 30 in. in diameter. The volume can be calculated for any pipe size by the equation given in the notes to Table 22. The velocity for the P_v reading can be obtained from Table 22 and the area of the pipe from Table 23.

Table 22
VOLUME OF AIR DELIVERED BY VARIOUS VELOCITY PRESSURES (P_v) AND PIPE DIAMETERS

Velocity pressure (in H_2O)	Velocity (ft/min)	Air volume (ft^3/min) for pipe diameter (in.)																				
		5	6	7	8	9	10	11	12	13	14	15	16	17	18	19	20	21	22	24	26	28	30
0.1	1,154	157	227	308	403	510	630	762	907	1,064	1,234	1,416	1,612	1,819	2,040	2,273	2,518	2,776	3,047	3,626	4,256	4,936	5,666
0.2	1,632	223	321	436	570	721	890	1,099	1,282	1,505	1,745	2,003	2,279	2,573	2,885	3,214	3,561	3,926	4,309	5,128	6,018	6,980	8,013
0.3	1,999	273	393	534	698	883	1,090	1,319	1,570	1,843	2,137	2,453	2,791	3,151	3,533	3,936	4,362	4,809	5,277	6,281	7,371	8,549	9,813
0.4	2,308	315	453	617	806	1,020	1,259	1,523	1,813	2,128	2,468	2,833	3,223	3,639	4,079	4,545	5,036	5,553	6,094	7,252	8,511	9,871	11,332
0.5	2,581	352	507	690	901	1,140	1,408	1,703	2,027	2,379	2,759	3,167	3,604	4,068	4,561	5,082	5,631	6,208	6,813	8,108	9,516	11,036	12,669
0.6	2,827	386	555	756	987	1,249	1,542	1,866	2,221	2,606	3,022	3,470	3,948	4,456	4,996	5,567	6,168	6,800	7,463	8,882	10,424	12,090	13,878
0.7	3,054	416	600	816	1,066	1,349	1,666	2,015	2,398	2,815	3,265	3,748	4,264	4,814	5,397	6,013	6,662	7,345	8,061	9,594	11,259	13,058	14,990
0.8	3,265	445	641	872	1,140	1,442	1,781	2,154	2,564	3,009	3,490	4,006	4,558	5,146	5,769	6,428	7,122	7,852	8,619	10,256	12,037	13,960	16,025
0.9	3,463	472	680	925	1,209	1,530	1,889	2,285	2,720	3,192	3,702	4,249	4,835	5,458	6,119	6,818	7,554	8,329	9,141	10,878	12,767	14,807	16,997
1.0	3,650	498	717	975	1,274	1,613	1,991	2,409	2,867	3,364	3,902	4,479	5,096	5,753	6,450	7,187	7,963	8,779	9,635	11,467	13,458	15,608	17,917
1.1	3,828	522	752	1,023	1,336	1,691	2,088	2,526	3,007	3,529	4,092	4,698	5,345	6,034	6,765	7,537	8,352	9,208	10,106	12,026	14,114	16,369	18,791
1.2	3,998	545	785	1,069	1,396	1,766	2,181	2,639	3,140	3,686	4,274	4,907	5,583	6,302	7,066	7,873	8,723	9,617	10,555	12,561	14,742	17,097	19,627
1.3	4,162	567	817	1,112	1,453	1,839	2,270	2,746	3,269	3,836	4,449	5,107	5,811	6,560	7,354	8,194	9,079	10,010	10,986	13,074	15,344	17,795	20,428
1.4	4,319	589	848	1,154	1,508	1,908	2,356	2,850	3,392	3,981	4,617	5,300	6,030	6,907	7,632	8,503	9,422	10,388	11,401	13,568	15,923	18,467	21,200
1.5	4,470	610	878	1,195	1,560	1,975	2,438	2,950	3,511	4,121	4,779	5,486	6,242	7,046	7,900	8,802	9,753	10,752	11,801	14,044	16,482	19,115	21,944
1.6	4,617	630	907	1,234	1,612	2,040	2,518	3,047	3,626	4,256	4,936	5,666	6,446	7,277	8,159	9,090	10,073	11,105	12,188	14,504	17,023	19,742	22,663
1.7	4,759	649	934	1,272	1,661	2,102	2,596	3,141	3,738	4,387	5,087	5,840	6,645	7,501	8,410	9,370	10,383	11,447	12,563	14,951	17,547	20,350	23,361
1.8	4,897	668	962	1,309	1,709	2,163	2,671	3,232	3,846	4,514	5,235	6,010	6,837	7,719	8,654	9,642	10,684	11,779	12,927	15,384	18,055	20,940	24,038
1.9	5,031	686	988	1,345	1,756	2,223	2,744	3,320	3,950	4,637	5,378	6,174	7,025	7,930	8,891	9,906	10,976	12,101	13,281	15,806	18,550	21,514	24,697
2.0	5,162	704	1,014	1,380	1,802	2,280	2,815	3,407	4,054	4,758	5,518	6,335	7,207	8,136	9,122	10,163	11,261	12,416	13,626	16,217	19,032	22,072	25,338
2.1	5,289	721	1,039	1,414	1,846	2,337	2,885	3,491	4,154	4,875	5,654	6,492	7,385	8,337	9,347	10,414	11,540	12,722	13,963	16,617	19,502	22,618	25,964
2.2	5,414	738	1,063	1,447	1,890	2,392	2,953	3,573	4,252	4,990	5,787	6,644	7,559	8,534	9,567	10,660	11,811	13,022	14,291	17,008	19,961	23,150	26,575
2.3	5,535	775	1,087	1,479	1,932	2,446	3,019	3,653	4,348	5,102	5,918	6,793	7,729	8,725	9,782	10,899	12,077	13,314	14,613	17,390	20,409	23,670	27,172
2.4	5,655	771	1,110	1,511	1,974	2,498	3,084	3,732	4,441	5,212	6,045	6,939	7,895	8,913	9,992	11,134	12,336	13,601	14,927	17,764	20,848	24,179	27,757
2.5	5,771	787	1,133	1,542	2,015	2,550	3,148	3,809	4,533	5,320	6,169	7,082	8,058	9,097	10,198	11,363	12,591	13,881	15,235	18,131	21,278	24,678	28,329
2.6	5,885	803	1,156	1,573	2,054	2,600	3,210	3,884	4,622	5,425	6,292	7,223	8,218	9,277	10,400	11,588	12,840	14,156	15,536	18,490	21,700	25,167	28,890
2.7	5,998	818	1,178	1,603	2,094	2,650	3,271	3,958	4,710	5,528	6,411	7,360	8,374	9,454	10,599	11,809	13,085	14,426	15,832	18,842	22,113	25,646	29,440
2.8	6,108	833	1,199	1,632	2,132	2,698	3,331	4,031	4,797	5,630	6,529	7,495	8,528	9,627	10,793	12,026	13,325	14,691	16,122	19,188	22,519	26,117	29,981
2.9	6,216	848	1,220	1,661	2,170	2,746	3,390	4,102	4,882	5,729	6,645	7,628	8,679	9,798	10,984	12,238	13,561	14,951	16,408	19,527	22,917	26,579	30,511
3.0	6,322	862	1,241	1,690	2,207	2,793	3,448	4,172	4,965	5,827	6,758	7,758	8,827	9,965	11,172	12,448	13,792	15,206	16,689	19,861	23,209	27,033	31,033
3.1	6,426	876	1,262	1,717	2,243	2,839	3,505	4,241	5,047	5,924	6,870	7,886	8,973	10,130	11,357	12,653	14,020	15,458	16,965	20,189	23,695	27,480	31,546
3.2	6,529	890	1,282	1,745	2,279	2,885	3,561	4,309	5,128	6,018	6,980	8,013	9,117	10,292	11,538	12,856	14,245	15,705	17,236	20,512	24,074	27,920	32,051
3.3	6,631	904	1,302	1,772	2,315	2,929	3,616	4,376	5,208	6,112	7,088	8,137	9,258	10,451	11,717	13,055	14,466	15,948	17,503	20,831	24,447	28,353	32,548
3.4	6,730	918	1,312	1,799	2,349	2,973	3,671	4,442	5,286	6,204	7,195	8,259	9,397	10,609	11,893	13,252	14,683	16,188	17,767	21,144	24,815	28,779	33,037
3.5	6,829	931	1,341	1,825	2,384	3,017	3,724	4,507	5,363	6,294	7,300	8,380	9,534	10,763	12,067	13,445	14,898	16,425	18,026	21,452	25,177	29,199	33,519
3.6	6,925	944	1,360	1,851	2,417	3,060	3,777	4,570	5,439	6,383	7,403	8,499	9,670	10,916	12,238	13,636	15,109	16,668	18,282	21,757	25,534	29,613	33,995
3.7	7,021	957	1,379	1,876	2,451	3,102	3,829	4,633	5,514	6,472	7,505	8,616	9,803	11,067	12,407	13,824	15,317	16,887	18,534	22,057	25,886	30,022	34,464
3.8	7,115	970	1,397	1,902	2,484	3,143	3,881	4,696	5,588	6,558	7,606	8,732	9,935	11,215	12,574	14,009	15,523	17,114	18,783	22,353	26,234	30,425	34,926
3.9	7,208	983	1,415	1,926	2,516	3,184	3,931	4,757	5,661	6,644	7,706	8,846	10,065	11,362	12,738	14,193	15,726	17,338	19,028	22,645	26,577	30,823	35,383

4.0	7,300	995	1,433	1,951	2,548	3,225	3,982	4,818	5,733	6,729	7,804	8,958	10,193	11,507	12,900	14,373	15,926	17,559	19,271	22,934	26,915	31,215	35,834
4.1	7,391	1,008	1,451	1,975	2,580	3,265	4,031	4,787	5,805	6,812	7,901	9,070	10,319	11,650	13,060	14,552	16,124	17,777	19,510	23,219	27,250	31,603	36,279
4.2	7,480	1,020	1,469	1,999	2,611	3,305	4,080	4,937	5,875	6,895	7,997	9,180	10,444	11,791	13,219	14,728	16,319	17,992	19,747	23,500	27,580	31,986	36,719
4.3	7,569	1,032	1,486	2,023	2,642	3,344	4,128	4,995	5,945	6,977	8,091	9,288	10,568	11,930	13,375	14,903	16,513	18,205	19,880	23,770	27,906	32,365	37,153
4.4	4,656	1,044	1,503	2,046	2,673	3,382	4,176	5,053	6,013	7,057	8,185	9,396	10,690	12,068	13,530	15,075	16,703	18,416	20,211	24,053	28,229	32,739	37,583
4.5	7,743	1,056	1,510	2,069	2,703	3,421	4,223	5,110	6,081	7,137	8,277	9,502	10,811	12,205	13,683	15,245	16,892	18,624	20,440	24,325	28,548	33,109	38,007
4.6	7,828	1,068	1,537	2,092	2,733	3,458	4,270	5,166	6,148	7,216	8,369	9,607	10,930	12,339	13,834	15,414	17,079	18,829	20,665	24,594	28,863	33,475	38,427
4.7	7,913	1,079	1,554	2,115	2,762	3,496	4,316	5,222	6,216	7,294	8,459	9,711	11,049	12,473	13,983	15,580	17,264	19,033	20,889	24,859	29,175	33,836	38,843
4.8	7,997	1,090	1,570	2,137	2,791	3,533	4,362	5,277	6,281	7,371	8,549	9,813	11,166	12,605	14,131	15,745	17,446	19,234	21,110	25,123	29,484	34,195	39,254
4.9	8,080	1,102	1,586	2,159	2,820	3,569	4,407	5,332	6,346	7,447	8,637	9,915	11,281	12,736	14,278	15,908	17,627	19,434	21,329	25,383	29,790	34,549	39,661
5.0	8,162	1,113	1,603	2,181	2,849	3,606	4,541	5,486	6,410	7,523	8,725	10,016	11,396	12,865	14,423	16,070	17,806	19,631	21,545	25,641	30,092	34,900	40,063
5.1	8,243	1,124	1,618	2,203	2,877	3,642	4,496	5,440	6,474	7,598	8,812	10,116	11,509	12,993	14,566	16,230	17,983	19,826	21,760	25,896	30,391	35,247	40,462
5.2	8,323	1,135	1,634	2,224	2,905	3,677	4,540	5,493	6,537	7,672	8,896	10,214	11,621	13,120	14,708	16,388	18,159	20,020	21,972	26,148	30,688	35,591	40,857
5.3	8,403	1,146	1,650	2,246	2,933	3,712	4,583	5,546	6,600	7,745	8,983	10,312	11,733	13,245	14,849	16,545	18,332	20,211	22,182	26,399	30,982	35,931	41,248
5.4	8,482	1,157	1,665	2,267	2,961	3,747	4,626	5,598	6,662	7,818	9,067	10,409	11,843	13,369	14,989	16,700	18,504	20,401	22,390	26,641	31,273	36,269	41,635
5.5	8,560	1,167	1,681	2,288	2,988	3,782	4,669	5,649	6,723	7,890	9,151	10,505	11,952	13,493	15,127	16,854	18,675	20,589	22,597	26,892	31,561	36,603	42,019
5.6	8,637	1,178	1,696	2,308	3,015	3,816	4,711	5,700	6,784	7,962	9,234	10,600	12,060	13,615	15,264	17,007	18,844	20,776	22,801	27,135	31,846	36,934	42,399
5.7	8,714	1,188	1,711	2,329	3,042	3,850	4,753	5,751	6,844	8,032	9,316	10,694	12,167	13,736	15,399	17,158	19,012	20,960	23,004	27,377	32,130	37,263	42,776
5.8	8,790	1,199	1,726	2,349	3,068	3,883	4,794	5,801	6,904	8,103	9,397	10,787	12,274	13,856	15,534	17,308	19,178	21,143	23,205	27,616	32,410	37,588	43,150
5.9	8,866	1,209	1,741	2,369	3,095	3,917	4,836	5,851	6,963	8,172	9,478	10,880	12,379	13,975	15,667	17,456	19,342	21,325	23,404	27,853	32,688	37,911	43,520
6.0	8,941	1,219	1,755	2,389	3,121	3,950	4,876	5,900	7,022	8,241	9,558	10,972	12,483	14,093	15,799	17,604	19,505	21,505	23,602	28,088	32,964	38,231	43,887

Based on standard air conditions (70°F. 29.92 in. Hg) and 1 center-of-the-pipe pitot tube reading (4,005×0.9113 = 3,650).

Velocity (ft/min) = $3{,}650\sqrt{\text{velocity pressure (in } H_2O)}$.

Volume of air (ft^3/min) = velocity (ft/min)×area of pipe (ft^2).

Table 23
AREA AND CIRCUMFERENCE OF ROUND PIPE

Diameter (in.)	Area in.²	Area ft²	Circumference (in.)
1	0.79	0.005	3.14
2	3.10	0.022	6.28
3	7.07	0.049	9.42
4	12.57	0.087	12.57
5	19.60	0.136	15.71
6	28.27	0.196	18.85
7	38.48	0.267	21.99
8	50.27	0.349	25.13
9	63.60	0.442	28.27
10	78.50	0.545	31.42
11	95.00	0.660	34.56
12	113.10	0.785	37.70
13	132.70	0.922	40.84
14	153.90	1.069	43.98
15	176.70	1.227	47.12
16	201.00	1.396	50.26
17	226.90	1.576	53.41
18	254.40	1.767	56.55
19	283.50	1.969	59.69
20	314.10	2.182	62.83
21	346.30	2.405	65.97
22	380.10	2.640	69.11
23	415.40	2.885	72.26
24	452.30	3.142	75.40
25	490.80	3.409	78.54
26	530.90	3.687	81.68
27	572.50	3.967	84.82
28	615.70	4.276	87.96
29	660.50	4.587	91.11
30	706.80	4.909	94.25
31	754.70	5.241	97.39
32	804.20	5.585	100.53
33	855.30	5.940	103.67
34	907.90	6.305	106.81
35	962.10	6.681	109.96
36	1017.80	7.069	113.10
37	1075.20	7.467	116.24
38	1134.10	7.876	119.38
39	1194.50	8.296	122.52
40	1256.60	8.727	125.66

Note: Area = 3.14×diameter²/4 = 0.7854×diameter².
Circumference = 3.14×diameter.

For example, assume a P_v reading of 1.6 in. of water in a 13-in.-diameter round pipe; from Table 22 we find that the velocity is 4617 ft/min, and the air volume is 4256 ft³/min. Since Table 22 is based on standard air conditions, a density correction factor must be considered if the fan handles air at other than standard temperature and altitude. Factors for correcting for temperature and altitude are given in Table 24.

Table 24
AIR VELOCITY CORRECTIONS FOR TEMPERATURE AND ALTITITUDE[a]

Temperature (°F)	Correction factor	Altitude (feet m.s.l.)	Correction factor
0	0.932	0[b]	1.000
20	0.952	500	1.009
40	0.971	1000	1.018
60	0.990	1500	1.028
70[c]	1.000	2000	1.037
80	1.009	2500	1.046
100	1.028	3000	1.056
120	1.046	3500	1.066
140	1.064	4000	1.075
160	1.082	4500	1.086
180	1.098	5000	1.095
200	1.114	5500	1.106
225	1.137		
250	1.157		
275	1.177		
300	1.197		
350	1.236		
400	1.273		
450	1.310		
500	1.345		

[a] For correcting Table 22 for other than standard conditions.
[b] Sea level. Standard condition.
[c] Standard condition.

Example

If the fan in the above example is handling air at 200°F and is located at an altitude of 3000 ft, the following corrections should be made.

Solution

1. From Table 24 we find that the correction factors for temperature and altitude are 1.114 and 1.056, respectively.
2. The composite correction factor = 1.114 × 1.056 = 1.176.
3. The corrected velocity = 4617 × 1.176 = 5430 ft/min.
4. The corrected volume = 4256 × 1.176 = 5005 ft³/min.

Centrifugal Fans

Two types of fans are in common use in gins, centrifugal and axial flow, but centrifugal fans are in the majority. The centrifugal fan has a single inlet and is fully enclosed in a cast iron or sheet metal housing in which bladed wheels revolve. Size designations are confusing because each manufacturer uses his own descriptive terms. Many fans, however, are designated simply by size from No. 20 to No. 70 or F-76 to F-238.

Rembert fans with perforated flat disks or cones permit the material to pass through the fan housing without damage from the wheel. These fans are generally used for transferring seed cotton from one location to another (such as from trailer to storage place) without use of a separator.

Fans should not be operated at speeds higher than necessary, and wheel tip speed should never exceed 18,000 ft/min. If it becomes necessary to speed up fans, resulting changes in performance may be calculated from the following basic laws:

1. Capacity (cubic feet per minute) varies directly with the speed (revolutions per minute)
2. Pressure (inches of water) varies as the square of the speed (revolutions per minute)
3. Power (horsepower) varies as the cube of the speed (revolutions per minute)

These laws are expressed in equations as follows:

$$\frac{\text{Original capacity}}{\text{Final capacity}} = \frac{\text{original speed}}{\text{final speed}} \tag{1}$$

$$\frac{\text{Original pressure}}{\text{Final pressure}} = \frac{\text{original speed squared}}{\text{final speed squared}} \tag{2}$$

$$\frac{\text{Original horsepower}}{\text{Final horsepower}} = \frac{\text{original speed cubed}}{\text{final speed cubed}} \tag{3}$$

Motor current can be substituted for horsepower in Equation 3, with only a slight error, so long as the voltage and frequency remain the same. This will be more convenient for ginners since motor current is easier to measure than actual horsepower. The results will be accurate enough for field calculations.

$$\frac{\text{Original motor current}}{\text{Final motor current}} = \frac{\text{original speed cubed}}{\text{final speed cubed}} \tag{4}$$

Many ginners do not realize what Equation 4 means to them in power consumption. If the speed of a fan is doubled, the capacity is also doubled; but the resistance pressure is four times as great and the horsepower consumed is eight times as great.

Assume, for example, that a fan operating at 1100 r/min and delivering 2500 ft^3/min of air against a P_s of 6 in. is using 5 hp. With all other conditions remaining the same, the speed is doubled to 2200 r/min. The volume of airflow will be doubled, to 5000 ft^3/min; the P_s will be increased to 24 in. of water resistance; and the power consumed will increase from the initial 5 to 40 hp.

Care should be taken to choose a fan that will be operating within the most efficient capacity range and within the speed range recommended by the manufacturer. Table 25 can be used as a guide for selecting fans for their most efficient operation. Most manufacturers base fan performance ratings on handling standard air at 70°F and 29.92 in. of mercury barometric pressure, and weighing 0.075 lb/ft^3. Corrections to pressure and horsepower for variations in temperature and altitude are given in Table 26.

Example

Select a fan with a capacity of 5000 ft^3/min at 140°F with a system P_s of 12 in. of water to be operated 2500 ft above sea level.

Solution

1. From Table 26 we find that the correction factors for temperature and altitude are 1.13 and 1.10, respectively.

Table 25
SELECTION OF FANS FOR MOST EFFICIENT OPERATION

	Most efficient		Static pressure[b]	
Fan size or equivalent	Capacity range (ft^3/min)	Speed range[a] (r/min)	Straight blade (in. H_2O)	Multiblade (in. H_2O)
Single 20	1,250—1,750	2,200—3,000	Up to 14	Up to 17
Single 25	1,500—2,000	1,800—2,800	Up to 16	Up to 19
Single 30	2,000—3,000	2,000—3,000	Up to 18	Up to 21
Single 35	3,000—4,000	2,000—2,600	Up to 18	Up to 21
Single 40	4,000—5,000	1,800—2,400	Up to 22	Up to 24
Single 45	5,000—6,000	1,600—2,200	Up to 23	Up to 25
Single 50	6,000—8,000	1,400—2,000	Up to 22	Up to 25
Parallel 40	8,000—10,000	1,800—2,400	Up to 22	Up to 24
Parallel 45	10,000—12,000	1,600—2,200	Up to 23	Up to 25
Parallel 50	12,000—16,000	1,400—2,000	Up to 22	Up to 25
Series 40	4,000—5,000	1,800—2,400	Up to 44	Up to 48
Series 45	5,000—7,000	1,600—2,200	Up to 46	Up to 50
Series 50	6,000—8,000	1,400—2,000	Up to 44	Up to 50

[a] Fan wheel tip velocity should not exceed 18,000 ft/min.

[b] Fan intake plus discharge static pressure. These values will vary slightly with manufacturer and number of blades on wheel.

Table 26
FAN PERFORMANCE CORRECTIONS FOR TEMPERATURE AND ALTITUDE

Temperature (°F)	Factor	Altitude (ft m.s.l.)	Factor
0	0.87	0[a]	1.00
20	0.91	500	1.02
40	0.94	1000	1.04
60	0.98	1500	1.06
70	1.00	2000	1.08
80	1.02	2500	1.10
100	1.06	3000	1.12
120	1.09	3500	1.14
140	1.13	4000	1.16
160	1.17	4500	1.18
180	1.21	5000	1.20
200	1.25	5500	1.22
225	1.29		
250	1.34		
275	1.39		
300	1.43		
350	1.53		
400	1.62		
450	1.72		
500	1.81		

[a] Sea level. Standard condition.

[b] Standard condition.

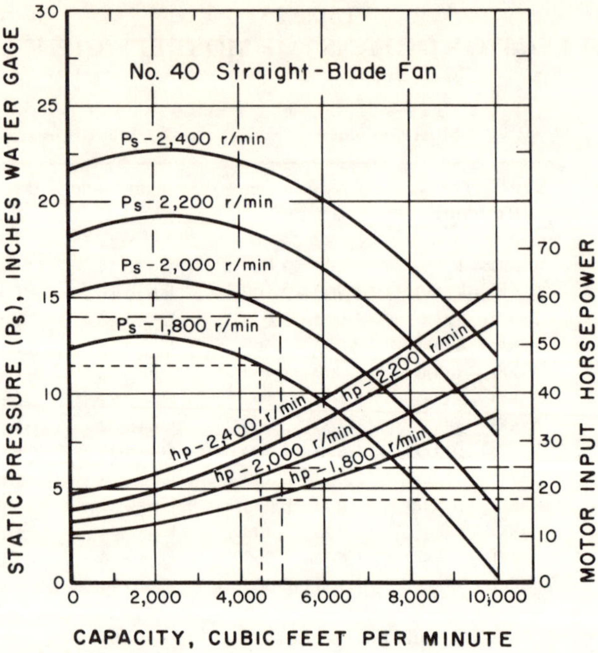

FIGURE 6. Typical performance characteristics curves for a No. 40 straight-blade fan.

2. Composite correction factor = 1.13 × 1.10 = 1.24.
3. 12 in. H_2O × 1.24 = 14.9 in. of water P_s at 70°F and sea level.
4. Fan performance tables of one manufacturer show a size 40 gin fan has a capacity of 5000 ft³/min against a P_s of 15 in. of water at a speed of 1800 r/min and will require 18.7 hp.
5. Correct the horsepower and pressure to operating conditions of 140°F and 2500 ft altitude.

$$18.7 \text{ hp} \div 1.24 = 15 \text{ hp}$$

$$14.9 \text{ inH}_2\text{O} \div 1.24 = 12 \text{ inH}_2\text{O } P_s$$

6. Final performance of altitude and temperature: 5000 ft³/min, 12 in. of water P_s, 1800 r/min, and 15 hp.

The performance characteristics curves of a fan provide a good means of determining its capabilities; the performance of fans of the same size designation will be slightly different for each manufacturer. Use the correct curves for manufacturer and fan size when checking fan performance. Figure 6 shows the performance characteristics curves for a size 40 straight-blade fan. If the fan capacity and speed are known, the P_s and horsepower can be determined from the fan curve as follows. Go to 4500 ft³/min on the X-axis and move vertically until you intersect the 1800-r/min curve for horsepower. Now move horizontally to the right and read 18 hp on the Y-axis. Return to 4500 ft³/min on the X-axis and move vertically again until you intersect the 1800-r/min curve for P_s. Now move horizontally to the left and read 11.5 in. of water on the Y-axis.

If you wish to increase the volume of air to 5000 ft³/min, the necessary speed of the fan can be calculated with Equation 1:

$$\frac{\text{Original capacity}}{\text{Final capacity}} = \frac{\text{original speed}}{\text{final speed}}$$

$$\frac{4500}{5000} = \frac{1800}{\text{final speed}}$$

$$\text{Final speed} = 2000 \text{ r/min}$$

The new horsepower and static pressure can now be determined from the performance curves in Figure 6, and as previously discussed.

If fan performance curves are not available, the new conditions can be calculated if the original speed, P_s, volume of air flow, and motor current are known. This is done by using the basic laws of fan performance.

Example

Original conditions for a size No. 40 straight blade fan: fan speed is 1800 r/min measured with a tachometer. P_s 11.5 in. measured with a pitot tube and manometer. Volume of airflow is 4500 ft³/min measured with a pitot tube and manometer. Motor current is 23 A measured with a loop-around ammeter.

Solution

Volume of airflow desired is 5000 ft³/min.

Using Equation 1:

$$\frac{4500 \text{ ft}^3\text{/min}}{5000 \text{ ft}^3\text{/min}} = \frac{1800 \text{ r/min}}{\text{final speed}}$$

$$\text{Final speed} = 2000 \text{ r/min}$$

Using Equation 2:

$$\frac{(1800 \text{ r/min})^2}{(2000 \text{ r/min})^2} = \frac{11.5 \text{ in. } H_2O}{\text{final pressure}}$$

$$\text{Final pressure} = 14.2 \text{ in. } H_2O$$

Using Equation 3:

$$\frac{(1800 \text{ r/min})^3}{(2000 \text{ r/min})^3} = \frac{23 \text{ A}}{\text{final motor current}}$$

$$\text{Final motor current} = 31.5$$

Table 27 can be used to determine whether the electric motor will be overloaded with this increase in speed.

Axial Flow Fans

Axial flow fans include those fans which produce motion of the air by the thrust effect of the inclined blades. The air enters axially at the leading edge of the blade, and the flow is essentially parallel to the impeller shaft. In addition, the air is given a rotating motion about the shaft axis as a center. Fans in this category are usually classified into tube axial, vane axial, and disk or propeller. Propeller fans are rarely used in cotton gins.

A tube axial fan consists of a relatively large hub with helical blades rotating within a cylinder, and will operate against resistance pressures of 2 or 3 in. of water. These are sometimes referred to in the ginning industry as clean-air or duct fans.

Table 27
AMPERE RATINGS OF THREE-PHASE, 60-CYCLE AC INDUCTION MOTORS[a]

Hp	Syn. speed (r/min)	Current in amperes 220 V	Current in amperes 440 V
1	3600	2.9	1.5
	1800	3.2	1.6
	1200	3.7	1.9
	900	4.1	2.1
1½	3600	4.3	2.1
	1800	4.6	2.3
	1200	5.2	2.6
	900	5.8	2.9
2	3600	5.6	2.8
	1800	6.0	3.0
	1200	6.5	3.3
	900	7.3	3.6
3	3600	8.2	4.1
	1800	8.7	4.4
	1200	9.2	4.6
	900	10.3	5.2
5	3600	13.3	6.7
	1800	14.0	7.0
	1200	14.6	7.3
	900	16.2	8.1
7½	3600	19.4	9.7
	1800	20.3	10.2
	1200	20.9	10.5
	900	23.0	11.5
10	3600	25.5	12.8
	1800	26.6	13.3
	1200	27.1	13.6
	900	29.8	14.9
	600	34.0	17.0
15	3600	37.7	18.9
	1800	39.0	19.5
	1200	39.3	19.7
	900	43.1	21.6
	600	49.1	24.6
20	3600	49.6	24.8
	1800	51.0	25.5
	1200	51.4	25.7
	900	55.8	27.9
	600	63.6	31.8
25	3600	61.5	30.8
	1800	62.8	31.4
	1200	63.3	31.7
	900	68.5	34.3
	600	77.8	38.9
30	1800	74.7	37.4
	1200	75.0	37.5
	900	80.7	40.4
	600	91.8	45.9
40	1800	98.0	49.0
	1200	98.4	49.2
	900	105.0	52.5

Table 27 (continued) AMPERE RATINGS OF THREE-PHASE, 60-CYCLE AC INDUCTION MOTORS[a]

Hp	Syn. speed (r/min)	Current in amperes 220 V	440 V
	600	119.0	59.5
50	1800	122.0	61.0
	1200	123.0	61.5
	900	130.0	65.0
	600	146.0	73.0
60	1800	145.0	72.5
	1200	146.0	73.0
	900	153.0	76.5
	600	172.0	86.0
75	1800	179.0	89.5
	1200	180.0	90.0
	900	188.0	94.0
	600	210.0	105.0
100	1800	235.0	118.0
	1200	236.0	119.0
	900	247.0	124.0
	600	273.0	137.0

[a] The ampere ratings of motors vary somewhat, depending upon the type of motor, but the values given here can be considered average for the various types. In the case of high-torque squirrel cage motors, the ampere ratings will be at least 10% greater than the value given.

Vane axial fans consist of an axial flow impeller within a cylinder, combined with a set of guide vanes located either before or after the impeller. These guide vanes assist in recovering the energy imparted as tangential acceleration and therefore fans of this design are capable of operating against static pressures up to 6 or 7 in. of water.

Axial flow fans are sized by the diameter of the cylinder housing. The basic fan laws, which were discussed in connection with centrifugal fans, also apply to these fans. Performance curves can be used in the same manner as for centrifugal fans. One of the distinguishing characteristics of axial flow fans is that they generally have a drooping horsepower curve, with the horsepower being a maximum at no flow. This is the opposite of centrifugal fans, which have minimum horsepower at no flow.

Unlike the centrifugal fan, axial flow fans cannot handle material through the fan impeller. They are used in cotton gins primarily on low-pressure systems such as lint cleaner condensers and battery condensers. An access door should be provided in the air duct on both sides of the fan to facilitate inspection for lint fly buildup, especially on the fixed vanes.

Mechanical Conveyors

The mechanical conveyors used in cotton gins fall into three main categories: (1) belt conveyors, (2) screw conveyors, and (3) chain conveyors.

Belt conveyors, although formerly used in seed-cotton distributors, are now used primarily for conveying cottonseed from under the gin stands to a dropper discharging into a small-pipe pneumatic conveyor line. The principal advantage of using the belt conveyor under the gin stands is that it is self-cleaning. In installations ginning cotton for planting seed, use of the belt conveyor enables the ginner to keep seed varieties separated with a minimum of cleaning. The belt conveyor is efficient, requires relatively small horsepower, and has a high carrying capacity. A properly designed belt system has a long service life, and although the initial cost is usually high, installation is recommended where amortization of the initial cost is assured through reduced cleaning costs or for other reasons. Seed belt conveyor capacities are flexible because of the relatively high belt speeds permissible. An 8-in. belt running at a speed of 100 ft/min in a 6-in.-deep trough will handle over 18 tons of seed per hour at 75% capacity.

Screw conveyors and screw lifts are the most common mechanical handling devices used in cotton ginning, being used to convey cottonseed, gin trash, and seed cotton. Screw conveyors are simple and relatively inexpensive but require more power than belt conveyors. The standard pitch screw has a pitch approximately equal to the diameter and is used on most horizontal installations and on inclines up to 20°. Horizontal screw conveyors are usually operated in a U-shaped trough with the screw supported at various standard spacings. Screw elevators are enclosed in cylindrical housings. The tube is operated full with no brackets used between ends. Screw elevators are normally used in the gin only to convey cottonseed. Concise formulas and data are not available for individual designs; however, the empirical equation given below can be used.

$$\text{ft}^3/\text{hr} = \frac{(D^2 - d^2)}{36.6/\text{min}} \text{ X P X shaft r/min}$$

where D = screw diameter (inch), d = shaft diameter (inch), and P = screw pitch (inch) — normally equal to D.

The actual capacity will be much less than the theoretical because of screw housing clearance, fluid characteristics of material, screw length, material head, and lift. When specific operating data are not available, an estimate of 50 to 60% of the theoretical capacity should be used.

In some areas of the Cotton Belt, chain or scraper conveyors are used to elevate cottonseed from the seed pile into trucks for transfer to the mill. These scraper conveyors are similar to grain elevator conveyors; however, flights are normally shallow and have widths ranging from 12 to 18 in. The capacity of such conveyors will depend on the incline needed. A rule of thumb for chain belt capacity is: horizontal, 115% of conveyor flight volume; 20° incline, 77% of flight volume; 30° incline, 55% of flight volume, and 40° incline, 33% of flight volume.

Theoretical horsepower requirements for flight conveyors can be determined by the following equation:

$$\text{Horsepower} = \frac{2VL_cW_cF_c + Q(LF_m + H)}{33{,}000}$$

where V = speed of conveyor (feet per minute), L_c = horizontal projected length of conveyor (feet), W_c = weight of chain and flights (pounds per foot), F_c = coefficient of friction for chains and flights, Q = pounds of material to be handled per minute, L = horizontal projected length of loaded conveyor (feet), F_m = coefficient of friction for material, and H = height of lift (feet).

The coefficients of friction range from approximately 0.25 for metal-to-metal contact to 0.50 for metal-to-wood contact. The materials coefficient of friction for cottonseed to metal is estimated to be approximately 0.80.

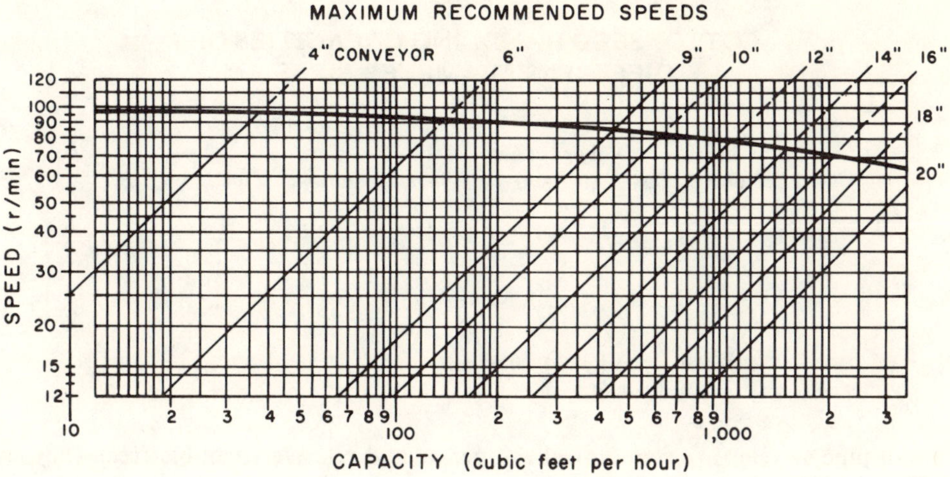

FIGURE 7. Capacity of screw conveyors.

The power requirement of a screw conveyor is a function of its length, elevation, type of hangers, type of flights, internal resistance or viscosity of the material, coefficient of friction of the material on the flights and housing, and weight of the material. Consideration should also be given to the extra power required to start a full screw and to free a jammed screw. An approximation of the power required to operate a normal horizontal screw can be determined by the following equation.

$$\text{Horsepower} = CLWF/33{,}000$$

where C = conveyor capacity (cubic feet per minute), L = length of conveyor (feet), W = bulk material weight (pounds per cubic foot) (cottonseed = 25 to 30 lb/ft^3, average trash weight = 10 to 12 lb/ft^3, seed cotton = 4 to 6 lb/ft^3), and F = material factor (cottonseed = 0.9 and gin trash = 0.9).

Screw conveyor size, capacity, and recommended speeds for normal gin operations are given in Figure 7.

Normally, a 16-in., 5-hp unit is capable of loading 30 to 40 tons of cottonseed an hour into an average-height truck and trailer. Some provision should be made for raising and lowering the incline section to adjust to the height of the sideboards.

Materials handling is a highly specialized enterprise. Its procedures and details have developed largely from usage and experience in cotton gins. Large installations should be designed and installed by a materials-handling engineer. Smaller, less involved installations can be designed and installed by local mechanics or by plant supervisors and engineers.

Cottonseed Handling

Seed at gins may be handled readily by gravity, belts, screw conveyors, or pneumatic piping. Belt and pneumatic conveying systems are self-cleaning between the ginning of different varieties if the seed is to be segregated by variety. The use of belt or pneumatic conveyors is therefore desirable if any appreciable quantity of seed is to be saved for planting.

Where vertical lifts are employed, screw elevators are generally satisfactory. In some instances seed may be conveyed by belts or low-pressure piping. Conveying seed by air in small pipes is now the general practice.

Table 28
COTTONSEED HANDLING CAPACITIES OF DIFFERENT DIAMETER PIPES

Pipe diameter (in.)	Volume of air[a] (ft^3/min)	Capacity[b] lb/min	tons/hr	bales/hr
4	436	128	3.8	9.6
5	682	200	6.0	15.0
6	982	288	8.6	21.6
8	1745	512	15.4	38.4

[a] Based on a velocity of 5000 ft/min.
[b] Based on 800 lb of seed per bale.

Small-pipe systems are economical to operate and relatively trouble-free. They enable the cotton producer and ginner to maintain pure varieties of cottonseed because the apparatus is self-cleaning. They have adequate capacity for moving as much as 500 lb of seed per minute, as fast as the cotton is ginned. They may also be used for carrying seed to storage bins and for moving it later for grading, sterilizing, and other pressures at rates to suit plant capacity. Cottonseed has been successfully handled by small-pipe flowing systems for distances up to 700 ft at operating costs lower than larger pipe and fan methods.

Air-Conveying Principles and Calculations

Pipes 4 to 8 in. in diameter have proved large enough for handling cottonseed in gins and delinting plants. Table 28 gives the seed-handling capacity of such pipes.

For preliminary calculations to determine total resistance pressures that the pump must overcome, it is customary to allow 16 oz of resistance (or 1 lb/in.^{2}) for each 200 linear feet of piping. In calculating, each short elbow and each valve must be considered equivalent to approximately 15 ft of straight pipe. For a more accurate estimate on which the factory can provide the blower unit and suggest its speed, the pressure losses for the individual elements that make up the system may be estimated as follows:

	Estimated pressure loss in oz/in.^{2}
4-in. piping, each 100 ft	6.0
6-in. piping, each 100 ft	5.0
6-in. piping, each 100 ft	3.0
8-in. piping, each 100 ft	2.0
For both 4- and 5-in. pipes	
Elbows	0.8
Base and tapered discharges from dropper	2.0
Cyclone collector and sacker at end of pipe	1.0
For both 6- and 8-in. pipes	
Elbows	0.5
Base and tapered discharges from dropper	2.0
Cyclone collector and sacker at end of pipe	1.0

Average air velocities in small seed pipes are in the range 4200 to 5200 ft/min, with 5000 ft/min as the common figure. Air volumes of about 4 ft^3/lb cottonseed have been used in short piping systems with an air pump, but it is advisable to alow 5 ft^3 where pipe length exceeds 250 ft.

For horizontal blowing of cottonseed, the air velocity in the pipe should not be less than 4000 ft/min. No seed should pass through the blower.

Blowers

Since the operating air pressure range for small-pipe systems is usually 1 to 4 lb/in^2, the two-lobe rotary air pump is the type most commonly used; it is known in the trade as a positive-pressure blower. These blowers can purge the piping to overcome minor chokages by a temporary increase in air pressure.

Performance data for the rotary positive-pressure blowers now being used successfully at cotton gins and seed establishments are given in Table 29.

A screened intake or air filter is imperative on cottonseed-handling blowers to protect the lobes and casing from excessive wear. Screened-intake filters may be either factory built or homemade. Large areas of close-mesh bronze screen wire are necessary for homemade intake filters; these filters should have at least 5 ft^2 of gross screen area, all accessible for daily or more frequent cleanings.

Relief valves are seldom used on positive-pressure blowers because they prevent purging the system at higher pressures when chokages threaten. If used, they should be set to pop off (release) at double the working pressures.

Feeding Seed into Pressure Pipes

Cottonseed may be satisfactorily fed into a small-pipe pressure system by means of a dropper, or rotary sealed wheel (sometimes called a vacuum wheel), that mechanically drops the seed into the air line on the discharge side of the pump or blower. ''Seed plugs'' of the auger type have also been used successfully.

Speeds of the rotary sealed wheel, or dropper, should be relatively low — 30 to 60 r/min — and internal seals at the ends and pocket divisions are necessary to prevent serious air leakage. It is customary to provide an independent drive for the hopper, because its speed is much slower than that of a rotary positive-pressure blower. A taper of 20 in. or more should be used on feeder base outlets or jet boxes to prevent chokage. Jet boxes must be set as close to the feeder as possible.

Piping

Aluminum is the most commonly used piping material, but standard 20-gage galvanized pipe or galvanized lightweight tubing may be used where the piping is to be exposed to the weather. Joints should be flanged wherever seed passes through the pipe. Six-bolt companion flanges with rubber gaskets are recommended. On the blower intake and discharge, or at points ahead of the seed dropper, standard threaded pipe and fittings can be used for handling the compressed air.

Seed-handling elbows should be 18-gage or heavier and must be of the long sweep type to give satisfactory service without chokage. A 36-in. radius is recommended for pipes 4, 5, and 6 in. in diameter, and a 48-in. radius for pipes 8 in. in diameter.

Risers or lifts in seed piping should preferably be on an incline rather than vertical, both to keep the angles at elbows as large as possible and to save the piping length that diagonals afford in comparison with right-angle runs. Pipe should not be allowed to slope downward in the direction of flow. Downward slopes may cause chokages.

Valves, Branches, and Discharges

Valves for small-pipe systems frequently give trouble in operation unless they are of good workmanship and are well fitted for tightness. Not more than two valves should be used in the ordinary cottonseed system, because leakage and careless adjustments invite trouble.

Table 29
PERFORMANCE DATA FOR ROTARY POSITIVE-PRESSURE COTTONSEED BLOWERS[a]

				Pressure[c]									
				1 lb/in²g		2 lb/in.²g		3 lb/in.²g		3½ lb/in.²g		4 lb/in.²g	
Seed capacity (lb/min)	Size no.[b]	Pipe diameter (in.)	Speed (r/min)	Capacity (ft³/min)	Brake hp	Capacity (ft³min)	Brake hp	Capacity (ft³/min)	Brake hp	Capacity (ft³/min)	Brake hp	Capacity (ft³/min)	Brake hp
Up to 100	615	4	575	292	1.9	256	3.5	228	5.2	217	6.0	—	—
			690	368	2.3	332	4.3	304	6.3	292	7.2	—	—
			1080	626	3.6	590	6.7	562	9.7	550	11.2	—	—
Up to 175	717	5	575	555	3.2	511	6.1	478	8.9	464	10.4	—	—
			690	687	3.8	643	7.3	610	10.4	596	12.5	—	—
			925	956	5.2	912	9.7	879	14.4	866	16.9	—	—
Up to 250	717	6	1200	1275	6.4	1225	12.8	—	—	1175	21.8	—	—
			1400	1500	7.5	1450	15.0	—	—	1400	25.2	—	—
			1620	1750	9.0	1650	17.0	—	—	1600	29.1	—	—
Up to 500[d]	820	8	1000	—	—	1520	14.0	—	—	—	—	1440	28.0
			1200	—	—	1680	17.0	—	—	—	—	1600	34.0
			1450	—	—	2070	20.0	—	—	—	—	2020	40.0

[a] Adapted from tables published by manufacturers.
[b] One of several trade designations.
[c] 4 ft³ of air/lb of seed may be allowed at 1 lb/in²g. At all higher pressures, allow 5 ft³
[d] At maximum capacity operating at a velocity of 5000 ft/min at a pressure of 2 lb_f/in.², limiting conveying distance will be 500 ft.

For lines containing the vane-type seed valves, the takeoff angle should not exceed 30°. The deflector vane should be of adequate thickness and should be well fitted into the body of the valve, with the seated end adjusted to prevent seed chokage by lint or by seed build-up at the valve intake.

The discharge funnel materially assists airflow through the pipe to open bins, but it should not be directed at a wall or any obstacle that might cause the seed to crack.

Cylcone collectors of seed for delivery to sackers may be used at the ends of small-type systems. The downspouts from these collectors should be 8 in. or more in diameter to prevent choking or bridging of seed at their bases.

Other Features of Small-Pipe Systems

Approximate distances for conveying cottonseed efficiently with small-pipe systems are 200 to 700 ft of piping. Since no two systems are exactly alike in length and number of risers and elbows, the limitations of pressure and volume peculiar to the type of blower must be carefully considered in the design of any small-pipe system. Power cost is usually about half that of ordinary cotton-gin fan systems. Initial cost of small-pipe systems using positive-pressure blowers is somewhat lower than that of the gin fan system. The efficiency of the small-pipe handling system depends on correct installation.

Measuring Equipment

Most modern gins are equipped with devices for measuring cottonseed from each bale before or during its discharge into the conveying system. A common type of cottonseed measuring device consists of a pocket wheel that receives the seed from the gins and dumps automatically when each succeeding pocket of the wheel has received a specified amount of seed. Depending on the size of the unit, each 10 or 20 lb of seed is measured as it is dumped, and the device can be equipped to signal the completion of each bale.

Cottonseed Aeration and Storage

Mechanical harvesting and modern high-capacity gins have increased the need for bulk storage of cottonseed. In recent years more and more emphasis has been placed on maintenance of cottonseed quality, for both planting and oil mill purposes. Aeration systems are now being made to engineering designs rather than by trial and error, helping to reduce seed losses. Properly designed aeration systems help prevent free fatty acid formation and maintain initial seed viability. With improved system designs, selection of proper equipment, and good operating procedures, cottonseed quality can be maintained efficiently and effectively.

Storage

Most of the cottonseed from the ginning process goes into storage either for oil mill processing or for planting. A small amount will be returned to the farm for cattle feed. Approximately 4.27 million tons of cottonseed were produced in the U.S. in 1978, of which about 3.16% (135,000 tons) was stored and processed for planting. The remainder was used primarily in the milling industry. Storage methods are similar for preserving quality for either purpose. Proper aeration is the key to minimizing storage losses and quality deterioration. Storage methods vary in different geographic areas.

Mechanically harvested cotton when ginned may have a seed moisture content (wet basis) as low as 6% and as high as 18%, with the majority within the range of 10 to 15%. These moisture contents, in most storage areas, present no problems for safe storage with proper aeration and cooling. However, cottonseed for planting purposes should not go into storage with a moisture content above 12%. The highest moisture content in seed will usually occur at the beginning of the harvest season. Most ginners and storage warehousemen, at the beginning of the season, will exercise their prerogative of sending this high-moisture seed to the oil mill. The drier seed will go into storage for planting-seed processing later in the year.

Most present-day cottonseed storages have moisture-proof concrete floors. The structural framing may be either wood or steel, with the outside clad with sheet steel. Some planting-seed storages may have walls faced on the inside. Facing gives some protection from extreme temperature changes that may occur in steel-clad buildings. However, cottonseed is a good insulator and wide temperature changes occur only in a relatively thin layer along the walls.

Cottonseed storage capacities may range from 400 to several thousand tons. In computing storage capacity, a density of 80 ft^3/ton is used for seed depths of 24 ft or less. A density of 75 ft^3/ton is used for seed depths or 24 to 50 ft. For depths of 50 ft or more, a density of 70 ft^3/ton is used. Table 30 gives the capacities of shallow, flat storages with depths to 24 ft, commonly used for planting seed.

Calculating the capacities of Muskogee-type (deep, peaked) storages is more complicated than calculating for shallow, flat storages. Following is an equation for computing the approximate capacities of Muskogee-type storages using the assumptions given:

1. All roof slopes are 45°
2. Head house is 10 ft wide
3. Any space lost to aeration system, filling and unloading systems (tunnel, etc.), and internal building structural losses is disregarded
4. House is filled to the roof
5. All dimensions are inside of building
6. Density is 70 ft^3/ton for tightly packed seed over 50 ft deep

Calculate as follows:

$$\text{Volume in ft}^3 = WL(S + H) - H^2 (L = W - 1.333H)$$

$$\text{Capacity in tons} = \frac{\text{volume in ft}^3}{70 \text{ ft}^3/\text{ton}}$$

where S = sidewall depth (feet), H = maximum depth of seed less depth of sidewall (feet), L = length of storage (feet), and W = width of storage (feet).

Aeration and Cooling

One difficulty encountered in the past in designing aeration systems for moving small amounts of air through stored cottonseed was the lack of information on the resistance that cottonseed offers to air flowing through it. Research in recent years has established some values for P_s losses or resistance to airflow by cottonseed at depths of 10 to 80 ft and airflow rates of 2 to 20 ft^3/min/ton of seed. Table 31 gives some comparative values for three depths and three airflow rates and the resulting P_s and approximate horsepower requirements. Additional P_s values and horsepower requirements at various depths are given in U.S. Department of Agriculture Marketing Research Report No. 1020, *Aeration of Cottonseed in Storage.*

An airflow rate not less than 10 ft^3/min/ton is recommended for cotton planting seed. At this rate, cottonseed having 12% moisture can be cooled sufficiently with ambient air to maintain seed quality. Many certified seed producers use an airflow rate of 15 ft^3/min/ton. However, at this rate, the P_s is approximately 65% greater and

Table 30
CAPACITIES OF FLAT COTTONSEED STORAGES

Inside width (ft)	Capacity/linear ft (tons)	Average depth of seed (ft)	Capacity in tons when length of house in feet is						
			80	100	120	140	160	180	200
30	4.50	12	360	450	540	630	720	810	900
	5.25	14	420	525	630	735	840	945	1050
	6.00	16	480	600	720	840	960	1080	1200
	6.75	18	540	675	810	945	1080	1215	1350
	7.50	20	600	750	900	1050	1200	1350	1500
	8.25	22	660	825	990	1155	1320	1485	1650
	9.00	24	720	900	1080	1260	1440	1620	1800
36	5.40	12	432	540	648	756	864	972	1080
	6.30	14	504	630	756	882	1008	1134	1260
	7.20	16	576	720	864	1008	1152	1296	1440
	8.10	18	648	810	972	1134	1296	1458	1620
	9.00	20	720	900	1080	1260	1440	1620	1800
	9.90	22	792	990	1188	1386	1584	1782	1980
	10.80	24	864	1080	1296	1512	1728	1944	2160
40	6.00	12	480	600	720	840	960	1080	1200
	7.00	14	560	700	840	980	1120	1260	1400
	8.00	16	640	800	960	1120	1280	1440	1600
	9.00	18	720	900	1080	1260	1440	1620	1800
	10.00	20	800	1000	1200	1400	1600	1800	2000
	11.00	22	880	1100	1320	1540	1760	1980	2200
	12.00	24	960	1200	1440	1680	1920	2160	2400
45	6.75	12	540	675	810	945	1080	1215	1350
	7.87	14	630	787	945	1102	1260	1417	1575
	9.00	16	720	900	1080	1260	1440	1620	1800
	10.12	18	810	1012	1215	1417	1620	1822	2025
	11.25	20	900	1125	1350	1575	1800	2025	2250
	12.37	22	900	1237	1485	1732	1980	2227	2475
	13.50	24	1080	1350	1620	1890	2160	2430	2700
50	7.50	12	600	750	900	1050	1200	1350	1500
	8.75	14	700	875	1050	1225	1400	1575	1750
	10.00	16	800	1000	1200	1400	1600	1800	2000
	11.25	18	900	1125	1350	1575	1800	2025	2250
	12.50	20	1000	1250	1500	1750	2000	2250	2500
	13.75	22	1100	1375	1650	1925	2200	2475	2750
	15.00	24	1200	1500	1800	2100	2400	2700	3000
60	9.00	12	720	900	1080	1260	1440	1620	1800
	10.50	14	840	1050	1260	1470	1680	1890	2100
	12.00	16	960	1200	1440	1680	1920	2160	2400
	13.50	18	1080	1350	1620	1890	2160	2430	2700
	15.00	20	1200	1500	1800	2100	2400	2700	3000
	16.50	22	1320	1650	1980	2310	2640	2970	3300
	18.00	24	1440	1800	2160	2520	2880	3240	3600

the horsepower requirement increases by about 140% over that for 10 ft^3/min/ton. Research results indicate that 10 ft^3/min/ton is sufficient to cool cottonseed to not less than 50°F, as recommended for planting seed.

In large Muskogee-type storages for oil mill processing, the depth of the cottonseed usually is a limiting factor in the amount of air that can be moved economically. For

Table 31
RESISTANCE OF COTTONSEED TO AIRFLOW AT SPECIFIED RATES OF FLOW AND DEPTH OF SEED IN LEVEL STORAGE

Depth of cottonseed (ft)	Airflow rate (ft^3/min/ton)	Static pressure (in. H_2O)	Hp required/ 1000 tons
20	5	2.8	3.5
	10	6.4	19.0
	15	10.6	43.0
30	5	6.5	10.0
	10	15.8	42.5
	15	26.5	100.0
40	5	12.3	17.0
	10	30.5	86.0
	15	52.5	210.0

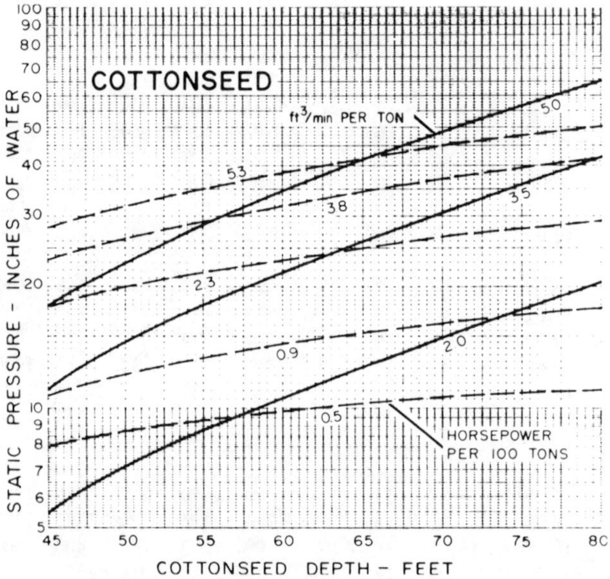

FIGURE 8. Fan horsepower and static pressure requirements for aerating cottonseed at three airflow rates at depths ranging from 45 to 80 ft.

example, with cottonseed 80 ft deep and an airflow rate of 2 ft^3/min/ton, the P_s is approximately 20 in. (water gage), as given in Figure 8. Research results also indicate that this airflow rate is sufficient to cool cottonseed 80 ft deep to temperatures of 40° to 45°F by late December or January.

Usually no additional aeration is necessary once the seed has been cooled to the desired temperature. Even so, seed temperatures should be monitored throughout the storage period, especially in planting seed, since hot spots occasionally develop. Usually a few hours of additional aeration will remove the heat from the trouble spot, provided it is discovered promptly.

Many planting-seed storages use commercially available systems for monitoring temperatures of stored cottonseed. Systems using thermocouples are probably the most common. Thermocouple cables can usually be fastened to some part of the roof or truss structure and suspended to near the floor. A breakaway cable hanger is recommended to prevent structural damage to the building in case of excessive force on the cable caused by seed setting. The sensing points in each cable may be spaced 3, 4, or 5 ft apart, depending on the depth of seed and the number of sensing points desired. The distance between horizontal cables is optional, but is usually 20 ft in each direction lengthwise and crosswise of the storage.

Measuring temperatures in deep-stored cottonseed is much more difficult. Thermocouple cables are sometimes used by placing them in the cottonseed horizontally during the filling process; these cables will tend to shift with the seed and make it difficult to determine where the sensing points are located. Another common method is to drive a pipe or heavy conduit vertically into the cottonseed at selected locations and hang mercury thermometers, tied to a heavy cord, at desired intervals inside the pipe or conduit. When temperature readings are needed, the cord is rapidly removed from the pipe and each thermometer read as it is removed. When all thermometers have been read and the temperatures recorded, the thermometers are returned to place and left until the next reading time. This method will give a reasonable estimate of the temperature in the seed surrounding the pipes.

Aeration system designs fall into two general types, manifold and central duct. The manifold system has several aeration ducts spaced on the floor and connected to a fan. This type of system makes it possible to select sections or combinations of sections of the storage to be aerated at any one time, since each aeration duct is provided with a gate or slide valve that can be opened or closed as needed. The central-duct system has one or two large aeration ducts in or near the center of the storage, lengthwise of the building. A supply pipe from the center of the central duct connects to a fan outside of the building. In most central-duct systems, the unloading tunnel is covered with perforated metal or some other material that will enable it to serve as part of the aeration duct. With this type of system, aeration may be restricted to that time when the tunnel is covered with cottonseed. Additional aeration ceases once the tunnel is uncovered for moving seed out of the storage.

Design of Aeration Systems

Most aeration systems have these principal parts: (1) one or more fans to supply the required volume of air at a specified static pressure, (2) ducts to move the air into or out of the cottonseed, (3) supply pipes to connect the fans and ducts, and (4) a motor to drive the fan.

The primary factors that influence design of a system and selection of aeration equipment are (1) the size and type of structure in which the system is to be installed, (2) the depth of cottonseed through which the air will be moved, (3) the airflow rate per cubic foot (or per ton) to be provided, and (4) the quantity of seed to be served by each fan.

After these points have been settled, the following may be determined: (1) total air volume to be supplied, (2) the P_s against which the fan must operate, (3) the type and size of fans and motors, and (4) the type and size of aeration ducts and supply pipes.

Most aeration systems are designed with the airflow downward through the cottonseed. Downward airflow eliminates moisture condensation and accumulation in the top layers of cottonseed, which can occur when warm, moist air moves upward into the cold top layers of seed. Downward airflow also counteracts any convectional air movement that might take place. In addition, temperature and odor of the exhaust air from the fan can give some indication of seed quality to the experienced warehouse operator.

Airflow rates used herein are based on research reported in U.S. Department of

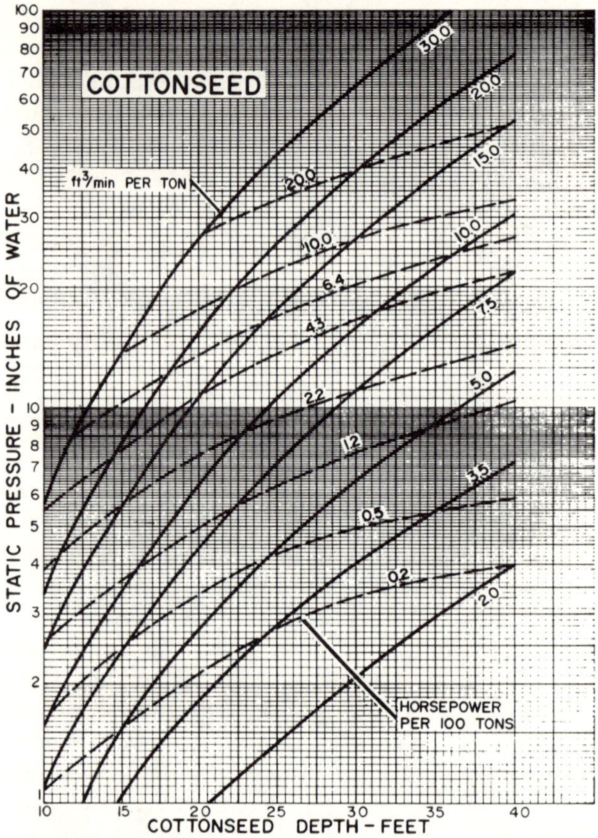

FIGURE 9. Fan horsepower and static pressure requirements for aerating cottonseed at various airflow rates and at cottonseed depths from 10 to 40 ft.

Agriculture Market Research Report No. 1020. In designing an aeration system it should be recognized that uniformity of airflow is impossible to achieve in most aerated storages, especially in Muskogee-type or peak-loaded storages, where seed depths may vary from 10 ft at the sidewalls to 100 ft or more at the peak of the mass. Since air moves along the path of least resistance, the ducting should be so designed to equalize the air paths as nearly as possible and to assure at least the desired airflow through the seed at all locations.

The following design criteria are used in selecting the aeration ducts, supply pipes, fan sizes, and motor sizes. To determine the required aeration duct surface area, a velocity of 10 to 20 ft/min at the duct surface is assumed. To calculate supply pipe sizes, an air velocity of 1500 to 2000 ft/min is used, except that a velocity of 2500 ft/min is permissible in short pipes (5 to 10 ft in length). Fan sizes are determined by the total amount of air each fan is to deliver and the P_s, depending on the airflow rate and depth of seed (Figures 8 and 9). The approximate horsepower required for each fan can also be estimated from Figures 8 and 9, once the airflow rate and cottonseed depth are known. Fan manufacturers' capacity tables will give the horsepower required for their fans under specific conditions.

For example, a flat storage for planting seed, 60 ft wide and 140 ft long, with cottonseed 16 ft deep has a capacity of 1680 tons at a density of 80 ft^3/ton. Using a design airflow rate of 10 ft^3/min/ton, the total volume of air required is 16,800 ft^3/min. The amount of aeration duct surface area required, using an air velocity at the duct surface of 10 ft/min, is 1680 ft^2. Any combination of duct diameters and lengths that provides this amount of area would be satisfactory. The distance between ducts on the floor should not be greater than 1½ times the cottonseed depth. In this storage 36-in. diameter, half-round, perforated, corrugated ducts were selected. All ducts are 24 ft long and spaced 17.5 ft apart on the floor.

With this system, either two or four fans manifolded on each side of the building may be used. Using two fans, each is required to deliver 8400 ft^3/min. The P_s against which the fans must operate determined from Figure 9, for a depth of 16 ft at an airflow of 10 ft^3/min/ton is approximately 3.9 in. (water gage). The horsepower required (Figure 9) is approximately 1.2/100 tons, or 10 hp per fan, using two fans. Whether using two or four fans, the P_s that the fans will operate against is the same — 3.9 in. (water gage). Using four fans, each fan will be required to deliver only one half the volume required when using two fans.

In designing an aeration system for a Muskogee-type or peak-load storage, the same calculations are used as for the flat storage. As an example, a Muskogee storage building that is 120 × 150 ft, with 20-ft sidewalls and is filled with cottonseed to a maximum depth of 75 ft has a capacity of approximately 11,000 tons, calculated as above. Because of the cottonseed depth and the P_s and horsepower requirements, the system in this storage is designed to deliver 2 ft^3/min/ton. The total required volume of air is 22,000 ft^3/min at a P_s of approximately 17.5 in. (water gage) (Figure 8). One or two fans may be used; two are used in this design. Each fan is required to deliver 11,000 ft^3/min at 17.5 in. (water gage) of pressure and will require approximately 50 hp. The center 90 ft of the unloading tunnel is used as the aeration duct. This part of the tunnel is covered with perforated or expanded metal to allow airflow through it. Tunnel doors provide for starting aeration as soon as the center portion of the tunnel is covered with seed. On either end of the tunnel, 30 ft are sealed against any air movement to eliminate short-circuiting of the air through the shallow seed directly over the tunnel near the outside wall.

Types of Fans

The most commonly used fans for aerating cottonseed are of the centrifugal (radial flow) type. Propeller (axial flow) fans may be used where the P_s is not more than 3 or 4 in. (water gage). Centrifugal fans must be used to aerate deep storages.

Three types of centrifugal fans are used in aeration of cottonseed. The forward-curve fan has a large number of blades and operates at relatively low speed. One disadvantage of this fan is that the motor may be overloaded if the P_s is decreased, causing an increase in air delivery; it is not commonly used for aerating cottonseed when the motor load will vary considerably between the beginning of the season, with a partially filled storage, and the end of the season, when the storage is full.

The backward-curve centrifugal fan is a high-speed fan that usually has 12 blades. It is slightly more efficient than a forward-curve or straight-blade fan. One of its advantages is a self-limiting horsepower characteristic; when used near the point of maximum efficiency with a motor of adequate size, there is no danger of motor overload.

A centrifugal fan with straight blades, sometimes called a pressure fan, or more commonly an industrial exhauster, is widely used in aerating cottonseed. It has some overloading characteristics when the P_s is reduced, but will not overload as much as a forward-curve fan. A slide gate or damper placed in the supply pipe leading to the fan can be adjusted to maintain a relatively constant P_s on this fan, as the amount of cottonseed varies.

Axial flow fans are basically low-pressure fans, but certain designs are suitable for up to medium-pressure operation.

The centaxial fan, relatively new in design, is being used in some of the latest aeration systems. The combination design has the characteristics of both the basic axial and centrifugal fan. It has a low noise level and is nonoverloading. This fan offers the installation simplicity of an axial flow design and the performance of the centrifugal. Its efficiency is very good, and it will perform in pressure ranges of up to 17 in. (water gage) at relatively low speeds.

Fans should be selected on the basis of performance ratings supplied by the manufacturers. A performance rating specifies the volume of air delivered by the fan at various static pressures. Reliable performance ratings are assured if the fans have been rated in accordance with the testing code of the Air Moving and Conditioning Association.

Handling Planting Cottonseed

Cottonseed to be used for planting requires some additional consideration during storage. Seed from mechanically harvested cotton will usually go into bulk storage without any additional cleaning at a moisture content of 12% or less. It may be conveyed into storage by mechanical methods with a belt or screw conveyor or pneumatic system. Less damage will occur to the cottonseed in a belt conveyor than in either of the other two methods. To minimize seed damage in a pneumatic conveyor, all interior pipe surfaces should be as smooth as possible. Impact damage can be kept low by making all pipe turns with a radius of at least 4 ft.

The nonflowing characteristics of cottonseed cause a major problem in moving it out of storage. Considerable hand labor in conjunction with portable conveyors is used. Pneumatic removal is being used in some storages. In most of these cases, however, the size and weight of the suction pipes are so great that very little if any reduction in labor is realized.

Treating and Delinting Cottonseed

Processing and treating of planting seed will start after the harvest season, usually in December. The cottonseed will be moved from the storage and into the processing plant where it will be cleaned. After cleaning, it may be delinted by one of several processes. The best planting methods require a delinted seed that can be planted with precision accuracy.

Several methods for delinting cottonseed are in use. In machine delinting, the cottonseed is reginned to remove some additional linters from the original ginned cottonseed. Following machine delinting, flame delinting may be used to remove additional linters and fuzziness and help to make the seeds flow more easily. Close machine delinting removes more of the linters than either of the above two methods. Acid delinting has gained popularity in recent years, because it removes all linters, leaving the black seed, which will flow like grain. The most precise planting control in existing planters can be obtained with acid-delinted seed. Since acid-delinted seed cannot be stored as long as seed delinted by other methods, acid delinting is usually done as near planting time as possible.

After the cottonseed has been delinted, it is treated with the necessary chemicals to protect seed viability and to reduce bacterial activity in the soil during the early growing stages of the plant. The treated seed is bagged, labeled, and stored for distribution and sale. Because of the controls and precautions necessary in the treatment and delinting processes, most planting seed is processed in commercial seed facilities.

Collection and Disposal of Gin Waste

In the research program of the U.S. Department of Agriculture, an effort is made to anticipate the future needs of the ginning industry. Currently, one of the most press-

Table 32
AVERAGE WEIGHT OF FOREIGN MATTER REMOVED DURING GINNING A 500-LB BALE (GROSS WEIGHT) FROM COTTON HARVESTED BY VARIOUS METHODS

	Method of harvest			
	Hand-picked	Machine-picked	Machine-stripped	Hand-snapped
Percentage of crop[a]	<0.5	63	36	<0.5
Avg waste weight, lb[b]	107	203	740	615

[a] Percentages from Charges for Ginning Cotton, Costs of Selected Services Incident to Marketing, and Related Information, Season 1973-74, U.S. Department of Agriculture Econ. Res. Serv. (Rep.) ERS-2, Washington D.C., 1974.

[b] Averages calculcated from seed-cotton weights, using the seed-lint ratio furnished by cotton breeders and correcting it for 3% seed-cotton moisture removal during ginning. Weighted waste averaged for the crop: 322 lb/bale.

ing needs is for air pollution abatement equipment for waste collection and disposal. This need has been in the forefront since the passage of the Federal Clean Air Act of 1970.

At most gin plants, the only pollutant emitted into the atmosphere is particulate matter, either exhausted from some type of abatement or escaping from some type of abatement equipment. Additional pollutants in the form of smoke and fly ash are emitted from gin plants that incinerate the collected waste material. At the present time, all cotton-producing states are operating under some form of process-weight standard having an allowable emission level based on the total weight of material being processed.

The actual levels of gin emissions vary, depending on the time and method of harvest. During the 1978—1979 cotton harvesting season, 65% of the crop was machine-picked and 34% was machine-stripped. The amount of foreign matter removed by gin equipment is quite variable, ranging from an average low of 107 lb per bale for hand-picked cotton to an average high of 740 lb per bale for machine-stripped cotton (Table 32).

The conveying systems in modern cotton gins are predominately pneumatic and require large volumes of air to move the cotton through the processing sequence. A gin with a capacity of 12 to 15 bales per hour requires approximately 100,000 ft^3 of air per minute to operate properly. Trash removed from the seed cotton by the cleaning equipment of the gin is usually carried out of the gin by these pneumatic systems.

The ginner's waste problem is twofold. First, he must collect the material, and second, he must dispose of it. Both operations must be performed without causing excessive air pollution.

High-Efficiency Cyclones

Cyclones, the most widely used collection equipment on high-pressure discharges, collect the bulk of the waste being discharged. Cyclones are used extensively because they are effective, inexpensive, and require little maintenance. They are essentially cylindrical, with a long, conical hopper bottom and a center cylinder through which the air escapes. The trash-laden air enters tangentially near the top. Centrifugal force caused by the whirling action of the trash and air pushes the trash outward and into the conical hopper below. The air passes up and out through the center cylinder.

A properly sized high-efficiency cyclone collects 99.9% of the total trash introduced and virtually 100% of the trash that is larger than 30 μm in diameter (Figure 10).

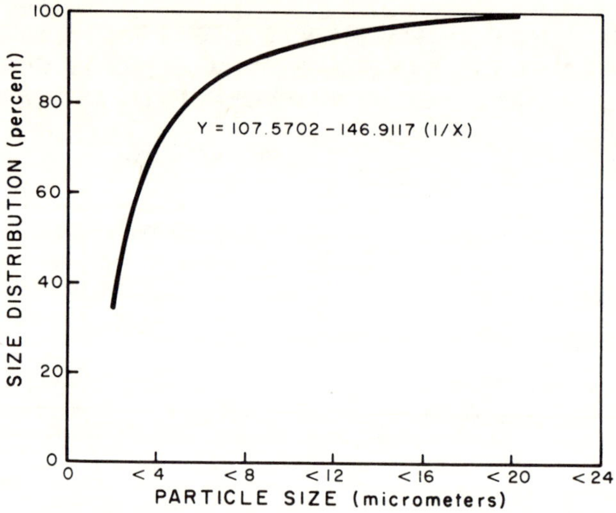

FIGURE 10. Percentage of various sizes of gin trash particles emitted from exhaust of a small-diameter cyclone.

Cyclone size and arrangement depend on several variables, the most important of which is the volume of air to be handled. This can be measured by the use of a pitot tube and manometer. In general, low-volume fans require only a single cyclone installation; however, as air volume is increased, two to four parallel-mounted cyclones are required to ensure high efficiency and keep the sizes within practical limits. A diameter of 34 in. is considered optimum. Figure 11 illustrates relative dimensions for small-diameter cyclone designs. Table 33 is a guide to the selection of cyclone size and arrangement based on air volume, and Figures 12, 13, and 14 illustrate typical cyclone arrangements.

The pressure resistance the fan is capable of overcoming is another important item in cyclone sizing. A minimum inlet air velocity of 3000 ft/min is recommended for efficient removal of the trash and dust. The calculated pressure drop in cyclones designed for an inlet air velocity of 3000 ft/min is 4.3 in. of water, which is within the normal operational limits of straight-blade centrifugal fans.

Figure 15 shows the optimum velocity range for operating cyclone collectors and gives approximate sizes required for various applications. Selection should be made from this graph between the dotted lines to provide a good compromise of collecting efficiency to horsepower load or pressure drop. These limits are from 3 to 5 in. of pressure as measured on the water gage and correspond to the recommended inlet velocities of 3000 ft/min.

Space and height limitations will be factors in choosing a multiple mount. A 4-ft-diameter cyclone will be a minimum of 17 ft high (8-ft body, 4-ft cone and at least 1 ft for the clean-air exit pipe), whereas a 2-ft-diameter cyclone will only be 9 ft high.

For cyclones discharging into screw conveyors, care should be taken to provide sufficient conveyor size and speed to prevent stoppages. Conveyors mounted beneath trash-collecting cyclones should be equipped with a hopper to accommodate temporary overloads.

Since the splitter inlet transition on multiple-cyclone arrangements is a potential source of trouble, a single cyclone should be used on each lint-cleaner waste fan whenever possible. If the cyclones discharge into a screw conveyor, the lint-cleaner waste cyclone should be located near the outlet end of the conveyor to minimize conveyor chokage.

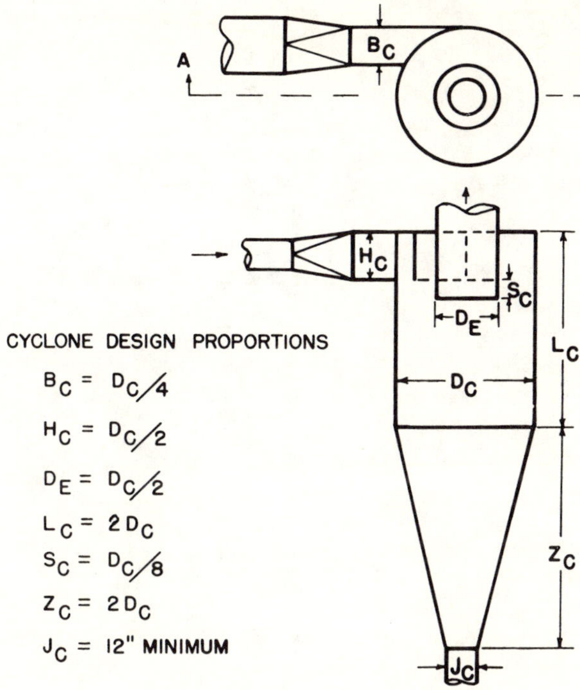

FIGURE 11. Relative dimensions for a small-diameter design cyclone.

Inline Filters

Inline filters were developed specifically for controlling fly lint and dust emissions from condenser exhausts. Three designs of inline filters are available (Figures 16 and 17). In one of these, the stationary-screen design, a fixed filtering screen is mounted in an arc to match a rotating brush reel. In another, the filtering screen covers a revolving drum, which is cleaned by a rotating brush cylinder located near the bottom of the drum. In the third design, called a round air filter, a flat circular screen is cleaned by a brush attached to a rotating arm.

All three designs operate on the same basic filtering principle. Fly lint and small-leaf trash are strained from the exhaust air by a fine-mesh filtering screen. Fly lint accumulates on the screen, forming a filtering batt for the separation of fine dust particles. Periodically, the accumulated material must be removed from the screen to prevent the buildup of excessive back pressure.

Inline filters are most efficient when operated on a collecting-cleaning cycle controlled by a pressure-differential switch. The switch automatically starts the screen-cleaning action when the pressure reaches a predetermined level and then stops cleaning action when the pressure is relieved. This system allows an efficient dust-filtering batt to form before the screen is cleaned, which improves the dust-collecting ability of the inline filter. In those instances in which inline filters are operated without a pressure-differential switch, the filter-cleaning mechanism is allowed to operate continuously, simplifying the system but significantly lowering dust-collection efficiency.

Switches should be adjusted to start the cleaning action before the pressure reaches a level that would critically affect exhaust fan performance. The critical pressure varies from installation to installation. However, experience has shown that vaneaxial fans can usually operate satisfactorily against pressures of 1 to 1½ in. of water over normal operating pressure. Tubeaxial fan systems usually operate satisfactorily when pressures

Table 33
CYCLONE SIZING AND ARRANGEMENTS RECOMMENDED FOR AVERAGE GINNING CONDITIONS

Air	Single		Double		Triple		Quadruple	
volume (ft^3/min)	Diameter, D_c (in.)	Approximate height (ft)	Diameter, D_c (in.)	Approximate height (ft)	Diameter, D_c (in.)	Approximate height (ft)	Diameter, D_c (in.)	Approximate height (ft)
1,500	24	9	—	—	—	—	—	—
2,000	28	10	20	8	—	—	—	—
2,500	31	11	22	8	—	—	—	—
3,000	34	12	24	9	20	8	—	—
4,000	40	14	28	10	23	8	20	8
5,000	44	16	31	11	26	10	22	8
6,000	48	17	34	12	28	10	24	9
7,000	—	—	37	13	30	11	26	10
8,000	—	—	40	14	32	12	28	10
9,000	—	—	42	15	34	12	29	11
10,000	—	—	44	16	36	13	31	11
11,000	—	—	46	16	38	14	32	12
12,000	—	—	48	17	40	14	34	12
14,000	—	—	—	—	42	15	—	—
16,000	—	—	—	—	45	16	40	14
18,000	—	—	—	—	48	17	42	15
20,000	—	—	—	—	—	—	44	16
22,000	—	—	—	—	—	—	46	16
24,000	—	—	—	—	—	—	48	17

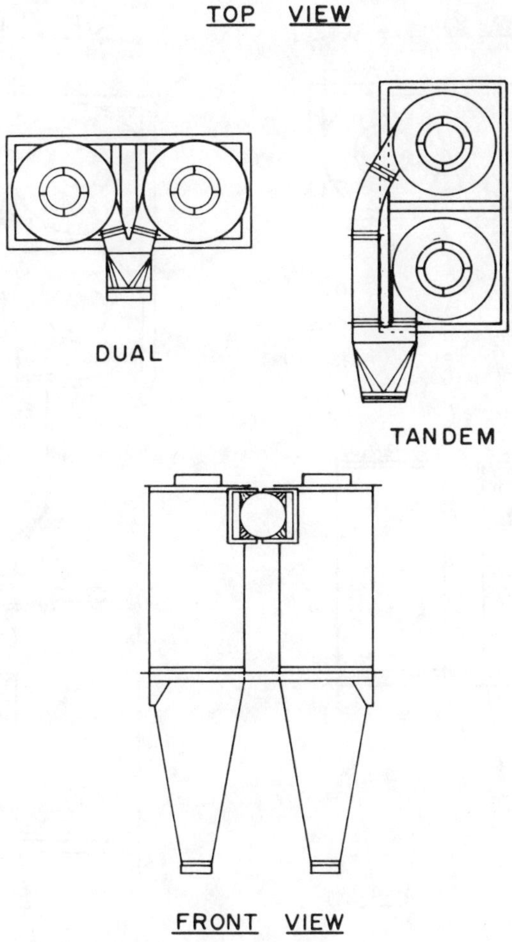

FIGURE 12. Double cyclone arrangement.

do not exceed 3/4 in. of water. Final adjustment of the pressure-differential switch should be made under actual operating conditions.

The collection efficiency of an inline filter also depends upon the amounts of fly lint and dust in the exhaust air. Tests in New Mexico using machine-picked cotton showed an overall collection efficiency of 87% for a stationary screen inline filter on the exhaust of a first-stage lint cleaner. Efficiencies of approximately 81% were obtained in Texas when ginning machine-stripped cotton. The Texas tests also showed no significant differences in collection efficiency among the three inline filter designs. Generally, an inline filter can be expected to collect all of the fly lint and up to about 70% of the fine dust.

Inline filters require fine-mesh filtering screens with high open area percentages. Bolting-grade stainless steel or Monel® wire cloth has proved to be effective. Dimensions of wire cloth that have been successfully employed in inline filters are given in Table 34.

Inline filters should be sized according to the volume of condenser exhaust air to be filtered. Screen face velocities of approximately 750 ft/min have been satisfactorily employed. This velocity results in initial added pressure (screen resistance and shock losses) of 0.3 in. of water or less. Lower face velocities can be used, but the size of the filter must be increased. High face velocities create higher pressures. Recommended screen sizes for various air volumes are given in Table 35.

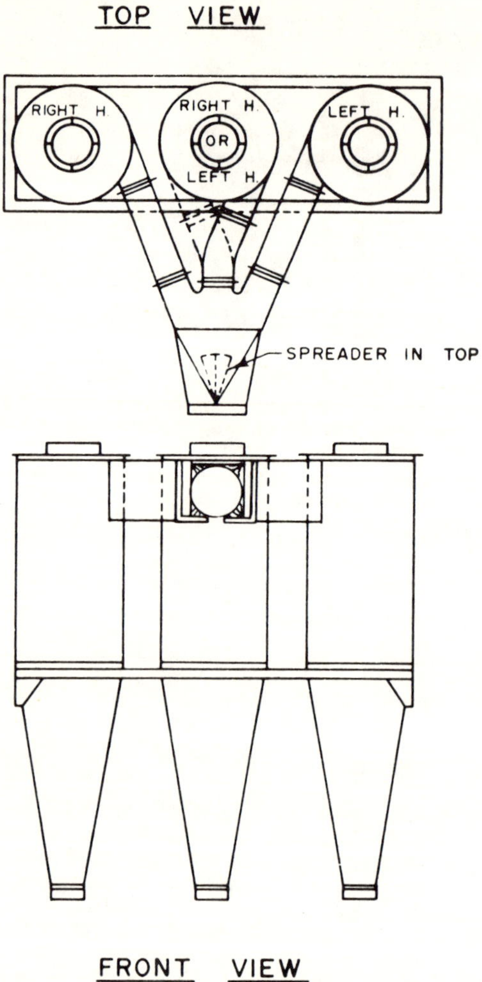

FIGURE 13. Triple cyclone arrangement.

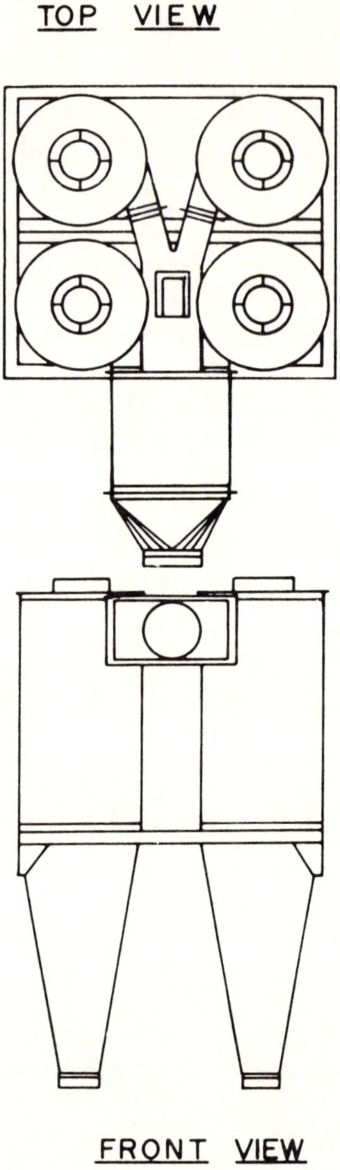

FIGURE 14. Quadruple cyclone arrangement.

Proper selection and adjustment of brushes are required for effective cleaning of the filtering screen. Soft bristles, such as horsehair, are not stiff enough; nylon bristles approximately 0.02 in. in diameter and 1 in. long have the necessary stiffness for satisfactory cleaning of filter screens. The brush should be adjusted so that the tips of the bristles firmly contact all of the screen surface area. Any part of the screen that the brush fails to touch will not be cleaned well and will tend to become clogged. Brush reels for stationary screen filters and brush arms for round air filters should be driven at speeds of about 12 r/min. Brush cylinders on revolving drum filters should operate at approximately 30 r/min.

Inline filters should be inspected daily throughout the active ginning season. Accumulations of lint on the brushes should be removed, and the waste exit should be kept

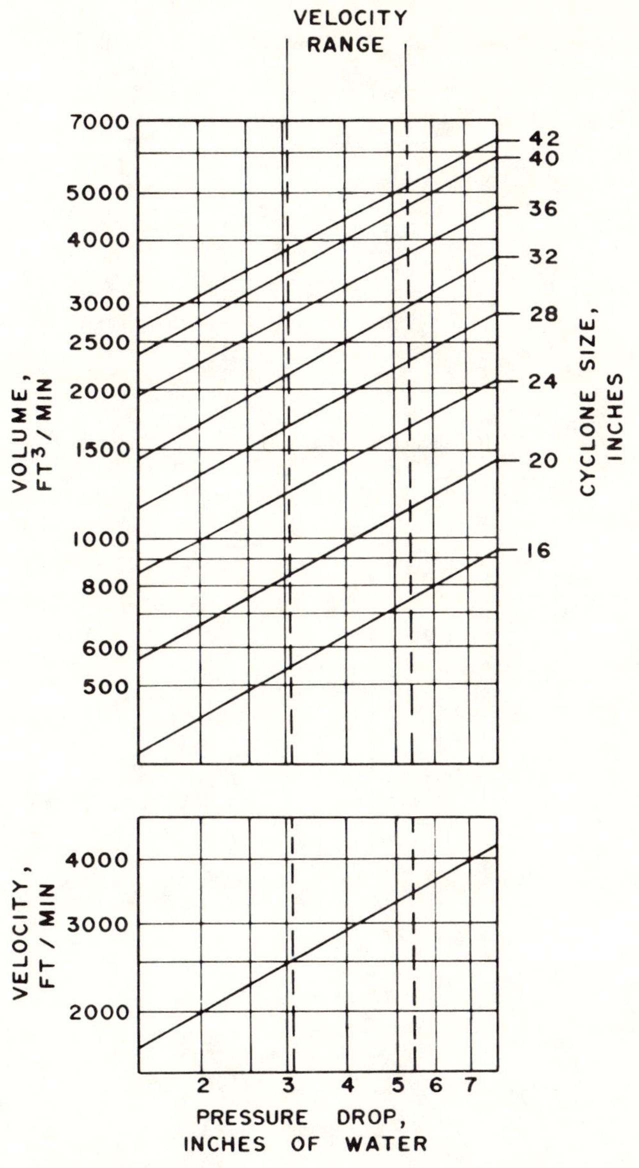

FIGURE 15. Small diameter cyclone selection chart.

clean of obstructions. Any part of the filtering screen which has become clogged should be cleaned with an airhose. Each year, before the ginning season starts, inline filters should be thoroughly cleaned and inspected for (1) good brush contact, (2) screen condition, (3) clearance through waste exit and mote lines, (4) correct tension and alignment of chain drives, and (5) accumulations of lint material in the electric motors.

Condenser Drum Covering

Short fibers and dust exhausted from condenser fans are probably the emissions most objectionable to gin employees and residents of the immediate area. This lint fly settles on electric lines, the gin yard, rooftops, and surroundings, causing not only a housekeeping problem but a serious fire hazard. Tests at the U.S. Cotton Ginning Research Laboratory at Stoneville, Miss. have shown that the lint fly problems can be

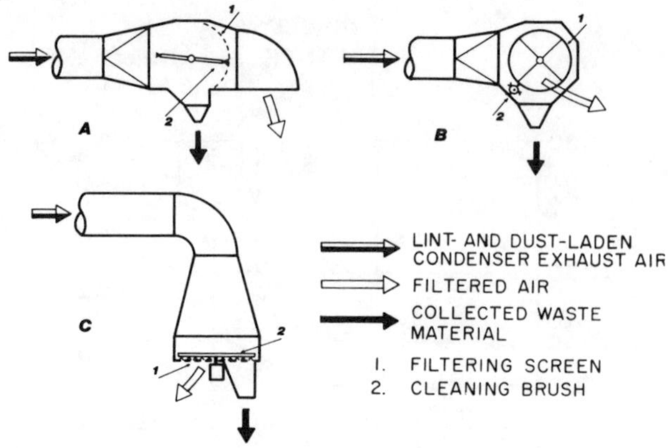

FIGURE 16. Schematic diagrams of three designs of in-line air filters. A, Stationary screen type B, Revolving drum type. C, Round air filter.

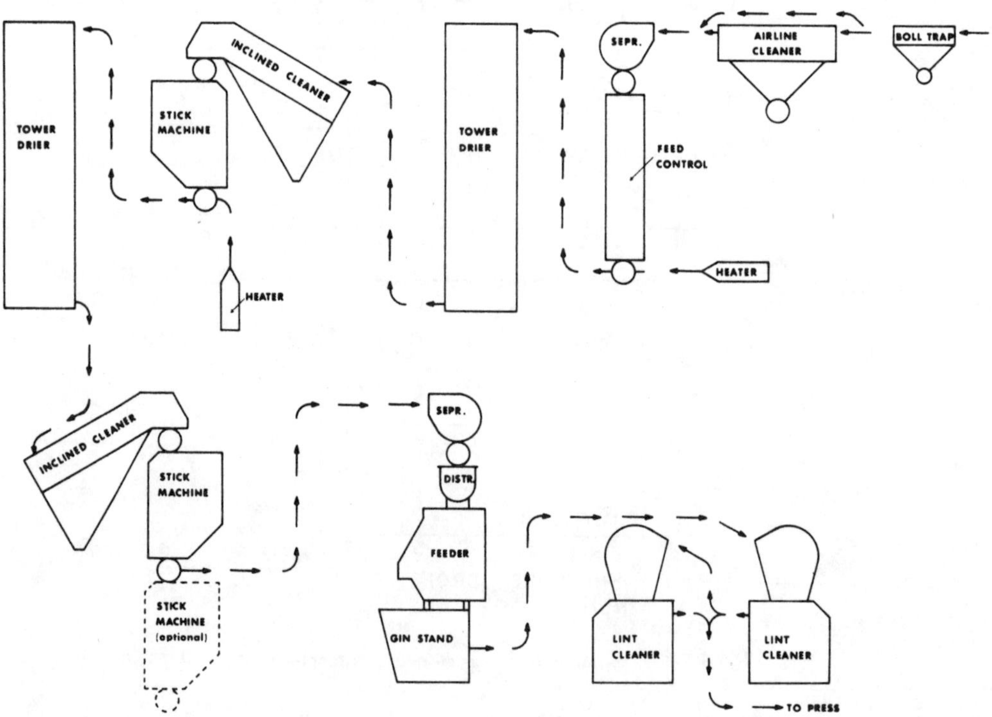

FIGURE 17. Three designs of inline air filter. Left to right: round air filter, stationary screen type, and revolving drum type.

eliminated, for all practical purposes, by covering condenser drums with fine-mesh stainless-steel screen wire or replacing its covering with fine, perforated metal. The screen wire can be either 100 × 100 mesh with approximately 31% of open area or 70 × 70 mesh with approximately 54% of open area. The fine, perforated metal should give approximately 20% of open area (Table 36). The greater open area will result in less pressure on the system. The lint fly is eliminated, with only the very fine dust being discharged with the exhaust air. The slight increase in air resistance is not a problem where exhaust fans are used. The screen is stretched smooth over the con-

Table 34
SPECIFICATIONS FOR BOLTING-GRADE WIRE CLOTH

Mesh of screen (wires/in.)	Wire diameter (in.)	Size of opening (in.)	Open area (%)	Pressure drop at 750-ft/min face velocity (in. H_2O)
40× 40	0.0065	0.0185	54.8	0.04
70× 70	0.0037	0.0106	54.9	0.05
105×105	0.0030	0.0065	46.9	0.10

denser drum and soldered around both ends and longitudinally across the drum where seams are necessary. It will be necessary to adjust the doffing rollers to ensure clearance to the drum.

Caution — Condenser drums that are blanked off with no more than one fourth of the drum area exposed to the inlet air and lint should not be covered with screen. Ample exposed drum area is necessary to compensate for the reduction in open area when fine screen is applied.

Skimmers

Skimmers are occasionally used instead of small-diameter cyclones for those fan exhausts that do not discharge large trash. Exhausts from unloader fans, overflow-suction fans, and separator-pull fans contain small-leaf trash, dust particles, and fly lint. Some of this small material can be collected by skimmers, but large trash such as burs and sticks clog the skimmer's narrow skimming slot.

Exhaust air enters the skimmer and follows the curved sidewalls along a circular arc 90 to 180°. The circular motion creates centrifugal forces that act on the entrained trash particles, pushing them toward the outside wall, where an adjustable vane and slot skims the particles and some of the air from the main airstream. The skimmed air must then be discharged into another trash collection device, such as a cyclone collector, for final separation of the trash. This remainder of the main airstream is exhausted to the atmosphere.

The collection efficiency and pressure characteristics of a skimmer depend upon its radius of curvature, the air velocity, and the size of the skimming-slot opening. The size and weight of the particulate matter entrained in the air also affect collection efficiency. Higher collection efficiencies and pressure are associated with short radius of curvature, high velocity, and wide skimming slot. Thus the design of a skimmer requires a compromise between collection efficiency, operating pressure, and the volume of skimmed air that requires further cleaning.

Bur Hoppers

It is usually necessary for cotton gins to store gin trash temporarily prior to final disposal. Elevated bur hoppers provide a convenient means of accumulating truckloads of gin trash, and also expedite the loading operation. Loading the trucks is accomplished by simply opening the bottom of the bur hopper and allowing the accumulated trash to fall into the truck.

Bur hoppers are available in lengths of 20, 30, and 40 ft. The hopper size required depends upon the type of cotton being ginned, the ginning rate, and the number of trucks being used to haul trash. Low capacity gins processing less than eight bales of stripped cotton per hour or gins processing picked cotton usually find 20-ft hoppers satisfactory. Higher capacity gins, especially those ginning stripped cotton, require the 30- and 40-ft hoppers.

Table 35
SCREEN-SIZING TABLE FOR INLINE FILTERS

Condenser capacity (bales/hr)	Exhaust volume[a] (ft^3/min)	Exhaust fan diameter and type[b] (in.)	Gross screen area required[c] (ft^2)	Stationary screen[d]		Revolving drum diameter[e] (in.)	Round filter diameter (in.)
				Radius (in.)	Width (in.)		
Up to 2	Up to 3,000	V18	5.3	15	30	20	31
2—4	3,000—6,000	V18—V21	10.6	21	42	30	44
4—6	6,000—9,000	V21—T36	15.9	25	50	35	54
6—8	9,000—12,000	V24—T42	21.2	29	58	41	62
8—10	12,000—15,000	V26—T42	26.5	32	64	46	70
10—15[f]	15,000—20,000	T42	35.6	38	76	53	81
15—20[f]	20,000—25,000	T42	44.4	42	84	59	90

[a] Exhaust volume measured without lint in the system.
[b] Type of exhaust fan, vane axial or tube axial, indicated by letters V and T, respectively.
[c] Gross screen area based on 750 ft/min face velocity, assuming a 25% loss in area for screen supports.
[d] Based on a 120° screen arc.
[e] Length of revolving drum assumed to be equal to the diameter.
[f] Press condensers only.

Table 36
DUST AND LINT FLY IN EMISSIONS FROM A UNIT LINT CLEANER CONDENSER EXHAUST WITH VARIOUS TYPES OF COVERINGS

Type of drum covering and hole diameter	Holes/in.^{2}	Open area (%)	Lint fly and dust concentration (g/1000 ft^3)	Static pressure loss (in. H_2O)
Perforated metal, 0.109-in. diameter (standard)	47	45	50.8	0.52
Perforated metal, 0.033-in. diameter	234	20	32.0	0.70
100×100 mesh screen (31.4% open area) over standard drum (45% open area)[a]		14	21.6	0.75
70×70 mesh screen (54.9% open area) over standard drum (45% open area)[b]		25	—	—

[a] 0.0056 in. between mesh.
[b] 0.0106 in. between mesh.

Trash can be blown directly into a bur hopper by one or more air lines or dropped into the hopper by cyclone collectors. The use of cyclone collectors is the preferred method. All of a gin's cyclone collectors may be mounted on top of the hopper. However, this practice has the disadvantages of inaccessibility to the collectors and a high risk of wind damage. In recent years it has become common for gins to group most of their cyclones in a bank located close to the ground and to concentrate the separations from each collector into a single air line discharging into a single set of collectors on top of the hopper.

Livestock Feed

In a few areas of the Cotton Belt, gin trash has become important as a livestock feed. On the high plains of Texas, where concentrated cattle-feeding operations have been developed, the use of gin trash as a feed roughage has become commonplace. Gin trash is usually ground and mixed with the other rations at feed lots before feeding. However, some stockmen feed unground gin trash to their cattle as a supplemental roughage, usually allowing them to feed at will from piles of trash deposited on the ground.

Several cotton gins have installed elaborate grinding and pelleting equipment for gin trash. These systems produce pellets about 1/4 in. in diameter that can be easily handled, stored, and transported. Gin trash pellets, because of their high bulk density, can be economically shipped long distances; also, pellets are easier to handle and to mix with other feedstuffs than untreated gin trash. Storage is also simplified since pellets do not constitute as much of a fire hazard as untreated gin trash. Molasses or other materials can be added to gin trash during pelleting to increase its nutritional value and palatability.

Unifilter Collection System

A totally new system for collecting all waste from a cotton gin has been developed at the U.S. Cotton Ginning Research Laboratory, Stoneville, Miss. The unifilter trash-collection system is composed of a settling chamber equipped with an impinging baffle

Table 37
OVERALL FILTER EFFICIENCY OF THE UNIFILTER COLLECTION SYSTEM FOR THREE GINNING RATES

Ginning rate (bales/hr)	Trash input (lb/hr)	Overall efficiency (%)
4	571	99.23
6	851	99.36
8	1097	99.55

and a revolving drum covered with a filter medium. Light material and dust collect on the filter medium and are removed by vacuum nozzles. Cleaned air is discharged from both ends of the filter drum. After feasibility studies on small models indicated that the principle was sound, a model that would handle 25,000 ft^3/min was built and evaluated for overall collection efficiency (Table 37).

A full-size trash collection system has been designed, fabricated, and installed for operational studies at a commercial gin whose capacity is 15 to 20 bales per hour. The enclosure is 13 ft wide, 24 ft long, and 16 ft high, with a trash hopper underneath. A 12-in.-diameter screw conveyor is located in the bottom of the hopper to carry the heavy trash to a vacuum dropper, which removes the trash from the enclosure. Two filter drums, each 8 ft in diameter and 13 ft long, are located near the top of the enclosure. The filtered air is discharged from both ends of the drums to the atmosphere. The drums are covered with a nonwoven polyester medium which has a higher efficiency than polyurethane foam. The line carrying low-pressure condenser air enters the enclosure above the filter drums, whereas the line for trash-laden, high-pressure air enters below the drums and is deflected down into the hopper to protect the medium. The low-pressure condenser air can be introduced into the enclosure below the filter drums if desired, without affecting system efficiency.

All of the air discharged from the gin is piped into the enclosure for cleaning. A total of 111,000 ft^3/min of contaminated air is cleaned through 500 ft^2 of filter. This volume of airflow results in a face velocity of 222 ft/min on the filter.

The light material and dust collect on the filter medium, which is continuously cleaned by ten nozzles, each 24 in. wide. A No. 50 fan furnishes 700 ft^3 of air per nozzle per minute. This fan also handles the trash from the hopper.

The system was in operation for the entire 1973 season, during the ginning of over 7000 bales of machine-picked cotton. It was found necessary to clean out the nozzles and remove the trash buildup on the nozzle supports every 200 to 300 bales, a job accomplished by two men in about 10 min on each occasion. Lint catching on parts of the nozzles can cause a problem, especially if lint-cleaner waste is collected with this system. Those portions of the nozzles which cause problems should be removed if possible and the nozzles sanded smooth, especially around the ends where tags normally start to form. The flexible plastic hose between the nozzle and the suction manifold should be a materials-handling type with a rubber liner.

Several rib fires occurred during the ginning season, and at least four major fires went through the filter to the trash pile. These fires caused no damage to the filter enclosure and only minor damage to the filter medium — not enough to require replacement.

The proper sizing of a unifilter system for a gin plant is dependent on the volume of air to be filtered and the face velocity at the filter medium. The efficiency goes up as the face velocity goes down. A face velocity of 150 to 250 ft/min is satisfactory.

Table 38
FILTER SELECTION FOR UNIFILTER COLLECTION SYSTEMS, BASED ON AIR VOLUME AND FACE VELOCITY AT FILTER MEDIUM

Drum diameter (ft)	Filter area (ft^2)	Number of nozzles	Air volume (ft^3/min) at face velocity of 100 ft/min	150 ft/min	200 ft/min	250 ft/min
4	50.4	2	5,040	7,560	10,080	12,600
	75.6	3	7,560	11,340	15,120	18,900
5	62.8	2	6,280	9,420	12,560	15,700
	94.2	3	9,420	14,130	18,840	23,550
	125.6	4	12,560	18,840	25,120	31,400
6	75.2	2	7,520	11,280	15,040	18,800
	112.8	3	11,280	16,920	22,560	28,200
	150.4	4	15,040	22,560	30,080	37,600
7	88.0	2	8,800	13,200	17,600	22,000
	132.0	3	13,200	19,800	26,400	33,000
	176.0	4	17,600	26,400	35,200	44,000
	220.0	5	22,000	33,000	44,000	55,000
8	150	3	15,000	22,500	30,000	37,500
	200	4	20,000	30,000	40,000	50,000
	250	5	25,000	37,500	50,000	62,500
	300	6	30,000	45,000	60,000	75,000
9	169.8	3	16,980	25,470	33,960	42,450
	226.4	4	22,640	33,960	45,280	56,600
	283.0	5	28,300	42,450	56,600	70,750
	339.6	6	33,960	50,940	67,920	84,900
10	188.4	3	18,840	28,260	37,680	47,100
	251.2	4	25,120	37,680	50,240	62,800
	314.0	5	31,400	47,100	62,800	78,500
	376.8	6	37,680	56,520	75,360	94,200

Table 38 gives the capacity at various face velocities for several drum sizes. The enclosure should be designed to fit the size and number of drums chosen for a particular installation.

Research on Incinerator

Of the many disposal methods being tried, incineration is the simplest for the ginner; however, no incinerator installed at gins before 1975 meets the strict emission standards now in force. Incinerators approved for all types of standard waste are available on the market, but none have been approved specifically for gin waste.

The Cotton Ginning Research Laboratory at Stoneville, Miss. has installed a small, multichamber incinerator to study ability to dispose of gin waste. This system is composed of an automatic loader, two burning chambers, an automatic ash removal system, and a heat exchanger in the stack.

The incinerator operates on the controlled-air principle. Waste is heated in the lower chamber by controlling the introduction of air which, in turn, controls the temperature. This system results in very low air velocities so that ash particles are not entrained and carried to the upper chamber. Only smoke, unburned gases, and very small particles pass into the upper chamber. There the smoke is reheated and additional air is introduced so that the gases and smoke particles are oxidized rapidly. The high-temperature stack gases leave the upper chamber and enter the heat exchanger, where they are cooled by air passing over the wall of the exchanger. As the waste moves across the lower chamber, it is reduced to ash and discharged from the incinerator by the ash removal cylinder.

A pound of gin trash contains approximately 7000 Btu of heat, depending on composition and moisture content. Based on an average of 203 lb of trash per bale from machine-picked cotton and 740 lb per bale for machine-stripped, the total potential heat value per bale would be 1,421,000 Btu for machine-picked cotton and 5,880,000 Btu for machine-stripped cotton.

If the heat exchanger is capable of operating at 40% efficiency, a ten-bale/hour gin plant could possibly reclaim 5,684,000 Btu/hr from machine-picked cotton and 23,520,000 Btu/hr from machine-stripped cotton. This amount of heat should be adequate for drying cotton under almost all conditions when ginning machine-picked cotton and more than adequate when ginning machine-stripped cotton. Reclaiming this heat would help offset the initial investment in an incinerator equipped with a heat exchanger.

Pink Bollworm Control in Gin Operations

Some cotton-production areas, primarily in the southwestern U.S., are infested with pink bollworm, *Pectinophora gossypiella* (Saunders). The pink bollworm is an insect that can cause severe damage to cotton. The U.S. Department of Agriculture, cooperating with state departments of agriculture, requires that certain regulations be observed to prevent the buildup and spread of this destructive cotton pest. These regulations pertain to the treatment and movement of cotton, cottonseed, and their byproducts within and from regulated areas. Producers, ginners, and seed processors are affected by these regulations.

Pink Bollworm Quarantine and Regulations

Quarantines and regulations are enforced by states and the Animal and Plant Health Inspection Service (APHIS), Plant Protection and Quarantine Program (USDA). Assistance with specific problems or information on current regulations may be obtained by calling local APHIS representatives listed under U.S. Government, Department of Agriculture, in the telephone book. Table 39 shows the quarantine requirements regarding the specifications of fans used to treat pink bollworm-infested gin trash.

POWER: MOTORS AND ENGINES

As elsewhere in industry, energy shortages and high fuel prices have focused attention on ways to reduce the cost of power in cotton gins. The advantages and disadvantages of the most common sources of power are presented here to assist ginners in evaluating these factors, so as to select the most economical and efficient sources of power for a given installation.

The principal characteristics required of cotton-gin power units are constant speed, ample capacity to handle overloads, durability, and economy in operation and maintenance. The most common sources of power are electric motors and diesel, natural gas, and liquified petroleum (bottled) gas engines. Each has significant advantages and disadvantages.

A 1978 survey of gin equipment in the U.S. by the U.S. Department of Agriculture Agricultural Marketing Service showed that the percentages of each source of power for the main engine or motor were as follows:

Table 39
FEDERAL PINK BOLLWORM QUARANTINE REQUIREMENTS REGARDING WHEEL DIAMETER, INLET SIZE, AND SPEED OF OPERATION OF SINGLE FANS FOR TREATING GIN TRASH

Maximum inlet size (in.)	Fan wheel diameter (in.)	Minimum speed (r/min)	Maximum safe speed (r/min)
10—10½	19	2760	3020
	19½	2690	2940
	20	2620	2860
10½—11	20½	2560	2790
	21	2490	2730
	21½	2430	2660
	22	2380	2610
	22½	2330	2550
	23	2280	2490
11½—12	23½	2230	2440
	24	2180	2390
	24½	2140	2340
	25	2090	2290
	25½	2100	2250
	26	2060	2200
	26½	2020	2160
12—12½	27	1980	2120
	27½	1940	2080
	28	1910	2050
	28½	1880	2010
	29	1840	1980
	29½	1810	1940
	30	1790	1910
	30½	1750	1880
	31	1725	1850
	31½	1700	1825
	32	1700	1790
	32½	1700	1760

Source of power	%
Electricity	87.5
Diesel fuel	4.6
Natural gas	5.9
Bottled gas	2.0

Several factors should be considered in selecting a power unit for a given installation. Among these are amount of power required, initial cost of equipment, cost of fuel or electric energy, cost of lubrication, kind of labor needed, expense of maintenance, probable service life, and decrease in power capacity resulting from use.

Diesel, natural gas, and LPG engines have much to recommend them for the seasonal service of cotton ginning. In selecting such engines, the purchaser should allow for a liberal margin in power over that which is normally required, to accommodate peak loads and future expansion. Most manufacturers have a continuous rating for their engines. This rating should be applicable for continuous-duty cotton-gin service without further derating.

Table 40 lists the brake horsepower and other characteristics of several engines designed for continuous cotton-gin service.

The amount of electrical energy required varies considerably, depending on how constant and intermittent loads are handled and on how much fans and other accessories are used. In electrically operated gins, it is highly advisable to have separate motors to carry the intermittent work loads of pumps, presses, and seed loading rather than have the main motor carry these loads. (In large gins, this tends to result in numerous motors for individual drives. Electric power companies usually offer consulting and testing services to their patrons, to assist them in achieving the most economical use of electricity.)

Manufacturers of cotton ginning machinery publish data sheets giving the recommended horsepower for their machines, such as gin stands, separators, cleaners, and condensers, operating at varying capacities. These data sheets usually show both the recommended size for an individual electric motor and the horsepower to be used for calculating total horsepower requirements for a central engine-driven plant. The horsepower for purposes of calculation in a central engine-driven plant is less than the recommended horsepower for individual electric motors. This is due to the fact that normally an individual machine will require much less horsepower than is required to overcome periodic surges of flow through the equipment. In a central engine-driven plant, the surges will average out at any one time so that the overall load on the central engine will never need to be as great as the sum of the horsepowers of the individual electric motors.

Electric motors for powering cotton gins must be able to operate in air heavily laden with dirt and lint, which is common to cotton gins. Also, the ambient temperatures, particularly in the upper areas of a cotton gin, often become very high. In addition, it is impossible to obtain uniform power requirements for the various machines in the gin plant, even though bulk feed control units have done much to level out extreme surges in power requirements. Uniformity in these conditions is difficult for any one type of electric motor to attain.

Totally enclosed motors are protected from dust and lint, but even fan-cooled enclosed motors do not have the overload capacity of open motors, which can dissipate heat more easily. The totally enclosed motor may give more trouble because of overheating than the open motor will give because of contamination from dust and lint. A compromise may be a motor built in a large frame that has large openings to prevent clogging by lint and dust. The large frame enables this motor to dissipate heat more readily. Thus, the motor can operate under overloads of up to 25% for short periods of time to overcome the incalculable surges that sometimes develop in gin machinery. The cost of this large-frame, open motor is between the costs of a standard drip-proof motor and the totally enclosed fan-cooled motor, which is the most expensive. Despite apparent shortcomings, many standard drip-proof motors are giving satisfactory service.

GINNING RECOMMENDATIONS

Gins to Handle Specific Types of Cotton

Numerous tests by the U.S. Department of Agriculture cotton ginning research laboratories and cooperating agencies have formed the basis for recommendations on the sequence and amount of gin machinery to be used with cotton harvested by various methods, to achieve satisfactory bale value for the producer on one hand and to preserve the inherent quality of the cotton on the other.

As a result of these findings, and considering premiums and discounts of the mar-

Table 40
CHARACTERISTICS OF SEVERAL ENGINES DESIGNED FOR CONTINUOUS COTTON-GIN SERVICE

Model[a]	Configuration— no. of cylinders	Bore		Stroke		Displacement		Maximum rating[b]			Continuous rating[c]			Peak torque			Altitude capability[d]	
		in.	mm	in.	mm	in.³	ℓ	Hp	(kW)	r/min	hp	(kW)	r/min	lb-ft	(N·m)	r/min	ft	m
V-378-P	V-6	4 5/8	117	3¾	95	378	6.2	145	(108)	3,000	125	(93)	2,800	280	(380)	1,900	500	152
V-504-P	V-8	4 5/8	117	3¾	95	504	8.3	195	(145)	3,000	170	(127)	2,800	375	(508)	1,900	500	152
N-855-P	In line—6	5½	140	6	152	855	14.0	235	(175)	2,100	195	(146)	1,800	647	(877)	1,500	500	152
NT-855-P	In line—6	5½	140	6	152	855	14.0	335	(250)	2,100	265	(198)	1,800	1,006	(1,364)	1,500	11,000	3,335
NTA-855-P	In line—6	5½	140	6	152	855	14.0	360	(269)	2,100	280	(209)	1,800	1,080	(1,564)	1,500	12,000	3,660
NTA-855-P	In line—6	5½	140	6	152	855	14.0	400	(298)	2,100	325	(242)	1,800	1,150	(1,559)	1,600	8,000	2,440
KT-1150-P	In line—6	6¼	159	6¼	159	1,150	18.9	450	(336)	2,100	365	(272)	1,800	1,350	(1,831)	1,500	9,000	2,745
KTA-1150-P	In line—6	6¼	159	6¼	159	1,150	18.9	525	(392)	2,100	425	(317)	1,800	1,575	(2,138)	1,500	12,000	3,660
KTA-1150-P	In line—6	6¼	159	6¼	159	1,150	18.9	600	(448)	2,100	470	(351)	1,800	1,650	(2,240)	1,600	6,500	1,980
VTA-1710-P	V-12	5½	140	6	152	1,710	28.0	700	(522)	2,100	545	(407)	1,800	1,925	(2,610)	1,500	12,000	3,660
VTA-1710-P	V-12	5½	140	6	152	1,710	28.0	800	(597)	2,100	620	(463)	1,800	2,200	(2,983)	1,550	6,000	1,830
KT-2300-P	V-12	6¼	159	6¼	159	2,300	37.8	900	(671)	2,100	700	(522)	1,800	2,475	(3,355)	1,500	10,000	3,050
KTA-2300-P	V-12	6¼	159	6¼	159	2,300	37.8	1,050	(783)	2,100	850	(634)	1,800	2,890	(3,919)	1,600	9,500	2,900
KTA-2300-P	V-12	6¼	159	6¼	159	2,300	37.8	1,200	(895)	2,100	940	(701)	1,800	3,300	(4,475)	1,600	7,500	2,290
KTA-3067-P	V-16	6¼	159	6¼	159	3,067	50.3	1,600	(1,194)	2,100	1,250	(933)	1,800	4,400	(5,966)	1,550	6,000	1,800

[a] Models with T in designation are turbocharged; those with TA are turbocharged and aftercooled; all others are naturally aspirated.

[b] Maximum output without engine-driven fan. Rating is applicable to intermittent applications and conforms with SAE standard J816b; British Standard overload BS 649; and DIN 6270 continuous rating B.

[c] Continuous rating without engine-driven fan. Rating is applicable to continuous 24-hr-a-day service and conforms with BS 649 and DIN 6270 continuous rating A.

[d] Maximum power is delivered up to the altitudes shown. Above these altitudes, naturally aspirated engines should be derated 3% per 1000 ft and turbocharged engines should be derated as noted on certified curves for specific models. Continuous ratings of naturally aspirated engines are attainable to approximately 6500 ft (1980 m).

From Cummins Engine Co., Inc., Columbus, Ind., 1974. With permission.

keting system, recommendations for ginning machine-picked and machine-stripped cotton have evolved. These recommendations are primarily for ginning 15/16 to 1 1/8 in. White and Light-Spotted cotton, but the principles apply to all cotton. Even though recommendations were developed using experimental gin setups handing three to eight bales per hour, they have also been found applicable for new high-capacity ginning plants installed in the past 10 or 12 years. Some variation from, or modification of, these recommendations will be necessary in certain areas of the Cotton Belt at certain times of the year or for special harvesting conditions. High grades will result from cotton coming from clean fields and harvested by machines in good condition; while grassy, weedy fields and poor harvesting practices will result in lower grades.

Machine-Picked Cotton

Research has shown that machine-picked Upland cotton varieties from across the Cotton Belt have similar cleaning characteristics. The basic ginning recommendation for sequence and amount of machinery for machine-picked cotton is as follows:

1. Rock and green-boll trap
2. Feed control
3. Tower drier or equivalent
4. Cylinder cleaner: five to seven cylinders
5. Stick and green leaf machine
6. Tower drier or equivalent
7. Cylinder cleaner: five to seven cylinders
8. Cylinder cleaner: five to seven cylinders (optional)
9. Feeder, extractor type
10. Saw gin stand
11. Saw-cylinder lint cleaner
12. Saw-cylinder lint cleaner
13. Press

The optional cylinder cleaner (item 8) is recommended for the difficult-to-clean cottons of the irrigated west. A gin flow chart for machine-picked cotton is shown in Figure 18.

Machine-Stripped Cotton

Because machine-stripped cotton contains six to ten times as much foreign matter as machine-picked cotton, ginning systems in stripper areas have to be more elaborate than those in picker areas. Additional extraction equipment is required to cope with large amounts of burs and sticks which, unless removed, will seriously lower gin stand performance. Provisions also have to be made for removing green bolls and sand from stripped cotton. Green-boll traps and airline cleaners located early in the cleaning sequence perform these functions.

Even though the cleaning characteristics of stripped cottons vary slightly from variety to variety, the following array of gin machinery has been found to be near optimum for most varieties:

1. Green-boll trap
2. Air-line cleaner
3. Feed control
4. Tower drier or equivalent
5. Cylinder cleaner: five to seven cylinders
6. Stick machine

FIGURE 18. Recommended gin machinery arrangements for machine-picked cotton.

7. Tower drier or equivalent
8. Cylinder cleaner: five to seven cylinders
9. Stick machine
10. Optional stick machine
11. Feeder, extractor type
12. Saw gin stand
13. Saw-cylinder lint cleaner
14. Saw-cylinder lint cleaner
15. Press

The optional stick machine (item 10) is recommended only for stripped cotton containing in excess of 32% foreign matter. This range of foreign matter content is approximately equivalent to a lint turnout of 22% or less. A flow diagram of the recommended machinery for a stripper gin is shown in Figure 19.

The predominant varieties of American-Pima cotton in the U.S. today are Pima S4 and Pima S5. These varieties are normally machine picked and arrive at the gin with variable moisture and trash content. The following sequence of machinery has been found to be optimum for most conditions encountered.

1. Rock and green-boll trap
2. Feed control
3. Tower drier or equivalent
4. Cylinder cleaner: five to seven cylinders
5. Stick machine
6. Cylinder cleaner: five to seven cylinders
7. Tower drier or equivalent
8. Cylinder cleaner: five to seven cylinders
9. Feeder

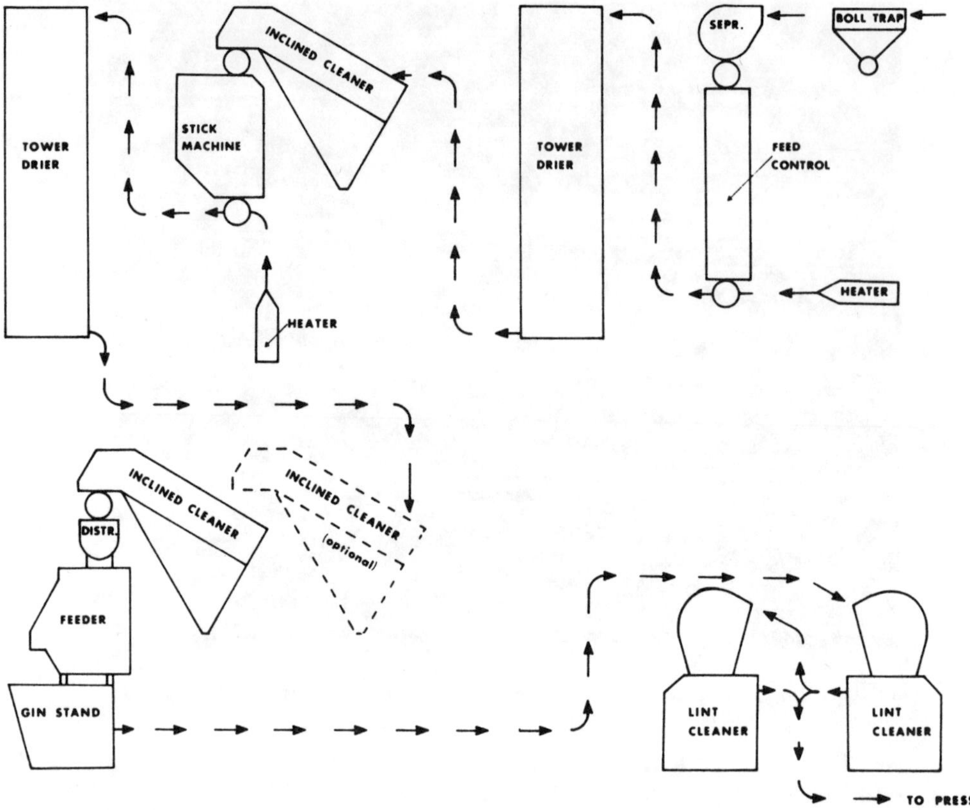

FIGURE 19. Recommended gin machinery arrangements for stripped cotton.

10. Roller gin stand
11. First stage opener-beater-cleaner and air type lint cleaner
12. Second stage opener-beater-cleaner and air type lint cleaner
13. Press

The total number of cylinders of cleaning should not exceed 18. A flow diagram of the recommended machinery is shown in Figure 20.

Bypasses

The recommendation above represents the maximum amount of machinery that should be needed. Obviously, any machinery which is not necessary for the particular lot of cotton should be bypassed. Driers, seed-cotton cleaners and extractors, and lint cleaners should be provided with bypasses to allow selection of less drying, cleaning and lint cleaning when extra-clean, dry cotton is brought to the gin. For example, hand-picked cotton now represents less than 1% of the total crop, but may be significant in a few areas. Appropriate bypasses should be used in handling this cotton, especially if it is clean and dry. Also, in irrigated areas, certain varieties have better cleaning characteristics, which often require use of only one lint cleaner to produce satisfactory quality. Special market conditions will sometimes influence the amount of equipment used also.

A necessary part of the ginning machinery recommendation is that the driers should be adjusted to produce lint at the gin stand with a moisture content from 5 to 8% — 5 to 6% for roller ginning and 7 to 8% for saw ginning. Research has shown that

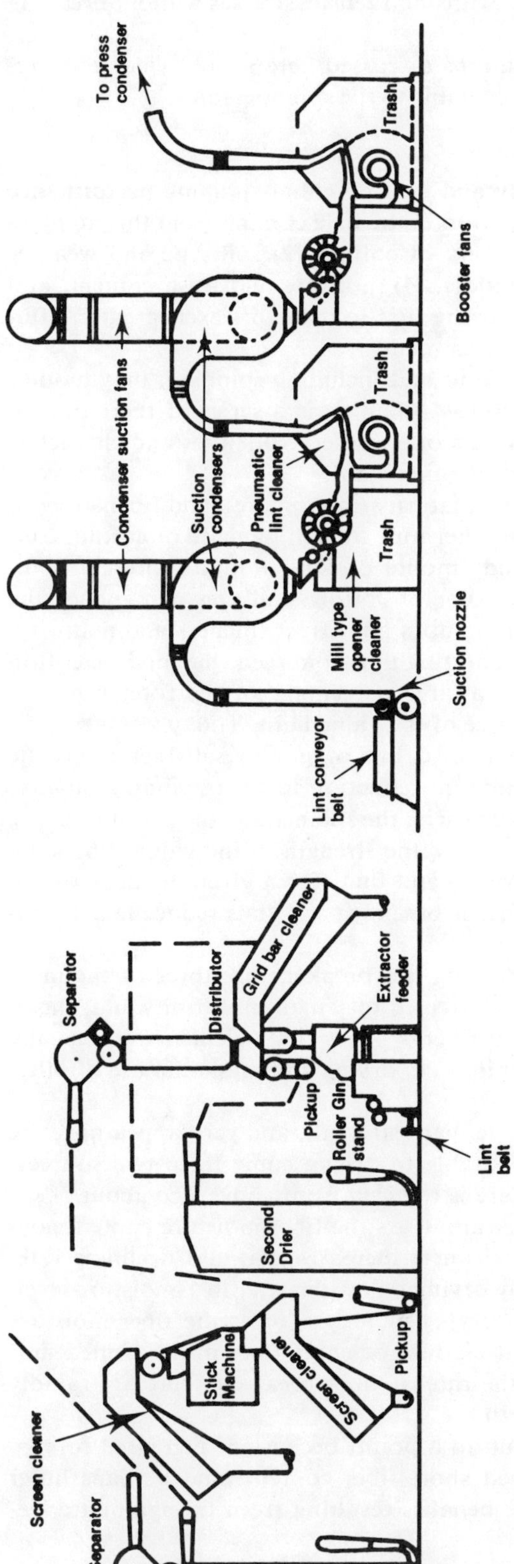

FIGURE 20. Recommended gin machinery arrangement for roller ginning American-Pima cotton.

cotton at this moisture level is more able to withstand tensile stresses without breaking than cotton at lower moisture contents.

Judicious use of gin machinery combinations discussed here should yield good returns to the producer and provide grades acceptable to the spinning mills.

Effects of Gin Machinery on Cotton Quality

Cotton possesses its highest fiber quality and best potential spinning performance when it is on the stalk. Lint quality of the cotton in the bale coming from the gin press depends on many factors, including (1) variety of cotton, (2) soil type and weather conditions, (3) cultural and harvesting practices, (4) moisture and trash content, and (5) ginning treatments and processes. Usually the first four factors exert greater influence on lint quality than extremes in ginning treatment.

Nevertheless, any mechanical handling, up to and including spinning, may modify the natural qualities or characteristics of cotton. Ginning is a series of thermopneumatic and mechanical processes and at best can only preserve qualities and characteristics inherent in the cotton when it enters the ginning plant.

Lint grade, staple length, fiber fineness and maturity (micronaire), and fiber strength — in the order listed — are the factors that determine the selling price of cotton. Lint grade is determined principally by color and amount of foreign matter present. Ginning processes may do little to change the color of cotton, and they can change the staple length to only a limited extent. Ginning does not affect fineness and maturity.

In addition to its principal function of separating lint from seed, the modern cotton gin is equipped to remove from the cotton a large percentage of the foreign matter that would significantly reduce the selling price of the ginned lint. Today's ginner must have a twofold objective: (1) to produce ginned lint of grades satisfactory to the grower, and (2) to gin the cotton with minimum reduction in fiber spinning quality, so that it will meet the demands of its ultimate users, the spinner and the consumer.

Research has shown that, above certain limits, the strength of individual fibers decreases as the fibers become drier and warmer, and that, for a given quantity of gin machinery, increased drying may increase fiber breakage and thus reduce length uniformity.

Yarn strength, yarn appearance, and spinning-end breakage are three of the most important spinning quality elements. All are affected by length uniformity and, therefore, by the proportion of short or broken fibers. These three elements are usually best preserved when cotton is ginned with minimum drying and minimum use of other gin machinery.

The relations between gin drying, lint grade, yarn strength, and yarn appearance are shown in Figure 21. Grade increases attributable to drying come from two sources. The grade increase that occurs when moisture is reduced from above 7 to about 7% is due to moving from rough to smooth preparation. As the fiber moisture content goes below 7%, foreign matter removal by the cleaners increases, but at a declining rate, so there is no real advantage to lint grade by drying below the 5% fiber-moisture level.

Yarn strength continuously decreases as drying increases (or as the fiber moisture content decreases) because of the increased amount of short fiber in the ginned lint. Below the 5% moisture level at ginning, the rate of fiber breakage increases rapidly with corresponding decreases in yarn strength.

Yarn appearance improves with drying up to a point, because of increased foreign matter removal. But the effect of increased short fiber content (and perhaps other factors) on yarn appearance outweighs the benefits resulting from foreign matter removal.

The effects of seed cotton cylinder cleaners on lint grade, yarn strength, and yarn appearance are shown in Figures 22 and 23. As the number of cleaning cylinders in-

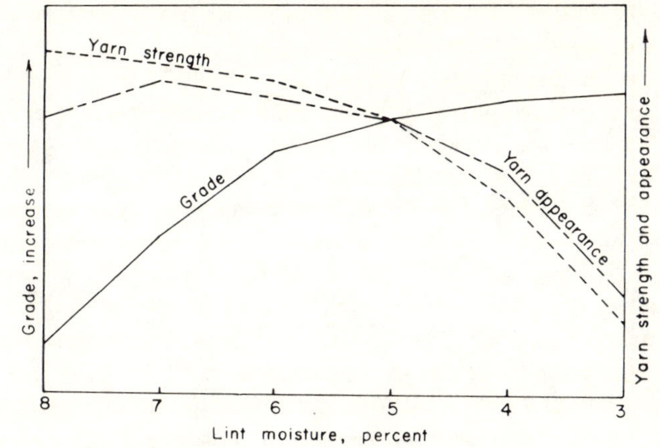

FIGURE 21. Effects of gin drying on lint grade, yarn strength, and yarn appearance.

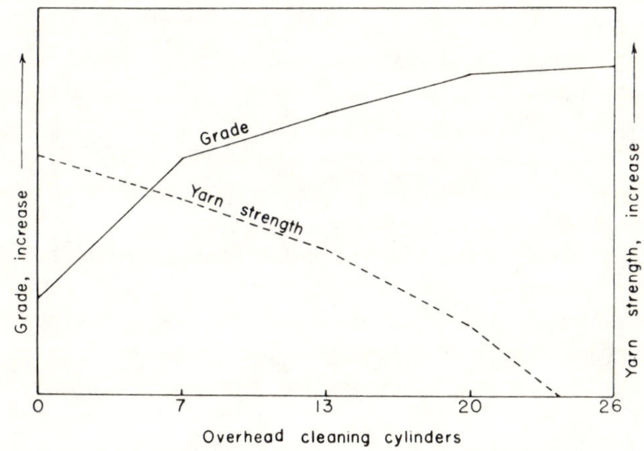

FIGURE 22. Effects of seed-cotton cleaners on lint grade and yarn strength.

creases, the lint grade usually increases. The cleaning efficiency of each cylinder is relatively constant; less foreign matter is removed and less grade improvement is expected from each succeeding cylinder. When two lint cleaners are used and the cotton is dried to the 5 to 7% fiber moisture level, 12 to 14 cylinders are considered adequate. The first cleaner should not have more than seven cylinders, to prevent roping of inadequately dried seed cotton. The yarn strength decreases steadily as the number of cleaning cylinders increases.

For cleanly picked cotton, the yarn appearance curve also shows a steady downward trend because of the machined effect, whereas the machine-picked cotton curve shows an upward trend because of foreign matter removal, until the machined effect overrides the foreign matter effect at about 13 cylinders.

The effect of lint cleaning on lint grade, yarn strength, and yarn appearance is shown in Figure 24. Lint grade classification shows some grade increase as the number of lint cleaners increases, but each succeeding lint cleaner gives less grade improvement than the preceding one. In addition to removing fine trash, the lint cleaners often blend Light-Spotted cotton so that it passes into the White grades. The lint cleaners also remove some short fiber. Hand stapling and machine measurements often show a slight

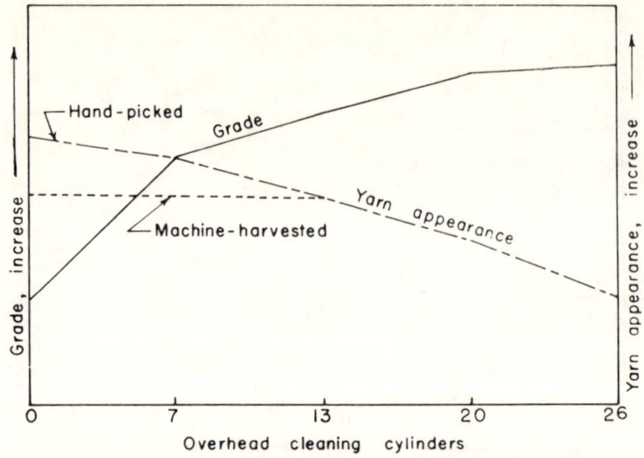

FIGURE 23. Effects of seed-cotton cleaners on lint grade and yarn appearance for hand-picked and machine-picked cotton.

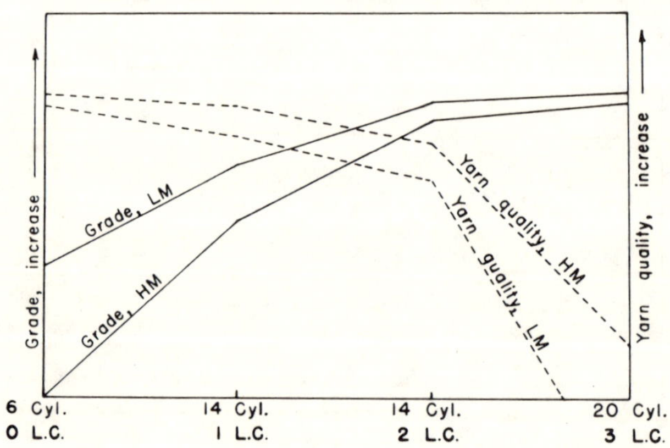

FIGURE 24. Effect of lint cleaners on lint grade, yarn strength, and appearance.

increase in length. This length increase usually reduces manufacturing waste but has little, if any, effect on yarn strength.

Yarn appearance is usually affected adversely by lint cleaner treatment. The effect is usually slight when only one or two lint cleaners are used, but appearance declines rapidly as additional cleaners are used.

The overall effect of ginning processes on grade and yarn quality is shown by Figure 25. This chart shows the effects of gin drying and four different combinations of machinery on lint grade and yarn quality. Because of the usually rapid decline in lint grade improvement and in yarn quality after the 14-cylinder and 2-lint-cleaner combination is used, that combination is recommended as the maximum ginning machinery compatible with acceptable yarn quality. Use of this machinery combination results in little grade improvement in low-moisture cotton as compared to high-moisture cotton, whereas the spinning quality of low-moisture cotton rapidly decreases as additional machinery is used.

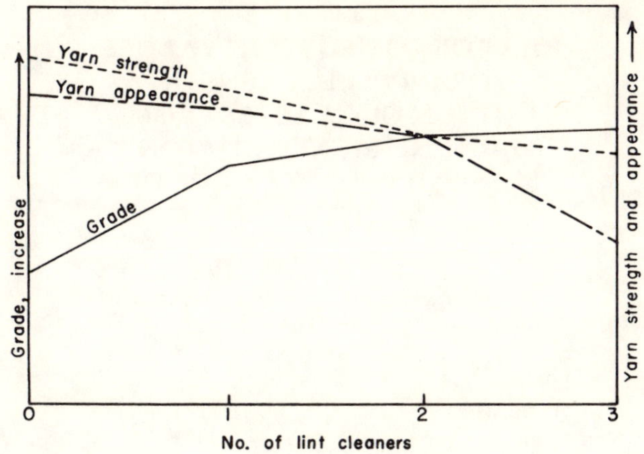

FIGURE 25. Effects of seed-cotton cleaning and lint cleaning on lint grade and yarn quality for cotton of two fiber-moisture levels during gin processing. Yarn quality includes yarn strength, yarn appearance, and spinning breakage.

MISCELLANEOUS CONSIDERATIONS

Fuel Value of Gin Wastes

The fuel value of a material is also known as its calorific value, heat content, or heat of combustion. All these terms refer to the amount of heat liberated by a unit mass of a material when it oxidizes. The gross heat of combustion may be obtained by using an oxygen bomb calorimeter; the gross value is then converted to a net heat of combustion value at normal atmospheric pressure. The net heat of combustion value is thus considered the same as fuel value.

Heat of combustion values were determined in 1973 at the U.S. Cotton Ginning Research Laboratory, Stoneville, Miss. for gin wastes from machine-picked cotton harvested from mid-October to mid-December (Table 41). These values may be considered typical for Upland cotton, and the mean value of 7928 Btu/lb is useful as an overall average for dry ginning wastes.

To apply these data, one must take into account the moisture content of the material at the time it is incinerated. For example, if the boll hulls have a moisture content of 16% when cotton is ginned, their fuel value would be less than the tabular value of 8194 Btu/lb. The proper value may be easily found by converting the moisture content (wet basis) to moisture regain (dry basis) and making a correction to the tabular value by using Equations 5 and 6.

Moisture content (M) may be converted to moisture regain (R) by the formula

$$R = 100M/(100 - M) \tag{5}$$

where R = moisture regain (dry basis) in percent and M = moisture content (wet basis) in percent. For example, when $M = 16$, $R = 1600/84.0 = 19$.

The fuel value of ginning wastes at a specified moisture content may be calculated from Table 42 by using the formula

$$H_R = 100H_D/(100 + R) \tag{6}$$

where H_R = fuel value at specified moisture regain in Btu per pound, H_D = fuel value of dry material (Table 42) in Btu per pound, and R = moisture regain in percent.

Table 41
HEAT OF COMBUSTION VALUES FOR DRY SPECIMENS OF COTTON-GIN WASTES FROM MACHINE-PICKED COTTON, MISSISSIPPI DELTA CROP OF 1973[a]

Type of waste	Heat of combustion[b] (Btu/lb)
Leaf	7831
Lint	7758
Sticks, stems, and hulls	8194
Mean	7928

[a] Average of samples from three harvestings collected at locations within the gin.
[b] Net value equivalent to fuel value.

Table 42
FUEL VALUE OF COTTON GIN WASTES AT SPECIFIED MOISTURE CONTENTS, MISSISSIPPI DELTA MACHINE-PICKED COTTON, CROP OF 1973

Type of waste	Moisture content (%)	Fuel value (Btu/lb)
Leaf	12	6893
Sticks, stems, and hulls	16	6886
Lint and motes	11	6902

For example, when boll hulls have a moisture content (wet basis) of 16%, their fuel value will be $H_{19} = 819{,}400/119.0 = 6886$ Btu/lb. Thus, the moisture content of hulls may have a considerable effect on their fuel value.

The fuel value of other types of ginning wastes at the time of incineration may also be calculated by using Equations 5 and 6. If we arbitrarily assume certain moisture contents for the materials listed in Table 41 and calculate their fuel values on that basis, more realistic values are obtained (Table 42).

Because ginnery production rates are customarily measured in units of bales per hour, it is useful to calculate ginning waste fuel values in units of Btu per bale so that these data may be readily converted to Btu per hour by simply multiplying the per-bale heat content by the ginning production rate. Thus, it is necessary to take into account the actual quantities of gin wastes as well as their type and moisture content. For example, a typical machine-picked bale in the Mississippi Delta area might contain about 100 lb of waste, of which about 1/3 would be sticks, stems, and hulls at a 16% moisture content; whereas, a stripper-harvested bale might contain 700 lb of waste, with about 570 lb being sticks, stems, and hulls at 12% moisture (Table 43). By applying the data of Table 42 to typical ginning wastes and using Equations 5 and 6, the wastes from first-pick spindle-harvested and stripper-harvested bales may be shown to have fuel values of approximately 675,700 and 5,017,600 Btu, respectively.

A quick method of estimating the fuel value of ginning wastes is to multiply the

Table 43
TYPE AND QUANTITY OF GINNING WASTES AND THEIR FUEL VALUES OBTAINED FROM MISSISSIPPI DELTA SPINDLE-PICKED COTTON AND TEXAS HIGH PLAINS STRIPPER-HARVESTED COTTON PER 500 LB OF LINT GINNED

	Spindle-picked			Stripper-harvested		
Type of waste	Quantity[a] (lb)	Moisture content[a] (%)	Fuel value[b] (Btu)	Quantity[a] (lb)	Moisture content[a] (%)	Fuel value[b] (Btu)
Leaf	21	12	144,800	100	11	696,700
Sticks, stems, and hulls	35	16	241,000	570	12	4,111,400
Motes and lint cleaner waste	42	11	289,900	30	10	209,500
Total	98	—	675,700	700	—	5,017,600

[a] Estimated from various laboratory observations.
[b] Rounded to nearest hundred.

estimated quantity of ginning waste by 0.87 (to adjust for moisture content) and by 7900 (an overall value for the fuel value of dry ginning wastes). Thus, 98 × 0.87 × 7900 gives 673,600 Btu as the fuel value for gin trash from a bale of first-pick spindle-harvested cotton, and 700 × 0.87 × 7900 gives 4,811,100 Btu as the fuel value of ginning waste from a bale of stripper-harvested cotton.

Sampling Cotton

In the U.S. cotton industry the producer normally retains title to his cotton until after it has been ginned. Trading procedures require that the bale be sampled and that the lint be classified according to staple length and grade to establish its value. The Agricultural Marketing Service (AMS), Cotton Division, U.S. Department of Agriculture maintains staple and grade standards and offers cotton classing as a service to U.S. Department of Agriculture cut samples and mechanically collected samples. Cut samples are taken from the bale after it is removed from the gin press and the mechanically collected samples are collected during the ginning of the bale.

Cutting of samples has long been an established procedure and historically was done by hand at the gin. Today most cut samples are collected at the warehouse. They are cut from the bale with a mechanical cutter and are manually removed and placed in bags for delivery to AMS. In either case, the bale is left with open sample holes. This has been the source of much criticism of the U.S. cotton bale, particularly in foreign markets.

The other method of sampling, known as automatic sampling, involves taking a number of subsamples from the bale during the ginning cycle. The segments of the sample are taken from the lint flue by means of a sampling valve that is actuated by an automatic timer. These segments are condensed and pressed into a small bale and inserted into a paper wrapper. Equipment for this process was originally developed at the U.S. Cotton Ginning Research Laboratory, Stoneville, Miss. Automatic sampling equipment has been commercialized and is utilized today in a high percentage of modern gins, particularly in the west. The 1978 Cotton Gin Equipment Survey conducted by the Agricultural Marketing Service of the U.S. Department of Agriculture indicated that a total of 329 gins were equipped with automatic samplers. Almost 3/4 of these were located in California, Arizona, and Texas.

Automatic samplers are also gaining acceptance in the southwest and midsouth in modern high capacity gins utilizing gin universal density presses. Much of the automatically sampled cotton is shipped directly to the mill, and the bales arrive at the mill in good condition with no open sample holes.

In 1979 three companies were manufacturing automatic sampling equipment. These companies are Lummus Gin Co., Columbus, Ga.; Bush Hog/Continental Gin Co., Prattville, Ala.; and Superior Cotton Sampling, Inc., Bakersfield, Calif. In all three lines transistorized solid state control circuitry is used, and the number of moving mechanical parts has been minimized for trouble-free operation. Since this equipment furnishes the sample which serves as the basis for bale value determination, uninterrupted dependable operation is imperative. To accomplish this, a well organized operational plan should be developed by the ginner. The important parts of this plan are: (1) installation of trouble-free dependable equipment, (2) proper maintenance and supervision of the equipment, and (3) adequate records to assure that each bale is sampled and that the sample is preserved. This plan should place a high priority on specialized operator training and maintenance at the gin of an adequate inventory of replacement parts and supplies for the sampling system. A member of the gin crew on each work shift should be specially trained to operate the sampler and should be given the responsibility of seeing that each bale is properly sampled.

Moisture Meters

An electric portable cotton moisture meter is valuable to the ginner if it is properly used and its limitations are understood. It is useful both on the gin yard and in the gin. On the yard it can be used to determine the approximate average moisture content of the seed cotton on trailers waiting for ginning. If grouping is practiced, it can be used for grouping loads or modules of seed cotton having similar moisture content. Some ginners mark the moisture percentage on the back of each trailer ticket, and some mark it on the trailer with chalk. When taking moisture readings of loads of seed cotton, two or more readings should be taken at different locations within the load. These readings should be averaged, because wide differences in moisture within the same load are not unusual. Readings should be taken at least several inches below the surface, because the outer portions of the load change rapidly with changes in the ambient air relative humidity.

Within the gin, seed cotton moisture measurements at the feeder aprons and lint measurements at the lint slide are the most common. Here, also, at least two samples should be measured and the results averaged. The size of the sample for testing should be no smaller than that recommended by the instrument manufacturer. In most electrical resistance-type moisture meters, the chance of error is greater when the sample is smaller than the recommended minimum size than when it is larger than the recommended maximum size.

Electrical moisture meters suitable for use at gins are of two types — the electrical resistance type (sometimes called the conductance type), and the capacitance type (also called the dielectric type). The electrical resistance type operates on the principle that the resistance of a mass of cotton to the flow of an electric current increases as the cotton becomes drier. The capacitance type operates on the principle that the capacitance of an electrical condenser varies with the moisture content of a mass of cotton placed between the condenser plates.

The methods of taking moisture measurements by means of the electrical type of meters are known as "secondary" measurement methods, and the meters must originally be calibrated by using a "primary" method such as the oven method (sometimes called the gravimetric method). The electrical resistance type has been widely accepted in the ginning industry, and a number of makes and models are commercially available. The resistance type has an advantage because it responds more rapidly to variations in moisture conditions than does the capacitance type.

The capacitance-type meters currently available are generally more bulky than the electrical resistance-type meters and require that the sample be weighed for accurate readings. Although for greatest accuracy the samples used in resistance-type meters should be weighed, this is generally not necessary and is seldom done.

Manufacturers of resistance-type meters suitable for gin use offer either multiple use probe or sample chambers, or a number of accessory sample electrodes that make the instrument useful for moisture measurement of seed cotton, cottonseed, and lint.

The use of a moisture meter by the ginner brings the importance of moisture to the attention of the producer-customer and should assure better bale values.

Moisture meters should be given reasonable care. Although they are generally ruggedly constructed, the meter movement and internal components can be damaged by severe mechanical shocks. Meters with Bakelite® cases should be protected from cracking or breaking.

Battery maintenance is important with battery-operated instruments. Most difficulties encountered with this type of instrument are due to battery troubles. Corrosion is serious and may result from battery leakage. Failure to remove batteries when the meter will not be used for prolonged periods usually results in battery wall failure which permits liquids to escape and attack the internal components of the meter. At the close of the ginning season, batteries should be removed and discarded because they usually will not be suitable for further use. Table 44 lists several makes and models of moisture meters and gives their power requirements. When changing batteries, polarity must be observed.

The temperature of the sample affects the reading of a resistance-type meter. For accurate readings, temperature corrections should be made according to the manufacturer's recommendations. The electrical resistance of a mass of cotton decreases as the temperature of the cotton increases. Therefore, the moisture content of the sample will appear higher if the reading is taken while the cotton is hot. This is generally not serious if the sample temperatures are within the range of 60 to 80°F.

Static electricity can cause reading errors and erratic deviations in the meter needle. This generally occurs only in dry weather, and particularly in dry, cold weather. Static charges on the meter can be removed by wiping with a damp cloth. Sometimes electrical grounding of an instrument will improve its operation when static charges are present.

Some suggestions for using moisture meters are

1. Be sure the batteries are good.
2. Keep electrodes and insulation clean and free of moisture.
3. Maintain proper sample size.
4. Maintain sample density according to manufacturer's recommendations.
5. Use an average of several samples for greater accuracy.
6. Correct for atmospheric or cotton temperature to obtain greatest accuracy.
7. Remove the batteries during long periods of storage.
8. Remember that the moisture meter is an aid and not a cure-all for moisture control during ginning.

The commercial gin machinery manufacturers have incorporated the electric moisture meter into continuous moisture sensing and drier control systems. These electronic systems are built into the overhead machinery sequence and give the gin operator automatic control of the drying system with the assurance that the cotton is ginned within

Table 44
POWER REQUIREMENTS FOR COTTON MOISTURE METERS

Moisture meter	AC line operated — 50—60 cycle	Power requirements — battery operated (battery size and type)				
		1½ V	9 V	22½ V	45 V	67½ V
Accra-Temp®,[a] Bush-Hog/ Continental	115 V	None	None	None	None	None
Delmhorst	None	Penlight	None	Burgess® U-15, Eveready® 412, and RCA® VS084	None	None
Fiber-Rite®,[a] Murray	115 V	None	None	None	None	None
Fiber-Rite®,[b] Murray	None	Penlight and C	None	None	None	Burgess® UX45, Eveready® 416, and RCA® VS318
Fuel Miser®,[a] Cliff Granberry Corp.	115 V	None	None	None	None	None
Hart® K-101[a]	None	D	None	None	Burgess® Z30NX, Eveready® W350, Ray-O-Vac® 711, and RCA® ZS114	None
Hart® K-103	115 V	None	None	None	None	None
Hart® CU-2[b]	None	D	None	None	Burgess® Z30NX, Eveready® W350, Ray-O-Vac® 711, and RCA® ZS114	None
Hart® R-41[b]	None	C or D	None	None	Burgess® XX30, Eveready® 455, and RCA VS055	None
KPM Aqua Boy®[b]	None	None	None	Burgess® U-15, Eveready® 412, and RCA® VS084	None	None
Moisture-Matic®,[b] Lummus	115 V	None	None	None	None	None
U.S. Testing Co., ACCO®[b]	115 V	None	None	None	None	None
WR-7000 Verity®[b]	None	None	Burgess® AU6, Neda 1604, Eveready® 216	None	None	None

[a] Built into the ginning system for continuous moisture monitoring.
[b] Portable meter.

the proper moisture range — 5 to 8%. When this type of system is used, a properly calibrated portable electric meter should be available to check the performance of the automatic control system.

ACKNOWLEDGMENTS

Grateful acknowledgment is hereby made of the indebtedness to our colleagues and associates in the cotton ginning industry for their criticisms and suggestions which served as assistance toward the preparation of this chapter on cotton ginning. We used *Cotton Ginners Handbook* (Agriculture Handbook No. 503, Agricultural Research Service, U.S. Department of Agriculture, Washington, D.C., 1977) as our principal reference in compiling the information presented; therefore, we hereby express particular appreciation for the contributors to that handbook for their efforts: D. M. Alberson, former Agricultural Engineer, Agricultural Research Service. W. Stanley Anthony, Agricultural Engineer, U.S. Cotton Ginning Research Laboratory, SEA-AR*, Stoneville, Miss. R. V. Baker, Agricultural Engineer, South Plains Ginning Research Laboratory, SEA-AR, Lubbock, Tex. R. F. Colwick, Agricultural Engineer, Cotton Production Research Unit, SEA-AR, Mississippi State, Miss. G. N. Franks, former Agricultural Engineer, Agricultural Reserach Service. W. E. Garner, Agricultural Engineer, U.S. Cotton Ginning Research Laboratory, SEA-AR, Stoneville, Miss. Anselm C. Griffin, Jr., physicist, U.S. Cotton Ginning Research Laboratory, SEA-AR, Stoneville, Miss. I. W. Kirk, former Agricultural Engineer, Southwestern Cotton Ginning Research Laboratory, Mesilla Park, N.M.; at present Assistant Director, Southern Regional Research Center, New Orleans, La. J. W. Laird, Agricultural Engineer, South Plains Cotton Ginning Research Laboratory, SEA-AR, Lubbock, Tex. Oliver L. McCaskill, former Agricultural Engineer, U.S. Cotton Ginning Research Laboratory, SEA-AR, Stoneville, Miss. Gino J. Mangialardi, Jr., Agricultural Engineer, U.S. Cotton Ginning Research Laboratory, SEA-AR, Stoneville, Miss. Vernon P. Moore, former cotton technologist, Agricultural Research Service. J. H. Tony Price, Executive Vice President, Texas Cotton Ginners' Association, Dallas, Tex. S. T. Rayburn, Jr., Agricultural Engineer, U.S. Cotton Ginning Research Laboratory, SEA-AR, Stoneville, Miss. C. S. Shaw, former cotton technologist, Agricultural Research Service. L. L. Smith, former Agricultural Engineer, Agricultural Research Service. V. L. Stedronsky, former Agricultural Engineer (deceased), agricultural Research Service. W. E. Taylor, Agricultural Engineer, Oklahoma State University, Stillwater, Okla. J. E. Vaughn, manager, Fire Prevention and Engineering Bureau of Texas, Dallas, Tex. R. A. Wesley, Agricultural Engineer, Field Crops Mechanization Research Laboratory, SEA-AR, Stoneville, Miss. T. E. Wright, Agricultural Engineer, Southwestern Cotton Ginning Research Laboratory, SEA-AR, Mesilla Park, N. M.

In the roller ginning information we drew heavily on *Roller Ginning American-Egyptian Cotton in the Southwest,* Agriculture Handbook No. 257, ARS, USDA, and other works of the U.S. Department of Agriculture Southwestern Cotton Ginning Research Laboratory, Mesilla Park, N.M. We are especially indebted to Clarence Leonard, Physicist, SEA-AR, and Marvis Gillum, Agricultural Engineer, SEA-AR, of the USDA Southwestern Cotton Ginning Research Laboratory, Mesilla Park, N.M., for their assistance in compiling the information on roller ginning.

Additional references used are gratefully acknowledged.[1-11]

The authors are grateful for the cooperation of the U.S. Department of Agriculture cotton ginning research laboratory staffs and the gin machinery manufacturers who

* Science and Education Administration-Agricultural Research (SEA-AR), USDA, was formerly Agricultural Research Service, USDA.

made possible the compilation of information presented in the figures and the tables of technical data that support the text. Without their help this information could never have been documented.

Trade names are used in this publication solely for the purpose of providing specific information. Mention of a trade name does not constitute a guarantee or warranty of the product by the U.S. Department of Agriculture or an endorsement by the Department over other products not mentioned.

REFERENCES

1. U.S. Department of Agriculture, Handbook for Cotton Ginners, *U.S. Dep. Agric. Agric. Handb.* No. 260, 1964.
2. Arkansas and Missouri Ginner's Association, *Handbook for Ginners*, Arkansas and Missouri Ginners' Association, West Memphis, Ark., 1953.
3. U.S. Department of Agriculture, Roller Ginning American-Egyptian Cotton in the Southwest, *U.S. Dep. Agric. Agric. Handb.* No. 257.
4. **Johnston, T. J.,** *Cotton Ginners Handbook*, Southern Cotton Ginners' Association, West Memphis, Ark., 1956.
5. Arkansas-Missouri Ginners' Association, *Guidectory, A Guide to Better Ginning*, Arkansas-Missouri Ginners' Association, West Memphis, Ark., 1961.
6. **Pendleton, A. M. and Moore, V. P.,** Cotton Ginning to Preserve Fiber Quality, ESC-560, U.S. Department of Agriculture, Washington, D.C., 1967.
7. **Bennett, Charles A.,** *Roller Cotton Ginning Developments*, Texas Cotton Ginners' Association, Dallas.
8. **Bennett, Charles A.,** *Saw-Toothed Cotton Ginning Developments*, Texas Cotton Ginners' Association, Dallas.
9. **Bennett, Charles A.,** *Cotton Ginning Systems in the U.S. and Auxiliary Developments*, Texas Cotton Ginners' Association, Dallas.
10. **Brown, H. B.,** *Cotton: History, Species, Varieties, Morphology, Breeding, Culture, Diseases, Marketing, and Uses*, McGraw-Hill, New York, 1938.
11. **Anon.,** Cotton Gin Equipment, Agricultural Marketing Service, U.S. Department of Agriculture, Washington, D.C., 1978.

STORAGE, HANDLING, AND QUALITY EVALUATION OF COTTONSEED

Marion E. Whitten

Cottonseed is marketed in a simple and direct manner. The cotton harvest begins in early July in south Texas, and spreads north, east, and west as the crop matures. Completion of the harvest varies from December 1 in areas of central Texas and Oklahoma to mid-January in areas west of there. Most of the crop today is picked by mechanical harvesters (90%) as compared with almost none in 1945. The harvesting period is much shorter since machines can pick the crop much faster than by hand.

With the rapid harvesting of the crop, storage of the seed cotton before ginning has become a problem since ginning capacity cannot handle the crop as fast as it is harvested. About one third of the supply of seed for a year reaches the mills within 1 month during peak harvest time. From September through December the mills receive more than 80% of the year's supply; by the end of January, 94%.[1]

Growers and ginners have resorted to storing seed cotton in the fields or on the gin yard. The seed cotton is compressed into free-standing modules or ricks. Harvesters can operate at full capacity without having to wait for trailers to be returned from the gin. The ginner can spread his operation over a longer period of time. There are about 3000 gins in the U.S.

After the lint is baled at the gin, it is usually sold by the grower to cotton merchants. The cottonseed is usually bought by the ginner who accumulates seed from many bales and sells them to oil mills.

Since cottonseed are bulky and are subject to rapid deterioration under certain conditions, storage of large quantities of seed on the farm or at gins is not practical. Ginners usually ship seed to oil mills as rapidly as they are separated from the seed. Seed storage capacity at gins is usually very limited. Most seed move by truck which the ginner may own or have under contract. Some mills maintain their own trucks for collection of seed.

Cottonseed may be bought by oil mills from gins on an "as is" basis or on grade. When seed are bought "as is", the price is based on average quality of seed produced in that area rather than the quality of each individual lot. The same price is paid for all seed delivered to the mill at that particular time. Seed are regularly sampled and analyzed by the oil mill to determine the average quality of the seed. This practice does not give growers and ginners an incentive to produce better quality seed since every farmer and gin receives the same price for his seed.

Cottonseed may vary greatly in quality from lot to lot, in the quality of products that are available as well as the quantity of products. For this reason about 70% of the crop is bought by oil mills from gins on grade.

SAMPLING COTTONSEED AT OIL MILLS

When cottonseed are received from gins, they are sampled by mill personnel before the seed are unloaded. A study revealed that ten random probes obtained from a gross lot of 100 tons would satisfactorily represent that lot, resulting within three grade points values of the true grade 95% of the time.[2] Since the average truck moves approximately 20 to 25 tons and small trucks 10 to 12 tons, it is necessary to probe the large trucks two or three times and the small trucks once to obtain ten probes for each 100-ton gross lot. Only ½ lb of seed per ton is obtained for the gross samples by this procedure. The 50 to 60 lb sample is then passed over a shaker-cleaner which removes the foreign material. The shaker-cleaner equipment usually has a divider or splitter from which the laboratory sample is obtained.

After the trucks are sampled, the cottonseed are unloaded and placed in storage silos or storage houses. These storage facilities are usually equipped with cooling equipment which pulls air through the pile when required of seed with high moisture content. (In areas where high moisture is no problem, mills may not have this equipment.)

QUALITY EVALUATION OF COTTONSEED

The Official Standards for Grading Cottonseed

The cottonseed grading system now used was developed cooperatively by the cottonseed processing industry and the U.S. Department of Agriculture. The present Official Standards Grades for Cottonseed were established in 1932 and have been used continuously since that time.[2]

The Cotton Division, Agriculture Marketing Service, U.S. Department of Agriculture, is charged with administration of the standards under direction of the Supervisor of Cottonseed Grading. Cotton Division personnel inspect sampling facilities and method of sampling at oil mills, making suggestions and/or corrections where necessary.

The supervisor of cottonseed grading inspects independent cottonseed laboratories which have been licensed by the Department to grade cottonseed. Periodic analyses of cottonseed samples are also used to determine the accuracy of the laboratory. Cost of this supervision is covered by a fee per sample analyzed. The official cottonseed chemists forward a copy of each officially analyzed sample certificate to the Cotton Division, Marketing Statistics group at Memphis, Tenn. This group analyzes the data and issues biweekly reports during the harvesting season. These reports include the average, high, and low grades for the period as well as prices paid for seed and for cottonseed meal and hulls. These data are broken down to the county level.

Grading and Evaluation of Cottonseed

Under the present system for grading cottonseed, basis grade cottonseed was set at 18.5% oil and 3.50% ammonia content. For the Quality Index, prime cottonseed is prescribed as containing no more than 12% moisture, 1.8% free fatty acids and 1% foreign matter. The Quantity Index is based on the price relationship of 1 lb of oil to 5 lb of meal.

Quality Index

The grade of cottonseed is divided into two parts, the Quality Index which is generally related to those elements associated with quality of the product and the Quantity Index that reflects the value of the quantity of the products that can be obtained from the cottonseed. The Quality Index is 100 when the chemical analysis indicates that prime products can be made from the cottonseed. When the chemical analysis indicates production of lower quality products or difficulty in cleaning or storage of the cottonseed, the Quality Index reflects this situation.

Discounts for Foreign Matter

Industry trading rules require the deduction of weight of foreign matter from the gross weight of cottonseed to discourage the practice of incorporation of foreign matter with seed. Because foreign matter stimulates deterioration and is difficult and costly to remove, foreign matter is also included as an element of grade. The standards require a discount of 0.1 units in the Quantity Index for each 0.1% foreign matter in excess of 1.0%.

Moisture Factor

Moisture content is the primary cause of deterioration in cottonseed. Prime cottonseed of low moisture content may be stored for well over a year and yield prime products. Excess moisture in seed may cause rapid deterioration. A moisture content of 12% was established as the critical point in the storage of cottonseed beyond which cooling or drying is necessary to preserve quality. To offset cooling or drying expense necessitated by excess moisture content, the standards require a discount of 0.1% in the Quality Index for each 0.1% moisture in excess of 12.0%.

Free Fatty Acids Factor

Cottonseed produced under favorable weather conditions usually contain from 10 to 12% moisture and from 0.5 to 1.0% free fatty acids. Data on which the standards were based indicated that cottonseed with 1.8% or less free fatty acids would produce prime products. To compensate for loss in product value a discount of 0.4 unit of Quality Index for each 0.1% free fatty acids in excess of 1.8% is assessed. Recent studies by the author indicate that product value loss break point is at 1.5% instead of 1.8% free fatty acids.

Quantity Index

Price studies revealed that the value of 1 lb of crude cottonseed oil was equivalent to as much as 6 and as little as 4 lb of cottonseed meal (41% protein). An oil to meal ratio of 1:5 occurred most frequently and that ratio was selected as the basis of the Quantity Index. When value of products was plotted against the oil-ammonia reciprocal, it was found that the following simple formula would fit this relationship: oil content × 4 + the ammonia content × 6 + a constant of 5. This formula is used in calculating the Quantity Index.

The grade of cottonseed is obtained by multiplying the Quantity Index by the Quality Index. This grade is multiplied by the value of the Basis Grade 100 cottonseed to obtain the value of that lot of cottonseed.

STORAGE OF COTTONSEED

Cottonseed are stored for an average of 120 to 130 days at oil mills. Cottonseed of questionable keeping quality are processed immediately on arrival at oil mills or as soon as possible. Seed containing above 12% moisture may "heat" (a rise in temperature caused by the deterioration) unless steps are taken to cool the cottonseed. With close control and adequate air cooling, it is possible to store seed containing 2.5 to 5.0% free fatty acids and 10 to 11% moisture.[3] One cannot stress too much the need for extreme watchfulness when storing high fatty acids seed above 10% moisture content. These seed will heat in pockets and seed temperatures must be observed daily in order to air-cool selectively. Unless these seed are watched daily, the entire mass may char.

Experimental lots of high moisture seed (19%) have been stored successfully after treatment with 1.5% propionic acid as a preservative showing no perceptible increase in the fatty acids content at the end of 180 days storage.[4]

When cottonseed are received at the oil mill, they are cleaned of foreign material before storage when possible since this material will contribute to more rapid deterioration of the stored seed.[3]

BUYING COTTONSEED ON OFFICIAL GRADE

When cottonseed are bought "on grade", they are sampled at the oil mill and analyzed and graded by licensed cottonseed chemists according to methods issued by the U.S. Department of Agriculture. Samples usually represent from 50 to 100 tons. Duplicate laboratory samples are prepared. One is sent to the chemist, the other retained on file.

Payment for the lot of cottonseed represented by the sample is made on the basis of 100 grade seed (or otherwise if agreed upon by the mill and gin) when the seed are received. When the grade certificate is received from the official chemist, the oil mill corrects the actual value as determined by the grade certificate and accordingly pays the ginner for his seed.

QUARANTINE

When cottonseed are contaminated with the pink boll worm, that area is declared under quarantine to prevent spread of the insect. Seed may not be sold for shipment outside the area unless treated. Several means have been used: steam; heat; chemicals. Usually the seed are processed in the area under quarantine since the oil, meal, and linters produced may be shipped outside the area without treatment. Treatment of the seed to kill the pink boll worm moth larvae, and egg materially lowers the value of the seed.

COTTONSEED USED FOR PLANTING

Cottonseed which are used for planting are generally handled more carefully than "gin run" seed. More care is exercised in picking the cotton and in drying the seed cotton if that is necessary. Excess heat will materially affect the viability of the seed. Most large seed breeders have their own gins where they carefully gin each seed variety to maintain its identity. Seed breeders often contract with growers to grow their new varieties, paying them a premium for the necessary extra care. Planting seed must be of low moisture as well as prime cottonseed. Breeders further remove all or most of the linters remaining on the cottonseed either by reginning or by chemical means. They are then further treated with chemicals to inhibit fungus or bacteria growth when the seed are planted.

REFERENCES

1. National Cottonseed Products Association, Cottonseed and Its Products, Memphis, Tenn., 1978.
2. **Whitten, M. E., Doughtie, R. T., Jr., and Coleman, W. T.,** Improved factors in grading cottonseed, *U.S. Dep. Agric. Mark. Res. Rep.*, No. 884, 1971.
3. **Alderks, O. H.,** in Cottonseed and Cottonseed Products, Bailey, A. E., Ed., New York, 1958.
4. **Rayburn, S. T., Jr., Griffin, A. C., Jr., Whitten, M.E.,** Storing cottonseed with propionic acid, *Trans. ASAE*, 2 (5), 1, 1978.

Index

INDEX

A

C

D

E

F

G

H

I

J

K

L

M

N

O

P

Q

R

S

T

U

V

X

Y

Z